MW00814811

Reviews of Accelerator Science and Technology

Volume 6

Reviews of Accelerator Science and Technology

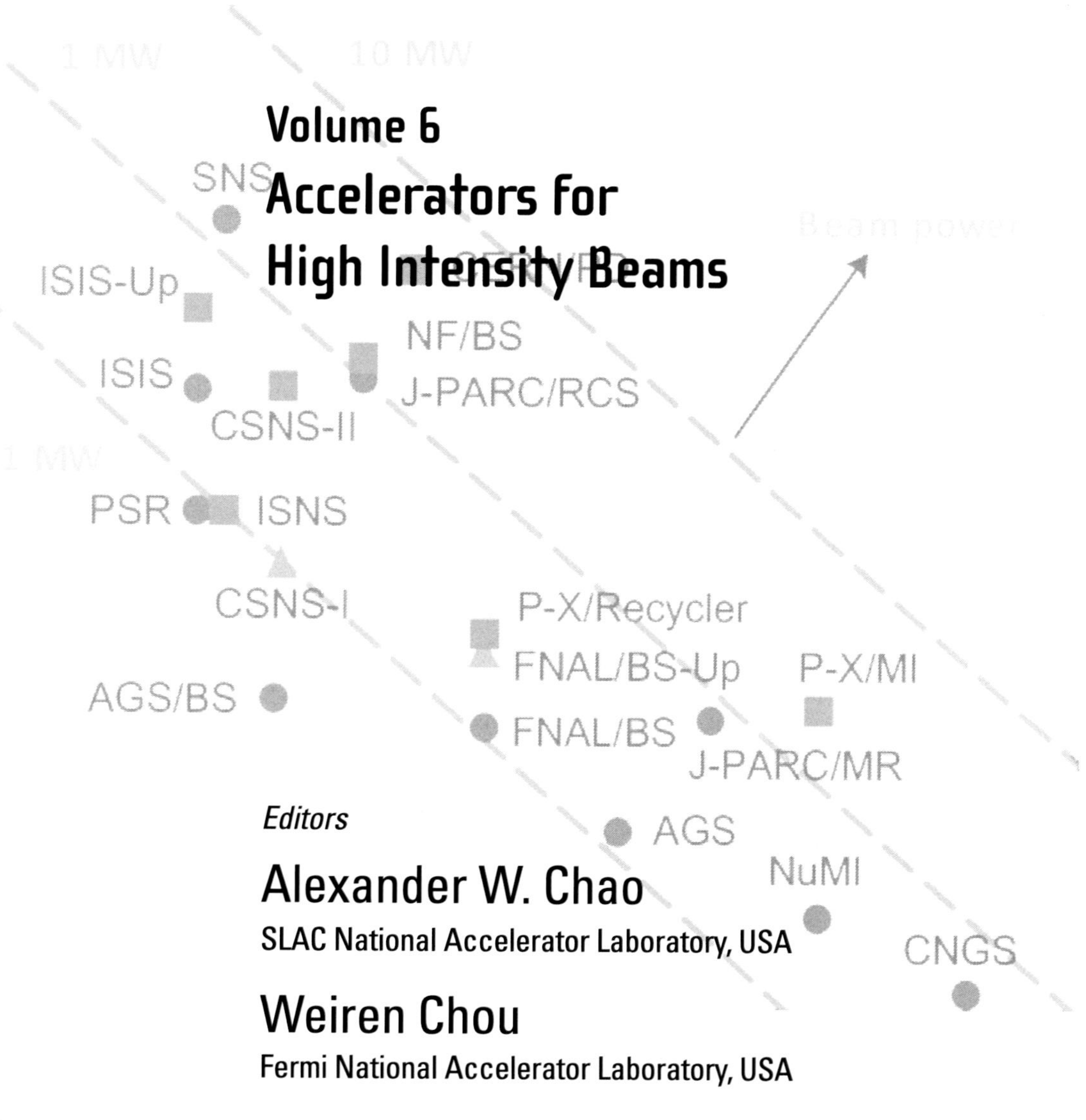

Volume 6

Accelerators for High Intensity Beams

Editors

Alexander W. Chao
SLAC National Accelerator Laboratory, USA

Weiren Chou
Fermi National Accelerator Laboratory, USA

World Scientific

NEW JERSEY · LONDON · SINGAPORE · BEIJING · SHANGHAI · HONG KONG · TAIPEI · CHENNAI

Published by

World Scientific Publishing Co. Pte. Ltd.
5 Toh Tuck Link, Singapore 596224
USA office: 27 Warren Street, Suite 401-402, Hackensack, NJ 07601
UK office: 57 Shelton Street, Covent Garden, London WC2H 9HE

British Library Cataloguing-in-Publication Data
A catalogue record for this book is available from the British Library.

REVIEWS OF ACCELERATOR SCIENCE AND TECHNOLOGY
Volume 6: Accelerators for High Intensity Beams

ISBN 978-981-4583-24-4

Printed in Singapore by Mainland Press.

Reviews of Accelerator Science and Technology
Vol. 6 (2013) v

DOI: 10.1142/S1793626813010017

Editorial Preface

As particle accelerators strive for ever-increasing performance, high intensity particle beams become one of the critical demands requested across the board by a majority of accelerator users (proton, electron and ion) and for most applications. Much effort has been made by our community to pursue high intensity accelerator performance on a number of fronts. Recognizing its importance, we devote Volume 6 of RAST to Accelerators for High Intensity Beams.

High intensity accelerators have become a frontier and a network for innovation. They are responsible for many scientific discoveries and technological breakthroughs that have changed our way of life, sometimes taken for granted. A wide range of topics is covered in the fourteen articles in this volume. The first two articles are overviews of applications in two major areas: the high intensity frontier in particle physics by Robert Tschirhart, and the high intensity frontier in nuclear physics by Kenichi Imai. Several other applications and their associated accelerators are then discussed in subsequent articles: radioactive ion beams and radiopharmaceuticals by Laxdal, Morton and Schaffer, spallation neutron sources and accelerator driven systems by Henderson, accelerators for inertial fusion by Bangerter, Faltens and Seidl, particle beam radiography by Peach and Ekdahl, rapid cycling synchrotrons and accumulator rings by Tang, and superconducting hadron linacs by Ostroumov and Gerigk.

Key accelerator subsystems that allow high intensity operation are also covered in this volume. These include ion injectors by Stockli and Nakagawa, ion charge strippers by Nolen and Marti, targets and secondary beams by Noah, neutron beamlines by Bentley, Cooper-Jensen and Andersen, and beam-material interactions by Mokhov.

In this assembly of scholarly articles, there are several topics that we decided not to cover because they have appeared in previous RAST volumes. These include fixed field alternating gradient (FFAG) accelerators and cyclotrons (Volume 1), energy recovery linacs (Volume 3) and high performance electron guns (Volume 3). The readers are referred to these previous publications for complete coverage of the current theme. Another major topic, high intensity beam dynamics, will be covered in a future volume.

It is a RAST tradition that in each volume we feature one or two articles that concern the accelerator community worldwide but are not necessarily related to the theme. In this volume, we dedicate an article, on personal recollections by Giorgio Brianti and David Plane, to a great pioneer, John Adams, and his many contributions to CERN as well as to the field of scientific research.

RAST has entered its sixth year. We are grateful that this journal continues to receive strong support from our community. It reaches a broad audience. Several volumes have become bestsellers of World Scientific Publishing. From feedback received, the success is attributed to review articles that readers can trust and use, and to the stories of individuals who have shaped our field and inspired new generations. We express our sincere gratitude to the board of advisors, authors, referees and many colleagues, whose dedication, contributions and suggestions are essential and vital to the success of this journal.

Alexander W. Chao
SLAC National Accelerator Laboratory, USA
achao@slac.stanford.edu

Weiren Chou
Fermi National Accelerator Laboratory, USA
chou@fnal.gov

Contents

Reviews of Accelerator Science and Technology
Vol. 6 (2013) 1–18

DOI: 10.1142/S1793626813300016

Beams for the Intensity Frontier of Particle Physics

Robert S. Tschirhart
Fermi National Accelerator Laboratory,
Batavia, IL 60510, USA
tsch@fnal.gov

Advances in high intensity beams have driven particle physics forward since the inception of the field. State-of-the-art and next generation high intensity beams will drive experiments searching for ultrarare processes sensitive through quantum corrections to new particle states far beyond the reach of direct production in foreseeable beam colliders. The recent discovery of the ultrarare B meson decay $B_s \to \mu\mu$, with a branching fraction of 3×10^{-9} for example, has set stringent limits on new physics within direct reach of the Large Hadron Collider. Today, even in the context of the Higgs boson discovery, observation of finite neutrino masses is the only laboratory evidence of physics beyond the Standard Model of particle physics. The tiny mass scale of neutrinos may foretell and one day expose physics that connects quarks and leptons together at the "grand unification" scale and may be the portal through which our world came to the matter-dominated state so different from conditions we expect in the early universe. Here we describe next generation neutrino and rare processes experiments that will deeply probe these and other questions central to the field of particle physics.

Keywords: Particle physics intensity frontier.

1. Themes in Particle Physics

Particle physics has made enormous progress in understanding the nature of matter and forces at a fundamental level and has unlocked many mysteries of our world. The development of the Standard Model of particle physics has been a magnificent achievement of the field. Many deep and important questions have been answered, and yet many mysteries remain. The discovery of neutrino oscillations, discrepancies in some precision measurements of Standard Model processes, observation of matter–antimatter asymmetry, and the evidence for the existence of dark matter and dark energy, all point to new physics beyond the Standard Model.

The pivotal developments of our field, including the discovery of the Higgs boson, have progressed within three interlocking frontiers of research — the energy, intensity, and cosmic frontiers — where discoveries and insights in one frontier powerfully advance the other frontiers as well. Themes spanning these frontiers that can be addressed with experiments driven by high intensity beams include:

- Are there new forces in nature?
- Do any new properties of matter help explain the basic features of the natural world?
- Are there any new (normal or fermionic) dimensions to space–time?

In pursuit of these themes, the mainstays of laboratory particle physics are high energy colliding beam experiments on the one hand, and intense beams on fixed targets on the other. Although one usually thinks of the former as the place to discover new particles, and the latter as the place to tease out rare and unusual interactions, history provides several examples of precise measurements at high energy colliders (for example, the mass of the W boson and the B_s oscillation frequency) and unexpected discoveries at high intensity experiments (for example, flavor mixing in quarks and in neutrinos).

The experimental research opportunities discussed in the following sections address these deep questions in several ways:

- *New forces.* Experiments have established flavor-violating processes in quarks and neutrinos, so it seems conceivable that charged leptons, for example, can violate flavor too. High intensity muon sources can search for these phenomena via muon-to-electron conversion and related processes, and the anticipated large sample of tagged τ-lepton decays at the high luminosity e^+e^- collider at Super KEK-B in Japan will likewise have high sensitivity. Many of the theoretical ideas unifying forces and flavor violation anticipate baryon number violation, and high intensity neutron sources can extend the limits on neutron–antineutron oscillations by orders of magnitude, and large underground detectors will continue to advance limits on proton decay. These same ideas posit measurable flavor-changing neutral currents, thereby mediating rare decays such as $K \to \pi\nu\bar{\nu}$ and $B \to \mu\mu$ (B_u and B_s).
- *New properties of matter.* According to the Sakharov conditions, the baryon asymmetry of the universe requires *CP*-violating (matter–antimatter-asymmetric) interactions, but their strength in the Standard Model is insufficient to account for the observed excess. It is not known whether the missing *CP* violation takes place in the neutrino sector or the quark sector. Advances in high intensity beams will aid both searches by increasing the reach of neutrino oscillation experiments and by enabling a new suite of searches for nonzero electric dipole moments (EDMs). The latter program is broad and overlaps the domain of nuclear physics — using high intensity particle sources to search for an EDM of the neutron, proton and muon directly, and the electron, exploiting amplification in short-lived isotopes such as ^{225}Ra, ^{223}Rn and ^{211}Fr [1, 2].
- *New dimensions.* Many extensions of the Standard Model introduce extra dimensions: in the case of supersymmetry, the dimensions are fermionic. The space of non-Standard interactions opens up possibilities for the interactions mentioned above: quark and neutrino *CP* violation and quark-flavor-changing neutral currents with supersymmetry, and flavor-changing neutral currents from a warped fifth spatial dimension. Rare K and B meson decays, EDMs, and neutron–antineutron oscillations are closely tied to these possibilities.

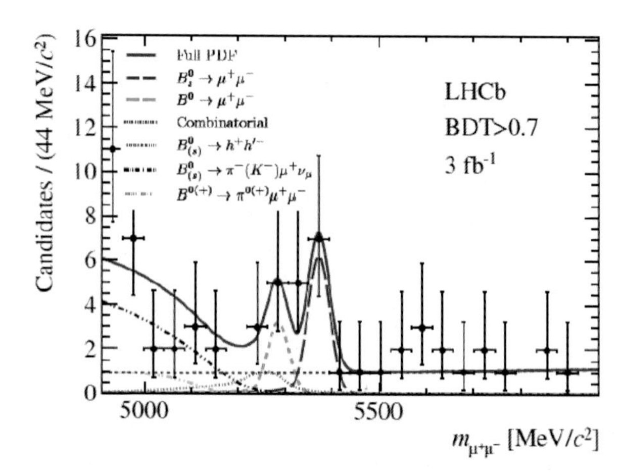

Fig. 1. Observation of the $B_s \to \mu\mu$ process by LHCb with a measured rate of $[2.9^{+1.1}_{-1.0}\ (\mathrm{stat})^{+0.6}_{-0.4}(\mathrm{syst})] \times 10^{-9}$ [3] and a 4-σ significance. The CMS experiment at the LHC has reported [4] a similar rate with a 3-σ significance.

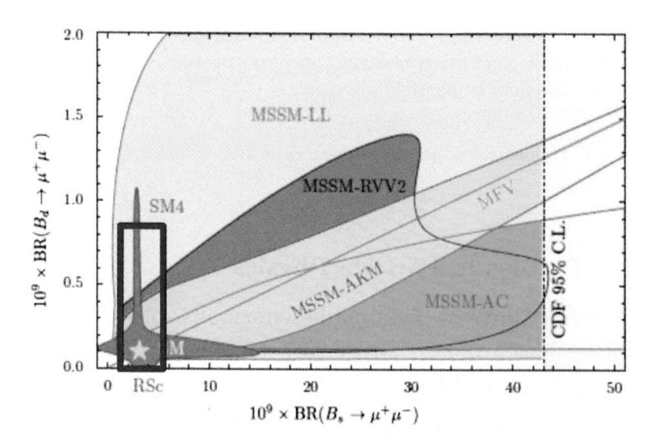

Fig. 2. Recent measurements [3, 4] of $B_d \to \mu\mu$ and $B_s \to \mu\mu$ overlaid on the range of enhancements beyond the Standard Model, which is indicated by a star [5]. Minimal supersymmetric theories, in particular, are highly constrained.

The recent observation of $B_s \to \mu\mu$ shown in Fig. 1 and limits on $B_d \to \mu\mu$ severely constrain the parameter space of supersymmetry, as illustrated in Fig. 2.

2. The Intensity Frontier Research Program

Research on rare processes, precision measurements, and neutrino physics is proceeding at high intensity accelerator facilities worldwide, with beam energies spanning 20 orders of magnitude from 300 neV ultracold neutrons to the 8 TeV LHC proton–proton collider. As noted previously, ultrarare processes are incisive probes of new particles beyond the Standard Model that can induce quantum fluctuations in these processes. Likewise, precision measurements of processes well understood in the Standard Model, such as the anomalous magnet moments of electrons [6] and muons [7]or precision measurement of Moller e^-–e^- scattering [8], are deep probes of physics beyond the Standard Model. An emerging experimental program at existing high intensity electron beam facilities [9] can also search for "dark sector" [10] physics at relatively low mass scales but coupling very weakly to known particles. A comprehensive review of the worldwide Intensity Frontier research program is beyond the scope of this article. Recently, as part of the planning process for US particle physics, this field was broadly reviewed at the Intensity Frontier Workshop and in associated proceedings [11]. In this article we will explore a few probes in some depth where future accelerator facilities hold particular promise of driving these experiments.

3. Neutrino Physics

A leading experimental pursuit in particle physics today is the search for physical phenomena beyond the Standard Model. Even in the context of the recently discovered Higgs boson and the compelling evidence for the existence of dark matter, neutrino research is the only laboratory setting today that has revealed phenomena beyond the Standard Model through the existence of massive neutrinos. As with to quark mixing, the existence of finite neutrino masses implies the existence of a mixing matrix that connects the mass eigenstates to the weak eigenstates. By analogy with quarks, considering the mixing as a unitary 3×3 matrix allows for an undetermined imaginary component of the mixing (a complex phase) where matter–antimatter asymmetries can be manifested. However, in contrast to charged quark mixing, the neutrino mixing matrix includes the possibility of "Majorana" [12] mixing, and another complex mixing phase, where neutrinos can transform into antineutrinos through a complex Majorana mixing term possibly corresponding to another manifestation of matter–antimatter asymmetry. Establishing the 3×3 paradigm and associated unitarity is of great experimental and theoretical interest, since deviations can be induced by sterile neutrinos that can mix with the three known mass eigenstates.

The very small mass scale of neutrino mixing illustrated in Fig. 3 (0.001–1 eV), which is a factor of a million below the quark mass scales, is yet another hint and avenue for physics beyond the Standard Model, with so-called "see-saw" [13] models relating the small neutrino mass scale to the Grand Unified Theory (GUT) scale (10^{15} GeV) where quarks and leptons may unify.

An intense worldwide program to measure the mixing properties of neutrinos has continued to surprise us, and the emerging pattern of large mixing between neutrino species is now providing further intriguing hints of possibly other new physics beyond the Standard Model. Recent reactor-based experiments [14] have established through precision disappearance measurements the large mixing of electron neutrinos with other neutrino species. With existing

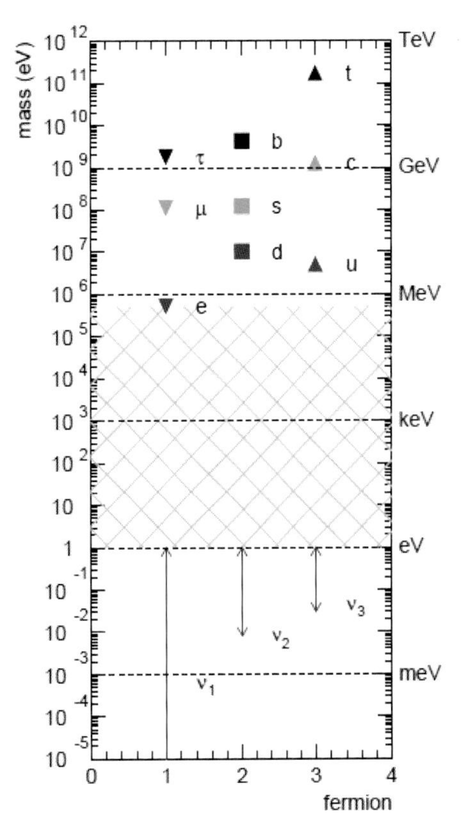

Fig. 3. The mass of known fundamental fermions, illustrating the very small mass scale of neutrinos with respect to all known charged leptons and quarks [17].

data to date, the long-baseline (300–1300 km) experiments, most recently the T2K experiment [15], have provided clear evidence for the corresponding appearance of electron neutrinos in an intense muon neutrino beam, which is another important validation of the 3×3 mixing model of neutrinos. Through the remainder of the decade the T2K [15] and NOvA [16] long-baseline experiments and other experiments will improve the precision of mixing parameters and will also begin to probe the mass hierarchy (sign of Δm mass differences) of the neutrino states, which is currently unknown. Another clue for new physics that may have an associated symmetry evident in neutrino mixing is the apparent maximal mixing of muon and tau neutrinos. Determining if this mixing is maximal or not is an important goal for next generation mixing experiments.

The large observed mixing among neutrinos also facilitates [18] the sensitivity of next generation experiments to matter–antimatter-asymmetric (*CP*-violating) mixing between neutrino states. The observed predominance of matter in our world and the Sakharov conditions [19] require the existence of *CP*-violating processes, which among neutrinos through the process of leptogenesis [20] can provide a framework for understanding how the baryon asymmetry (baryogenesis) of our matter-dominated world emerges. However, the observation of *CP* violation in neutrinos does not by itself establish leptogenesis. *CP* violation can only be observed in neutrinos through long-baseline mixing experiments. *CP* violation asymmetries can indeed be large, $O(1)$, but there are certain values of the *CP* violation phase that would be unobservable due to interference from matter effects along a long flight baseline through the earth. The fraction of *CP*-violating phase (δ) accessible as a function of measurement error $\Delta(\delta)$ is shown in Fig. 4 [21], which illustrates the worldwide interest in pursuing *CP* violation in neutrino mixing. Theories and models inspired by the grand unification of leptons and quarks motivate a target sensitivity for *CP* violation among neutrinos that is comparable to the quark sector (CKM 2011) illustrated in Fig. 4.

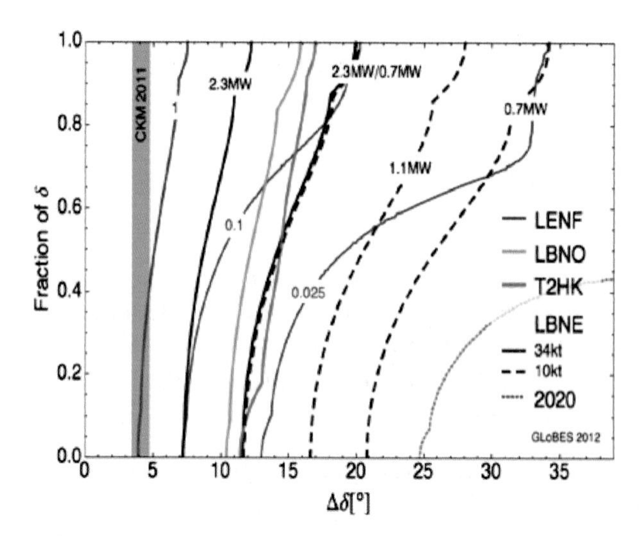

Fig. 4. Measurement sensitivity of next generation long-baseline neutrino mixing experiments to *CP*-violating asymmetries evident in the energy spectrum of neutrinos and antineutrinos from a well-known source after they propagate a long distance (300–2600 km).

3.1. *Superbeams*

The next generation pursuit of mass-hierarchy and *CP*-violating asymmetries is based on high power pulsed proton sources (700–2500 kW) driving target and magnetic horn systems to prepare intense muon neutrino "superbeams" born of pion decay which illuminate distant (300–2600 km) massive detectors (5–500 kt). The optimum muon neutrino energy is between 300 and 3000 MeV for mass-hierarchy and *CP*-violating asymmetry experiments. Detector and superbeam designers strive to optimize sensitivity by balancing constraints of beam power, beam energy, and detector mass — all cost drivers in these large scale experiments. The *CP* violation sensitivity of the Long Baseline Neutrino Experiment (LBNE) [23] for a range of (power) $\times$ (detector mass), for example, is shown in Fig. 5, where the thickness of bands shows the systematic understanding of the neutrino content within the superbeam and mixing parameter variation.

The far-detector technology of choice for LBNE is a 34 kt (fiducial) liquid argon time projection chamber (TPC) that can reconstruct charged and neutral current neutrino interactions with high spatial granularity. This fine spatial reconstruction of neutrino interactions facilitates identifying electrons created in charged current interactions following the oscillation of a muon neutrino into an electron neutrino over the 1300 km on-axis baseline flight path from the Fermi National Accelerator Laboratory (Fermilab). The "bubble-chamber-like" spatial resolution of liquid argon TPCs can separate electron electromagnetic showers from the copious background of $\pi^0 \to \gamma\gamma$ electromagnetic showers

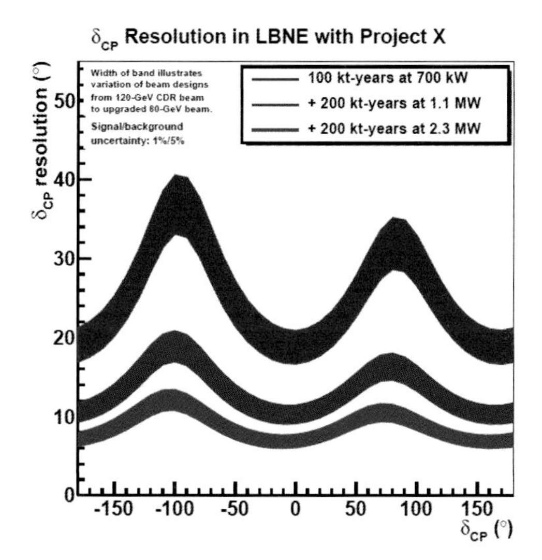

Fig. 5. The *CP* violation sensitivity of the Long Baseline Neutrino Experiment (LBNE) [23] for a range of (power) × (detector mass) as a function of the *CP*-violating phase δ. The width of the band is the estimated systematic error in understanding the incident neutrino flux and variation of neutrino mixing parameters.

with high efficiency over a broad range of incident neutrino energies.

In contrast, the HyperK initiative [24] in Japan is based on a massive 560 kt (fiducial) water Cerenkov detector that builds on the successes of the SuperK predecessor. The HyperK experiment strategy uses a 2.5° off-axis neutrino beam over a 300 km baseline from the Japanese Proton Accelerator Research Complex (JPARC), where the kinematics of $\pi \rightarrow \mu\nu_\mu$ decay constrain the ν_μ energy to a narrow, relatively low energy band at the expense of neutrino beam flux. The low energy of the incident beam of neutrinos strongly suppresses the production of $\pi^0 \rightarrow \gamma\gamma$ through neutral currents which are relatively difficult to distinguish with water Cerenkov detector technology from electrons produced from oscillated electron neutrinos in the beam. The very large mass of the detector compensates for the reduced neutrino beam flux of this off-axis technique.

The very massive (560 kt) water Cerenkov detector strategy of HyperK and the "modest" (34 kt!) detector strategy of LBNE have complementary strengths, but both rely critically on high power proton sources to prepare a high intensity broadband beam for LBNE and a lower intensity narrowband beam for HyperK.

For superbeam experiments such as T2K, MINOS [22], NOvA, Hyper-K, LBNE, and the Long Baseline Neutrino Observatory (LBNO) [25], the control of systematic errors will be a major issue, since neither the detection cross-sections nor beam fluxes are known within the required precision. Near detectors, together with hadron production data, will play an important role in nailing down the neutrino flux impinging on the massive remote detectors to a precision needed to measure mass-hierarchy and *CP*-violating effects.

However, this alone will not be sufficient to obtain percent level systematics, since the beam at the near detector is composed mostly of ν_μ and hence a measurement of the ν_e cross-section is not possible, but in the far detector the signal is ν_e; see for example Ref. 26. Unfortunately, there are no strong theory constraints on the ratio of muon-to-electron neutrino cross-sections either [27]. Here, experiments like MINERvA [28] and MicroBooNE [29] or a facility like νSTORM [30] will be important in modeling both production spectrum and neutrino interaction cross-sections. Better theory calculations of neutrino–nucleon interactions will also certainly be required. Such calculations are possible with lattice QCD [31] and will be carried out over the next several years. In this context, these calculations will help disentangle hadronic from nuclear effects in neutrino–nucleus scattering.

3.2. *Muon storage ring sources*

The questions of leptonic *CP* violation and the completeness of the three-flavor picture can only by addressed by very high precision measurements of neutrino and antineutrino oscillation probabilities, specifically including channels where the initial and the final flavor of the neutrino are different. Several neutrino sources have been conceived to reach high sensitivity and to allow the range of measurements necessary for removing all ambiguities in the determination of oscillation parameters. The sensitivity of these facilities is well beyond that of the presently approved neutrino oscillation program. Studies [21], illustrated in Fig. 6, have so far shown an intense high energy neutrino source based on a stored muon beam, which gives the best performance for *CP* measurements over the entire parameter space. The timescale and cost of neutrino factories based on stored muon beams are, however, not clear compared to the relative certainty of the cost and complexity of superbeam experiments. Second

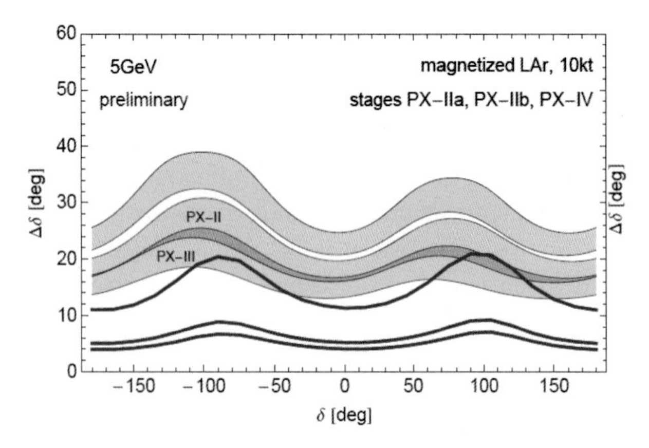

Fig. 6. The *CP* violation sensitivity ($\Delta\delta$) of the Long Baseline Neutrino Experiment (LBNE) for 10 kt of liquid argon detector mass and a range of beam power (stages PX-IIa to PX-IV) from the Project X power evolution (light blue). The sensitivity of a muon storage ring driven by different stages of Project X illuminating a *magnetized* 10 kt liquid argon detector is shown in dark blue.

generation superbeam experiments such as LBNE using megawatt proton drivers like Project X [1] may be an attractive option in certain scenarios, but eventually the issue of systematics control may limit the reach of superbeam experiments. In response to the measurement of large θ_{13}, "neutrino factory" designs has been reoptimized to a stored muon energy of 10 GeV and a single baseline of 2000 km using a 100 kt magnetized iron detector. It is possible to further reduce the stored muon energy to around 5 GeV and concomitantly the baseline to 1300 km without an overall loss in performance if one changes the detector technology to improve efficiency around 1–2 GeV.

Other possible design choices include a magnetized liquid argon detector or a magnetized, fully active plastic scintillator detector. If one of these technology choices can be shown to be feasible, there currently appears to be no strong physics performance reason to favor the 10 GeV over the 5 GeV option, or vice versa.

The low energy option seems attractive, due to its synergies with planned superbeam experiments like LBNE and because the detector technology would allow for a comprehensive physics program in atmospheric neutrinos, proton decay, and supernova detection. Within the low energy option, detailed studies of luminosity staging have been carried out, which indicate that even at 1/20 of the full-scale beam intensity and starting with a 10 kt detector, significant physics gains beyond the initial phases of a pion-decay-based beam experiment, like LBNE, can be realized [1]. At full beam luminosity and with a detector mass in the range of 10–30 kt, a 5 GeV neutrino factory — Low Energy Neutrino Factory (LENF) — offers the best performance of any conceived neutrino oscillation experiment, which is shown in Fig. 4. The Project X [1] staged evolution of the Fermilab accelerator campus affords opportunities to realize the low energy muon storage ring, with sensitivities per stage noted in Fig. 6.

3.3. *Cyclotron sources of neutrinos*

Another interesting concept studied to measure *CP* violation among neutrinos is the Daeδalus initiative [32]. The Daeδalus concept is an array of high power cyclotrons (1–7 MW) at three different baselines (1.5, 8.0, 20 km) from a massive (50–500 kt) water or scintillator gadolinium-doped detector sensitive to the decay-at-rest neutrino spectrum illustrated in Fig. 7.

The Daeδalus initiative is pursuing innovative accelerator technology to reduce space charge effects in high power cyclotrons and the production cost of high power cyclotrons that would produce neutrinos at the three different baseline stations. The sensitivity [32] of Daeδalus operating with the 560 kt water detector of HyperK compared to the superbeam sensitivity of HyperK and the combined sensitivity of Daeδalus with the nomial HyperK superbeam sensitivity is illustrated in Fig. 6.

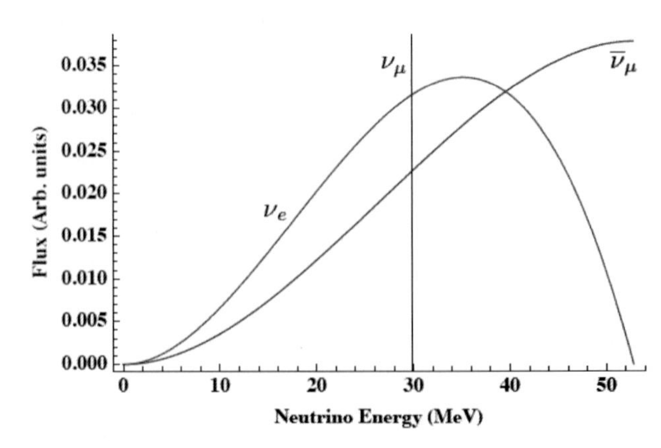

Fig. 7. The energy spectrum of neutrinos from charged pions in a stopping target where both the pion and decay muon decay at rest [32].

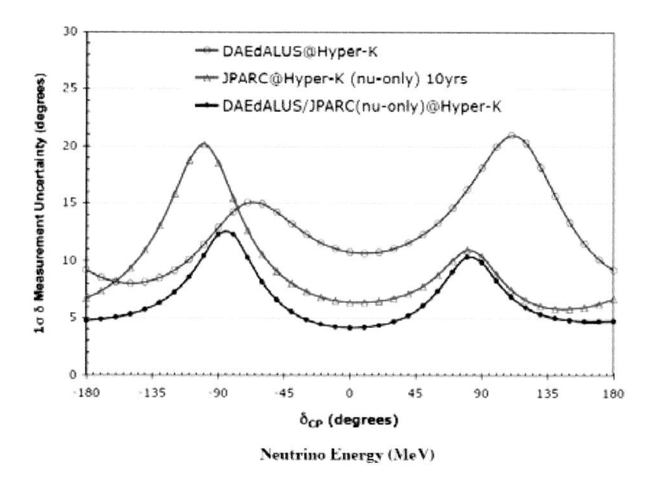

Fig. 8. $\Delta\delta$ sensitivity of the Daeδalus initiative for various combinations of the proposed HyperK experiment [24].

4. Rare Processes, Muons

The process of muon-to-electron conversion is just one example in the broader field of charged lepton flavor violation (CLFV). An excellent review of CLFV and the flavor physics of leptons can be found in Ref. 33. Two classes of diagrams can contribute to conversion. The first class includes magnetic moment loop diagrams with a photon exchanged between the loop and the nucleus; these diagrams can proceed with many different sorts of particles in the loop including but not limited to supersymmetric particles, heavy neutrinos, and a second Higgs doublet. This class of diagrams also produces nonzero rates for the process $\mu \to e\gamma$, which has been probed recently by the MEG experiment at PSI [34] to a limit of 5.7×10^{-13} at the 90% confidence level [35], which has highly constrained loop-induced models [36]. The second class includes both contact terms that parametrize compositeness and the exchange of a new heavy particle, perhaps a leptoquark or a Z^0. This class of diagrams does not give rise to the process $\mu \to e\gamma$. Through these processes, muon-to-electron conversion experiments have sensitivity to new-physics mass scales up to about 10,000 TeV/c^2, far beyond the scales that will be accessible to direct observation at the LHC. Here in this discussion we focus on the next generation muon-to-electron conversion experiments which can reach beyond the already impressive MEG sensitivity and will have better scaling properties with increasing muon beam intensity.

4.1. *The COMET and Mu2e experiments*

Techniques to search for coherent muon-to-electron conversion have been refined for many years at the PSI and TRIUMF laboratories [37], with the leading sensitivity from the SINDRUM program of measurements [38]. In the 1990s a new and powerful concept to dramatically increase the sensitivity of muon-to-electron experiments was developed for the Moscow Meson Factory [39], which inspired the MECO initiative in the US [40].

Today two initiatives are pursuing a major step in sensitivity, on the scale of ×10,000 in rate with respect to SINDRUM: the Mu2e experiment at Fermilab [41] and the COMET experiment at JPARC [42], which both aim for a sensitivity of $5 \times 10^{-17}(R_{\mu e})$, which is defined with respect to the standard model muon nuclear capture rate.

Both the Mu2e and COMET experiments are driven with high intensity negative muon beams produced by a proton beam striking a target that produces negative pions, and the subsequent backward decay muons are collected with solenoids into a low energy beam. The yield [43] of stopping muons as a function of the proton beam driver kinetic energy is shown in Fig. 10, which indicates a threshold of about 0.6 GeV and an optimum proton beam kinetic energy of 2–3 GeV. The Mu2e and COMET experiments will initially use existing 8 GeV proton synchrotrons to drive the experiment, with the first stage of COMET commencing in 2015 and nominal sensitivity running for both COMET and Mu2e around 2020. The Mu2e experiment can later be driven by the Project X linacs [1] at 1–3 GeV of proton drive beam energy at ×10–×100 the beam power, thereby increasing the sensitivity accordingly.

As illustrated in Fig. 9 for the instance of Mu2e, the backward-going muon beam is collected and

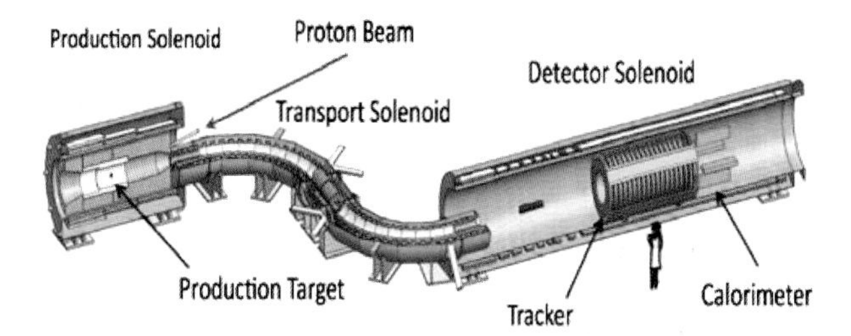

Fig. 9. The Mu2e detector at Fermilab [39] is composed of a production solenoid, a transport solenoid, and a detector solenoid. The COMET experiment at JPARC has similar design elements configured in detail differently.

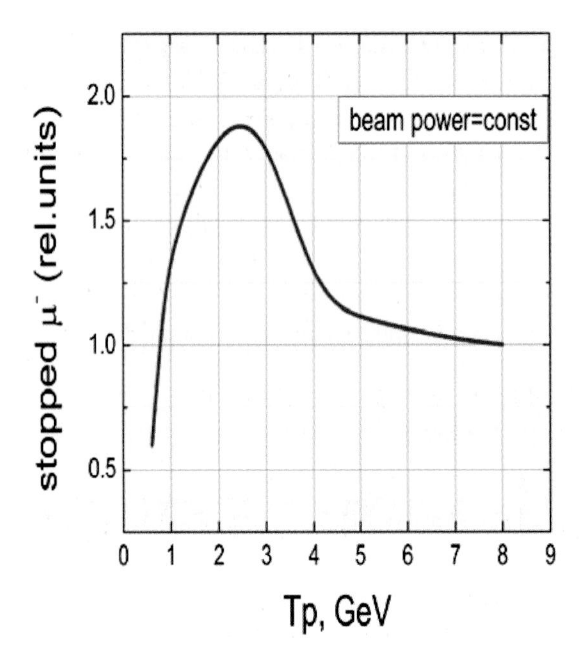

Fig. 10. The stopped muon yield at constant power for the Mu2e experiment geometry as a function of proton beam kinetic energy [43].

transported to a set of thin Al target stopping foils, and muons are captured into the Al atomic inner K shell, forming a muonic atom. The Bohr radius of the K shell of muonic Al is about 20 fm and the nuclear radius of Al is about 4 fm, which provides a large overlap between the muon wavefunction and that of the nucleus.

The two major decay modes of muonic Al atoms are muon decay in orbit (DIO), which occurs about 40% of the time, and normal muon capture (NMC) on the nucleus, which occurs about 60% of the time. DIO produces electrons with a continuous energy spectrum, which is essentially the Michel decay spectrum, modified by the orbital motion of the muon and the form factor and recoil of the nucleus. In an extreme configuration, both neutrinos are at rest and the electron recoils against the intact Al nucleus. This is the configuration in which the electron has the maximum energy in the lab frame and can generate background to the conversion signal region (105 MeV) for muonic Al. The energy spectrum falls to this end point roughly as $(E–E_{\mathrm{max}})^5$. In contrast, NMC produces protons, neutrons, and photons which drive the hit activity in the detector but produce reconstructible high momentum electrons only via secondary processes. The μ-to-e conversion produces a monoenergetic electron with an energy — ignoring neutrino masses — equal to that of the end point of the continuous spectrum from DIO (105 MeV for Al). In summary, the technique is to carefully measure the momentum spectrum from electrons emitted from the target foils and to search for an excess at the end point.

The muon beam that drives Mu2e is initially produced by 8 GeV pulsed protons from the Fermilab proton source and, in the case of COMET, from a special configuration of the Rapid Cycling Synchrotron (RCS) and the Main Ring (MR) at JPARC.

For both COMET and Mu2e, a bunch of protons with a full width of about 200 ns is extracted onto a thin cylindrical gold target located in the middle of a high field solenoid — the production solenoid (PS), shown in Fig. 9. In the production target, p–Au interactions produce pions that are captured into helical trajectories in the field of the solenoid; these pions decay into muons that are also captured by the field of the solenoid. The relevant pion momenta are produced in the backward direction, and the graded PS magnetic field (5 T at the proton-downstream end falling to about 2.2 T at the proton-upstream end) drives these low momentum pions toward the experiment. The backward-going muons exit the PS and enter the S bend graded field transport solenoid (TS), also shown in Fig. 9. The bend in the TS induces a dipole term which allows, by appropriate placement of absorbers and collimators, the sign selection of the muon beam and the stopping of any antiprotons accompanying the muon beam. The TS transmits the μ^- beam into the detector solenoid (DS), where it encounters the foils that make up the stopping target, where about 30% of the muons are stopped.

Downstream of the stopping target is a tracking system, and downstream of that is an electromagnetic calorimeter (ECal). In both of these devices, the inner annular region to a radius of about 38 cm is empty. This allows those muons that do not stop in the stopping target to pass through the detector to a beam dump. The DS magnetic field is also graded to form a magnetic mirror that reflects half of the conversion electrons back toward the tracker. In the volume occupied by the tracker and ECal, the DS magnetic field is highly uniform at 1.0 T. When a conversion or DIO electron is emitted from the stopping target, it travels in a helical trajectory and, if it has sufficient transverse momentum

(p_T), its trajectory will be measured by the tracker. Only those electrons with $p_T > 55\,\text{MeV}/c$ will reach the tracker and only those with $p_T > 80\,\text{MeV}/c$ will intersect enough of the tracker to form a reconstructible track.

Because almost all tracks from DIO have $p_T < 1/2\ m_\mu$, they will never reach the tracker. This is the key to making a measurement of $R_{\mu e}$ with a sensitivity of $O(10^{-17})$: the apparatus is only sensitive to the tail of the DIO energy distribution.

High momentum electrons that pass through the tracker will eventually intersect the ECal, which will provide an independent energy measurement and a position measurement, both of which can be used to confirm track candidates. The μ^- beam that reaches the stopping target is contaminated by many e^- and some π^-, both of which can produce false signals when they interact with the stopping targets. These backgrounds occur promptly. To defeat them, as illustrated in Fig. 11, the experiment exploits the lifetime of muonic Al, about 864 ns: Mu2e waits for 700 ns following the arrival of the proton bunch at the production target and then begins counting electrons that are emitted from the Al foils of the stopping target. By this time, all of the beam from the production target has passed through the stopping target and the prompt backgrounds have died away. After a total of 1694 ns the cycle is repeated. It is also critical that few protons arrive at the production target between the bunches. If protons arrive out of time,

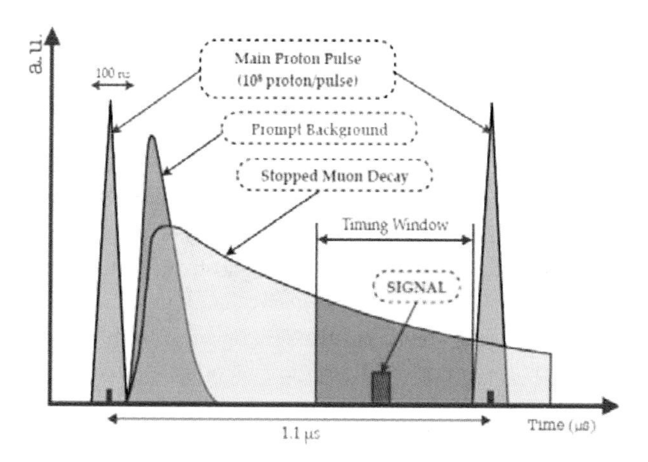

Fig. 11. The time structure of proton beam incident on the COMET experiment at JPARC and the consequent muon decay time structure [44]. The period is optimized to minimize pion decays ($\tau = 24\,\text{ns}$) while maintaining a high efficiency for stopped muon decays ($\tau = 800\,\text{ns}$) for an aluminum stopping target.

they can produce e^- and π^- that arrive at the stopping target within the live gate. To reduce this background Mu2e requires a beam extinction of 10^{-9}; that is, for every 10^9 protons that arrive at the production target within the bunch, there should be no more than one proton (on average) between bunches. The dominant background sources are expected to be poorly measured DIO electrons, radiative π^- capture on the target foils, scattered beam electrons, μ decay in flight, cosmic-ray-induced, and other relatively minor background processes.

5. Rare Processes, Kaons

By analogy with CLFV in muon decays, physics beyond the Standard Model can be observed in loops that affect the rate of rare kaon decays where the Standard Model contribution is highly suppressed as in the case of $K \to \pi\nu\bar{\nu}$ decays or forbidden in the case of $K_L \to \mu e$, for example. The $K^+ \to \pi^+\nu\bar{\nu}$ and $K_L \to \pi^0\nu\bar{\nu}$ processes are of great interest, since the Standard Model contributions can be calculated to high precision (3–5%) [45], allowing physics beyond the Standard Model enhancements to the rate of just 25% to be established with high confidence (5σ). The $K_L \to \pi^0\nu\bar{\nu}$ process is manifestly *CP*-violating and hence is a particularly sensitive probe of new physics that is *CP*-violating.

To date, no evidence of physics beyond the Standard Model has been observed in rare quark and lepton decays, which has correspondingly constrained new physics to very high mass scales (>100 TeV) or highly constrained the couplings of new physics that may be present at 1–10 TeV in order to conceal their presence. One can consider the extreme limit of couplings identical to the CKM matrix, with the associated GIM suppression that would minimize the presence of new particles in loops, and this ansatz is referred to as minimum flavor violation (MFV) [46]. The $K^+ \to \pi^+\nu\bar{\nu}$ and $K_L \to \pi^0\nu\bar{\nu}$ pair of processes is a powerful probe of MFV, which, as is evident from Fig. 12, strictly constrains the relationship between these two rates. Any measured deviation from this MFV constraint would be a clear indication that new physics beyond the Standard Model exists.

State-of-the-art rare decay kaon experiments have probed branching fractions in the 10^{-11}–10^{-12} range; these decays include the rarest particle decay ever observed, $B(K_L \to e^+e^-) = 9 \times 10^{-12}$ [48], and

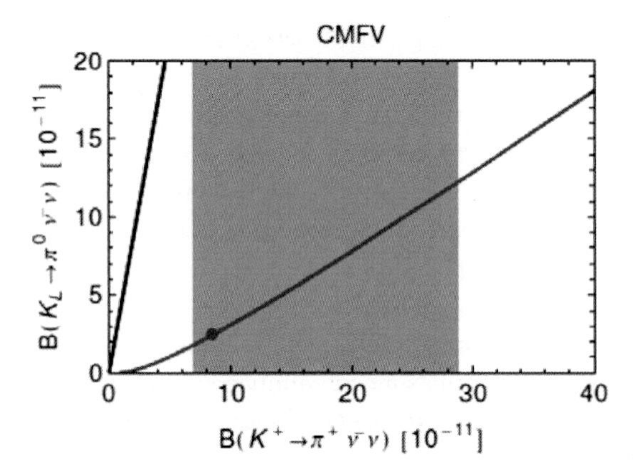

Fig. 12. The highly constrained relationship between the charged and neutral $K \to \pi\nu\bar{\nu}$ rates in minimum flavor violation models, indicated as the blue line; the red dot indicates the Standard Model rates. The gray band illustrates the current measurement of the charged mode, and the black line illustrates the corresponding limit on the neutral mode from isospin analysis of the charged mode measurement [47].

the discovery [49] of the long-sought-after process $K^+ \to \pi^+\nu\bar{\nu}$. These measurements were achieved with 20–50 kW of slow-extracted proton beam power from proton synchrotrons at Brookhaven National Laboratory (BNL).

Current experiments at J-PARC and the European Organization for Nuclear Research (CERN) aim to reach the SM level of sensitivity for $K_L^0 \to \pi^0\nu\bar{\nu}$ and to improve the measurement of $K^+ \to \pi^+\nu\bar{\nu}$ by an order of magnitude, respectively. Next generation experiments at Fermilab could reach the 1000-event SM level for the $K \to \pi\nu\bar{\nu}$ processes, which will require branching fraction sensitivities at the 10^{-14} level; the continuous beam from the proposed Project X accelerator [1] at Fermilab will facilitate balancing (a) the competing requirements of high rates necessary for reaching ultrahigh sensitivity in a finite time and (b) the backgrounds incurred due to accidental effects. In this section, we discuss the status and prospects of the most promising rare kaon experiments and initiatives.

5.1. *Experimental study of $K^+ \to \pi^+\nu\bar{\nu}$*

The E787 experiment at the BNL Alternating Gradient Synchrotron (AGS) reported first evidence of $K^+ \to \pi^+\nu\bar{\nu}$ in 1997 [49] with a rate of $B(K^+ \to \pi^+\nu\bar{\nu}) = 4.2^{+9.5}_{-3.5} \times 10^{-10}$. In 2008, the BNL E949 experiment [50] reported a combined E787–E949 final result of

$$B(K^+ \to \pi^+\nu\bar{\nu}) = 1.73^{+1.15}_{-1.05} \times 10^{-10},$$

based on the observation of a total of seven events. Compare this result with the SM prediction [43]:

$$B(K^+ \to \pi^+\nu\bar{\nu}) \\ = (0.781 \pm 0.075 \pm 0.029) \times 10^{-10}.$$

E787 and E949 represented the culmination of a long series of experiments using stopped kaons. Today, the CERN experiment NA62 [51] is pursuing the next step beyond discovery with a promising new technique, driven by the SPS proton facility, which aims for 100-event sensitivity at the SM level. The proven techniques developed at the AGS could further be exploited with the existing Fermilab accelerator complex to ultimately reach 1000-event sensitivity. These experimental programs are discussed in turn below.

5.1.1. *BNL experiments E787 and E949*

The BNL E787 and E949 experiments were driven by ~40 kW of 24 GeV protons from the AGS that impinged on a platinum target. A magnetic channel selected 700 MeV/c particles, which were filtered with electrostatic separators to establish a 70%-pure 700 MeV/c K^+ beam [52]. This low-energy-separated K^+ beam was transported to a stopping target where 20% of the kaons in the beam stopped and decayed with the characteristic lifetime of 12 ns. The basic experimental principles for these experiments were to measure as much as possible about the incident K^+ and the decay π^+, which are the only observable particles, and to ensure that no extra particles occurred simultaneously. Each kaon was identified and tracked, and had its energy measured. For pions, the momentum (p), energy (E), depth in a range stack (R), and the entire $\pi \to \mu \to e$ decay sequence were determined with large solid angle detector systems surrounding the stopping target. Suppression of muons due to $K^+ \to \mu^+\nu(\gamma)$ decays was crucial; combined particle identification from the observation of the $\pi \to \mu \to e$ decay sequence and relative kinematic tests resulted in a suppression factor $>10^6$ for muons. The region of phase space with the charged track momentum above the two-body $K^+ \to \pi\pi$ ($K_{\pi 2}$) mass peak is the principal measurement region in which potential backgrounds from other kaon decays could be confidently eliminated; a lower region of phase space was more problematic, due to additional background sources from pion interactions.

5.1.2. *The future CERN program to measure* $K^+ \to \pi^+\nu\bar{\nu}$

The NA62 collaboration proposes [51] measuring the $K^+ \to \pi^+\nu\bar{\nu}$ process with a sensitivity of 80 SM events with less than 10% background. This proposal benefits from the succession of experiments at the CERN North Area (NA) that have culminated in the precision measurement of the $K^+ \to e^+\nu$ decay. The NA62 collaboration is now preparing the detector to measure $K^+ \to \pi^+\nu\bar{\nu}$ using decay-in-flight techniques refined in long series of in-flight experiments at the CERN North Area. The NA62 design has also benefited from developments for the "Charged Kaons at the Main Injector" (CKM) Fermilab proposal [53]. The in-flight approach focuses on the lower region of phase space and may have higher $\pi^0 \to \gamma\gamma$ detection efficiency than the stopped K^+ technique, and does not require tagging the $\pi \to \mu \to e$ decay chain, which could permit operating in a higher rate environment. The CERN Superconducting Proton Synchrotron (SPS) will drive the NA62 experiment with 400 GeV protons. In common with the stopping K^+ experiments, the NA62 initiative relies critically on high resolution timing, kinematic rejection, particle identification, hermetic vetoing, and redundancy of measurements. To realize the necessary sensitivity with an in-flight technique, NA62 plans to perform low mass tracking of the incident unseparated beam with an ~1 GHz total rate, 40 MHz/cm^2, achieve positive kaon identification in this high rate environment by means of a differential Cherenkov counter insensitive to pions and protons with minimal accidental mistagging, achieve a muon rejection of at least 10^5 with a sampling hadron calorimeter, achieve two-or-more-standard-deviation π/μ separation up to 35 GeV/c momentum with a ring-imaging Cherenkov (RICH) counter system, and veto the charged particles originating from three- and four-body kaon decays. Initial prototype running in the $K^+ \to \pi^+\nu\bar{\nu}$ configuration with several subdetectors in place occurred in 2013, and a 2–3-year run to reach a sensitivity of 80 Standard Model events is expected to commence in 2015.

5.1.3. *The future Fermilab initiative to measure* $K^+ \to \pi^+\nu\bar{\nu}$

The ORKA ("Golden Kaon") experiment [54] aims to enhance the basic techniques developed at BNL to reach a sensitivity 100 times greater than that achieved previously and 10 times that proposed by NA62 at CERN, corresponding to 1000 standard model events. ORKA will use a 95 GeV proton beam extracted from the Fermilab Main Injector (MI) to produce a 600 MeV/c separated K^+ beam with a particle ratio of $K/\pi \sim 3$. The favored location for ORKA is the former CDF hall at Fermilab, where the existing CDF solenoid can be employed; a primary beam transport line will be constructed, running from the MI to the CDF hall. The kaons will be stopped at the center of a detector assembly (shown in Fig. 13) similar in concept to BNL E949. The pions arising from the decays $K^+ \to \pi^+\nu\bar{\nu}$ will be tracked in a uniform solenoidal magnetic field and identified using the $\pi \to \mu \to e$ decay sequence and measurements of their range, momentum, and energy. The absence of other coincident activity will also be required. Due to running at 600 MeV/c, a substantially higher kaon stopping fraction can be realized for ORKA with little increase in accidental rates compared to E949. All the ORKA detector systems will be improved versions of those used in BNL E949 described above.

Overall, an order-of-magnitude improvement in acceptance, with finer segmentation, increased resolutions, and reduced backgrounds are expected compared to E949. The CDF solenoid magnet will be used with a 1.25 T magnetic field, to allow improved momentum resolution. Detector system improvements include 4× finer segmentation of the pion stopping region "range stack" (RS) and an enhanced photon veto detector with 23 radiation lengths, compared to 17 in E949. The length of the drift chamber,

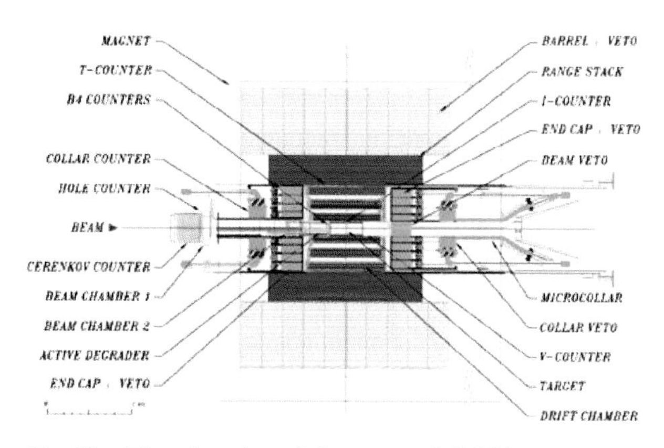

Fig. 13. Elevation view of the proposed ORKA experiment at Fermilab [54].

range stack, and barrel veto will be extended from 50 cm to 80 cm in the beam direction to increase the solid angle acceptance. ORKA is expected to have the sensitivity to collect about 210 events/year at the SM level, i.e. two orders of magnitude greater than achieved at BNL. Including background uncertainties, the experimental precision will approach that of the SM prediction after about three years of operation.

5.2. *Experimental pursuit of* $K_L \to \pi^0 \nu\bar{\nu}$

The first dedicated experiment in pursuit of this process was KEK E391a [55], which has evolved into the KOTO experiment [56] at JPARC. The E391a experiment established a limit $<2.6 \times 10^{-8}$ [57] at the 90% confidence level, compared to the SM prediction [45] of

$$B(K_L \to \pi^0 \nu\bar{\nu}) = (2.43 \pm 0.39 \pm 0.06) \times 10^{-11}.$$

Measuring this highly suppressed process at the Standard Model level requires very intense kaon sources and is a driver of the J-PARC research program in Japan. The KOTO experiment at J-PARC is pursuing $K_L^0 \to \pi^0 \nu\bar{\nu}$ discovery with an initially proposed sensitivity of a few events at the Standard Model level; a higher sensitivity experiment is planned for the future. The very high beam power available with the Project X [1] evolution of the Fermilab complex allows consideration of experiments with much higher sensitivity, at the 1000-event level in the SM. Pursuit of this challenging measurement is complicated by the fact that all particles in both the initial and final states are neutral and consequently difficult to detect. The prospects of the KOTO experiment and Project X next generation concepts will be discussed in turn.

5.2.1. *The future* $K_L \to \pi^0 \nu\bar{\nu}$ *experiment in Japan (KOTO)*

The KOTO experiment is based at J-PARC, which is a new high intensity proton accelerator research complex in Japan designed to deliver 2×10^{14} protons at 30 GeV every 3.3 s. The complex began operation in 2009, and is gradually increasing beam intensity for users. The KOTO experiment is a major evolution of the KEK PS E391a experiment, with large improvements of beamline and detector components. At J-PARC the 30 GeV proton beam impinges on a single target shared by multiple secondary beam lines. A neutral K_L beamline is formed 16° with respect to the incident proton beam. The KOTO beamline was redesigned to reduce the neutron halo/core ratio to 3×10^{-5}, one order of magnitude smaller than KEK E391a. Based on GEANT simulation studies, the number of collimators was reduced to just two and the beam aperture is determined by only three surfaces, thereby minimizing the rate of scattered neutrons out of the beam and into the detector. The beamline was constructed in 2009, and the beam shape and halo component were measured to be consistent with the GEANT simulation, and the kaon yield was also measured in the 2009 run to be ×2.3 higher than assumed in the KOTO proposal. The pure CsI calorimeter has been substantially improved with respect to E391a by incorporating the longer and smaller transverse-size crystals from the Fermilab KTeV experiment [58].

In order to cope with the higher KOTO beam intensity and reject accidental activity, all detector element analog wave forms are sampled at 125 MHz with a system that provides both a large dynamic range (14 bits) and excellent timing performance (<1 ns). Decay photons that escape through the calorimeter beam hole are tagged with very high speed detector modules located inside the neutral beam. The beam–hole tagger is built from modules that consist of a lead plate followed by an aerogel Cherenkov counter. The electron pairs produced by incident photons leave hits in multiple modules, which is less likely for pions produced by neutrons. The detection inefficiency of the beam–hole system is expected to be less than 10^{-3} for photons with energies larger than 2 GeV.

In 2010, KOTO performed an engineering run with 60% of the calorimeter in place which was characterized with momentum-analyzed electrons. In 2012, the detector was completed, and commission data was acquired in late 2012, followed by the first production running in 2013 at 10% of the nominal beam intensity, which will be sufficient to probe rates at the Grossman–Nir limit [59], above which the $K_L \to \pi^0 \nu\bar{\nu}$ rate is excluded through a model-independent interpretation of the measured $K^+ \to \pi^+ \nu\bar{\nu}$ rate. Beyond KOTO, the collaboration is now exploring a following experiment to collect several hundred SM events. Techniques being considered now are a new optimized beamline extracted at 5° from a new target station to increase the kaon yield, and increasing the size of the decay volume and the calorimeter.

5.2.2. *Considerations for a Project X $K_L \to \pi^0\nu\bar{\nu}$ experiment*

The KOPIO initiative [60] at the BNL AGS (which was not realized) proposed measuring the $K_L \to \pi^0\nu\bar{\nu}$ process with a Standard Model sensitivity of 100 events, which would have required ~10,000 h of upgraded BNL AGS proton beam (100×10^{12} 24 GeV protons every 5 s on target). In the KOPIO technique, the kaon momentum is determined with time of flight (TOF) techniques in the momentum range of 300–1200 MeV/c to suppress dominant backgrounds from $K_L^0 \to \pi^0\pi^0$ decays in which two photons are unobserved. The neutral beam incident on the KOPIO detector was designed with a large targeting angle ($\theta = 42°$) from the production target to produce the low momentum neutral kaons critical to the TOF strategy of the experiment [60]. The KOPIO K_L beam had an average kaon momentum of 800 MeV/c, with ~1000 neutrons ($E_n > 10$ MeV) for every K_L in the beam acceptance, which requires that the beam propagate through an excellent vacuum.

This approach is well matched to the Project X kaon momentum spectrum shown in Fig. 14. The high precision timing properties of Project X continuous wave proton linac technology provide experimental tools (TOF techniques) for strengthening the experimental signature and rejecting background processes to the required level. The projected TOF performance of KOPIO at the AGS was limited by achievable proton beam bunching of the AGS of approximately 250 ps.

The Project X beam pulse timing, including target time slewing, is expected to be less than 50 ps, which would substantially improve the momentum resolution and background rejection capability of the $K_L \to \pi^0\nu\bar{\nu}$ experiment driven by the Project X beam illustrated in Fig. 15. The AGS K_L/p yield from 24 GeV protons is 20 times as high as the Project X K_L/p yield for 3 GeV protons. Project X compensates for this relative yield with a proton flux that is 150 times greater than the AGS KOPIO goal of 100×10^{12} protons every 5 s. Therefore, the Project X neutral kaon flux into the nominal KOPIO beam acceptance is eight times the AGS kaon flux. A nominal five-year Project X run is 2.5 longer than the duration of the KOPIO AGS initiative; thus, the reach of a Project X $K_L \to \pi^0\nu\bar{\nu}$ experiment is 20 times that of the KOPIO goals.

A TOF-based $K_L \to \pi^0\nu\bar{\nu}$ experiment driven by Project X would need to be reoptimized for the Project X K_L momentum spectrum, TOF resolution, and corresponding background rejection. The very high K_L beam flux, the potential of breakthrough TOF performance, and improvements in calorimeter

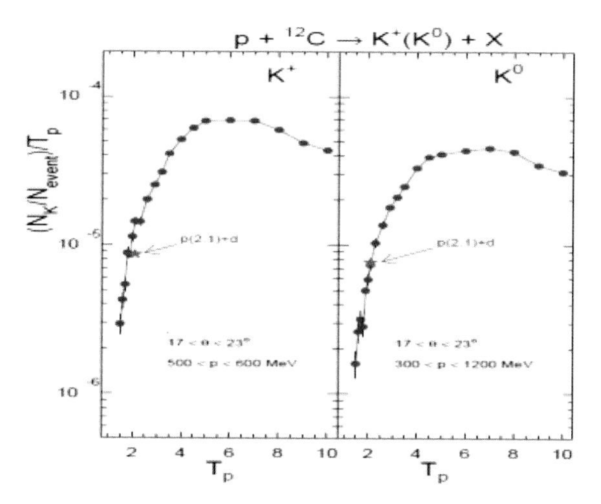

Fig. 14. The simulated (LAQGSM/MARS15) kaon yield [61] at constant beam power (yield/T_p) for experimentally optimal angular and energy regions as a function of T_p (GeV).

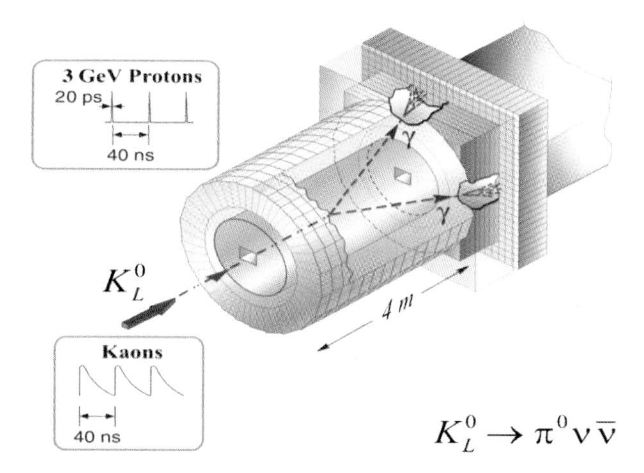

Fig. 15. Illustration of the key elements of the KOPIO technique implemented in a Project X [1] $K_L \to \pi^0\nu\bar{\nu}$ experiment: time-of-flight measurement of the K_L momentum, measurement of ($\pi^0 \to \gamma\gamma$), and veto of all other background-process particles.

detector technology support the plausibility of a $K_L \to \pi^0 \nu\bar{\nu}$ experiment with a sensitivity of ~1000 SM events.

6. Spallation Sources of Ultracold Neutrons and Isotopes for Intensity Frontier Research

Most theories beyond the Standard Model can induce matter–antimatter asymmetries (through "*CP*-violating phases") in known phenomena. The requirement of "*CPT* symmetry," an invariance required by quantum field theory to the product of charge, parity, and time inversion operators in sequence, means that *CP* violation requires *T* violation and consequently manifestly *T*-odd effects. Electric dipole moments (EDMs) are manifestly *T*-odd in coupling to electric and magnetic fields, and hence have been pursued for many decades as evidence for matter–antimatter asymmetry. Next generation spallation targets can enable both a new generation of EDM experiments and neutron–antineutron oscillation experiments (NNbar) which probe another Sakharov condition: baryon number (B) violation mediated by $\Delta B = 2$ physics beyond the Standard Model.

6.1. *Enhanced atomic EDMs*

Spallation targets optimized for particle physics in the megawatt (MW) class can produce copious quantities of isotopes for EDM research in Pb-eutectic or thorium targets [62] with the relevant light isotopes extracted by isotope separation online (ISOL) techniques. Short-lived light isotopes of interest include Ra, Fr, and Rn isotopes (^{219}Rn, ^{223}Rn, ^{211}Fr, ^{221}Fr, ^{223}Fr, ^{223}Ra, ^{225}Ra, $^{225-229}$Ac) to support fundamental searches for physics beyond the Standard Model. Francium isotopes are of interest due to a large relativistic enhancement of heavy nuclei where "Schiff screening" (atomic electrons effectively null any nuclear EDM in the nonrelativistic limit) is reduced. These radon and radium isotopes have favorable nuclear and atomic properties for enhanced EDM searches. The ^{225}Ra isotope, for example, is predicted to have a large octopole deformation, which can enhance EDMs induced by new physics by about 1000 times with respect to the same induced EDM in ^{199}Hg, currently the most sensitively probed atomic EDM. The octopole deformation of ^{224}Ra has recently been established at experiments at the CERN ISOLDE facility [63]; it is illustrated in Fig. 16. Experiments are currently underway with radon isotopes at TRIUMF [64] and radium isotopes at Argonne National Laboratory (ANL) and Kernfysisch Versneller Instituut (KVI) [65] to exploit the octopole enhancement, which has now been established.

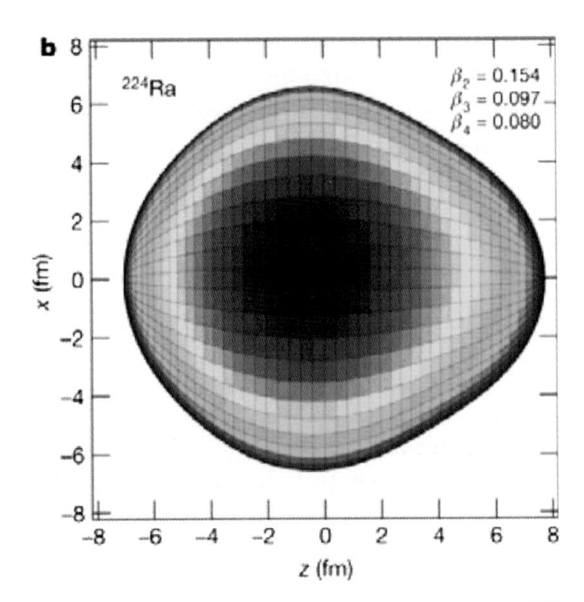

Fig. 16. Illustration of the octopole deformation of ^{224}Ra observed at the CERN ISOLDE facility [63].

6.2. *Ultracold neutrons*

Spallation targets optimized for particle physics in the MW class can also produce copious quantities of cold neutrons (CNs), very cold neutrons (VCNs), and ultracold neutrons (UCNs).

UCNs have the following properties:

- Can be stored in material bottles for hundreds of seconds and piped around corners;
- Typical velocities 0–8 m/s (0–350 neV) (kt < 4 mK);
- Wavelengths >50 nm;
- 100% polarizable with magnetic fields;
- Lifetime <1000 s.

Primary spallation neutrons are too fast to be useful for most nuclear physics applications, such as an NNbar search or n-EDM search. Creation of UCNs requires moderation to thermal energies with a moderator such as heavy water, cooling to VCNs

Table 1. Current cold neutron facilities worldwide [66].

Source	Type	Ecrit (neV)	UCN/ cm^3	Status	Purpose
LANL	Spallation/D2	180	35	Operating	UCNA/users
ILL	Reactor/turbine	250	40	Operating	n-EDM/users
Pulstar	Reactor/D2	335	120	Constructing	Users
PSI	Spallation/D2	250	10^3	Constructing	n-EDM
TRIUMF	Spallation/He-II	210	10^4	Planning	n-EDM/users
Munich	Reactor/D2	250	10^4	R&D	Gravity
SNS	N beam/He-II	130	400	R&D	n-EDM

using cryogenic materials such as solid methane, and then to UCN temperatures using liquid helium.

Radiative heating of the moderator and heat removal are design challenges. Optimization of the moderator configuration is needed to provide maximum yield of cold neutrons, which can then be enhanced for VCN and UCN production. Channeling of the VCN–UCN for NNbar experiments might utilize high-m superreflectors and graphite. Table 1 lists UCN projects operating or under construction around the world.

UCNs can be produced using a D_2O moderator tank, thermal radiation shields to maintain temperatures of $\sim$4 K, and a cold source such as liquid H_2, liquid ^4He, solid D_2, or solid CH_4 to get temperatures of $\sim$0.8 K. CH_4 is the brightest known cold neutron moderator, but is not usable at high power sources due to radiation damage. Other reflectors, such as high albedo materials like diamond nanoparticles, might be used as radiation-hard reflectors near the moderator to improve cold/VCN brightness [67]. Multilayer mirrors might also improve UCN populations provided to experiments.

6.2.1. *Neutron–antineutron oscillation (NNbar) searches*

A next generation NNbar oscillation search requires: (1) observing a sample of neutrons in a vacuum in the absence of a magnetic field for as long as possible, (2) observing as many neutrons as possible, (3) detecting the neutron–antineutron transition by the annihilation reaction, and (4) measuring the probability of appearance or setting a limit. Project X [1] can meet these criteria for a next generation experiment. There is a strong experimental motivation for developing a next generation experiment, since there is a possibility of increasing the detection probability by a factor of >100 with respect to the previous background-free experiment at the ILL research reactor [68]. There is a strong theoretical motivation, in that new theories of neutrino mass, low scale quantum gravity, low scale baryogenesis models, etc., accommodate neutron–antineutron transition probabilities that could be testable by a new experiment.

A next generation oscillation experiment requires:

Lower neutron velocities, larger time (VCN, UCN)

- A larger source to target flight distance;
- Use more of the 4π geometry;
- Use diffusive reflection for fast neutrons;
- Use supermirror reflectors;
- Use neutron focusing ellipsoidal mirrors;
- Use gravity for neutron manipulation.

6.2.2. *Neutron electric dipole moment (n-EDM)*

The worldwide search for a neutron electric dipole moment is highly motivated, and broadly pursued, as is evident from Table 1. Within the United States, for example, nuclear science advisory bodies have concluded [69]: "The successful completion of a *n*-EDM experiment, the initiative with the highest scientific priority in US neutron science, would represent an impressive scientific and technical achievement for all of nuclear physics, with ramifications well beyond the field."

Table 1 describes some of the facilities that can drive n-EDM experiments around the world. A next generation n-EDM experiment that focuses on an integrated design of source and detector could in principle make large gains in UCN density (e.g. 10^4–$10^5/cm^3$). In addition to UCN density, progress in detector capability and systematic control including higher electric fields and control of geometric phase distortions is critical for the next step beyond the current generation of experiments [70].

7. Summary

Particle physics research advances in a close dance with advances in technology. As observed by the visionary theorist Freeman Dyson [71]: "New directions in science are launched by new tools much more often than by new concepts. The effect of a concept-driven revolution is to explain old things in new ways. The effect of a tool-driven revolution is to discover new things that have to be explained."

High intensity beams in particle physics have a distinguished history of providing tools for discovery, and have considerable promise for the future. Indeed, future research on the recently discovered Higgs boson can reasonably be considered as "intensity frontier" research, since next generation experiments at the LHC and future lepton collider concepts rely more critically on luminosity upgrades rather than energy upgrades. Advances in superconducting RF technology, low emittance e^+e^- colliding beam technologies, and next generation high power cyclotron technologies are within our grasp and can provide the next generation tools for reaching far beyond the Standard Model through neutrino research, rare process, and precision measurements. Where these tools will lead us we cannot say today, but there is now a compelling nexus of theoretical motivation and technological capability to take these steps forward.

References

[1] A. Kronfeld and R. Tschirhart (eds.), *Project X: Physics Opportunities* (arXiv:1306.5009 [hep-ex]).

[2] J. Engel, M. J. Ramsey-Musolf and U. van Kolck, Electric dipole moments of nucleons, nuclei, and atoms: the Standard Model and beyond, *Prog. Part. Nucl. Phys.* (2013) (arXiv:1303.2371 [nucl-th]).

[3] Measurement of the $B_s \to \mu\mu$ branching fraction and search for $B_d \to \mu\mu$ decays at the LHCb experiment. LHCb collaboration; submitted to *Phys. Rev. Lett.* (July 2013) (arXiv:1307.5024 [hep-ex]).

[4] Measurement of the $B_s \to \mu\mu$ branching fraction and search for $B_d \to \mu\mu$ with the CMS experiment. CMS collaboration (July 2013) (arXiv:1307.5025 [hep-ex]).

[5] Figure courtesy of H. Jawahery, "Snowmass on the Mississippi" meeting (Aug. 2013); http://www.snowmass2013.org/tiki-index.php

[6] New measurement of the electron magnetic moment and the fine structure constant, *Phys. Rev. Lett.* **100**, 120801 (2008); Lepton dipole moments, B. L. Roberts (Boston Univ.), ed. W. J. Marciano (Brookhaven). Published in *Advanced Series on Directions in High Energy Physics* (2010).

[7] The new (g-2) experiment: a proposal to measure the muon anomalous magnetic moment to ± 0.14 ppm precision. (g-2) collaboration, Feb. 2009. FERMILAB-PROPOSAL-0989; Lepton dipole moments, ed. B. L. Roberts (Boston Univ.), ed. W. J. Marciano (Brookhaven). Published in *Advanced Series on Directions in High Energy Physics* (2010).

[8] K. S. Kumar *et al.*, Low energy measurements of the weak mixing angle (arXiv:1302.6263 [hep-ex]); The Moller experiment at Jefferson Laboratory, USA: http://arxiv.org/abs/1208.1260

[9] See for example the summary of the 2011 Intensity Frontier Workshop working group on "Hidden sector photons, axions, and WISPs": http://confluence.slac.stanford.edu/display/hspawg/Intensity+Frontier+Workshop

[10] See for example N. A. Hamed's remarks regarding hidden sector physics at the "Snowmass on the Mississippi" meeting in Aug. 2013: http://indico.fnal.gov/conferenceTimeTable.py?confId=6890#20130729

[11] See the proceedings of the Intensity Frontier Workshop (Washington DC, Dec. 2011): *Fundamental Physics of the Intensity Fronter*, http://xxx.lanl.gov/abs/1205.2671

[12] E. Majorana, Teoria simmetrica dell'elettrone e del positrone, *Nuovo Cimento* (in Italian) **14**, 171 (1937); B. Kayser, *Phys. Rev. D* **30**, 1023 (1984). For a recent review see: W. Rodehojann, *Int. J. Mod. Phys. E* **20**, 1833 (2011).

[13] M. Gell-Mann, P. Ramond and R. Slansky, in *Supergravity*, eds. D. Freedman and P. Van Nieuwenhuizen (North-Holland, Amsterdam, 1979), pp. 315–321; T. Yanagida, in *Proc. Workshop on Unified Theory and Baryon Number in the Universe*, eds. O. Sawada and A. Sugamoto (KEK, Tsukuba, Japan; 1979); R. Mohapatra and G. Senjanovic, *Phys. Rev. Lett.* **44**, 912 (1980); P. Minkowski, *Phys. Lett. B* **67**, 421 (1977).

[14] Y. Abe *et al.* (Double CHOOZ), *Phys. Rev. D* **86**, 052008 (2012); F. P. An *et al.* (Daya Bay), *Phys. Rev. Lett.* **108**, 171803 (2012); J. K. Ahn *et al.* (RENO), *Phys. Rev. Lett.* **108**, 191802 (2012); F. P. An *et al.* (Daya Bay), *Chin. Phys. C* **37**, 011001 (2013).

[15] The Tokai to Kamiokande (T2K) experiment in Japan: http://t2k-experiment.org

[16] The NOνA experiment in the US: http://www-nova.fnal.gov/how-nova-works.html

[17] Figure courtesy of Prof. Andre de Gouvea, Northwestern University, USA.

[18] See the discussion in Sec. II.2 of Ref. 1 regarding the relationship of *CP* violation and the magnitude of θ_{13} in long-baseline experiments.
[19] E. W. Kolb and M. S. Turner, *The Early Universe* (Perseus, 1994).
[20] M. Fukugita and T. Yanagida, *Phys. Lett. B* **174**, 45 (1986). Recent reviews are: S. Davidson, E. Nardi and Y. Nir, *Phys. Rep.* **466**, 105 (2008); W. Buchmuller, R. Peccei and T. Yanagida, *Annu. Rev. Nucl. Part. Sci.* **55**, 311 (2005).
[21] Figure from Ref. 1, courtesy of P. Coloma, P. Huber, J. Kopp and W. Winter, "Systematics in long-baseline neutrino oscillations for large θ_{13}" (2012), in preparation.
[22] The MINOS experiment in the US: http://www-numi.fnal.gov
[23] The Long Baseline Neutrino Experiment (LBNE) in the US: http://lbne.fnal.gov
[24] Letter of Intent: The Hyper-Kamiokande Experiment — detector design and physics potential (arXiv:1109.3262 [hep-ex]).
[25] Expression of Interest for a very long baseline neutrino oscillation experiment (LBNO), http://cds.cern.ch/record/1457543?ln=en, CERN-SPSC-2012-021; SPSC-EOI-007.
[26] P. Huber, M. Mezzetto and T. Schwetz, *JHEP* **0803**, 021 (2008) (arXiv:0711.2950 [hep-ph]).
[27] M. Day and K. S. McFarland, *Phys. Rev. D* **86**, 053003 (2012) (arXiv:1206.6745 [hep-ph]).
[28] The MINERνA experiment at Fermilab: http://minerva.fnal.gov
[29] The MicroBooNE experiment at Fermilab: http://www-microboone.fnal.gov
[30] nuSTORM: Neutrinos from STORed Muons (arXiv:1206.0294 [hep-ex]).
[31] P. Hagler, *Phys. Rep.* **490**, 49 (2010) (arXiv:0912.5483 [het-lat]).
[32] A. Adelmann *et al.*, Cyclotrons as drivers for precision neutrino measurements (arXiv:1307.6465 [physics.acc-ph]).
[33] R. Bernstein and P. S. Cooper, Charged lepton flavor violation: an experimenter's guide. Accepted for publication in *Phys. Rep.*, July 2013 (arXiv:1307.5787 [hep-ex]).
[34] J. Adam *et al.*, The MEG experiment at PSI, May 2013 (arXiv:1303.0754 [hep-ex]).
[35] MEG collaboration, New constraint on the existence of the $\mu^+ \to e^+\gamma$ decay, *Phys. Rev. Lett.* **110**, 201801 (2013).
[36] A. de Gouvea and P. Vogel, *Prog. Part. Nucl. Phys.* **71**, 75 (2013) (arXiv:1303.4097 [hep-ph]); W. J. Marciano, T. Mori and J. M. Roney, *Annu. Rev. Nucl. Part. Sci.* **58**(1), 315 (2008); http://www.annualreviews.org/doi/abs/10.1146/annurev.nucl.58.110707.171126
[37] An extensive review of the history of rare muon decay searchers can be found in Ref. 33.
[38] W. Bertl *et al.*, *The Eur. Phys. J. C–Part. Fields* **47**, 337 (2006).
[39] V. S. Abadjev *et al.*, MELC Experiment to search for the A to eA process. Tech. Rep. INR(786/92); http://mu2e-docdb.fnal.gov:440/cgibin/RetrieveFile?docid=76;filename=meco002.pdf; version=1.
[40] M. Bachman *et al.*, BNL Proposal E-940 (MECO) http://mu2e-docdb.fnal.gov/cgi-bin/ShowDocument?docid=284
[41] R. Abrams *et al.* (Mu2e collaboration) The Mu2e experiment (2012) (arXiv:1211.7019 [physics.ins-det]).
[42] D. Bryman *et al.*, The COMET experiment (COMET/PRISM/PRIME) (2006); J-PARC Letter of Intent; http://mu2e-docdb.fnal.gov/cgi-bin/RetrieveFile?docid=2202;filename = PRISM-PRIME.pdf;version=2
[43] D. Glenzinski, K. Knoepfel, N. V. Mokhov, V. S. Pronskikh and R. Tschirhart, On optimal beam energy for the Mu2e experiment for Project X stages. FERMILAB-TM-2559-APC-PPD.
[44] COMET beam timing figure courtesy of H. Nishiguchi (KEK), presented at Tau2012 in Nagoya, Sep. 2012.
[45] See Sec. III.2.1.1 of *Project X Physics Book* [1] for a recent discussion on $K \to \pi\nu\bar{\nu}$ rate predictions.
[46] G. D'Ambrosio, G. F. Giudice, G. Isidori and A. Strumia, *Nucl. Phys. B* **645**, 155 (2002); A. J. Buras, P. Gambino, M. Gorbahn, S. Jager and L. Silvestrini, *Phys. Lett. B* **500**, 161 (2001).
[47] A. Buras and J. Girrbach, Towards the identification of new physics through quark flavour violating processes (June 2013) (arXiv:1306.3775v1 [hep-ph]).
[48] D. Ambrose *et al.*, *Phys. Rev. Lett.* **81**, 4309 (1998).
[49] S. Adler *et al.*, *Phys. Rev. Lett.* **79**, 2204 (1997).
[50] S. Adler *et al.*, *Phys. Rev. D* **77**, 052003 (2008); A. V. Artamonov *et al.* (E949 Collaboration), *Phys. Rev. D.*
[51] NA62 experiment: (http://na62.web.cern.ch/NA62); http://greybook.cern.ch/programmes/experiments/NA62.html
[52] J. Doornbos *et al.*, Optics design and performance of LESB3, a two-stage separated 800 MeV/c kaon beamline, *Nucl. Instrum. Methods Phys. Res. A* **444**, 546 (2000).
[53] The Fermilab CKM proposal: http://www.fnal.gov/projects/ckm/documentation/public/proposal/proposal.html
[54] E. Worcester, ORKA, The Golden Kaon Experiment: precision measurement of and other ultra-rare processes, http://pos.sissa.it/cgi-bin/reader/conf.cgi?confid=181; http://www.fnal.gov/directorate/program_planning/Dec2011PACPublic/ORKA_Proposal.pdf
[55] The KEK E391a experiment: http://e391.kek.jp
[56] M. Togawa, The KOTO experiment at Japanese Particle Accelerator Research Center (J-PARC),

http://koto.kek.jp; Status and future prospects for the KOTO experiment, in *Proc. Sci.*, http://pos.sissa.it/cgi-bin/reader/conf.cgi?confid=181
[57] J. K. Ahn *et al.*, *Phys. Rev. D* **81**, 072004 (2010).
[58] KTeV experiment at Fermi National Accelerator Lab: http://ktev.fnal.gov/public
[59] Y. Grossman and Y. Nir, *Phys. Lett. B* **398**, 163 (1997) (arXiv:hep-ph/9701313 [hep-ph]).
[60] KOPIO initiative at Brookhaven National Lab: http://www.bnl.gov/rsvp/KOPIO.htm
[61] K. K. Gudima, N. V. Mokhov and S. I. Striganov, in *Applications of High Intensity Proton Accelerators*, eds. R. Raja and S. Mishra (2009); http://lss.fnal.gov/archive/preprint/fermilab-conf-09-647-apc.shtml
[62] B. Mustapha and J. A. Nolen, *Nucl. Instrum. Methods* **204**, 286 (2003).
[63] L. P. Gaffney *et al.*, *Nature* **497**, 199 (2013).
[64] The Radon EDM experiment at the TRIUMF ISAC facility (S-929); see associated discussion in Sec. V.3.1.1 of *Project X Physics Book* [1].
[65] J. R. Guest *et al.*, *Phys. Rev. Lett.* **98**, 093001 (2007); S. De, U. Dammalapati, K. Jungmann and L. Willmann, *Phys. Rev. A* **79**, 041402(R) (2009); see associated discussion in Sec. V.3.1.2 of *Project X Physics Book* [1].
[66] Compiled from the Project X Forum on Spallation Sources for Particle Physics (Mar. 2012, Fermilab); https://indico.fnal.gov/getFile.py/access?contribId=2&sessionId=1&resId=0&materialId=slides&confId=5372
[67] Diamond nanoparticle reflector technologies are discussed in "Neutron–antineutron oscillations with Project X" in Ref. 1.
[68] M. Baldo-Ceolin *et al.*, *Z. Phys. C* **63**, 409 (1994).
[69] US Nuclear Science Advisory Committee (NSAC) Long Range Plan (2007), DOE-OHEP/NSF P5 report: The particle physics roadmap (2008).
[70] A discussion on neutron-EDM systematics can be found in the Project X Spallation Target Forum presentations (Mar. 2012); http://indico.fnal.gov/conferenceDisplay.py?confId=5372
[71] F. Dyson, Imagined Worlds (Harvard University Press, 1995); F. Dyson, *The Sun, the Genome, the Internet: Tools of Scientific Revolutions* (Oxford University Press, 1997).

Robert Tschirhart is a senior staff scientist at the Fermi National Accelerator Laboratory (Fermilab) in Batavia, Illinois. He received his PhD from the University of Michigan in 1987. His research program focuses on precision and high sensitivity measurements in particle physics. He served as chief research program scientist for the Project X high intensity proton accelerator initiative, serves as spokesperson for the Kaons at the TeVatron collaboration (KTeV), and spokesperson for the ORKA initiative at Fermilab which is pursuing observation of rare kaon decay amplitudes. He was awarded an APS fellowship in 2004, and is the author and coauthor of more than 150 publications in peer-reviewed scientific and technical journals.

Reviews of Accelerator Science and Technology
Vol. 6 (2013) 19–36

DOI: 10.1142/S1793626813300028

Intensity Frontier of Accelerators for Nuclear Physics

Kenichi Imai
Advanced Science Research Center and J-PARC Center,
Japan Atomic Energy Agency,
Tokai-mura, Ibaraki-ken 319-1195, Japan
imai.kenichi@jaea.go.jp

High intensity accelerators for nuclear and hadronic physics are reviewed. The frontier of nuclear and hadronic physics with these accelerators is discussed with its perspectives. J-PARC is a world-leading accelerator of this kind and has just started its operation. As a good example, J-PARC and its physics program are reviewed.

Keywords: High-intensity accelerator; hadron beam; exotic nuclei.

1. Introduction

The term "high intensity frontier" has often been used in relation to "high energy frontier," especially in particle physics. Physics beyond the standard model, such as lepton flavor violation, can be studied with a high intensity muon beam at low energies. It can also be explored through various rare processes studied by a high luminosity B-factory such as Super KEK-B, which is under construction at KEK. The effect of new physics such as supersymmetry (SUSY) can be studied through these rare processes. On the other hand, SUSY particles have been directly searched for at high energy accelerators such as Tevatron and now LHC and ILC in future. The high intensity frontier is, therefore, complementary to the high energy frontier in particle physics.

In nuclear physics, a typical example of the high energy frontier is quark–gluon plasma (QGP) physics studied by relativistic heavy ion collisions at RHIC and LHC. The phase diagram of nuclear matter is one of the key issues in nuclear physics. Nuclear matter in the high temperature region has been studied with high energy heavy ion collisions. The formation of QGP has been confirmed at RHIC and LHC through several observables, such as "jet quenching" and elliptic flow of particles from QGP. It is believed that the produced QGP is strong interaction quark–gluon matter with almost zero viscosity [1]. The property of the quark–gluon matter is being extensively studied at RHIC and LHC.

Although the term "high intensity frontier" is not widely used in nuclear physics, most of the nuclear frontiers other than QGP need high intensity accelerators today. A typical example is the physics of nuclei far from the stability line. It is expected that more than 8000 nuclear species are bound. However, so far only less than half of them have been observed. Finding new isotopes and studying their structure is certainly a frontier of nuclear physics. These new isotopes have been produced through beam fragmentation of heavy ion beams or target fragmentation bombarded by high intensity protons. There have been unexpected discoveries in nuclear structure physics in the studies of neutron-rich isotopes such as the neutron halo and skin [2]. Therefore, a lot of efforts have been made to construct frontier accelerators so as to produce new unstable nuclei far from the stability line. In order to gain access to the new region of isotopes, one needs higher intensity beams since the production cross section sharply decreases as the neutron number increases.

Compared to heavy ion accelerator facilities, not many intense proton accelerator facilities for nuclear physics are being operated. However, the high intensity proton accelerator has a long history. Such accelerators were first constructed as meson factories in the 1970s. They were LAMPF at Los Alamos (now called LANSCE), of 800 GeV; TRIUMF in Canada, of 500 MeV; and SIN in Switzerland (now PSI), of 590 MeV. The energy was determined to produce many pions and muons efficiently. Nuclear physics with high intensity pion beams as well as polarized proton beams was performed. Particle physics with high quality pion and muon beams was also carried out. As a natural extension of the "π-meson" factories, a kaon factory was proposed first at Los Alamos, then at TRIUMF for both nuclear and particle physics in the 1980s. They were not funded, unfortunately. Following the failure of these proposals, the Japanese nuclear physics community proposed such a high intensity proton accelerator facility at higher energy called JHF (Japan Hadron Facility). It later evolved to J-PARC (Japan Proton Accelerator Research Complex) with the help of the neutron science and particle physics community in Japan. J-PARC provides high intensity secondary beams such as kaon, pion, muon and neutrino beams for nuclear and particle physics, as well as intense pulsed neutrons and muons for material and biological sciences and applications. In nuclear physics, one of the major subjects is strangeness nuclear physics. The intense kaon beams can produce nuclei with strangeness. With intense kaon beams, not only nuclei with single strangeness but also those with double strangeness can be produced. They are a frontier of the nuclear world with a new flavor. The study of these nuclei with strangeness and difference from ordinary nuclei will provide a deeper understanding of the physics of nuclei and nuclear force.

Chiral symmetry is a key concept of QCD. The symmetry is spontaneously broken in QCD vacuum and in nuclear matter at low temperature and low baryon density, such as ordinary nuclear matter. The chiral symmetry, however, is expected to be restored at high temperature or high baryon density. It is closely related to the phase diagram of nuclear matter. To understand the phase diagram, one should study the behavior of chiral symmetry, chiral phase transition or symmetry restoration. The properties of hadrons, especially mesons in nuclear matter, are expected to be related to the chiral symmetry restoration and are another important subject in nuclear physics at J-PARC.

The existence of exotic hadrons is related to the QCD confinement problem and has been studied for a long time. Recent observations of X, Y, Z mesons at mainly B-factories attracted much attention to hadron spectroscopy, because they are very difficult to be interpreted as qq-bar mesons. They cannot be assigned as expected cc-bar or bb-bar mesons. They are unexpected particles. The present situation is sometimes called the Renaissance of hadron spectroscopy. Hadron spectroscopy, including charm with hadron beams, is also a subject of J-PARC nuclear physics.

In this article, high intensity accelerators are reviewed with the emphasis on the proton accelerator and nuclear physics. There are three kinds of high intensity accelerator for nuclear physics, namely electron, proton and heavy ion accelerators. The electron accelerator is described in Sec. 2 with the emphasis on CEBAF at Jlab. Heavy ion accelerators are briefly summarized in Sec. 3. Proton accelerators are described in Sec. 4. Since J-PARC has recently started its operation, the J-PARC accelerator and its physics program are described in separate sections, namely Secs. 5 and 7. Fair at GSI is described in Sec. 6, since it will be another large facility for nuclear physics in the near future.

2. Electron Accelerator

There are several electron accelerators for nuclear physics. They are ELPH at Tohoku, ELSA at Bonn, MAMI at Mainz and CEBAF at Jlab. Table 1 briefly summarizes the electron accelerators for nuclear physics.

Among them, MAMI [3] and CEBAF provide high intensity continuous electron beams. The intensity frontier of the large electron accelerator for nuclear physics is CEBAF at Jlab. CEBAF is now being upgraded from 6 GeV to 12 GeV by adding superconducting cavities and an additional arc, as shown in Fig. 1. A new experimental area called Hall D has also been constructed to accommodate a new detector system for 12 GeV beams. It provides continuous electron beams of up to 85 μA. The beam power will reach 1 MW at 12 GeV.

Electron beams have been an excellent probe for studing the structure of a nucleon and nuclei. One of the major physics subjects with the upgraded CEBAF is the precise and complete study of nucleon

Table 1. Summary of electron accelerators.

ELPH (Tohoku)	Synchrotron	1.3 GeV	
ELSA (Bonn)	Synchrotron	3.5 GeV	
MAMI (Mainz)	Microtron	1.5 GeV	150 KW
CEBAF (Jlab)	Recirculation	12 GeV	1 MW

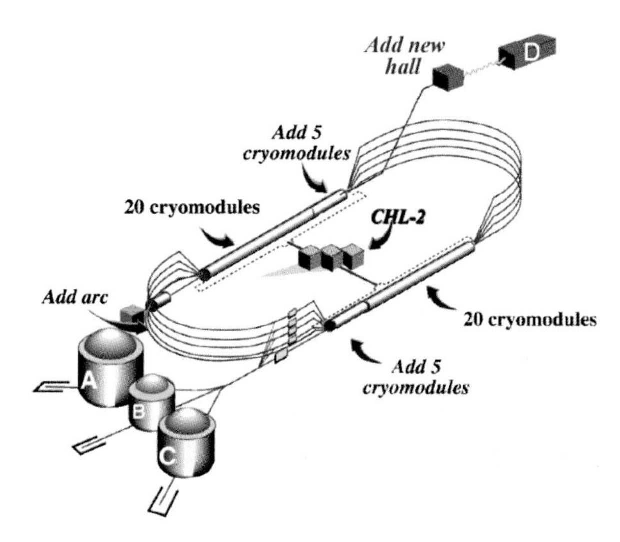

Fig. 1. The Jlab Continuous Electron Beam Accelerator Facility, showing the components needed for the 12 GeV upgrade [4].

structure, such as generalized parton distribution (GPD) and spin structure functions. The gluonic degree of freedom of hadrons is another major subject that can be studied by hadron spectroscopy with a photon beam. They are a natural extension of the hadron and nuclear physics programs carried out with the 6 GeV accelerator.

The spectroscopy of hypernuclei has been performed by the $(e, e'K^+)$ reaction with a high intensity electron beam at Jlab. This method is complementary to hypernuclear spectroscopy with π^+ and K^- beams, since different species of hypernuclei are produced. The study will be continued at Jlab as well as MAMI.

The nuclear interaction cross section for an electron is 100 times smaller than for a proton. Therefore, even an electron beam of 1 MW cannot compete with a proton beam for the production rate of secondary particles such as the pion, kaon and neutron. However, as for the production of photon beams, electron beams are far more superior to proton beams. The photon behaves like a hadron with no specific flavor at high energies. It is, therefore, complementary to hadron spectroscopy with a hadron beam, which always has some specific flavor. An overview of the Jlab 12 GeV upgrade project can be found in Ref. 4.

A photon beam for nuclear and hadron physics is also provided at SPring-8 (8 GeV synchrotron light source) by laser Compton backscattering (LEPS). The present LEPS, which provides 10^6 photons/s up to 2.4 GeV, is being upgraded to provide 10^7 photons/s (LEPS II) [5].

3. Heavy Ion Accelerator

Heavy ion accelerators have mainly been used for the study of unstable nuclei, except for the high energy accelerators such as RHIC and LHC. The unstable nuclei are produced as projectile fragments or target fragments. The projectile fragments, which have the same velocity as the beam, are separated by their mass and charge, and used as a radioisotope (RI) beam for various experiments. To increase the intensity of secondary RI beams and reach a new region of unstable nuclei, a heavy ion accelerator with a higher beam intensity and a capability to accelerate as many ions as possible up to U is required.

To utilize the target fragments, they are ionized and separated and accelerated at low energies for further experiments. ISOLDE at CERN is well known. For a system of this type, proton, deuteron and light ions are often used. The difficulty of an ion source of the fragments has somewhat limited the beam intensity. The planned facilities, however, expect 200–400 kW beams for the production of unstable nuclei as the target fragments. The heavy ion accelerators for the study of unstable nuclei are summarized in Table 2.

Table 2. Summary of the heavy ion accelerator facilities for the study of unstable nuclei.

RIBF (RIKEN)	3 cyclotrons	440 MeV/u light ion 350 MeV/u U ion 1 pμA (~80 kW)
SPIRAL2 (GANIL)	Linac	40 MeV d 5 mA (200 kW) 14.5 MeV/u HI 1 mA
HIRFL (Lanzhou)	2 cyclotrons CSR	10–50 MeV/u 100–400 MeV/u 10^9–10^{14} pps
FAIR (GSI)	SIS100	2.7 GeV/u U^{28+} 5×10^{11} U ions/p
	SIS300	34 GeV/u for U^{92+} 3×10^{11} U ions/s
RAON (S. Korea)	Linac	100 MeV/u U^{79} 8.3 pμA 600 MeV p 660 μA (400 kW)
FRIB (USA)	Linac	200 MeV/u U 0.7 emA (400 kW)

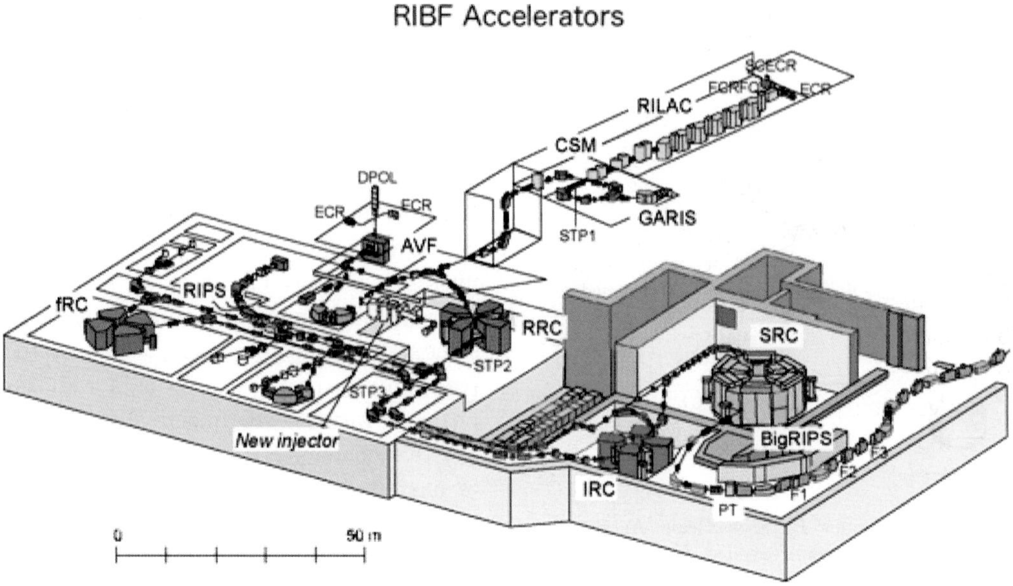

Fig. 2. Layout of the accelerators at RIBF (RIKEN).

The RI Beam Factory (RIBF) at RIKEN, Japan is one of the largest facilities to produce unstable nuclei which are operated today [6]. The layout of the accelerators is shown in Fig. 2. The injector is either a linac (RILAC) or an AVF cyclotron of $K = 70\,\text{MeV}$. They are followed by three ring cyclotrons, depending on the ion species: $K = 570\,\text{MeV}$ fixed frequency ring cyclotron (fRC), $K = 980$ intermediate ring cyclotron (IRC) and $K = 2500$ superconducting ring cyclotron (SRC). It accelerates light ions up to 440 MeV/u and very heavy ions including U ions to 350 MeV/u. The goal of the beam intensity is $1\,\text{p}\mu\text{A}$ and the beam power is 80 kW for the U ion beam. The intense RI beams via projectile fragmentation are separated with the BigRIPS separator and used for various experiments.

SPIRAL1 is a heavy ion accelerator facility which has two cyclotrons for the production of RI beams of up to 20 MeV/u at GANIL, France. It will be upgraded to SPIRAL2, which will facilitate a linac to provide high intensity beams: a 40 MeV deuteron beam of 5 mA (200 kW) and a 14.5 MeV/u heavy ion beam of 1 mA [7]. High intensity RI beams from the target fragmentation will be provided.

At Lanzhou, China, a heavy ion accelerator facility which has two cyclotrons accelerates heavy ions to 10–50 MeV/u to provide RI beams. A new cooling storage ring (CSR) has been constructed to provide 100–400 MeV/u heavy ion beams of 10^9–10^{14} pps [8].

GSI has operated the heavy ion accelerator facility at higher energy (~GeV/u) for a long time. A high intensity linac and a synchrotron, SIS18, are major accelerators providing heavy ion beams up to GeV/u for the production of RI beams as well as high energy nucleus–nucleus scattering experiments. New heavy ion synchrotrons, SIS100 and SIS300, are to be constructed, as well as various storage rings, as the FAIR project in the near future. The physics of unstable nuclei is an the important part of FAIR as the NUSTAR area [9]. Other nuclear physics, such as dense nuclear matter and hadrons and QCD, is also important. FAIR is separately described in Sec. 6.

A new RI beam facility (RAON at RISP) is to be constructed in South Korea. A new large facility for RI beams (FRIB) is planned in Michigan, USA. As shown in Table 2, the major accelerator of both facilities is a heavy ion linac for which the goal of the beam power is 400 kW for U ions. Thus, in the field of physics of unstable nuclei, many accelerator facilities are being operated, constructed and planned. New facilities always aim to provide higher intensity beams so as to gain access to new unstable nuclei.

The study of nuclear structure of nuclei far from the stability line and astronuclear physics will be further extended at these facilities. The details of the physics programs with these heavy ion facilities are out of the scope in this article. For example, the physics programs at RIBF can be found in Ref. 10.

4. Proton Accelerator

The intensity frontier of the proton accelerators was meson factories in 1970–1990 and these were built mainly for nuclear physics and also particle physics. They are LAMPF, TRIUMF and SIN. The accelerator of LAMPF (Los Alamos Meson Physics Facility) is an 800 MeV linac and its highest current was 1 mA. Many nuclear and particle physics experiments were performed with high intensity proton, pion, muon and neutrino beams. At LAMPF, a proton storage ring was constructed in 1985 to compress the proton beam pulses to less than 1 μs for the pulsed neutron production. In 1995, LAMPF was renamed LANSCE (Los Alamos Neutron Science Center) [11]. Although some nuclear and particle physics experiments, such as neutron β-decay asymmetry with ultracold neutrons, are performed, the major research subject of the facility is now material and biological science.

The accelerator of TRIUMF is a cyclotron and it accelerates H^- ions to 500 MeV. The nuclear physics programs with proton and pion beams were completed many years ago. For nuclear physics, it is now used to produce rare isotopes by target fragmentation with a high intensity proton beam of 50–75 kW. One of such RI beam facilities is ISAC. DC muon beams are also provided for material science.

The accelerator of SIN (Swiss Institute for Nuclear Physics) was also a cyclotron, accelerating protons up to 590 MeV. After completion of the initial nuclear physics programs, SIN was reorganized to PSI (Paul Scherrer Institute) in 1988. It produces intense cw neutron and muon beams, mainly for material science, although some particle and nuclear physics experiments are performed. The proton beam current has reached 2.2 mA, which corresponds to the beam power of 1.3 MW. This proton beam power is now highest at this energy region. Because of this high intensity proton beam, low energy muon beams of high intensity are available. The highest sensitivity of $\mu^+ \to e^+\gamma$ decay, which violates lepton flavor conservation, was achieved. Now the upper limit of the branching ratio has been obtained by the MEG Collaboration at PSI as 5.7×10^{-13} (90% confidence level) [12].

In England, after NIMROD, which was used for particle physics in its early days, a spallation neutron source named ISIS was constructed at Rutherford Appleton Laboratory. The main accelerator is a rapid cycling (50 Hz) synchrotron of 800 MeV. It provides 180 kW proton beams for neutron and muon sciences.

The neutron beams from the accelerator, especially pulsed neutron beams, attracted many users of material and biological science and other applications. In the USA, the Spallation Neutron Source (SNS) was constructed at ORNL. The proton beam is accelerated by a linac (this consists of drift tube, coupled cavity and superconducting cavity linacs) to 1 GeV. The accelerated beam is then stored and compressed to less than 1 μs with an accumulator ring (AR) to produce pulsed neutrons. The repetition rate is 60 Hz. They had already achieved a beam power of 1 MW. Although the major users are in material and biological science, nuclear reaction with neutron beams and fundamental properties of the neutron such as β-decay asymmetry and EDM will be studied.

In Europe, the European Spallation Source (ESS) is to be constructed in Lund, Sweden in the near future. The proton will be accelerated to 2.5 GeV by a linac. The expected beam power is 5 MW.

The Main Injector (MI) is a 150 GeV proton synchrotron at FNAL which was constructed to increase the luminosity of the Tevatron collider, as well as for the fixed target experiments and for intense neutrino beams. It is now operated for neutrino experiments and a few other experiments at 120 GeV. A beam power of about 400 kW was achieved for neutrino experiments and will be increased to 700 kW [13]. Although the major programs are for particle physics, SeaQuest is studying antiquark distribution in a nucleon and nuclei by using the Drell–Yan process.

J-PARC was constructed as a joint project for both neutron science and nuclear and particle physics. The accelerator consists of a linac, a 3 GeV rapid cycling synchrotron and a 50 GeV synchrotron. The pulsed proton beam from the 3 GeV synchrotron is used for the pulsed neutron source as well as the muon source. Nuclear and particle physics are mainly performed with the 50 GeV synchrotron. In this sense, it is a unique facility in the world. J-PARC started its operation in 2009. The details of the accelerator will be given in the next section.

A high intensity proton accelerator of 30 GeV (SIS100) is now under construction at GSI, Germany

Table 3. Summary of high intensity proton accelerators.

LANSCE	Linac + AR	800 MeV	100 kW
TRIUMF	Cyclotron	500 MeV	75 kW
PSI	Cyclotron	590 MeV	1.3 MW
ISIS	Synchrotron	800 MeV	180 kW
SNS	Linac + AR	1.0 GeV	1.0 MW
ESS	Linac	2.5 GeV	5.0 MW
MI	Synchrotron	8.0 GeV	
	Synchrotron	150 GeV	0.70 MW
J-PARC	Linac + RCS	3.0 GeV	1.0 MW
	Synchrotron	50 GeV	0.75 MW
FAIR (SIS100)	Synchrotron	29 GeV	0.20 MW

as a part of the FAIR project. It will provide high energy heavy ion beams as well as an upgrade of existing accelerators. Antiproton accumulator rings are a unique feature for the usage of proton beams compared to J-PARC. FAIR is also described in Sec. 6.

Table 3 briefly summarizes high intensity proton accelerators. The high intensity proton accelerators are used for various sciences, including particle and nuclear physics. The medium energy accelerators (the first six facilities in the table) are, however, heavily used for material and biological science with intense neutron and muon beams. At the last two facilities, J-PARC and FAIR, nuclear and particle physics play a major role.

5. J-PARC

J-PARC (Japan Proton Accelerator Research Complex) was constructed in Tokai, Japan. The accelerator complex consists of an injector linac, a 3 GeV rapid cycling synchrotron (RCS) and a 50 GeV main synchrotron (MR). The linac accelerates H^- ions up to 180 MeV, which are injected into the RCS by using the multiturn injection. The energy of the linac will be upgraded to 400 MeV in 2013 by adding cavities. The RCS accelerates protons up to 3 GeV. The repetition rate is 25 Hz. It can provide a proton beam of 1 MW as a goal and produce high intensity pulsed neutrons and muons for mainly material and biological science. Some of the 3 GeV beams are accelerated by the main synchrotron up to 50 GeV. At present, the maximum energy is limited to 30 GeV for the slow extraction and 40 GeV for the fast extraction, because of the limitation of power supplies. The goal of beam power at 50 GeV is 0.75 MW at a repetition rate of 0.3 Hz. The fast-extracted proton beam is used to produce a high intensity neutrino beam for the long baseline oscillation experiment (T2K). The slow-extracted proton beam is used to produce high intensity hadron beams such as kaon, pion and others for nuclear and particle physics experiments. The details of the accelerator design can be found in Ref. 14. An overview of the J-PARC accelerator complex is shown as a schematic view (Fig. 3) and also as a photograph (Fig. 4). One can see a straight building for the linac and a circular structure for the main synchrotron in the picture.

The linac consists of a volume-production-type H^- ion source, a 50 kV beam transport, a 3 MeV radio frequency quadrupole (RFQ), a 50 MeV drift tube linac (DTL) and a 180 MeV separated drift tube linac (SDTL). The RF of the RFQ and linacs is 324 MHz. The repetition rate is 25 Hz, which is equivalent to that of the 3 GeV synchrotron. However, the linac is designed to be operated at 50 Hz so as to provide the additional beam to the Accelerator-Driven nuclear waste transmutation System (ADS) in the near future. It is indicated with a blue box in Fig. 3, as a phase 2 program. The peak current of the linac is 50 mA.

The H^- beam from the linac is injected into the 3 GeV synchrotron by the multiturn injection with a stripper foil. The protons are accelerated to 3 GeV at a repetition rate of 25 Hz. The average beam current is 333 μA and the output beam power is 1 MW. For the spallation neutron source, the linac + accumulator compressor ring can be a choice as LANSCE and SNS. The neutron flux is roughly proportional to the proton beam power. At J-PARC, by accelerating to 3 GeV, 1 MW beam power is achieved with less beam current than for SNS. The 3 GeV synchrotron is indeed necessary as an injector for the 50 GeV synchrotron for nuclear and particle

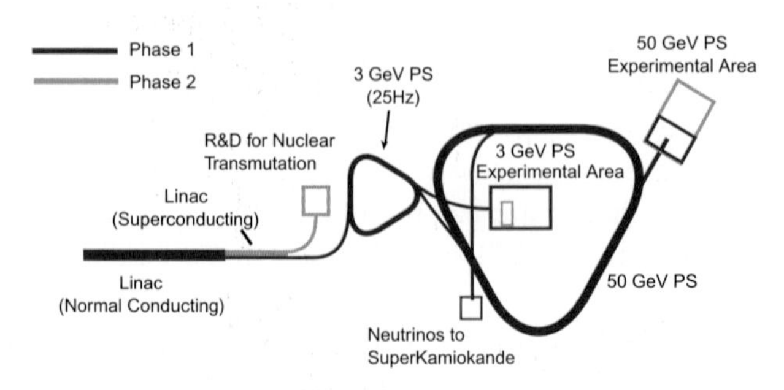

Fig. 3. Schematic overview of the J-PARC accelerator complex. The phase 2 project is shown in blue.

Fig. 4. Photograph of the J-PARC site. This is located in Tokai, Japan, very near the Pacific Ocean.

physics. The shape of the ring is triangular. The characteristic of the accelerator is high repetition (25 Hz) and large acceptance to provide the high power beam of 1 MW. There is no transition below 3 GeV. The main parameters of the 3 GeV synchrotron are summarized in Table 4.

The beam pulse length of less than 1 μs for neutron production limits the circumference of the ring. The threefold symmetry was chosen to have one long straight section for sufficient RF cavities.

For the high energy rapid cycling synchrotron, the RF cavity has been a difficult problem. A lot of R&D has been done at KEK to develop a new accelerating cavity loaded with a magnetic alloy such as FINMET [15]. This new cavity can provide a field gradient of much more than 50 kV/m, which is several times higher than for conventional ferrite-loaded cavities. It made possible the realization of the 3 GeV rapid cycling synchrotron with the parameters shown in Table 4. This type of cavity is also employed for the 50 GeV synchrotron.

The size of the magnet is determined to accept the high beam current. The collimator acceptance is 1.5 times larger than the beam emittance, to allow emittance growth up to 1.5 times after the beam injection. The beams are fast-extracted mostly to the muon production target and neutron production target located in the Material and Life Science Experimental Hall. Every 3 s, the beams are extracted to the 50 GeV synchrotron. The two buckets among the nine in the MR accept the two bunches from the RCS at once. This is repeated to

Table 4. The main parameters of the 3 GeV synchrotron.

Energy	3 GeV
Beam intensity	8.3×10^{13} ppp
Repetition	25 Hz
Average beam current	333 μA
Beam power	1.0 MW
Circumference	348.33 m
Magnetic rigidity	3.18–12.76 Tm
Lattice cell structure	(3-cell FODO × 2-module arc + 3-cell straight) × 3
Typical tune	(6.68, 6.27)
Mom. compaction factor	0.012 (no transition)
Bending magnets	
Number	24
Magnetic field	0.27–1.1 T
Field gap	210 mm
Quadrupoles	
Number	60
Field gradient	Max. 4.6 T/m
Bore diameter	290 mm, 330 mm
RF cavities	
Number	11 (+1)
RF voltage	42 kV/cavity
Radio frequencies	1.36–1.86 MHz
Emittance at injection	216 π mm.mrad
Emittance at extraction	81 π mm.mrad
Collimator acceptance	324 π mm.mrad
Physical aperture	486 π mm.mrad

Table 5. The main parameters of 50 GeV synchrotron.

Energy	50 GeV
Beam intensity	3.3×10^{14} ppp
Repetition	0.3 Hz
Average beam current	15 μA
Beam power	0.75 MW
Circumference	1567.5 m
Magnetic rigidity	12.8–170 Tm
Lattice cell structure	(3-cell DOFO $\times$ 8-module arc + 3-cell straight) $\times$ 3
Typical tune	(22.3, 17.3–22.3)
Mom. compaction factor	−0.001 (imaginary γ_T)
Bending magnets	
Number	96
Magnetic field	0.14–1.9 T
Quadrupoles	
Number	216
Field gradient	Max. 18 T/m
RF cavities	
Number	6
RF voltage	47 kV/cavity
Radio frequencies	1.67–1.72 MHz
Emittance at injection	54 π mm.mrad
Emittance at extraction	10 π mm.mrad (30 GeV)
Physical aperture	81 π mm.mrad

fill eight buckets, and then the acceleration starts in the MR.

The main parameters of the 50 GeV synchrotron are summarized in Table 5.

The MR lattice is designed to have a negative momentum compaction factor so that there is no transition which causes the beam loss. The three-fold symmetry is also chosen for the MR. The beam from the RCS is injected into one straight section and fast and slow extractions are made at the other two straight sections, as shown in Fig. 3. The most serious problem is a beam loss during the slow extraction. In the past, there were beam losses of 2% during the slow extraction from synchrotrons such as BNL-AGS and KEK-PS. At a high intensity accelerator such as the J-PARC 50 GeV synchrotron, the beam loss should be much less than 1% in order to avoid radioactivation of accelerator components. At the 20 kW operation, a slow extraction efficiency of 99.5% was achieved recently. For the operation of slow extraction for experiments at the Hadron Hall, a beam spill time of 2 s has been used. To keep a sufficient time period of the flat top for this spill time, the duration of the MR is 6 s.

The most serious problem of the high intensity accelerator is a radiation problem. The beam loss at injection and extraction and during acceleration causes radioactivation of accelerator components, which have to be accessed for maintenance. That should be avoided, especially in the early stage of the commissioning of accelerators. The beam intensity has been carefully increased step by step and it takes time to increase the beam intensity to its designed value. The commissioning and operation of the accelerators can be found in Refs. 16–18. Figure 5 shows the achieved and expected beam powers of the 3 GeV and 50 GeV synchrotrons as a function of time. However, it should be noted that the beam power of 400 kW was already achieved for muon and neutron production by the RCS and that of 200 kW was achieved for the MR fast-extracted beam for the neutrino experiment. The design goal of 1 MW beam power at 3 GeV will be achieved after upgrading the linac to 400 MeV in 2013.

The high intensity pulsed neutron source is a major facility of J-PARC. The neutrons are produced with a mercury target and there are 23 beam lines. Most of the beam lines are already equipped with experimental instruments and used for experiments [19]. Figure 6 shows the present status of the neutron source and beam lines. Most of the beam lines are used for material and biological science. A beam line (BL04) was constructed for nuclear reaction studies and another (BL05) was constructed to study fundamental properties of the neutron. Measurements of the lifetime, decay asymmetry and electric dipole moment of the neutron are planned. Those are the nuclear and particle physics with neutron beams.

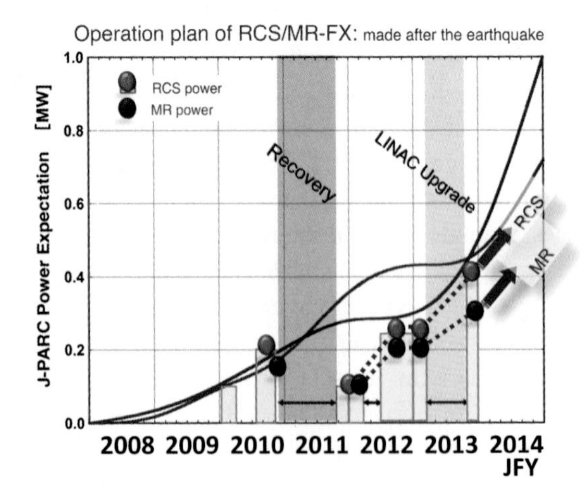

Fig. 5. Achieved and expected beam powers of the 3 GeV and 50 GeV synchrotrons as a function of time. The original plan is shown by solid lines, and circles show achieved beam power.

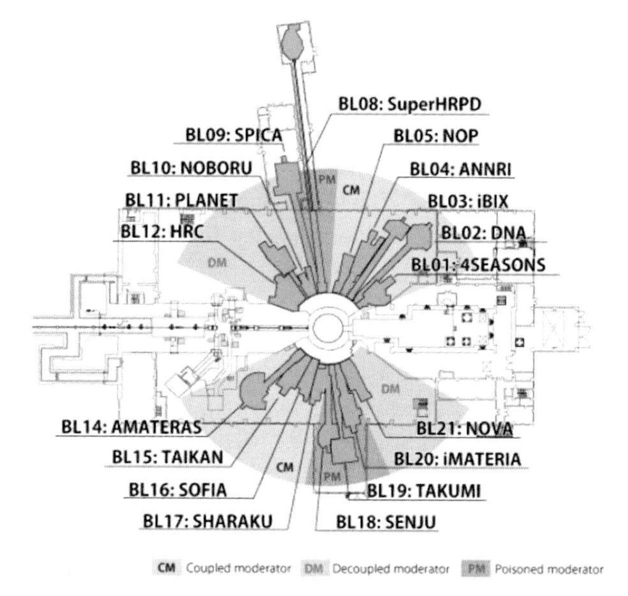

Fig. 6. Neutron beam lines of J-PARC. Eighteen beam lines are used for experiments.

In the material and life science facility, high intensity muon beams are also produced with a tungsten target just upstream of the neutron source. Three beam lines were constructed and used mostly for material science. Programs of particle physics such as a lepton-flavor-violating μe conversion experiment and muon g-2 measurement are expected to be carried out. For the g-2 experiment, an additional muon beam line followed by a muon linac is necessary. As with other high intensity accelerators at medium energies, with the high intensity pulsed neutron and muon beams, selected particle and nuclear physics experiments are foreseen [20].

6. FAIR

FAIR (Facility for Antiproton and Ion Research) is an international project located at GSI, Germany. It is a large scale accelerator complex utilizing existing facilities. The layout of the facility is shown in Fig. 7. A superconducting double synchrotron (SIS100 and SIS300) with magnetic rigidities of 100 and 300 Tm, respectively, is the central accelerator of FAIR. The circumference is 1100 m. Following an upgrade for high intensities, the existing GSI accelerators UNILAC and SIS18 serve as an injector for FAIR.

HESR is a high energy storage ring for storing antiprotons of up to 14 GeV. CR is a collector ring

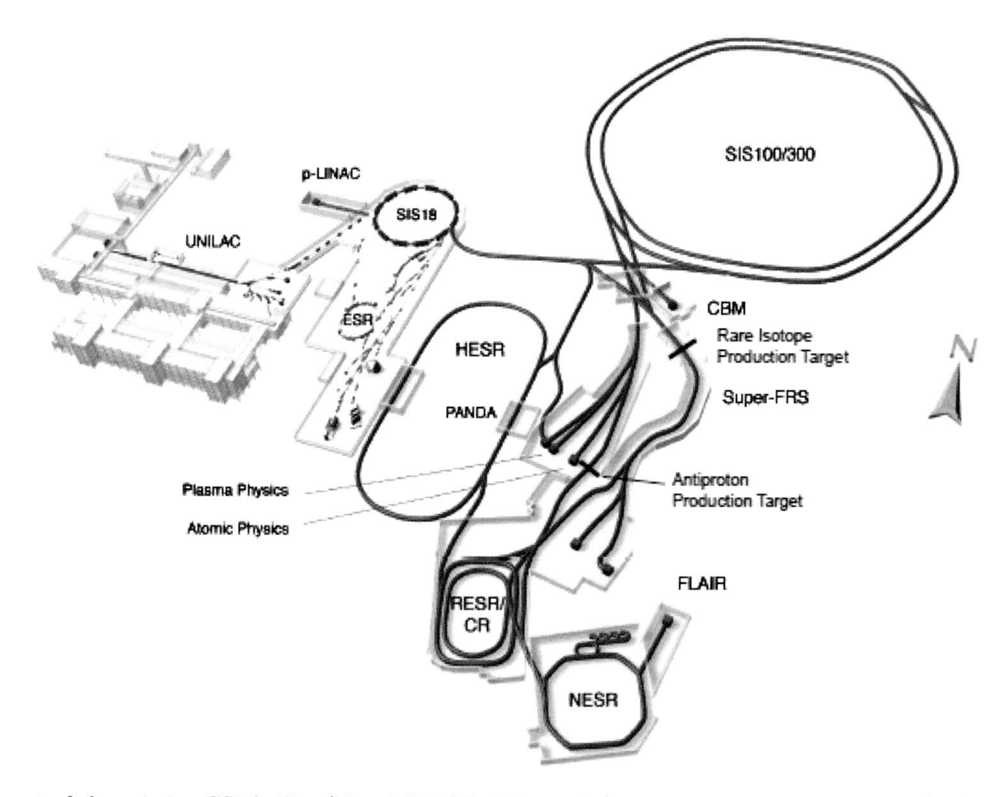

Fig. 7. Layout of the existing GSI facility (blue: UNILAC, SIS18, ESR) and the planned FAIR facility (red). The rings are the superconducting synchrotrons SIS100 and SIS300, the collector ring CR, the accumulator ring RESR, the new experimental storage ring NESR and the high energy antiproton storage ring HESR [21].

for collecting antiprotons of 3 GeV and radioactive ions with fast stochastic cooling capability. RESR is an accumulator ring for antiprotons which are precooled at CR. NESR is a new experimental storage ring which has an electron cooling capability and can cool and decelerate radioactive ions and antiprotons. Super-FRS is the superconducting fragment separator for the rare isotopes produced by heavy ion beams. Heavy ions up to U are accelerated and used for a nucleus–nucleus collision experiment (CBM) to study dense matter in the phase diagram. Details of the project can be found in Ref. 21.

The accelerator complex provides a variety of beams for various sciences. The accelerators and their basic parameters are summarized in Table 6.

The SIS100 accelerates protons and heavy ions up to 29 GeV and 2.7 GeV/u for U^{28+} ions. The expected beam intensities are 4×10^{13} protons per pulse and 5×10^{11} U ions per pulse, respectively. For the high intensities, the repetition rate is 1 Hz, with bending magnet ramp rates of 4 T/s. The output beam powers are about 200 kW for protons and 65 kW for U ions. The beams are extracted with either fast or slow extraction for various uses. The fully stripped U ions can be further accelerated with SIS300 up to 34 GeV/u.

The antiprotons are produced with the 29 GeV proton beam from SIS100, and accumulated and cooled with the CR and RESR rings and sent to NESR or HESR. NESR accommodates experiments with low-energy-cooled antiprotons. Hadron physics with the antiproton beam is mainly performed at HESR.

Heavy ion beams extracted from either SIS100 or SIS300 are used for the production of rare isotopes as beam fragments, which are further stored, cooled and decelerated by CR, PESR and NESR. Nuclear structure and reaction and astro-nuclear physics will be studied with RI beams (NUSTAR). The heavy ions from SIS300 are also used for the study of dense nuclear/quark matter in the phase diagram (CMB experiment). The energy was chosen to realize the maximum baryon density in U + U central collisions.

Table 6. Summary of rings.

SIS100 (synchrotron)	2.7 GeV/u for U^{28+} 29 GeV for protons	5×10^{11} U ions/p 4×10^{13} p/p
SIS300 (synchrotron)	34 GeV/u for U^{92+}	3×10^{11} U ions/s
CR (collector ring)	0.74 GeV/u for U^{92+} 3 GeV for antiprotons	
RESR (accumulator)	0.74 GeV/u for U^{92+} 3 GeV for antiprotons	
NESR (storage ring)	0.74 GeV/u for U^{92+} ~3 GeV for antiprotons	
HESR (storage ring)	0.9–14 GeV for antiprotons	

Hadron physics with antiprotons at HESR is prepared by the PANDA collaboration. In proton–antiproton annihilation many gluons and quark pairs are produced. At 14 GeV, charm quark pairs can also be produced. Hadron spectroscopy including a gluonic degree of freedom and charm quarks will be performed. Hypernuclei with more than single strangeness will be studied with antiprotons. CP violation is also included in the proposal. For these studies, a large scale detector is proposed as the PANDA detector, which is shown in Fig. 8. A pellet target of frozen hydrogen will be used. The PANDA detector is almost hermetic and consists of a forward spectrometer and a target spectrometer with calorimeters. Details of the detector design can be seen in Vol. 3 of Ref. 21. The facility will be also used for atomic physics, plasma physics and applications.

7. Nuclear Physics with High Intensity Accelerator J-PARC

At present, a high intensity proton accelerator that is heavily used for nuclear physics is J-PARC. We describe nuclear physics programs at J-PARC as a good example of nuclear physics with a high intensity accelerator.

The proton beam was successfully accelerated by the main synchrotron up to 30 GeV in 2009. The first extracted beam is used to produce a high intensity neutrino beam. It is used for the long baseline neutrino oscillation experiment, T2K ("Tokai to Kamioka"), together with the Super-Kamiokande detector, which is located about 295 km from J-PARC, as shown in Fig. 9. The mixing angles and mass differences of neutrinos have been measured by solar, atmospheric, reactor and accelerator neutrino oscillation experiments [22]. But, among them, Θ_{13} was not determined well. The major goal of the T2K experiment is to determine this mixing angle with high precision. The neutrino "beam" is produced from π (and K) decays in flight. The neutrino

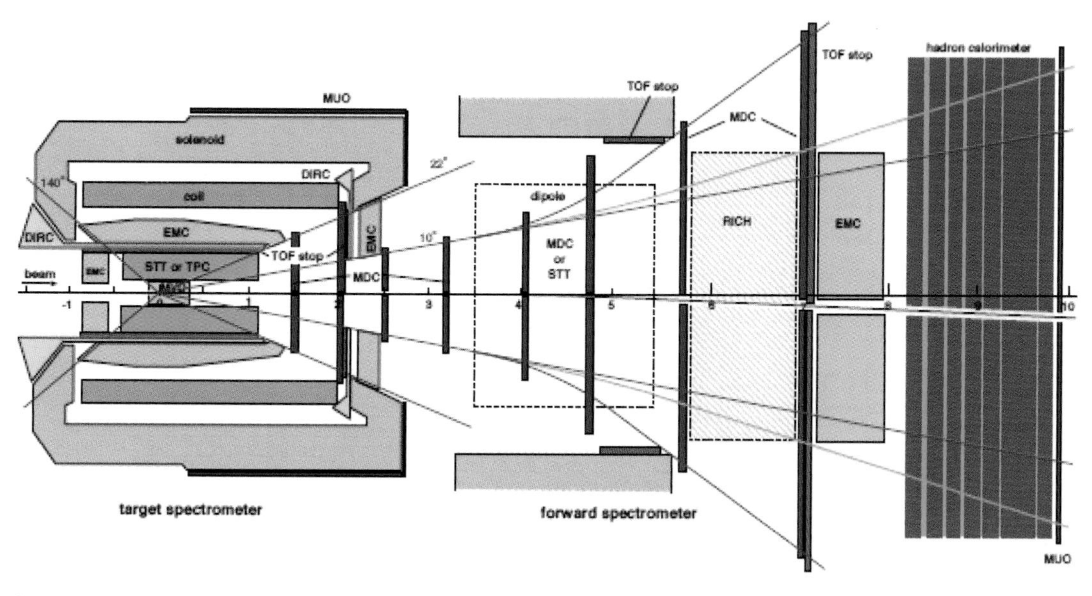

Fig. 8. Setup of the PANDA detector at FAIR. It consists of forward and target spectrometers, and covers almost 4π acceptance.

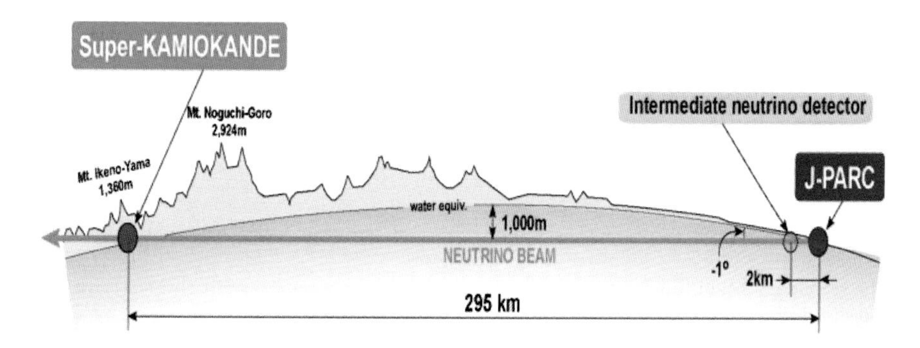

Fig. 9. Schematic view of the T2K long baseline neutrino oscillation experiment.

beam (ν_μ) off-axis of the π beam direction is sent to Kamioka. The mean energy of neutrinos is about 600 MeV, which is optimized for the oscillation during the flight from J-PARC to the Super-Kamiokande detector.

T2K started data-taking at the beginning of 2010. They accumulated data with a total proton beam of 6.4×10^{20} protons on target, and observed 28 ν_e neutrino interactions at the Super-Kamiokande detector by the summer of 2013. Figure 10 shows that the time structure of the neutrino events observed at Super-Kamiokande is consistent with that of the extracted beam from the 50 GeV synchrotron [23].

They have obtained the mixing angle, Θ_{13}, as $\sin^2(2\Theta_{13}) = 0.150 + 0.039 - 0.034$ for the normal mass hierarchy with $\delta cp = 0$ and $\Delta m^2_{23} = 2.4 \times 10^{-3} \mathrm{eV}^2$. The measurement of the CP violation parameter, δcp, in the neutrino mass matrix is

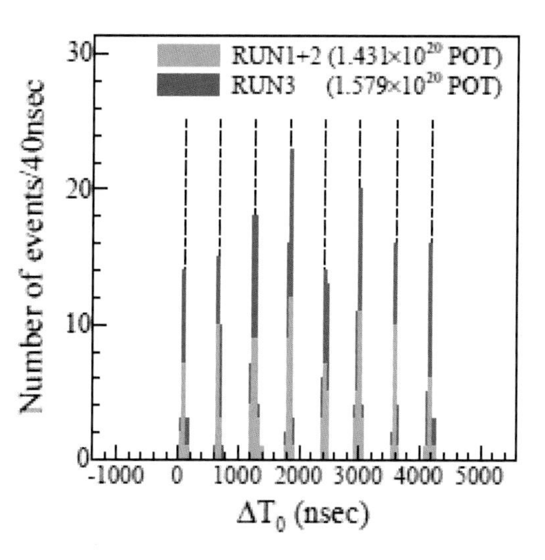

Fig. 10. The time structure of the neutrino events observed at Super-Kamiokande. It is consistent with the eight-beam bunch structure of the fast extraction from the main synchrotron [23].

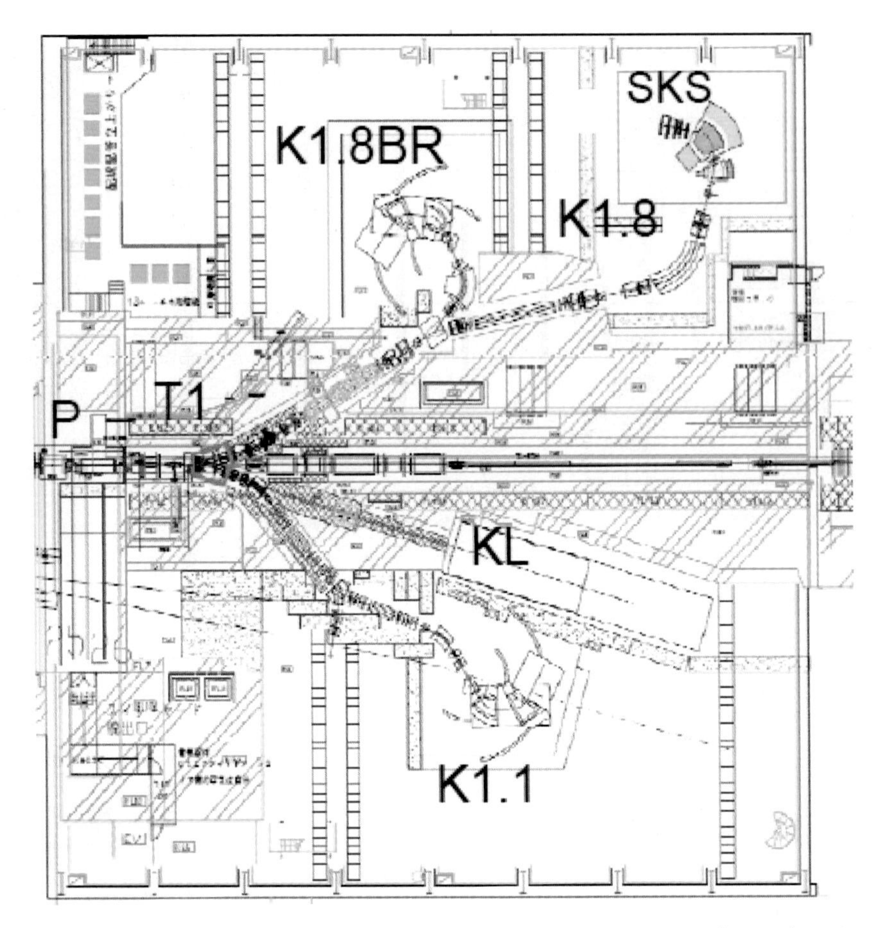

Fig. 11. Layout of the secondary beam lines at the Hadron Hall in 2013. K1.8 and K1.8BR (branch) are located above. The KL line is the straight line shown below. The K1.1 beam line (below) is under construction.

expected to be a future goal of neutrino oscillation experiments.

The slow-extracted proton beam is transported to the Hadron Hall, as shown in Fig. 3. So far there is only one proton beam line and one target for secondary beams, because of budget limitation. The layout of the secondary beam lines at the Hadron Hall at present is shown in Fig. 11. At the Hadron Hall, there are three secondary beam lines, K1.8, K1.8BR and KL, from the T1 target for the initial experiments [24].

K1.1, K1.1BR and high momentum beam lines of primary protons and secondary pions of about 20 GeV/c are to be constructed in a few years. The layout of the beam lines at the Hadron Hall in the near future is shown in Fig. 12. The beam line bent out the present Hadron Hall is for the μe conversion experiment (COMET). The K1.8 beam line can provide high-intensity-separated kaon beams up to 2 GeV/c. It has double-stage electric separators to obtain highly separated kaon beams. It was designed to carry out high resolution spectroscopy with the use of the SKS spectrometer [25]. K1.8BR is a shorter beam line which provides lower momentum kaon beams. The KL beam line provides neutral particles, typically K^0 mesons. K1.1 and K1.1BR will provide low energy kaon beams up to 1.1 GeV/c. One can also use not only kaon but also pion and antiproton beams.

Many proposals of particle and nuclear physics experiments have been submitted to J-PARC, and discussed at the PAC for nuclear and particle physics. So far 25 experiments have been approved and some of them have started. The list of proposals which have been approved by the PAC can be seen at the J-PARC website [26]. Out of the 25 experiments, several large scale experiments for particle physics, including T2K, are being or will be performed. There are 19 experiments for nuclear and hadronic physics [27]. Among them 14 experiments are for strangeness

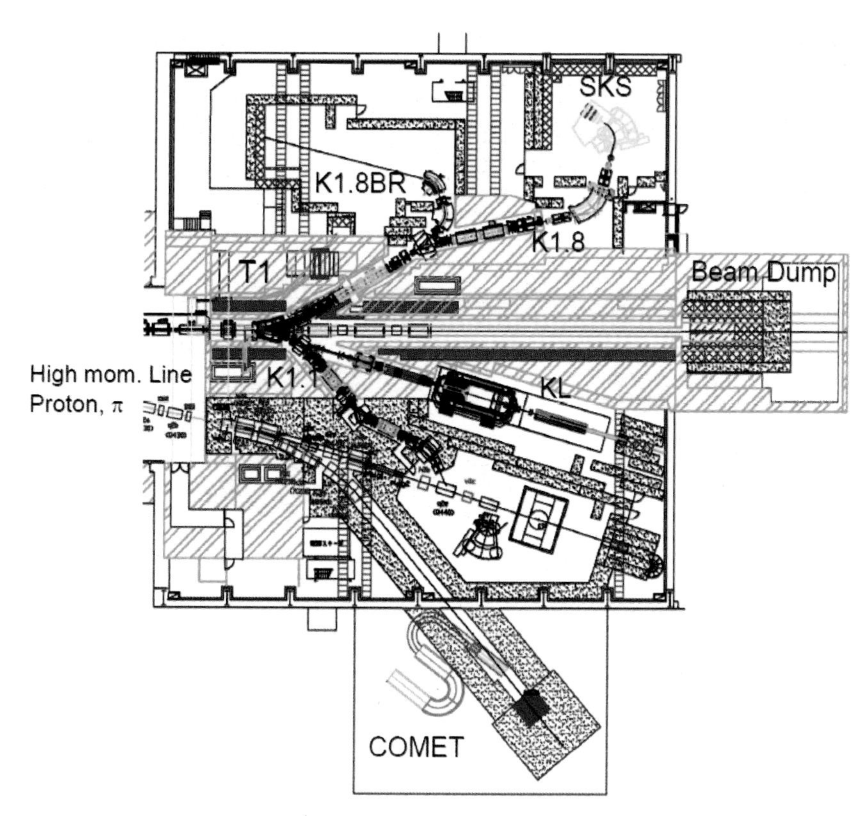

Fig. 12. Layout of the beam lines including the high momentum beam line whose target is far upstream and the COMET experiment which sticks out from the Hadron Hall. It will be realized in the near future (2015).

nuclear physics. We describe several of them here, with the emphasis on strangeness nuclear physics.

Nuclei with strangeness can exist much more than ordinary nuclei. However, only 35 single Λ nuclei and a few double hypernuclei are known. Nuclei with strangeness are, therefore, the frontier of the nuclear world. In the case of multistrangeness, very little experimental information is available. For example, the existence of Ξ hypernuclei is not known. A nuclear chart including strangeness is shown in Fig. 13. This world is explored at J-PARC.

The advantage of high intensity kaon beam is that nuclei with double strangeness can be produced with rather high statistics. The experiment E05 is one of the first priority experiments at the Hadron Hall [28]. The goal of E05 is to find Λ hypernuclei by (K^-, K^+) missing mass spectroscopy with the use of the K1.8 beam line spectrometer and SKS (Superconducting Kaon Spectrometer) spectrometer, which are shown in Fig. 14.

At BNL-AGS, from the analysis of the missing mass spectrum of the $^{12}C(K^-, K^+)$ reaction, an attractive potential of about 14 MeV for the Ξ hyperon and ^{12}B nucleus was suggested, although no peak structure was observed due to the poor energy resolution [29]. Thanks to the good energy resolution of the beam line and SKS spectrometers (2 MeV FWHM), a peak structure of the Ξ hypernucleus for the ^{12}C target is expected to be observed for the first time, and the Ξ potential is determined.

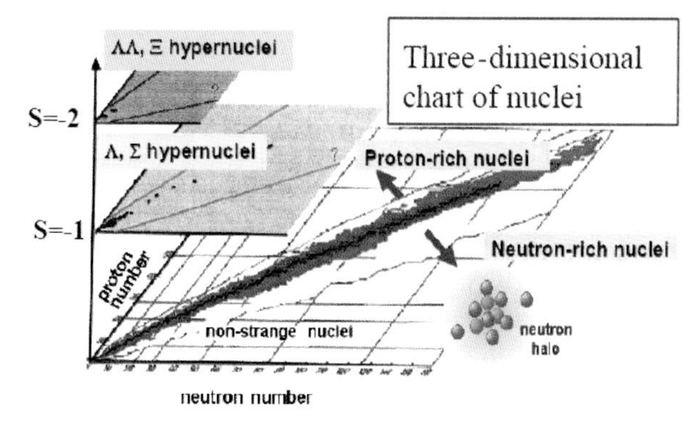

Fig. 13. Nuclear chart with strangeness. Very few nuclei are known for $s = -2$ nuclei.

Fig. 14. The K1.8 beam line and SKS spectrometer at the Hadron Hall. The green magnet is the dipole magnet of the beam line spectrometer. The yellow one is SKS.

The ΞN interaction is important by itself and also for the study of neutron stars. It is now believed that Λ hyperons exist in the core of neutron stars due to much information from Λ hypernuclei. There is, however, a conflict between the recent observation of a neutron star of 2 solar mass and the present knowledge of nuclear matter with hyperons. The experimental study of nuclear matter with hyperons is important for this reason.

Another experiment for nuclei with double strangeness is E07, a hybrid emulsion experiment to search for double Λ hypernuclei [30]. This hybrid emulsion technique was developed and successfully employed in the previous experiments at KEK-PS E176 and E373. In the latter experiment, a clean event of a double Λ hypernucleus, the Nagara event, was found. It was identified as $^{6}\mathrm{He}_{\Lambda\Lambda}$, which is called "Lamdpha," and its binding energy was uniquely determined [31]. The event is shown in Fig. 15.

From this event, the ΛΛ interaction ($\Delta B_{\Lambda\Lambda}$) is determined to be weakly attractive. The lower bound of the H dibaryon mass was also obtained. In these experiments, a few more double hypernuclei were found; however, the Nagara event is only one event which was uniquely identified.

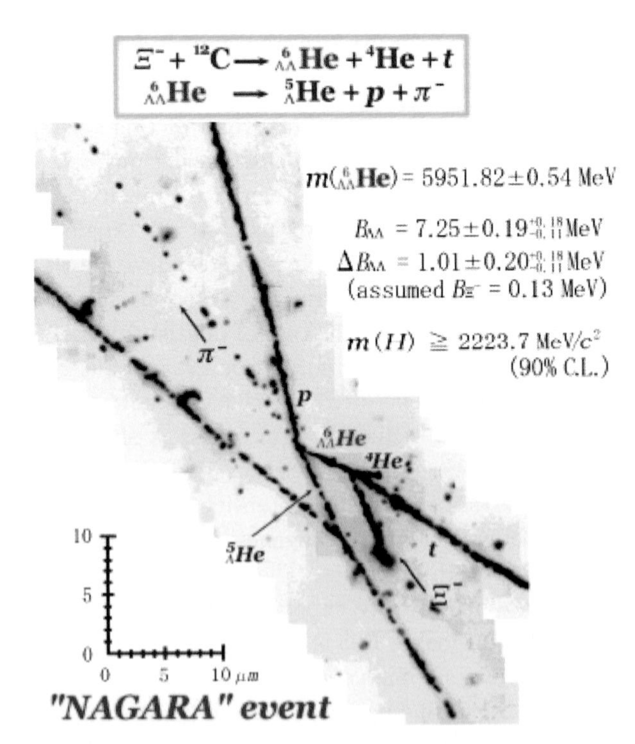

Fig. 15. Production and cascade decays of the double hypernucleus $^{6}\mathrm{He}_{\Lambda\Lambda}$ observed in the emulsion. $^{6}\mathrm{He}_{\Lambda\Lambda}$ was produced from the Ξ^- stopping point, and decays to a pion, a proton and $^{5}\mathrm{He}_{\Lambda}$. Then $^{5}\mathrm{He}_{\Lambda}$ decays again. From this event, the lower bound of the H dibaryon mass was given [31].

At J-PARC, by using the double-stage separators, one can obtain almost pure kaon beams at the K1.8 beam line. In the proposed E07 experiment, observation of 10-times-more double hypernuclei is expected. It will enable us to make a minichart of double hypernuclei and study their systematics. A high speed emulsion scanning system has been developed to analyze a larger amount of emulsion data.

The experiment E03 tries to measure X-rays from Ξ atoms for the first time [32]. The Ξ^- hyperons are produced again by the (K^-, K^+) reaction and stopped in a target where Ξ atoms are formed. Cascade X-rays from the Ξ atoms are detected with high efficiency Ge detectors, called Hyperball-J. The shifts of the X-ray energies due to the Ξ nucleus strong interaction potential can be measured. One can, therefore, study the Ξ nucleus potential. The X-rays from Ξ atoms of emulsion materials such as Ag and Br are also measured in E07. However, E03 expects higher statistics. Iron is used as a target to produce Ξ^- hyperons, and also as the stopping material. E03 and E07 do not need a high resolution spectrometer such as SKS. They plan to use the KURAMA spectrometer, which has a larger acceptance than SKS.

High precision spectroscopy of Λ hypernuclei with a large Ge detector called the Hyperball was one of the most productive experiments at KEK-PS and BNL-AGS. Hypernuclei were produced by (π^+, K^+) reactions at KEK-PS and by (K^-, π^-) reactions at BNL-AGS. The γ rays from most p shell hypernuclei were then detected with the Hyperball surrounding the targets, and their energy levels were precisely determined [33]. From these studies, spin-dependent ΛN interactions such as spin–spin, spin–orbit and tensor interactions were determined for p shell hypernuclei. One of the striking discoveries is the smallness of the spin–orbit interaction of hypernuclei. It is smaller almost by two orders of magnitude than the NN case. Figure 16 summarizes the energy levels previously measured by the Hyperball. The level splitting due to the ΛN spin–orbit interaction for $^9Be_\Lambda$ is only 43 keV. By using the Doppler method, B(E2) was measured for the E2 transition of $^7Li_\Lambda$ hypernuclei and a charge radius of the hypernucleus was found to be reduced by 20% compared to that of ^{6}Li.

A high precision spectroscopy experiment was proposed to J-PARC and approved as E13 [34]. For the J-PARC experiment, the Hyperball was upgraded to Hyperball-J. The peak efficiency was greatly improved by using clover-type Ge detectors. The anti-Compton suppressor was improved from BGO to PWO, which is much faster and can be used for higher beam rates. The initial program is to complete the γ spectroscopy of p shell hypernuclei and extend to sd shell hypernuclei. The measurement of B(M1) of $^7Li_\Lambda$ is also tried in order to measure the magnetic moment of the Λ hyperon in nuclear matter.

There are other hypernuclear physics experiments, such as spectroscopy of neutron-rich hypernuclei and weak decays. By adding a Λ hyperon to nuclei, unbound nuclei can be bound with Λ, such as $^9Be_\Lambda$, where ^{8}Be is unbound. Nuclei beyond the neutron drip line can, therefore, be bound for hypernuclei. One of such nuclei is $^6H_\Lambda$, where ^{5}H or even ^{4}H is unbound. These neutron-rich hypernuclei can be studied by the (π^-, K^+) reaction with a high resolution spectrometer. The experiment E10 searching for $^6H_\Lambda$ has already taken some data recently [35].

The search for kaonic nuclei is another high priority experiment. T. Yamazaki and Y. Akaishi predicted the kaon–nucleus bound state, and the lightest one is the K^-pp state [36]. If $\Lambda(1405)$ is a K^-p bound state, it is a natural application to heavier nuclei. There are many theoretical papers and a few experiments to support its existence. At J-PARC, the K^-pp state is being searched through ^{3}He(K^-, n) reaction with neutron TOF detectors and a cylindrical detector surrounding the ^{3}He target at the K1.8BR line [37]. At the K1.8 line, the same state has been searched via the $d(\pi^+, K^+)$ reaction with the SKS spectrometer [38]. The results will be obtained soon.

The chiral symmetry is expected to be restored in high temperature or high density nuclear matter. Partial restoration is expected at even normal nuclear density. At the coming high momentum beam line, an experiment to study the partial restoration of chiral symmetry in cold nuclear matter has been approved. With the use of a high momentum proton beam, possible mass shifts of vector mesons such as ρ, ω and ϕ in nuclei will be measured by their e^+e^- decays. The previous experiment at KEK claimed an evidence of the mass shift of ρ and ϕ mesons in nuclei [39]. The planned experiment will provide data with higher statistics and a wider

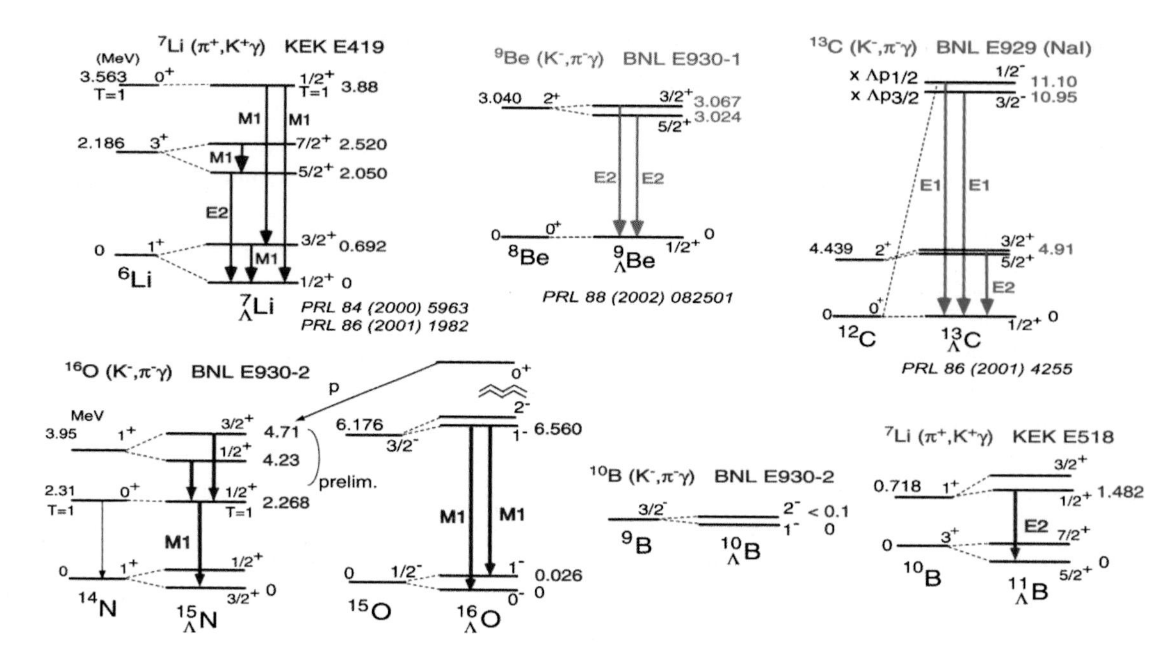

Fig. 16. Summary of γ transitions in p shell nuclei.

kinematic range as an extension of the previous KEK experiment [40].

An exotic hadron such as a pentaquark Θ^+ was also searched for by the $p(\pi^-, \mathrm{K}^-)X$ reaction at the K1.8 line as the first experiment at the Hadron Hall. It was the high statistics and high resolution (2 MeV FWHM) measurement. The result was, however, negative and gave the upper limit for its width as about 1 MeV [41]. Another exotic hadron recently suggested by lattice QCD calculations is the H dibaryon near the $\Lambda\Lambda$ threshold. A new experiment is being prepared to search for the H dibaryon with a hyperon spectrometer which is under construction [42].

In particle physics, there are several large scale experiments besides T2K. At the KL beam line, the $\mathrm{K}_L \to \pi^0\nu\nu$-bar decay experiment (KOTO) is being performed. It is the CP-violating rare decay and is expected to be sensitive to new physics [43]. The search for lepton flavor violation is proposed by measuring muon–electron conversion at the sensitivity of 10^{-16}. The pulsed proton beam is extracted from the MR to the Hadron Hall through the planned high momentum beam line. The muon beam is produced and transported to the COMET detector [44]. The measurements of g-2 and the electric dipole moment of the muon are also proposed with 10-times-better precision [45]. They look for the physics beyond the standard model.

In conclusion, J-PARC is the world's highest intensity proton accelerator for nuclear and particle physics. Much new physics is foreseen.

8. Summary

The high intensity proton accelerator has had a long history since the meson factory era. High intensity proton accelerators up to a-few-GeV energy are now widely used as the neutron and muon sources for material and biological science and applications. The field is expanding and a facility of higher beam power is planned.

For nuclear and particle physics, a high intensity proton accelerator of higher energies is awaited. The intense secondary particles, such as kaons, neutrinos, antiprotons and other hadrons at high energies, are quite useful for nuclear and particle physics. For particle physics, it is an intensity frontier complementary to the high energy frontier, like LHC. For nuclear physics, one can expand the frontier of nuclei, such as hypernuclei, kaonic nuclei and neutron-rich nuclei, with the use of high intensity accelerators. J-PARC has been constructed as such a facility, and it started its operation recently. The long baseline neutrino oscillation experiment (T2K) has already measured the mixing angle Θ_{13}. The frontier of the nuclear world will be fully explored at J-PARC. FAIR will join in, in the near future. The intensity frontier of the accelerator will thus expand the frontier of nuclear physics.

Acknowledgments

The author would like to thank Dr. S. Nagamiya and members of the J-PARC center and of the hadron physics group of Japan Atomic Energy Agency for preparing this article.

References

[1] K. Adcox *et al.* (PHENIX collaboration), *Nucl. Phys. A* **757**, 184 (2005).
[2] I. Tanihata, H. Savajols and R. Kanuga, *Prog. Part. Nucl. Phys.* **68**, 215 (2013).
[3] R. Heine *et al.*, in *Proc. iPAC 2010* (23–28 May 2010).
[4] V. D. Burkert, arXiv:1203.237301[nucl.ex] (2012).
[5] M. Yosoi, *AIP Conf. Proc.* **1388**, 163 (2011).
[6] H. Okuno, N. Fukunishi and O. Kamigaito, *Prog. Theor. Exp. Phys.* 03c002 (2012).
[7] S. Gales, *AIP Conf. Proc.* **1238**, 26 (2010).
[8] J. W. Xia *et al.*, *Nucl. Instrum. Methods Phys. Res. A* **488**, 11 (2002).
[9] NUSTAR Progress Report, eds. J. Gerl *et al.* (2012).
[10] T. Motobayashi and H. Sakurai, *Prog. Theor. Exp. Phys.* 03c001 (2012), *ibid.* 03c003–03c009.
[11] K. W. Jones and K. F. Schoenberg, *Proc. LINAC08* (2008), p. 88.
[12] J. Adam *et al.* (MEG collaboration), *Phys. Rev. Lett.* **110**, 201801 (2013).
[13] B. C. Brown *et al.*, *Phys. Rev. ST Accel. Beams* **16**, 071001 (2013).
[14] Accelerator Technical Design Report for J-PARC, KEK Report 2002-13, JAERI-Tech 2003-044, J-PARC 03-01 (2003).
[15] C. Ohmori *et al.*, High gradient cavity for JAERI–KEK joint project, in *Proc. 8th Eur. Part. Accel. Conf.* (2002), p. 257.
[16] M. Ikegami, *Prog. Theor. Exp. Phys.* 02B002 (2012).
[17] H. Hotchi *et al.*, *Prog. Theor. Exp. Phys.* 02B003 (2012).
[18] T. Koseki *et al.*, *Prog. Theor. Exp. Phys.* 02B004 (2012).
[19] MLF Annual Report 2011 (2011).
[20] Y. Arimoto *et al.*, *Prog. Theor. Exp. Phys.* 02B007 (2012).
[21] FAIR Baseline Technical Report, eds. H. H. Gutbrod *et al.* (2006).
[22] T. Schwtz, M. A. Tortola and J. W. F. Valle, *New J. Phys.* **10**, 113011 (2008).
[23] K. Abe *et al.* (T2K collaboration), *Phys. Rev. D* **88**, 032002 (2013); T. Sekiguchi, *Prog. Theor. Exp. Phys.* 02B005 (2012).
[24] K. Agari *et al.*, *Prog. Theor. Exp. Phys.* 02B008, 02B009 (2012).
[25] T. Takahashi *et al.*, *Prog. Theor. Exp. Phys.* 02B010 (2012).
[26] http://jparc.jp/researcher/Hadron/en/Experiments_e.html
[27] M. Naruki, *Prog. Theor. Exp. Phys.* 02B013 (2012); H. Tamura, *Prog. Theor. Exp. Phys.* 02B012 (2012).
[28] T. Nagae *et al.*, J-PARC E05 proposal.
[29] P. Khaustov *et al.* (AGS-E885), *Phys. Rev. C* **61**, 054603-1-7 (2000).
[30] K. Nakazawa, K. Imai and H. Tamura, J-PARC E07 proposal.
[31] H. Takahashi *et al.* (KEK-E373), *Phys. Rev. Lett.* **87**, 212502-1-5 (2001).
[32] K. Tanida *et al.*, J-PARC E03 proposal.
[33] H. Tamura *et al.*, *Nucl. Phys. A* **835**, 3 (2010).
[34] H. Tamura *et al.*, J-PARC E13 proposal.
[35] A. Sakaguchi *et al.*, J-PARC E10 proposal; H. Sugimura *et al.*, in *Proc. International Nuclear Physics Conference 2013* (Florence, Italy, 2013).
[36] T. Yamazaki and Y. Akaishi, *Phys. Rev. C* **76**, 45201 (2002).
[37] M. Iwasaki, T. Nagae, *et al.*, J-PARC E15 proposal; K. Agari *et al.*, *Prog. Theor. Exp. Phys.* 02B011 (2012).
[38] T. Nagae, J-PARC proposal E27.
[39] R. Muto *et al.*, *Phys. Rev. Lett.* **98**, 042501 (2007).
[40] S. Yokkaichi *et al.*, J-PARC proposal E16.
[41] K. Shiritori *et al.*, *Phys. Rev. Lett.* **109**, 132002 (2012).
[42] J. K. Ahn, K. Imai *et al.*, J-PARC proposal E42.
[43] T. Yamanaka *et al.*, J-PARC proposal E14; T. Yamanaka, *Prog. Theor. Exp. Phys.* 02B006 (2012).
[44] Y. Kuno *et al.*, J-PARC proposal E21.
[45] N. Saito, M. Iwasaki, *et al.*, J-PARC proposal E34.

Kenichi Imai obtained his Ph.D. (Nuclear Physics) from Kyoto University in 1975. He was Assistant Professor, Associate Professor and Professor at the Department of Physics, Kyoto University until 2010; Research Associate at Argonne National Laboratory from 1980 to 1982; Guest Senior Researcher at RIKEN (1995–2010) and Guest Professor at KEK (2002–2007). From 2010 to the present, he has been Group Leader at the Advanced Science Research Center and a member of the J-PARC Center, Japan Atomic Energy Agency. He is Professor Emeritus at Kyoto University. His major research field is spin physics and strangeness nuclear physics.

Reviews of Accelerator Science and Technology
Vol. 6 (2013) 37–57

DOI: 10.1142/S179362681330003X

Radioactive Ion Beams and Radiopharmaceuticals

R. E. Laxdal*, A. C. Morton† and P. Schaffer‡
TRIUMF, 4004 Wesbrook Mall,
Vancouver, BC, V6T 2A3, Canada
**lax@triumf.ca*
†morton@triumf.ca
‡pschaffer@triumf.ca

Experiments performed at radioactive ion beam facilities shed new light on nuclear physics and nuclear structure, as well as nuclear astrophysics, materials science and medical science. The many existing facilities, as well as the new generation of facilities being built and those proposed for the future, are a testament to the high interest in this rapidly expanding field. The opportunities inherent in radioactive beam facilities have enabled the search for radioisotopes suitable for medical diagnosis or therapy. In this article, an overview of the production techniques and the current status of RIB facilities and proposals will be presented. In addition, accelerator-generated radiopharmaceuticals will be reviewed.

Keywords: Radioactive ion beams; ISOL; in-flight; radiopharmaceuticals; accelerators.

1. Introduction

Modern accelerator technology has allowed nuclear physics to move to a new era where intense mass-selected radioactive ion beams (RIB) are available not only for studies of nuclear structure, but also for nuclear astrophysics and materials science. Many facilities are undergoing significant upgrades; as well, new facilities are being built or are being proposed, ranging from large global scale projects to small niche installations. In nuclear medicine, advances have come in step with the progress in RIB technology. Modern accelerators complement nuclear reactors in producing selected radionuclides used in the preparation of radiopharmaceuticals for diagnostic and therapeutic applications. Harvesting of these radionuclides using online production and separation techniques developed in the RIB community will allow access to new and promising imaging isotopes, and contribute to the optimization of radiotherapeutic isotopes.

2. Introduction to RIBs

In the last two decades a new scientific discipline has emerged from nuclear physics research worldwide: the production and study of energetic radioactive beams. The availability of energetic beams of short-lived radioactive ions, referred to as radioactive ion or rare isotope beams (RIBs), has enabled the study of the structure and dynamics of thousands of nuclear species never before observed in the laboratory [1]. These exotic beams, produced by two complementary techniques — in-flight fragmentation and isotope separation online (ISOL) — are now available at a number of facilities. The ISOL method was developed in the 1950s using some of the light ion accelerators that were available at that time. In the '70s and '80s several high energy heavy ion accelerators were built and as a consequence enabled production of RIBs through projectile fragmentation. A new generation of large scale RIB facilities is being built. The field of RIB physics is linked primarily to the study of nuclear structure under extreme conditions, but also to nuclear astrophysics, solid state physics, medical physics and the study of fundamental interactions. The advent of rare isotope beams has made it possible to explore how the properties of nuclei evolve with the ratio of neutrons to protons. These investigations have led to several major discoveries, such as the existence of halo nuclei, the modification of

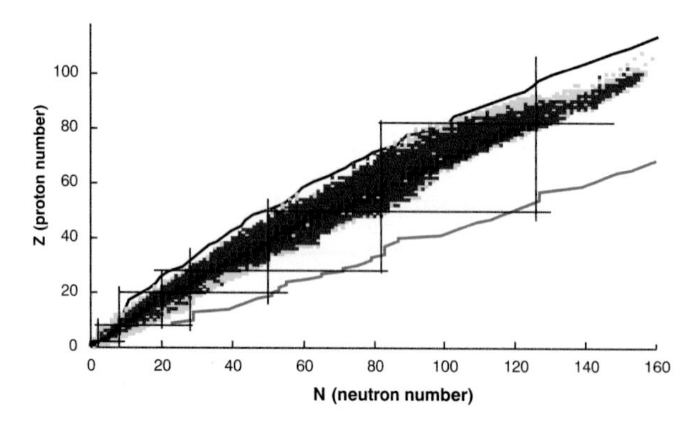

Fig. 1. Graph of the nuclides with predicted proton and neutron drip lines [6]. Black squares denote stable nuclei; pink, those for which excited states are known; and yellow, those with measured masses.

shell structure and magic numbers far from stability, decoupling of neutron and proton distribution, and proton and two-proton radioactivity.

Recent theoretical work predicts that there are roughly 7000 bound nuclei [2]. Of these, roughly 3000 either occur naturally or have been produced in laboratories (Fig. 1) [3–5]. RIB development continues to increase the number of nuclei available for study, particularly in regions far from stability.

2.1. *Historical overview*

In 1919 Rutherford demonstrated the transmutation of nitrogen into oxygen by impinging alpha particles from a radium source on a nitrogen gas cell and detecting protons from the ^{14}N(α, p)^{17}O reaction. In 1932 Cockcroft and Walton demonstrated an artificial nuclear reaction and transmutation by bombarding ^{7}Li with accelerated protons to produce alpha particles. The first artificial radioactive isotopes were produced by Curie and Joliet in 1934, by bombarding targets of boron and aluminum with alpha particles to produce ^{13}N and ^{30}P. Hahn and Strassmann discovered nuclear fission in 1938, producing barium after bombarding uranium with neutrons. Soon after, with the advent of nuclear fission reactors, many new species were created and studied. Hansen and Nielson are credited with the first demonstration of the ISOL method in 1951. Accelerated 11 MeV deuterons produced neutrons that were used to bombard a uranium target [7]. The fission products were thermalized in the target and delivered to an ion source with the ions subsequently selected in a mass separator. Klapisch and Bernas coupled an online isotope separator to the Orsay synchrocyclotron in France and produced beams of light lithium isotopes [8]. At CERN, the ISOLDE online facility, using the 600 MeV proton synchrocyclotron as a driver, began operation in 1964 and led the way for other modern ISOL facilities [9]. The first postaccelerator came online in 1989, when two cyclotrons were coupled together, one to produce radioactive species and one to accelerate and separate the products, at the Université Catholique de Louvain in Louvain-la-Neuve [10]. The Holyfield Radioactive Ion Beam Facility (HRIBF) at Oak Ridge National Laboratory came online in 1998 [11] and made use of a 50 MeV proton cyclotron as driver and a large 20 MV electrostatic tandem as postaccelerator. Finally, the ISAC ISOL facility at TRIUMF came online in 1999 [12] with an unprecedented driver power of 50 kW from the laboratory's 500 MeV main cyclotron.

The first in-flight separation was done at Oak Ridge, where a gas-filled magnetic separator was used to separate and identify fission products from a reactor [13]. Later, magnetic and electric fields were used at Garching to select fission products from a research reactor [14]. The idea of fragmenting fast heavy ions in a thin target was first demonstrated in the 1970s with the Bevalac accelerator at Lawrence Berkeley Laboratory (LBL) [15]. Later experiments in the 1980s led to the discovery of halo nuclei for very-neutron-rich ions [16, 17]. Many modern RIB facilities have grown out of existing infrastructure. Cyclotron-driven (NSCL, GANIL, RIKEN) and synchrotron-driven (GSI) heavy ion facilities developed in-flight fragmentation capabilities by adding target systems and either repurposing existing transport lines for filtration or installing dedicated fragment separators. Dedicated searches for superheavy elements were initiated at LBL [18], JINR [19] and GSI [20] during this period.

2.2. *RIB production methods*

The production of radioactive ions far from stability (i.e. those with neutron/proton ratios significantly different than those of their stable counterparts) is typically challenged by low cross-sections, isobaric contamination and short half-lives. A common feature is that the ion of interest is transported from the place of production to an isolated experimental setup where the nuclear properties can be studied with sufficient sensitivity. Beam transport serves to purify and to prepare the beam with respect to energy and energy spread, bunch length and ion optical properties for the experiments. Classical in-flight facilities are ideal for high energy beams of very-short-lived species, while classical ISOL facilities allow high quality, low emittance beams of lower energy and generally higher yields near stability. Hybrids of the two techniques are being proposed.

The main components of a classical in-flight facility [21, 22] are a high energy, heavy ion accelerator followed by a fragment separator. RIBs are produced by fragmentation of the projectiles in a thin target and are separated and selected in-flight. The incident beam is typically in the 50–200 MeV/u range and the radioactive fragments recoil in a forward angle cone with velocities similar to that of the incoming beam so that postacceleration is not required. The beam qualities compared to ISOL are significantly poorer due to the interactions in the target and in the degraders of the fragment separator. Since the flight time from target to experiment is typically of the order of hundreds of nanoseconds, the in-flight method is fast, enabling the study of ions with very short half-lives. Furthermore, the method is not sensitive to target/ion-source chemistry issues since the production mechanisms are largely dependent on the kinematics of high energy nuclear reactions.

In the classical ISOL technique [23, 24] an intense primary beam, generally of light hadrons, from a driver accelerator impinges on a thick target. In contrast to in-flight production, it is typically the target material rather than the projectile that is fragmented to create the RIB product. The driver beam initiates fission, spallation and fragmentation reactions in the target material, and the exotic isotopes that are produced diffuse and effuse out of the heated target through a transfer line to a neighboring ion source. The isotopes are ionized, accelerated from a biased platform and purified using an electromagnetic mass separator. The ions can then be delivered for experiments at low energy or accelerated to higher energies. Excellent beam qualities result due to the low initial transverse momentum, though the beam energies are typically lower than those at in-flight facilities. The range of species that can be produced is also limited, due to the chemical and transport properties of the atoms in the target, the half-life of the ion and the ease with which they are ionized. The two methods, shown schematically in Fig. 2, are complementary and several successful examples of each are in operation or under construction.

New facilities combining features of the two approaches are being proposed or built. Gas stoppers [25–27] can be used to thermalize fast-fragment beams (Fig. 2), which are then reaccelerated in a manner similar to that used at ISOL facilities to provide beams of comparable quality. Alternatively, ISOL facilities can be used to produce intense beams of neutron-rich fission products near stability (such as ^{132}Sn or ^{91}Kr) for use as primary beams for in-flight production. This hybrid approach allows the production of very-neutron-rich species at intensities higher than those that can be obtained with stable projectiles [28].

3. In-Flight Separation

3.1. *Drivers*

Important factors for the driver accelerator for in-flight facilities are the energy of the primary beam, primarily affecting beam quality; the intensity of the beam, determining the production yields; and the range of ion masses available, determining the range of final products. An advantage of the in-flight method is that the target dissipates only a fraction of the beam power; nonetheless, for high intensity facilities, the target area, beam dump and preseparator sections provide complex engineering challenges due to the beam power density and the associated activation and radiation from both the primary and the secondary beam.

Cyclotrons (e.g. at RIKEN) and, more recently, superconducting linacs (FRIB) are common technology choices. The two are similar, in that they can accelerate in continuous wave (CW) mode to deliver the maximum intensity for a given peak beam intensity limit. High energy cyclotron facilities utilize two or more cyclotrons in a cascaded way with stripping between stages to aid injection and to reduce the required total voltage. Superconducting magnet technology is used in high energy cyclotrons to reduce the weight of the magnet by allowing an increase in the field. Because of their multipass nature cyclotrons make efficient use of the RF structure but are complex to tune, increasing

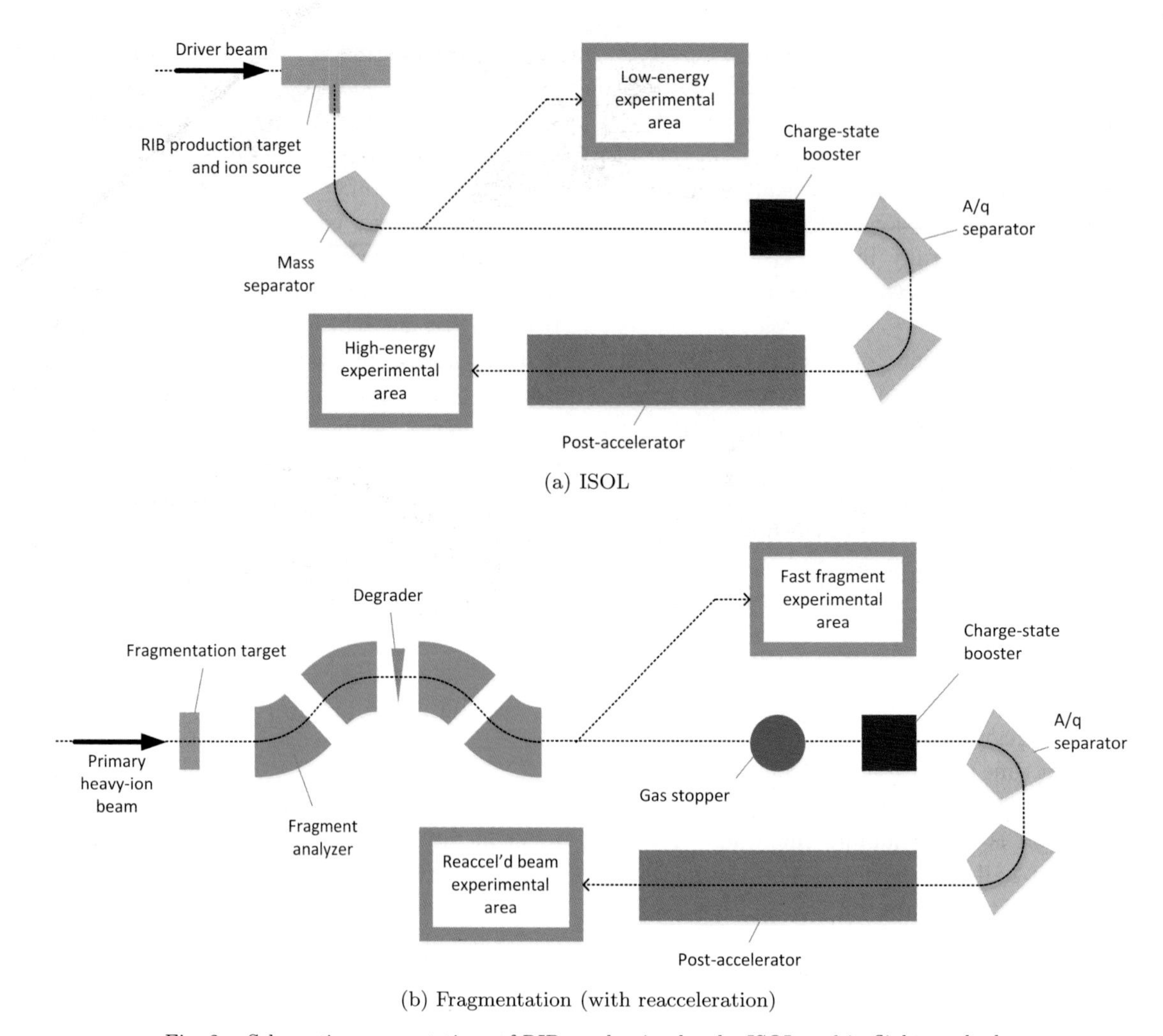

(a) ISOL

(b) Fragmentation (with reacceleration)

Fig. 2. Schematic representations of RIB production by the ISOL and in-flight methods.

tuning time, and typically experience beam loss at injection and extraction, reducing the final beam intensity. Hadron linear accelerators, on the other hand, have straightforward injection and extraction but are single pass devices. Superconducting radio frequency (SRF) technology has made acceleration in linacs much more efficient. It also allows operation at lower microwave frequencies and with larger apertures than would be typical for similar room temperature structures, since shunt impedance is not of paramount concern. The resulting large transverse and longitudinal acceptance allows the acceleration of high intensities and, importantly, the possibility of accelerating more than one charge state, reducing the impact to the final beam intensity from stripping [29]. SC driver heavy ion linacs cover a wide velocity range. A flexible design choice is to use several varieties of short double gap accelerating structures that each have a broad velocity acceptance so that ions from 500 keV/u to relativistic velocities can be covered with a few cavity types.

Synchrotrons (GSI) operate with linac drivers and can be cascaded to efficiently reach higher energies than those available to cyclotrons. For heavy ion facilities with $E > 400\,\mathrm{MeV/u}$, synchrotrons are the technology of choice due to their efficient use of RF and their compact annular footprint. Synchrotrons are cycling machines and so are pulsed by design. From this point the average current is much less than the peak intensity. The staccato beam delivery produces rapid thermal cycling of the target and can promote premature aging. Synchrotrons offer flexibility in extraction schemes: slow extraction over several seconds can be employed to approach a CW beam, while fast extraction allows for a well-defined production time of the secondary beam for injection into a storage ring. Furthermore, due to their higher energies, larger target thicknesses can be employed while secondary emittances are improved by increasing the transmission through a separator.

3.2. *Production methods*

Various production processes can be employed to produce beam for in-flight systems, including fragmentation, fission, direct reactions and Coulomb excitation. *Projectile fragmentation* involves an interaction of the projectile with a target nucleus where a small number of nucleons are removed. This is effective at producing neutron-deficient nuclei and some light neutron-rich nuclei. Both the Coulomb deflection and the recoil are small compared to the initial momentum. The observed fragment cross-sections are relatively constant, from approximately 40 MeV/u to 2 GeV/u [30]. A potentially useful feature of these reactions is that the nuclear spin of fragments produced at finite angles can be polarized. *Projectile fission* produces neutron-rich species. The fission process imparts reaction energies of the order of 1 MeV/u. As long as the forward-going energy is large with respect to the imparted recoil energy, the emerging beam can be efficiently transported in a downstream fragment separator. *Direct reactions* are processes complementary to the fragmentation process. These processes can have reasonable cross-sections and are selective so that a significant fraction of the projectile can be converted into a single product in a single nuclear state. This is advantageous since the final energy spread of the product can be kept small if only a single state is populated. In this case the energy spread is dominated by the thickness of the target. Efficient direct transfer reactions are limited to particular species near stability. *Fusion reactions* can occur if the projectiles are within a narrow energy window near the Coulomb barrier. This method is used to produce superheavy elements. *Coulomb excitations* can be used efficiently to produce neutron-deficient beams near stability. A high energy ion moving past a heavy target can excite the giant dipole resonance [31]. Subsequent decay is primarily by neutron emission. The secondary beams have recoil energies given by the decay energy of the neutron, $\sim 77/A^{1/3}$ MeV, so for incident beams of order 1 GeV/u the recoil spreading is a small effect. In principle, beams of specific heavy nuclei, near stability, can be produced with good emittances and with beam intensities of more than 10% of the primary beam intensity [21].

3.3. *Production targets*

The choice of target material is a balance between the nuclear cross-section and electronic interactions that impact energy loss, energy straggling and multiple scattering. Low Z target materials are preferred, due to the larger number of atoms for a given number of electrons. Beryllium and, to a lesser extent, graphite (carbon) targets are common. Small beam spots help improve separator resolution and acceptance, and also reduce emittance growth due to nuclear reaction kinematics and multiple scattering, but exacerbate target heating. Typically, the percentage of beam energy lost in the target is similar to the percentage momentum acceptance of the separator. Future high intensity facilities will drive the development of target technology, as more and more beam power will be deposited the target. Advanced designs for cooling targets have included a rapidly rotating graphite wheel in some cases with a multislice target to aid heat removal with power density in the range of tens of MW/cm^3 [32]. Liquid lithium is another choice for target material, partly because it is also an excellent cooling medium [33].

3.4. *Separation methods*

3.4.1. *Achromatic fragment separator*

The reaction products for fast projectiles are typically separated with an achromatic magnetic separator [34–36]. An in-flight separator typically has a $B\rho$ selection (proportional to A/Q if the velocity difference is not significant) in several dipole stages, with an optional ΔE degrader stage in between, complemented with higher order ion-optical devices. The degrader introduces a Z-dependent energy loss to create separation when several nuclei with the same A/Q are produced. The degrader can be shaped in a wedge to cancel the energy dispersion developed before the wedge to deliver all beams of a similar Z and Q to a focus at the image plane using a downstream achromat. A typical fragment separator with wedge degrader is shown in Fig. 3.

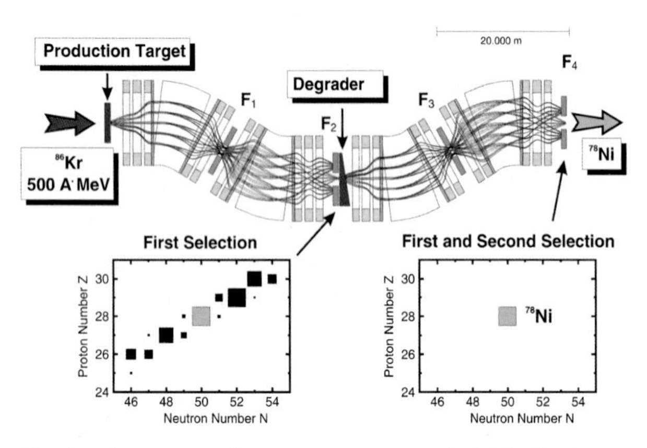

Fig. 3. A two-stage fragment separator with A/Q selection at F_2 and Z selection using a degrader to isolate a single isotope (in this case, ^{78}Ni) at F_4[37].

In-flight separators adapted to higher beam intensities can have further separation stages, including a preseparator, in order to safely separate the primary beam from the secondary species and dump the primary in a controlled manner in dedicated beam dumps. The secondary beam rate is directly related to the primary intensity and the relative separator acceptance. The total emittance of secondary beams is determined by the combination of the nuclear reaction kinematics and atomic processes such as multiple angular scattering and energy loss straggling in the production target and in any degrader. Large solid angle and large momentum acceptance are especially important if the device is to be used to separate light ions at 50–200 MeV/u.

3.4.2. *Solenoid*

For lighter ion beams of lower energy, 10–50 MeV/u, the in-flight technique has been used particularly with transfer reactions where highly forward-peaked exotic nuclei close to stability are produced [38]. A superconducting solenoid can act as an efficient fragment collector, especially at energies around the Fermi energy, where the kinematic focusing of the fragments is not as strong as in higher energy fragmentation reactions. However, the inherently small dispersion of the solenoid results in rather poor selection of RIBs of interest. In some cases the solenoids are cascaded with a slit and degrader in between for two-stage filtration, or a single solenoid with a conventional degrader behind can also be considered [39].

3.4.3. *Cooler*

The pulsed secondary beams available from synchrotron-based in-flight facilities during fast extraction are suitable for injection, beam manipulation and experiments in storage rings. For precision experiments the beam can first be cooled stochastically, then with electron cooling, in order to reach $\Delta p/p \sim 10^{-4}$ or better. The method of storing high energy RIBs has been most successful in measuring masses through Schottky mass spectrometry [40–42] on electron-cooled beams. Beams from CW facilities can also be considered for injection into rings (RIBF, HIE-ISOLDE).

3.4.4. *Gas-stopper*

An alternate method for handling the secondary beams from fragmentation targets is to slow down the ions in a buffer gas and then extract the ions at very low energies [25–27]. When the radioactive ions are thermalized in ultrapure helium gas, a large fraction remain in the 1+ or 2+ charge states, depending critically on the purity, pressure and residence time in the gas. The thermalized radioactive ions can be extracted in a few tens of milliseconds by drifting the ions to a supersonic nozzle. The beams can be studied at source potential or injected into a postaccelerator, as is done in the ISOL method, to achieve a low emittance ion beam. Various gas-stoppers have been built (e.g. at RIKEN, ANL, NSCL) or are under construction (FRIB).

4. ISOL Method

4.1. *Drivers and projectiles*

As at in-flight facilities, the range and yield of radioactive ions produced at ISOL facilities are dependent on the energy and intensity of the driver beams used. Since the ISOL technique employs thick production targets, driver beam intensities need to be chosen while considering the power-handling capability of the target. For high intensity facilities the target/source area, beam dump and preseparator sections contain the bulk of the activity and pose the most complex engineering challenges for the facility [43]. Remote handling, precise shielding, air handling, radiation monitoring, and licensing all are key issues for such a facility.

In contrast to in-flight production, it is the target material rather than the projectile that is typically used to produce the unstable products. ISOL projectiles are chosen for their ability to break up the target material in one or more production mechanisms. High energy ($\sim$500 MeV – 1.5 GeV) protons produce a wide variety of radioactive ions that can be either proton- or neutron-rich via both spallation and fission reactions. High energy protons have a large range in a target, thereby reducing the power density for a given current. Beam powers of 50 kW on target at 500 MeV have been demonstrated at ISAC [44]. A compact H^- cyclotron is a cost-effective driver accelerator and 70 MeV units are available commercially [45]. Lower energy proton drivers produce a limited range of (mostly) proton-rich radioactive ions through (p, n) and (p, 2n) reactions. The short range of 50 MeV protons in typical target materials leads to high power densities in targets, and limits the maximum current that can be used. Cyclotrons produce a CW beam that is optimal for reducing the thermal aging of the target. A synchrotron is used at CERN to deliver high energy, 1.4 GeV protons for ISOLDE. The cycling nature leads to a pulsed beam structure that may contribute to premature aging. Superconducting linacs can also be considered for use as high intensity proton drivers [46]. A new generation of high intensity proton drivers for accelerator-driven systems, neutron spallation sources and muon production will establish the technology for next-generation RIB drivers.

In some production schemes a two-stage process is envisaged where projectiles such as deuterons are stopped on a converter target and the energetic forward-going neutrons impinge on a thick production target of fissionable material [47]. Deuterons from cyclotrons (such as at HRIBF) or superconducting linacs (SPIRAL-II) near 40 MeV are effective drivers. Other proposals use neutrons produced from a reactor to produce fission isotopes, which are then ionized to produce radioactive ion beams (MAFF, Beijing ISOL).

An alternative is to use a high power electron linear accelerator to produce the radioactive beam species by bremsstrahlung-induced fission of a uranium target [48]. At low beam power the target material itself can create the photons (ALTO), while for higher intensities a cooled converter is employed (ARIEL). An example of the expected yields from a 500 MeV, 10 μA proton beam onto a 25 g/cm^2 uranium carbide (UC$_x$) target and those from a 50 MeV, 10 mA electron beam onto a Hg converter and 15 g/cm^2 ^{238}U target are shown in Fig. 4. Note that the production primarily by spallation reactions from high energy proton bombardment produces a wide range of ions while the production by photofission is dominated by neutron-rich products centered in two loci of N vs. Z.

Heavy ions with energies of 50–100 MeV/nucleon (GANIL) have significantly higher reaction cross-sections than protons and produce a large variety of both proton- and neutron-rich radioactive ions, usually on a light target, as with fragmentation facilities. Since the stopping power is much higher for heavy ions than protons, the target power density is a limiting factor.

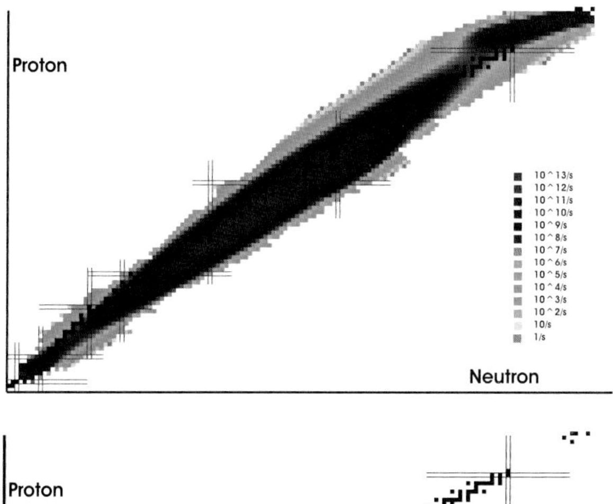

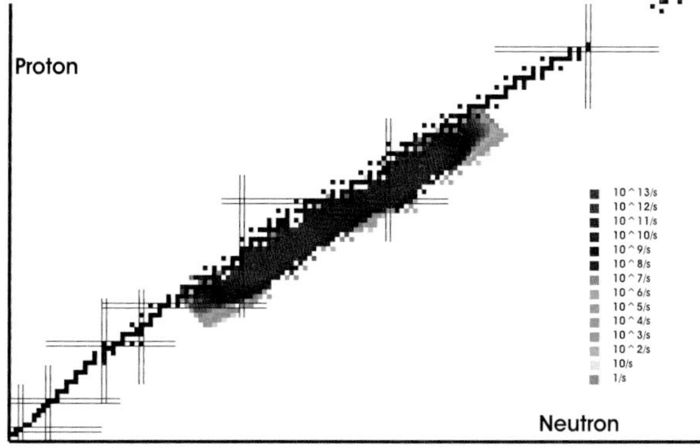

Fig. 4. Calculated in-target production for 10 μA, 500 MeV protons incident on a 25 g/cm^2 UC$_x$ target (top) and for 10 mA, 50 MeV electrons incident on a Hg converter and 15 g/cm^2 UC$_x$ target (bottom) [6]. The scales range from 1/s to 10^{13}/s.

4.2. *Production methods*

There are three main nuclear reactions typically used to produce rare isotopes in the ISOL target. *Spallation* is a breakup or fragmentation of the target material nuclei, with the most probable product being a few mass units lighter than the target nucleus. In this case the production of neutron-deficient nuclei is favored. *Fragmentation* is the counterpart of the spallation reaction and is favored for producing light, neutron-rich products from heavier target nuclei due to the neutron excess in the target. *Induced fission* occurs when the incoming projectile deposits sufficient energy in the target nucleus to induce a breakup into two roughly equivalent mass products. Again, because of the higher neutron-to-proton ratio in heavy nuclei, neutron-rich products in the medium mass region can be effectively produced by this reaction mechanism.

Determining the correct target material, target geometry and source for a particular beam requires careful development and study. After an isotope has been produced it is stopped in the target material matrix. In order to produce a RIB, the product must diffuse from the inside of the material to the surface and then effuse out of the target container to the ion source. Due to the stochastic nature of production decay losses can be significant for short half-life products. The diffusion speed is a function of the target size, the foil thickness and the temperature. The target material is maintained at a high temperature by a combination of beam heating and DC heaters. In the effusion process, the product atoms are in a gaseous state and randomly bounce around the heated target container and the target material surface until they enter the ion source through a heated tube. The effusion time and efficiency are dependent on the target temperature and surface chemistry. Isotopes can be significantly slowed or stopped altogether by colliding and sticking to target surfaces. A disadvantage with ISOL production in general is the difficulty in achieving high beam purity due to the many isobars of different elements produced simultaneously in the target. Furthermore, refractory elements are in general difficult to produce due to the high temperatures required to make them volatile. An important parameter is the lifetime of the target and the ion source unit from both thermal and radiation aging.

4.2.1. *He jet and ion guide methods*

An alternative to stopping the beam in a thick target is to stop the reaction products in gas. In the case of a gaseous catcher the radioactive atoms of interest stay in the gas phase, eliminating the need for high temperature systems. This results in a fixed holdup time independent of the element, making it possible to provide ion beams of essentially all elements. Two classes of gas stopping have been developed over the last 40+ years, and they are broadly categorized into He jets and ion guides. In the He jet concept the helium transports the reaction products to an ionizing cell. In one proposal [49], a high pressure helium jet is placed at a beam dump of a fragment separator. A secondary, normally waste beam can then be transported by a capillary to an ionizing cell to create a multi user facility.

In the ion guide method, nuclear reaction products recoiling out of a thin target are stopped in a gas (usually helium) and are transported by the gas flow through a differential pumping system directly into the acceleration stage of a mass separator. This process can be made fast enough for some reaction products to survive as singly charged ions. In this ion guide isotope separator on-line (IGISOL) [50, 51] technique, no ion source is used. Such a system is chemically insensitive and very fast (sub-millisecond). Technical advantages are its simplicity and durability; however, the system suffers from limited ion transport efficiency due to the relatively low stopping power of the helium gas. The ion guide was first used for light-ion-induced fusion–evaporation reactions, for which efficiencies of up to a few percent were obtained.

4.3. *Ion sources*

Ionization is required to convert the desired isotope atom into a charged particle beam. The beam can then be accelerated electrostatically from a biased platform, mass-analyzed, and transported to an experimental station, or be accelerated to higher energies. Different types of ion sources are available and the specific source is chosen to optimize the production of a certain species. *Surface ionization* [52] is a mechanism whereby an atom interacts with a heated surface. It can lose or gain an electron before leaving the surface as a positive or negative singly charged ion. Since the rate of ionization is dependent on the work function and the electron affinity surface, ionization is a highly element-selective method. The forced electron beam-induced arc discharge (*FEBIAD*) [53] ion source is well suited to RIB applications. Electrons, emitted from a heated cathode and accelerated into an anode chamber by means of a grid, ionize atoms in the anode chamber that have an ionizing potential less than the electron energy. The source operates efficiently in conjunction with high temperature thick target materials over a pressure range of 10^{-5}–10^{-4} Torr. A FEBIAD has good efficiency, particularly for heavier, slower-moving atoms. The resonant ionization laser ion source (*RILIS* [54]) is a powerful tool for the online production of RIBs. Here, selective ionization can be accomplished by using the nuclear spin dependence of the hyperfine splitting of the atomic levels with inherent suppression of unwanted isobaric contaminants. Two or three laser frequencies can be used to resonantly excite the electron into a level from which ionization proceeds.

4.4. *Mass separation*

The goal of the ISOL method is to achieve the highest beam intensity and the highest beam purity possible. The main beam purification is accomplished by passing the extracted beam through a high resolution magnet dipole. For high intensity installations an initial preseparator magnet with a resolution of $m/\Delta m \sim 300$ can be used to eliminate the bulk of the activity in a shielded area serviced by remote techniques. This is followed by a high resolution mass spectrometer. Typical resolving powers specified for modern separators are $m/\Delta m = 20{,}000$. In order to achieve this precision with reasonable transmission, the transverse emittance and energy spread can first be cooled in a radio frequency quadrupole (RFQ) cooler. In addition, the mass separator system can be floated on a high voltage platform to increase the ion velocity and hence reduce the unnormalized emittance before separation.

In RFQ coolers [55, 56], ions are manipulated and stored using electrical DC and RF fields. A buffer gas like helium or argon is introduced and through collisions between the buffer gas atoms and the radioactive ions energy is lost and cooling occurs. This cooling technique has been developed over recent years, yielding excellent performance. Efficiencies, defined as the beam intensity of the cooled beam versus that of the injected beam, of over 50% have been reached.

4.4.1. *Multiple-reflection time of flight*

An interesting alternative to magnetic separation is separation by time-of-flight mass spectrometry with the MR-TOF technique. This technique enables a fast, compact and inexpensive isobar separator [57]. In a demonstration, ions transmitted from a gas-filled stopping cell were guided into the system via a curved RFQ, cooled, and bunched into ion packets in a linear RF trap (injection trap) with a turnaround time of several nanoseconds. A prototype time-of-flight analyzer, in which ions from a dedicated test ion source traveled in the device for a selectable number of turns before being ejected, reflected in a planar ion mirror and flown through a pulsed ion gate for high resolution separation, was shown separately to have a resolving power of 10^5 [58]. After separation, the energy width of the resulting ion beam can be reduced in an energy buncher such that the ions can be injected efficiently into an accumulation trap to match the beam pulses to the downstream stage. Recent results at RIKEN have demonstrated using the MR-TOF technique as a mass analyzer for short-lived ions with a reported accuracy of $m/\Delta m \sim 10^7$ [58].

5. Postaccelerators

Many experiments take the beam directly from the ion source at energies up to ~60 keV, where properties such as ground state decay and masses of radionuclides can be measured. Solid state physics experiments use short-lived ions to probe materials. RIB postaccelerators are required to increase the energy of RIB ions from source potential to energies useful for a variety of other physics explorations. There are niche zones of experimental interest with particular energy ranges. Acceleration offering variable energies from 150 keV/u to 1.5 MeV/u is most interesting for nuclear astrophysics experiments. Typical nuclear physics experiments have interest near the Coulomb barrier 3–10 MeV/u. Recently there has been interest in accelerating RIBs to fragmentation energies (~100 MeV/u) in order to allow fragmentation of high intensities of RIBs to get to even more exotic species. Many postaccelerator facilities utilize linear accelerators that allow a staged installation with takeoff points along the linac chain to deliver RIBs to experimental facilities customized to exploit certain ion energies. The main emphases in all designs are to maintain high efficiency in order to achieve the maximum product at the end station, to control the beam emittance so as to allow precision experiments, and to control the beam purity since the exotic beam is in most cases of low intensity.

Typically, radioactive ions emerge from the online source singly charged. In principle it is possible to accelerate the particles in the 1+ charge state in order to achieve a high efficiency, but to reach the energies of interest this is prohibitively expensive when one is considering a reasonable range of interesting masses. To reach a given energy requires

$$E_{\text{out}}(\text{Mev/u}) = E_{\text{in}}(\text{Mev/u}) + V_{\text{eff}}(\text{MV})\frac{Q}{A}, \quad (1)$$

where V_{eff} is the effective voltage of the accelerator unit. In order to make acceleration more efficient, two strategies have been employed: one is to boost the charge by passing the beam through a charge-stripping medium (gas or solid), and the other is to use an ion source to remove additional electrons from the ion. Each of these processes reduces the beam transport efficiency since the ions emerge from the stripping or boosting device with a distribution of charge states and ion sources are never 100% efficient.

5.1. 1+ *acceleration*

The 1+ acceleration scheme can be effectively used for a modest mass range ($A < 50$), where a low frequency RFQ — and possibly a drift tube linac (DTL) — accelerates ions to a stripping energy of 100–300 keV/u (depending on the final energy) and a thin carbon foil 5–10 μg/cm^2 is used to increase their charge state. For a higher mass range, $A < 150$, the beam is accelerated in a very low frequency RFQ capable of accelerating a broad mass range to an energy of 10–20 keV/u, where gas stripping can be employed to produce ions with A/Q less than about 50, followed by the system described above. As the velocity increases, the requirements for transverse focusing are reduced and a DTL structure can be used with periodic focusing to provide more efficient acceleration. Although it is a very efficient method for the production of fully stripped light ions and provides excellent purity dependent only on the high resolution mass separator, it is expensive, requiring a high total accelerator voltage. As well, the efficiency for heavy ions is reduced since the poststripping charge state distribution is wide and multiple stripping stages have to be used.

5.2. *Charge state breeder*

A charge state breeder [59, 60] performs a transformation from a singly charged ion beam into a multiply charged one. Two types of charge state breeding ion sources are used: the electron beam ion source (EBIS) and the electron cyclotron resonance (ECR) ion source. Both rely on the principle of intense bombardment of the ions with energetic electrons, with electron impact ionization yielding ions in higher charge states. The plasma of ions and electrons is confined through electric and magnetic fields.

5.2.1. *EBIS*

In an EBIS a strong solenoidal magnetic field restricts a high intensity electron beam along the field axis to a very small radius. This results in electron beam densities over 100 A/cm^2. Ions are injected at low energy into the EBIS and are captured in an axial potential well created by an electrode structure, and are radially confined by the negative space charge of the electrons. The electron beam also serves to ionize the heavy ions. When the desired charge state is reached, the potential is lowered on the ejection side, allowing the highly charged ions to escape from the EBIS.

5.2.2. *ECR*

In ECR ion sources a plasma is confined in a bottle-type magnetic structure, usually consisting of a solenoid and a radial multipole field. Electrons are confined by the magnetic field and ions through the plasma potential. The electrons of the plasma are heated by injecting RF power at a frequency to obtain a resonant excitation of the electrons. When used as a charge state breeder, a 1+ ion beam is directly injected into the ECR source and subsequently slowed down and captured by the dense plasma of the ECR source. The beam contamination due to stable isotopes can, however, be the limiting factor for the postacceleration of very weak radioactive ion beams.

Many articles have compared the performance of ECRIS and EBIS charge breeders (e.g. [59]). In brief, an ECR source can capture ions in continuous mode and deliver continuous beams, while an EBIS yields a pulsed beam with a width of about $100\,\mu$s. An ECRIS has a higher intensity limit than an EBIS but this is not a significant problem for RIB applications. On the contrary, an EBIS typically has a higher efficiency than an ECRIS and delivers a cleaner beam since any residual background species in the ECR plasma chamber is also ionized. This last point is moving the RIB community away from the choice of ECRIS to EBIS for charge breeding, as RIB beams are typically of low intensity and can easily be overwhelmed by a stable component from the background.

5.3. *Accelerators*

5.3.1. *Linear accelerators*

The linear accelerator is the accepted standard for the postacceleration of radioactive beams. Linacs deliver high quality beams with low emittance growth and with high efficiency. They are typically operated CW to maximize efficiency or are pulsed with a repetition rate compatible with the pulse structure provided by the charge breeding system. With the rise in performance of SRF technology, linacs can be operated in CW mode with reasonable efficiency. Modern facilities are designing custom ISOL SC linac drivers with SC linac postaccelerators that utilize the same cavity geometries, RF ancillaries and cryomodule design to reduce engineering time.

Linacs are typically designed with a large acceptance. This is not critical in a low intensity post accelerator, so some selectivity can be built into the design [61]. For example, RFQs can be designed with a reduced longitudinal acceptance and utilize a multiharmonic prebuncher to achieve a small longitudinal emittance with reasonable capture [62]. Since ions are accelerated from a high voltage platform with a velocity difference given by the range of A/Q values, the prebuncher offers some selectivity based on time of flight and the reduced acceptance of the RFQ. Beam transport between linac sections can be designed with achromatic bends with dispersed focus at the symmetry point to provide further selectivity. A stripper foil can be employed both to increase the charge so as to reach a higher final velocity and to provide a further cut by altering the A/Q range of the poststripper ion cocktail. In effect, an entire linac chain can be used as a fragment separator.

5.3.2. *Cyclotrons*

Because of their limited A/Q acceptance, cyclotrons need highly charged ions and so require a charge breeder. At Louvain-la-Neuve and GANIL, an ECR source is directly coupled to the target through a cold transfer line. Because of the magnetic field and multiple turns used in the accelerating process of a cyclotron, the mass-analyzing power of these accelerators is very good. This results in a strong suppression of isobaric contaminants without loss in transmission. Although more efficient, the cyclotron is less flexible than a linac chain, that can produce full energy variability with multiple takeoff points.

6. RIB Facilities

6.1. *ISOL facilities*

ISOLDE makes use of a large variety of thick targets, including uranium carbide, that are irradiated with a pulsed beam of protons at 1.4 GeV from the Proton Synchrotron Booster at an average current of up to $2\,\mu$A and a rep rate of 0.4 Hz. The RIB can be postaccelerated with a compact, pulsed, normal-conducting linac up to an energy of 3 MeV/u. A Penning trap system, REXTRAP, cools and bunches the incoming ions from ISOLDE for subsequent transport to the charge breeder REXEBIS. The ion charge state is boosted to $2.5 < A/Q < 4.5$. A 101 MHz four-rod RFQ accelerates the ions from 5 to 300 keV/u. The beam is then rebunched into the first 101 MHz interdigital drift tube (IH) structure accelerating to 1.2 MeV/u. Three split ring cavities are used to give further acceleration to 2.2 MeV/u, and finally a 203 MHz nine-gap IH cavity is used to boost and to vary the energy between 2 and 3 MeV/u.

At the ISAC-I facility 500 MeV protons from TRIUMF's main cyclotron at currents of up to $100\,\mu$A impinge on one of two production targets to produce radioactive isotopes. Targets of SiC, Ta, ZrC, UC_x and Nb are typically used. The isotopes are ionized and the resulting beam is mass-separated and transported in an electrostatic beamline to either the low energy experimental area or, through a series of room temperature accelerating structures

(RFQ, DTL) operating CW, to the ISAC-I medium energy experimental area. For ions with $A \leq 30$ the ISAC postaccelerator utilizes a 1+ system for maximum transport efficiency to a medium energy area. The 35 MHz RFQ accelerates ions with $A/Q \leq 30$ to 150 keV/u and the poststripper variable energy 106 MHz DTL accelerates ions with $A/Q \leq 7$ up to 1.8 MeV/u.

The accelerated beam can also be transported to the ISAC-II superconducting linac for acceleration above the Coulomb barrier to the ISAC-II high energy area. The SC linac provides 40 MV of accelerating potential. Typical final energies are from 6 MeV/u to 15 MeV/u. An ECR charge breeder (CSB) is available to allow the acceleration of ions with $A > 30$. Here ions with $A/Q \sim 6$ are delivered from the CSB to the postaccelerator and in-flight filtration of CSB background is accomplished by a Nier-type spectrometer, TOF selection into the RFQ and a stripping stage after the DTL [61].

SPIRAL makes use of the GANIL coupled cyclotrons as a heavy ion driver. A thick graphite target is used and radioactive species are produced by fragmentation of the incident heavy ion beam, which is stopped in the target. The beams are ionized in a permanent magnet ECR source. This scheme has the advantage of simplicity but strongly limits the variety of beams available. A new ion source complex with a variety of 1+ sources delivering to an ECR charge breeder is currently under development (GANISOL) [63]; it should increase the number of elements available. The postaccelerator is the CIME cyclotron, which also serves as a high resolution mass separator. The final energy, which can reach 25 MeV/u, is the highest of current ISOL facilities.

The Californium Rare Isotope Breeder Upgrade (CARIBU) [64] at ANL has been built to supply ion beams of ^{252}Cf fission fragments which are thermalized in a gas catcher. The singly and doubly charged ions extracted from the gas catcher are mass-separated and either delivered to a low energy experimental area or charge-bred with a modified ECR source and subsequently reaccelerated by the ATLAS linac. An EBIS is being added as part of an upgrade to achieve purer beams.

Several smaller facilities exist around the world, doing a limited range of experiments using repurposed drivers or postaccelerators at nuclear physics institutes.

6.2. *Fragmentation facilities*

Due to increasing scientific interest, the major high energy heavy ion facilities in operation in the 1990s, NSCL-MSU, GANIL, GSI and RIKEN, have evolved to focus more on RIB production and experiments. The GANIL driver consists of two separated sector room temperature cyclotrons which produce heavy ions from C to Ar at energies up to 100 MeV/u and can accelerate masses up to U at 25 MeV/u. Large primary intensities (several μA) can be delivered but the final RIB intensities are limited by the lower forward momentum — GANIL has the lowest energies of the four facilities — and the limited acceptance of the beamlines which were not optimized for RIB transmission.

Production and selection efficiencies are larger at MSU/NSCL and RIKEN, where the fragment separators A1900 and RIPS were specifically built for efficient RIB selection. The NSCL/MSU facility is driven by coupled superconducting cyclotrons with $K = 500$ and $K = 1200$, while the initial RIKEN facility, still in operation today, has a $K = 540$ room temperature cyclotron. The SIS synchrotron at GSI can provide energies of up to 2 GeV/u but with intensities much lower than at the cyclotron-based facilities. The lower intensities are partially offset during the production stage by the higher forward momentum, the high efficiency of the FRS fragment separator, and the ability to use thicker targets in the experiments [65].

The FLNR facility at JINR (Dubna) operates several cyclotrons with the primary focus on accelerating heavy ions, in particular ^{48}Ca, onto heavy target materials for the synthesis of superheavy elements [19].

Japan has recently taken the global lead in high energy RIBs, as the first "next generation" in-flight facility, the Radioactive Ion Beam Factory (RIBF) in RIKEN, has been in operation since 2007 [66]. RIBF boasts a new high power heavy ion accelerator system consisting of three ring cyclotrons with K values of 570 (fixed frequency, FRC), 980 (intermediate stage, IRC) and 2500 (superconducting, SRC). A final energy of 350 MeV/u is obtained for up to the heaviest ions. RIBF has delivered in particular the world's most intense beam of ^{48}Ca at 200 particle-μA and uranium beams at an intensity of 10^9 pps. Radioactive beams are selected with a new superconducting fragment separator, BigRIPS [67, 68].

6.3. *Facilities under construction or proposed*

While a few proposals are for green field facilities, new RIB installations are in general building upon existing infrastructure. Major new projects are described below; however, given the tremendous interest in RIB physics, many new niche facilities are also being built or proposed.

6.3.1. *ARIEL*

The Advanced Rare Isotope Laboratory (ARIEL) [69] at TRIUMF is a ten-year initiative that will significantly expand the existing RIB program. The project goal is to make possible two, then three, simultaneous RIBs in a staged way to triple the RIB hours available per year, making ISAC/ARIEL the leading ISOL facility in the world. ARIEL will add a new electron driver at 10 mA and 50 MeV as a complement to the existing 500 MeV proton cyclotron driver to produce RIBs through photofission. The e-LINAC parameters were chosen to reach rates up to 10^{14} fissions per second. A new 500 MeV, 100 μA proton line will be added to serve a new target location in ARIEL, allowing three high power drive beams (two 50 kW protons and one 500 kW electron) to be used with three independent targets simultaneously. ARIEL will add a low energy front end with medium ($m/\Delta m \sim 5000$) and high resolution ($m/\Delta m \sim 20{,}000$) separators, an EBIS charge breeder and a new accelerator front end to 1.5 MeV/u. ARIEL/ISAC will support the simultaneous delivery of up to three RIBs with up to two accelerated beams, the delivery of new beam species, and increased beam development capabilities. The first phase is expected to be operational in 2015.

6.3.2. *SPIRAL-II*

The driver of the SPIRAL-II facility [47] is a high power, CW superconducting linac delivering up to 5 mA of deuterons at 40 MeV (200 kW) directed on a carbon converter and uranium target. Production of radioactive ion beams is based on fast-neutron-induced fission of a uranium target. The SPIRAL-II linac will also accelerate high intensities (up to 1 mA) of heavy ions to 14.5 MeV/u. The ions will be used to enlarge the range of exotic nuclei produced by the ISOL method. The construction of SPIRAL-II is split into two phases, with the linac available for stable beam acceleration expected in 2014 and RIB production from the SPIRAL-II linac in the latter part of the decade.

6.3.3. *HIE-ISOLDE*

The HIE-ISOLDE project [70] will see the addition of an SC linac driver to boost the final energy of the REX postaccelerator in several stages to reach 10 MeV/u for $A/Q = 4.5$ by 2018. In addition, the new CERN injector (LINAC4) expected to replace the current LINAC2 in 2018 will provide a major boost of the proton intensity onto the ISOLDE target to 10 μA. In the framework of HIE-ISOLDE, the target areas and ion sources are also being upgraded in order to make use of the more intense proton beams and to improve the efficiency of ion extraction and charge breeding.

6.3.4. *FRIB*

The FRIB [71] driver will be an SRF heavy ion linac that provides 400 kW beam power for all beams with uranium accelerated to 200 MeV/u and lighter ions with increasing velocity up to 600 MeV for protons. An upgrade path to extend the driver linac energy to 400 MeV/u for uranium and to add a light ion injector in order to accelerate beams for ISOL production is foreseen. Two ECR sources will be available to feed ions into the linac. The in-flight production target will be a rotating multislice graphite target. There will be a high momentum and angular acceptance three-stage fragment separator that can be tuned to optimize acceptance close to or far from stability. Fast in-flight beams or stopped and reaccelerated beams will be available. Three types of stoppers should also be available: a linear gas stopper for heavier and medium mass beams, a unique cyclotron gas stopper for light beams, and a solid stopper for certain elements at the highest intensities. These beams will be injected into the ReA3 postaccelerator with an EBIS charge breeder, a room temperature RFQ and an SC linac with plans to upgrade to 12 MeV/u operation. FRIB will make use of the current NSCL experimental areas, with the possibility of expansion to double the available area.

6.3.5. *FAIR*

The core of FAIR [72] is a new synchrotron, SIS100, which will deliver primary beams of 10^{12} $^{238}U^{28+}$ per second at 1.5–2 GeV/u to the NUSTAR facility (and other experimental areas). This corresponds to an increase in intensity of two to three orders of magnitude with respect to the current GSI synchrotron, SIS18. The heart of NUSTAR will be a fragment separator with improved efficiency, the Super-FRS, which will deliver a broad range of radioactive beams with an expected improvement in intensity over current values of up to 10^4. The beams will be used directly, degraded for low energy experiments, stopped, or injected into the NESR storage ring and decelerated and cooled.

6.3.6. *SPES*

SPES [45] is a new facility that adds a high intensity (750 μA), 70 MeV commercial cyclotron with two exit ports, a UC_x ISOL target and ion source, and a beam transport system with high resolution mass selection to the existing heavy ion linac PIAVE-ALPI complex in operation at INFN-LNL. To enable beam reacceleration with the linac, an RFQ cooler and a charge breeder are planned to be installed. The final energies are expected in the range of 5–15 MeV/u.

6.3.7. *ISOL@MYRRHA*

The ISOL@MYRRHA facility will be operated in parallel to MYRRHA-ADS. It proposes to use 200 μA of 600 MeV protons for the production of RIBs via the ISOL method. ISOL@MYRRHA aims to be complementary to existing facilities by focusing on experimental programs requiring long, uninterrupted beam times.

6.3.8. *EURISOL*

Present European planning includes the EURISOL [46] facility to complement FAIR NUSTAR as the world-leading ISOL facility. EURISOL will use a large CW SC linac to accelerate H^- ions to energies of 1 GeV. The option of using a pulsed beam at 50 Hz with a minimum pulse length of 1 ms has been kept open to enable possible sharing of the driver with other physics users. This beam of particles will deliver a power of up to 4 MW to one target station and, through a magnetic beam-splitting system some 100 kW, to three smaller target stations in parallel. Prior to postacceleration, the necessary charge breeding will be done in either an ECR source or in a high intensity CW EBIS. There will be at least one SC linac in which exotic ions will reach energies up to 150 MeV/u for $A/Q \leq 8$. The postaccelerated beams will have sufficient energy to undergo secondary fragmentation leading to neutron-rich nuclei further from stability, typical of this hybrid scheme.

6.3.9. *RISP*

The Korean nuclear physics community has launched an ambitious Rare Isotope Science Project (RISP). RISP is a multistage project with a commercial 70 MeV, 1 mA cyclotron as a 70 kW ISOL proton driver and a 200 MeV/u, 8.3 particle-μA ^{238}U heavy ion linac for in-flight production. An 18.5 MeV/u postaccelerator can accelerate stopped beams from the high energy driver or feed cyclotron-produced ISOL RIBs into the in-flight linac to produce high energy RIBs for the production of more exotic neutron-rich species.

6.3.10. *China*

The BRIF facility [73] is being installed at CIAE. The ISOL driver is a new 100 MeV cyclotron designed to provide up to 300 μA of protons. The postaccelerator is the existing 15 MV tandem with an ~4 MV superconducting booster cryomodule after the tandem. Plans also include the installation of cryomodules from the decommissioned Stony Brook facility, as well as the decommissioned K120 heavy ion cyclotron from PSI. The proposed Beijing ISOL [73] facility uses a neutron-rich fission fragment beam from both reactor-based thermal-neutron-induced ^{235}U fission and deuteron-linac-induced fast neutron ^{238}U fission. The CARR reactor at CIAE, online since 2012, can deliver a neutron flux rate of $8 \times 10^{14}/cm^2/s$ with a thermal power of 60 MW. The thermal neutrons from the CARR reactor will impinge on a ^{235}U target in an in-pile ion source. The deuteron linac will deliver 40 MeV, 5–10 mA deuteron beams and create fast neutrons in a converter to bombard a ^{238}U target. The fission fragments like ^{132}Sn and ^{91}Kr from both drivers will be mass-separated, charge-bred, and postaccelerated by an SC linac to an energy of 150 MeV/u, where in-flight production will be used to produce exotic species.

7. Introduction to Radiopharmaceuticals

Modern medicine is on the cusp of understanding disease at the molecular level. This awareness is enabled by tools developed in the field of nuclear medicine [74, 75] such as molecular imaging (MI) via radioactive probes (also known as radiopharmaceuticals) and is enabling progress toward personalized medicine, theranostics and targeted therapies via radiotherapeutics [76, 77]. The use of radioisotopes is unique, in that it provides a method for measuring biochemical processes *in vivo*, at a level where the tracer does not perturb the system under analysis, since the signal generators used (radioisotopes) make it possible to detect and localize quantities as low as 10^{-14}–10^{-15} mol/L [78].

MI has enabled health professionals and basic researchers to perform noninvasive studies of the changes that take place during the onset and progression of disease. MI using radiopharmaceuticals exhibits the highest degree of sensitivity, permitting earlier diagnoses as well as detection of minute changes in the behavior of a tissue or organ at the molecular level. By doing so health professionals are able to diagnose and stage a disease earlier than conventional technologies allow, as well as gauge the efficacy of a chosen treatment for individual patients.

Beyond imaging, there is active development in the use of targeted radiotherapeutics as a means of affecting systemic therapy of a disease when surgery is no longer feasible. The idea of a radioisotope used in therapy — radioisotope therapy (RIT) — is based on the idea of linking a suitable radionuclide which decays with radiation having a high linear energy transfer (LET) to a bioactive molecule that can be selectively directed to a specific tissue (i.e. a tumor) within the body. By doing so, one can deposit a lethal dose of radiation while avoiding a substantial dose to surrounding healthy tissue [79, 80]. This concept serves to avoid the side effects often observed when one is using external beams of photons or particles, such as those used in classical techniques of radiology.

Imaging applications typically seek shorter-lived isotopes and faster-clearing constructs to avoid an excessive patient dose; whereas therapeutic applications typically require longer-lived, stable constructs to ensure delivery of the nuclide to the site of interest and accumulation of an effective dose once there [81]. Present day radiopharmaceuticals are used for diagnostic purposes in the bulk of the cases, and the remainder are used in therapy. However, radiotherapeutic applications are undergoing significant development [82–84].

7.1. *A brief history*

The history of radiopharmaceuticals follows closely that of radioactive isotope production and study [85]. The first clinical tracer was investigated in Boston under Blumgart in 1925; ^{214}Bi activity from ^{226}Ra decay was registered in one arm after injection into the other. This enabled studies of circulation velocity, in particular as related to heart health [74, 75]. Later, John Lawrence investigated the therapeutic implications of ^{32}P for leukemia. In 1936 Hertz demonstrated that radioiodine could be made and used as a tracer, and in 1941 he used radioiodine in the treatment of hyperthyroidism [86]. Breakthroughs in detection were made in 1947 with the development of scintillation cameras. Widespread clinical use of nuclear medicine began in the early 1950s, as knowledge expanded about radionuclides, detection of radioactivity, and using radionuclides to trace biochemical processes. Pioneering works included the development of the first rectilinear scanner in 1951 and of the "Anger" gamma camera in 1958 [87]. ^{99m}Tc, first discovered in 1937 by C. Perrier and E. Segre [88, 89], became significant in the 1960s, after the development of a generator system by P. Richards in 1957 [90]. The concept of emission and transmission tomography, which began in the late '50s, progressed to the development of modern day positron emission tomography (PET) and single photon emission computed tomography (SPECT). In particular, the development of labeled ^{18}F-fluorodeoxyglucose (^{18}F-FDG) was a major factor in expanding the scope of the PET imaging modality [91].

8. Molecular Imaging

Imaging with radioisotopes is accomplished by using one of two methods: single photon (SPECT) or positron (PET) annihilation–based detection and algorithmic back projection to provide three- or four-dimensional, high contrast images with satisfactory spatial resolution using only a trace amount of the imaging agent. These methods differ from other radiology techniques such as CT and MRI, in that the latter provide mostly anatomical information. The merits of newer non-radionuclide-based modalities, such as fMRI, which measure changes in blood flow, temperature, pH, etc., are beyond the scope of this article and are not discussed in detail here.

SPECT imaging is best achieved with photons in the 150 keV range, so as to avoid giving an excessive dose to the patient while minimizing attenuation during transit to the detector and maximizing collimation once there [92]. ^{99m}Tc, used in over 80% of nuclear medicine scans around the world, has long set the preferred paradigm for isotope availability via the ^{99}Mo/^{99m}Tc generator, radioactive emission ($E_\gamma = 140$ keV) and radiopharmaceutical production (via commercial kits) [93]. Other important radionuclides showing suitable single photon emission for functional diagnosis are ^{201}Tl, ^{123}I, ^{67}Ga, ^{111}In and a few others [94, 95].

Despite the broad use of SPECT, PET continues to become more popular as it provides the ability to quantify physiological uptake and distribution of the nuclide injected. PET requires the coincident detection of 511 keV photons generated by the release and annihilation of positrons.

Development efforts for both SPECT and PET have traditionally been on the use of low molecular weight (i.e. small molecules of <1000 atomic mass units) targeting vectors to investigate the functional effects of disease on tissue. Newer efforts are focused on the development of higher (large molecule) molecular weight peptides, proteins, antibodies, etc. for the purpose of investigating the effects of disease on the expression of various biological markers [96].

9. Advances in Therapeutics

Traditional therapies consist of "regional" application of radiation either from an external beam (EB-XRT) or from a sealed local source (brachytherapy). Modern radiopharmaceutical therapies are developing unsealed active "systemic" agents with a high linear energy transfer (LET) associated with its decay products that would link to a biologically active molecule and selectively target a tumor site. Nuclides for therapy should selectively deposit high radiation doses on the target tissue (i.e. the tumor), mandating the use of isotopes that decay by particle emission. These radioisotopes emit particles, such as beta particles, alpha particles or Auger electrons. These are characterized by their range and decay energy, with typical ranges of 1–10 mm for beta emitters, <100 μm for alpha particles and a few μm for Auger electrons [97]. Established beta emitters are ^{90}Y and ^{131}I, with ^{177}Lu becoming recognized as a potentially useful isotope. Auger electron therapy isotopes are inherently at higher Z than $\beta+$ and γ ray–emitting nuclides. Radiometals, lanthanides and actinides, in particular, are under significant development [98–101]. Some isotopes are not yet available commercially and require unconventional production methods. A recent example is the production of four different terbium isotopes produced by thermal neutron capture in a high flux reactor (^{161}Tb) and by proton-induced spallation of tantalum targets and online mass separation at ISOLDE (^{149}Tb, ^{152}Tb, ^{155}Tb) [102, 103]. These four isotopes cover all nuclear medicine modalities (SPECT and PET as well as beta, alpha and Auger electron emission respectively for therapy). Since they have identical biochemical behavior they can be prescribed in concert: one isotope is used to diagnose patient-specific uptake while another is used for therapy — a so-called "theranostics" modality [77, 104, 105].

10. Producing Radiopharmaceuticals

Radioisotopes used in the production of radiopharmaceuticals are sourced from reactors and accelerators, depending on the isotope in question and the reaction used in its production. Reactors produce isotopes via (n, γ) or neutron induced fission reactions (n, F) and allow access to large quantities of certain isotopes with a range of specific activities [106]. The cost of reactor production is difficult to assess, due to heavy government subsidies toward the expense of the initial construction, the operation once critical, and maintenance during several decades of reactor life.

Accelerators, on the other hand, offer a decentralized production model. The advantages include the ability to produce a range of isotopes with a variety of decay mechanisms — alpha, beta (positron), Auger, etc. — for radiopharmaceutical or theranostic production with acceptable specific activities to allow a judicious choice of isotope (emission particle type, range, and energy loss characteristics) to provide optimized images or therapeutic effects. The disadvantages of accelerators for the routine production of medical isotopes include high (non-subsidized) operating costs and relatively low production capabilities when compared to reactors. Higher energy production machines also typically operate in a parasitic manner at selected research facilities around the world. Thus, the production schedule is often compromised by the scientific demands of the facility and may include significant planned downtime for maintenance or budget issues.

10.1. *Accelerator production of PET isotopes*

PET isotopes have been produced with commercial H- or proton cyclotrons located in hospitals or by facilities in close proximity to the imaging center. Proton energies typically range from 10 to 20 MeV, with production typically involving (p, n) and (p, α) reactions. Traditional isotopes of interest include a suite of light nonmetallic isotopes (such as ^{18}F, ^{11}C, ^{13}N and ^{15}O) [107].

More recent work has focused on metallic isotopes (e.g. ^{64}Cu, ^{68}Ga, ^{82}Rb, ^{86}Y, ^{89}Zr) [80, 108]. ^{64}Cu, ^{86}Y and ^{89}Zr can also be made with widely available 10–20 MeV machines, while ^{68}Ga has become increasingly popular due to its availability from ^{68}Ge/^{68}Ga generators [109, 110].

The interest in ^{82}Rb as a PET isotope for cardiology investigation is growing. This isotope is produced from an ^{82}Sr/^{82}Rb generator that requires 60–100 MeV protons for production [111]. In this case, production is centered in larger research centers: TRIUMF (110 MeV/100 μA), BNL (BLIP — 200 MeV/100 μA), LANL (100 MeV/200 μA), INR (160 MeV/120 μA), ARRONAX (70 MeV, 100 μA) and iThemba (66 MeV/250 μA). Only ARRONAX is a dedicated medical isotope production facility [112].

10.2. *Production of ^{99m}Tc*

Accelerator production of non-PET isotopes has recently undergone a renaissance, with much emphasis on alternative production of ^{99m}Tc. The ^{99m}Tc isotope (with a 6 h half-life) alone accounts for nearly 85% of all nuclear medicine imaging studies. It is typically extracted from a ^{99}Mo/^{99m}Tc generator ($t_{1/2} = 66$ h) that is sourced from a few large reactors around the world. Recent and recurring interruptions in the ^{99}Mo supply due to problems with one or more of the major production reactors are exacerbated by a global nonproliferation effort to cease production of highly enriched uranium (HEU) targets [113].

The present worldwide need for ^{99}Mo/^{99m}Tc generators totals approximately 10–12,000 six-day Ci (where a "six-day Ci" is a Curie of activity remaining six days after shipment) of ^{99}Mo delivery per week [73]. The predominant method of ^{99}Mo production, and the only method used in North America, is the ^{235}U(n, F)^{99}Mo reaction: fission of HEU by thermal neutrons in a reactor. A substantial effort is underway at most large capacity suppliers of ^{99}Mo to continue reactor-based alternatives by switching to low-enriched uranium (LEU) from HEU to reduce the global supply of enriched uranium.

Many accelerator-based approaches have been studied but, due to the subsidized availability of neutrons for the past several decades, few have been developed past the concept stage. Photofission of ^{238}U(γ, F)^{99}Mo has been considered at TRIUMF using a high intensity electron linac to produce intense gamma beams via a converter target. Estimates are that a 40 mA/50 MeV electron beam would produce 300 six-day Ci/week with ^{238}U targets [114]. Challenges to this approach include a very low production cross-section and technical issues associated with waste management and the development and operations of high power electron accelerators and targets. The photonuclear reaction ^{100}Mo(γ, n)^{99}Mo on separated ^{100}Mo would result in more efficient production, with an estimated 640 six-day Ci/week [114] for a 10 mA/50 MeV electron beam and less waste but with lower specific activity.

Another approach uses a proton accelerator to produce neutrons through the (p, n) reaction to fission ^{235}U(n, F)^{99}Mo. In this case a linac or a cyclotron produces protons between 150 and 500 MeV and up to 2 mA beam current which are directed onto a ^{235}U target. The goal is to produce an order of magnitude more secondary neutrons inside the target from ^{235}U fission. It is estimated that a proton beam of 350 MeV and 1 mA could produce 5000 six-day Ci/week [113].

The direct production of ^{99m}Tc by cyclotrons has been known since 1971, with laboratory scale studies performed by researchers at the University of Miami employing the ^{100}Mo(p, 2n)^{99m}Tc reaction [115]. The reaction has a reasonable cross-section at modest proton energies of 16–18 MeV, which is well within the range of small medical cyclotrons. This method would allow a fully localized distribution model where isotope production and patient could be under the same roof, a paradigm similar to what is done for ^{18}F-FDG in most major urban hospitals today. A consortium of institutions led by TRIUMF has been working on a full scale demonstration of this concept. To date, the TRIUMF consortium has publicly announced production capabilities of ~348 GBq (9.4 Ci) of ^{99m}Tc directly from ^{100}Mo in a single run [116]. The effort has required the development of new high power ^{100}Mo target production technology capable of withstanding the high current (up to 300 μA), prolonged irradiation (3–6 h) runs needed to produce this quantity of ^{99m}Tc.

10.3. *Medical isotopes from RIBs*

Given the enhanced capabilities of imaging and radionuclide therapy, present developments tend toward radioisotope production with higher radiochemical and radionuclidic purity, as well as enhancing the specific activity of the radiotracer. The role of isotope separators in providing online purification is being explored [117], particularly at ISOLDE.

Many therapeutic radioisotopes are produced in aging high flux reactors or by high energy particle accelerators in order to achieve appreciable production rates. However, these production routes may result in low specific activity [e.g. (n, γ)] or low purity (e.g. fragmentation, fission and spallation) product mixtures, necessitating purification techniques to produce useful samples. At lower accelerator energies the possible nuclear reaction channels tend to be more selective, leading to products of higher purity. With increasing particle energy the reactions become more and more nonspecific, giving a range of isotopic and isobaric mixtures. The result is a set of isotope products that have poor radionuclidic purity while still achieving high specific activity. The issue remains that, despite advances in nuclear and accelerator technology, the production of suitable isotopic products is restricted by subsequent chemical techniques.

Harvesting pharmaceuticals from RIB facilities is increasingly being pursued to enable greater availability of exotic or uncommon radioisotopes to drive new discoveries or improve applications currently in the clinical arena. RIBs offer unprecedented high purity products that are isotopically separated and carrier-free.

In order for radiopharmaceutical development from RIBs to proceed, mass separation of radioactive isotopes is needed. Work over the past decade has demonstrated that this is possible and affordable [85]. Spallation reactions and fission in combination with a mass separation process (offline or online) provide universal access to most radionuclides. The ability to implant high quality radionuclides has been demonstrated and may represent a preferred approach to the manufacture of stents and radioactive seeds for brachytherapy.

10.4. *Radiolanthanides from ISOLDE*

The radiolanthanide group of isotopes presents a number of interesting options for biomedical research with similar chemistry across the group of trivalent metallic radionuclides, many of which possess favorable radiation properties for use in PET, SPECT or radiotherapy. ISOLDE has demonstrated production of a full range of radiolanthanides through spallation from three standard target systems — tantalum (Ta), uranium carbide (UC_x) and niobium (Nb). UC_x is suitable for ^{225}Ra/^{225}Ac via spallation as well as ^{141}Ce, ^{153}Sm from fission, while neutron-deficient lanthanides are best obtained from spallation of Ta. Nb targets give access to four different Y isotopes suitable for PET and/or SPECT and therapy. All approaches use a surface ionization approach, with RILIS offering the possibility of higher efficiency or better chemical selectivity. After thick target ISOL production and magnetic separation, the ions are collected via a number of methods, including collection of the radioactive ions in ice kept at liquid nitrogen temperature (limited to short collection times), implantation into different salts (Na citrate, KNO_3 and others [118]), implantation into metallic foils (most suitable Al or Ta) or implantation into plastic foils (e.g. kapton or polyethylene). Since the lanthanides exhibit similar chemical behavior, a radiochemical separation [119] process is needed to obtain the chemical purity required for biomedical studies [120]. A number of *in vitro* and *in vivo* preclinical studies using radiolanthanides have been published over the last several years [102, 121].

11. Conclusion

Radioactive ion beam investigations performed over the last 50 years continue to expand as existing facilities are upgraded, new facilities prepare to come online and new proposals seek funding. Technical developments and innovative ideas are being applied to the two classical production approaches to increase yield, improve beam quality and enable a reach to ever-more-exotic species. Major advances in hadron superconducting linear accelerator technology have inspired new proposals with tremendous capability. The growth and sophistication inherent in RIB production parallels the advances in the application and development of radiopharmaceuticals. Accelerator-based production will continue to expand and decentralize compared to reactor-based production, while larger accelerator centers will allow, through online separation techniques, the development of an ever-more-sophisticated group of theranostic radiopharmaceuticals.

References

[1] M. Huyse, *Lect. Notes Phys.* **651**, 1 (2004).
[2] J. Erler *et al.*, *Nature* **486**, 509 (2012).
[3] A. H. Wapstra, G. Audi and C. Thibault, *Nucl. Phys. A* **729**, 129 (2003).
[4] G. Audi *et al.*, *Chin. Phys. C* **36**, 1287 (2012).
[5] M. Wang *et al.*, *Chin. Phys. C* **36**, 1603 (2012).
[6] *Five-Year Plan 2010–2015: Building a Vision for the Future*, eds. M. McLean and T. I. Meyer (TRIUMF, Vancouver, 2008).
[7] O. Kofoed-Hansen and K. O. Nielsen, *Phys. Rev.* **82**, 96 (1951).
[8] R. Klapisch and R. Bernas, *Nucl. Instrum. Methods* **38**, 291 (1965).
[9] H. L. Ravn, *Phys. Rep.* **54**, 201 (1979).
[10] P. Decrock *et al.*, *Phys. Rev. Lett.* **67**, 808 (1991).
[11] D. W. Bardayan *et al.*, *Phys. Rev. Lett.* **83**, 45 (1999).
[12] P. Schmor *et al.*, Initial commissioning of the ISAC RIB facility, in *Proc. 1999 Part. Accel. Conf. (PAC'99)* (New York City, 1999), pp. 508–512.
[13] B. L. Cohen and C. B. Fulmer, *Nucl. Phys.* **6**, 547 (1958).
[14] H. Ewald *et al.*, *Z. Naturforsch.* **19a**, 194 (1964).
[15] H. H. Heckman *et al.*, *Phys. Rev. Lett.* **28**, 926 (1972).
[16] I. Tanihata *et al.*, *Phys. Rev. Lett.* **55**, 2676 (1985).
[17] I. Tanihata *et al.*, *Phys. Lett. B* **160**, 380 (1985).
[18] G. T. Seaborg, W. Loveland and D. J. Morrissey, *Science* **203**, 711 (1979).
[19] Yu. Ts. Oganessian, *Pure Appl. Chem.* **78**, 889 (2006).
[20] G. Münzenberg, *Nucl. Instrum. Methods B* **70**, 265 (1994).
[21] D. J. Morrissey and B. M. Sherrill, *Phil. Trans. Roy. Soc. Lond. A* **356**, 1985 (1998).
[22] H. Geissel, G. Münzenberg and C. Riisager, *Ann. Rev. Nucl. Part. Sci.* **45**, 163 (1995).
[23] H. Ravn and B. Allardyce, *Treatise on Heavy-Ion Science: Volume 8, Nuclei Far from Stability*, ed. D. A. Bromley (Plenum, New York, 1989), pp. 363–439.
[24] P. Van Duppen, *Lect. Notes Phys.* **700**, 27 (2006).
[25] G. Bollen, D. J. Morrissey and S. Schwarz, *Nucl. Instrum. Methods A* **550**, 27 (2005).
[26] G. Savard, *J. Phys. Conf.* **312** (2011).
[27] M. Wada *et al.*, *Nucl. Instrum. Methods B* **204**, 570 (2003).
[28] I. Tanihata, *Nucl. Instrum. Methods B* **266**, 4067 (2008).
[29] P. N. Ostroumov and K. W. Shepard, *Phys. Rev. ST Accel. Beams* **3**, 030101 (2000).
[30] D. J. Morrissey and B. M. Sherrill, *Lect. Notes Phys.* **651**, 113 (2004).
[31] C. A. Bertulani and G. Baur, *Phys. Rep.* **163**, 299 (1988).
[32] F. Pellemoine *et al.*, *Nucl. Instrum. Methods A* **655**, 3 (2011).
[33] J. Nolan *et al.*, *A High Power Beam-on-Target Test of Liquid Lithium Target for RIA* (ANL-05/22, Argonne National Laboratory, 2005).
[34] H. Geissel *et al.*, *Nucl. Instrum. Methods B* **70**, 286 (1992).
[35] D. J. Morrissey, *Nucl. Instrum. Methods B* **126**, 316 (1997).
[36] H. Geissel *et al.*, *Nucl. Instrum. Methods B* **204**, 71 (2003).
[37] http://webdocs.gsi.de/~wolle/EB_at_GSI/FRS/GEISSEL/F09.pdf, used with permission.
[38] T. W. O'Donnell *et al.*, *Nucl. Instrum. Methods A* **422**, 513 (1999).
[39] A. Joubert *et al.*, The SISSI Project: an intense secondary ion source using superconducting solenoid lenses, in *Proc. 1991 Part. Accel. Conf. (PAC 1991)* (San Francisco, 1991), pp. 594–597.
[40] T. Radon *et al.*, *Nucl. Phys. A* **677**, 75 (2000).
[41] B. Franzke, H. Geissel and G. Münzenberg, *Mass Spectrom. Rev.* **27**, 428 (2008).
[42] K. Blaum, J. Dilling and W. Nörtershäuser, *Phys. Scripta* **T152**, 014017 (2013).
[43] P. Bricault *et al.*, *Nucl. Instrum. Methods B* **204**, 319 (2003).
[44] P. Bricault *et al.*, Progress in design of ISOL RIB ion sources and targets for high power, in *Proc. Cyclotrons and Their Applications 2007, 18th Int. Conf. (Cyclotrons 2007)* (2007), pp. 499–504.
[45] G. de Angelis *et al.*, *J. Phys. Conf.* **267**, 012003 (2011).

[46] Y. Blumenfeld *et al.*, *Int. J. Mod. Phys. E* **18**, 1960 (2009).
[47] M. Lewitowicz, *J. Phys. Conf.* **312**, 052014 (2011).
[48] W. T. Diamond, *Nucl. Instrum. Methods A* **432**, 471 (1999).
[49] J. J. Das *et al.*, *J. Phys. Conf.* **420**, 012165 (2013).
[50] J. Ärje *et al.*, *Nucl. Instrum. Methods* **186**, 149 (1981).
[51] P. Karvonen *et al.*, *Nucl. Instrum. Methods B* **266**, 4454 (2008).
[52] R. Kirchner, *Nucl. Instrum. Methods* **186**, 275 (1981).
[53] R. Kirchner and E. Roeckl, *Nucl. Instrum. Methods* **133**, 187 (1976).
[54] K. Blaum *et al.*, *Nucl. Instrum. Methods B* **204**, 331 (2003).
[55] R. B. Moore and G. Rouleau, *J. Mod. Optics* **39**, 361 (1992).
[56] G. Bollen *et al.*, *Nucl. Instrum. Methods A* **368**, 675 (1996).
[57] W. R. Plaß *et al.*, *Nucl. Instrum. Methods B* **266**, 4560 (2008).
[58] Y. Ito *et al.*, *Phys. Rev. C* **88**, 011306(R) (2013).
[59] P. Delahaye *et al.*, *Eur. Phys. J. A* **46**, 421 (2010).
[60] R. Vondrasek *et al.*, *Rev. Sci. Instrum.* **83**, 02A913 (2012).
[61] M. Marchetto *et al.*, In flight ion separation using a linac chain, in *Proc. XXVI Linear Accel. Conf.* (*LINAC12*) (Tel Aviv, Isreal, 2012), pp. 1059–1063.
[62] R. E. Laxdal *et al.*, Beam test results with the ISAC 35 MHz RFQ, in *Proc. 1999 Part. Accel. Conf.* (*PAC'99*) (New York City, 1999), pp. 3534–3536.
[63] P. Jardin *et al.*, *Rev. Sci. Instrum.* **83**, 02A911 (2012).
[64] J. A. Clark, *J. Phys. Conf.* **267**, 012002 (2011).
[65] J. Kurcewicz *et al.*, *Phys. Lett. B* **717**, 371 (2012).
[66] Y. Yano, *Nucl. Instrum. Methods B* **261**, 1009 (2007).
[67] T. Kubo, *Nucl. Instrum. Methods B* **204**, 97 (2003).
[68] T. Ohnishi *et al.*, *J. Phys. Soc. Jpn.* **79**, 073201 (2010).
[69] L. Merminga *et al.*, ARIEL: TRIUMF's Advanced Rare Isotope Laboratory, in *Proc. 2nd Int. Part. Accel. Conf.* (*IPAC 2011*) (San Sebastian, Spain, 2011), pp. 1917–1919.
[70] A. Herlert and Y. Kadi, *J. Phys. Conf.* **312**, 052010 (2011).
[71] M. Thoennessen, *Nucl. Phys. A* **834**, 688c (2010).
[72] *FAIR Baseline Technical Report*, eds. H. Gutbrot *et al.* (GSI, Darmstadt, 2006).
[73] W. Liu, *Nucl. Phys. News* **22**, 29 (2012).
[74] W. G. Myers and H. N. Wagner, Nuclear medicine: how it began, in *Nuclear Medicine: A Hospital Practice Book*, ed. H. N. Wagner (HP Publishing Co., New York, 1975).
[75] H. N. Wagner and G. T. Seaborg, *Nuclear Medicine: 100 Years in the Making, 1896–1996* (Society of Nuclear Medicine, Reston, VA, 1996).
[76] S. C. Srivastava, *Sem. Nucl. Med.* **42**, 151 (2012).
[77] J. A. Park and J. Y. Kim, *Curr. Top. Med. Chem.* **13**, 458 (2013).
[78] S. R. Cherry, *Phys. Med. Biol.* **49**, R13 (2004).
[79] M. J. Heeg and S. S. Jurisson, *Acc. Chem. Res.* **32**, 1053 (1999).
[80] M. D. Bartholomae, *Inorg. Chim. Acta* **389**, 36 (2012).
[81] J. A. O'Donoghue, M. Bardies and T. E. Wheldon, *J. Nucl. Med.* **36**, 1902 (1995).
[82] D. Sarko *et al.*, *Curr. Med. Chem.* **19**, 2667 (2012).
[83] Y. Liu and M. J. Welch, *Bioconjugate Chem.* **23**, 671 (2012).
[84] S. Del Vecchio *et al.*, *Q. J. Nucl. Med. Mol. Imaging* **51**, 152 (2007).
[85] D. L. Schlyer and T. J. Ruth, *Industrial Accelerators and Their Applications*, eds. R. W. Hamm and M. E. Hamm (World Scientific, 2012), pp. 139–181.
[86] H. N. Wagner, *Sem. Nucl. Med.* **28**, 213 (1998).
[87] E. Tapscott, *J. Nucl. Med. Technol.* **33**, 250 (2005).
[88] K. Yoshihava, *Top. Curr. Chem.* **176**, 2 (1996).
[89] J. R. Dilworth and S. J. Parrott, *Chem. Soc. Rev.* **27**, 43 (1998).
[90] P. Richards, *Radioactive Pharmaceuticals: Volume 6 of AEC Symposium Series*, ed. G. A. Andrews (US Atomic Energy Commission, 1966).
[91] T. Ido *et al.*, *J. Labelled Comp. Radiopharm.* **14**, 175 (1978).
[92] S. S. Jurisson and J. D. Lydon, *Chem. Rev.* **99**, 2205 (1999).
[93] *Technetium-99m Radiopharmaceuticals: Manufacture of Kits* (Technical Reports Series, International Atomic Energy Agency, 2008).
[94] P. J. Blower, *Annu. Rep. Prog. Chem. A* **95**, 631 (1999).
[95] A. M. Rey, *Curr. Med. Chem.* **17**, 3673 (2010).
[96] M. Fani *et al.*, *Eur. J. Nucl. Med. Mol. Imaging* **39**, S11 (2012).
[97] A. A. Tavares and J. M. Tavares, *Int. J. Radiat. Biol.* **86**, 261 (2010).
[98] G. J. Beyer *et al.*, *Eur. J. Nucl. Med.* **23**, 1132 (1996).
[99] G. J. Beyer *et al.*, *Proc. COST D8/D18 Workshop* (Prague, 2000).
[100] G. J. Beyer *et al.*, ^{142}Sm and ^{142}Pm have been used in *in vivo* studies, in *CERN Grey Book 1994*, pp. 243–245.
[101] G. J. Beyer *et al.*, *Eur. J. Nucl. Med.* **25**, 1157 (1998).
[102] U. Köster, New isotopes for medical applications, in *Proc. Int. Nucl. Phys. Conf.* (*INPC 2013*) (Firenze, Italy, 2013), in press.
[103] G. J. Beyer *et al.*, *Radiochim. Acta* **90**, 247 (2002).
[104] C. S. Cutler *et al.*, *Chem. Rev.* **113**, 858 (2013).

[105] G. R. Morais *et al.*, *Organometallics* **31**, 5693 (2012).
[106] G. C. Kriiger *et al.*, *Trends Biotechnol.* **31**, 390 (2013).
[107] G. Saha, *Basics of PET Imaging: Physics, Chemistry, and Regulations*, 2nd ed. (Springer, New York, 2010), pp. 117–189.
[108] C. J. Anderson and M. J. Welch, *Chem. Rev.* **99**, 2219 (1999).
[109] M. Fani *et al.*, *Contrast Media Mol. I* **3**, 53 (2008).
[110] I. M. Prata, *Curr. Radiopharm.* **5**, 142 (2012).
[111] S. M. Qaim *et al.*, *Appl. Radiat. Isotopes* **65**, 247 (2007).
[112] F. Haddad *et al.*, *Curr. Radiopharm.* **4**, 186 (2011).
[113] *Non-HEU Production Technologies for Molybdenum-99 and Technetium-99m* (Nuclear Energy Series, International Atomic Energy Agency, 2013).
[114] *Making Medical Isotopes: Report of the Task Force on Alternatives for Medical-Isotope Production*, eds. A. Fong, T. I. Meyer and K. Zala (TRIUMF, Vancouver, 2008).
[115] J. E. Beaver and H. B. Hupf, *J. Nucl. Med.* **12**, 739 (1971).
[116] Private communication.
[117] G. J. Beyer and T. J. Ruth, *Nucl. Instrum. Methods B* **204**, 694 (2003).
[118] G. J. Beyer *et al.*, *Medical Radionuclide Imaging 1980* (Heidelberg, Germany, 1–5 Sep. 1980) (Proceedings Series, International Atomic Energy, 1981), pp. 587.
[119] G. J. Beyer, E. Herrmann and H. Tyrroff, *Isotopenpraxis* **13**, 193 (1977).
[120] G. J. Beyer *et al.*, *J. Labelled Comp. Radiopharm.* **37**, 292 (1995).
[121] G. J. Beyer, *Hyperfine Interact.* **129**, 529 (2000).

Robert Laxdal is Head of the RF/SRF Department and Deputy Head of the Accelerator Division at TRIUMF and a Fellow of the American Physical Society. He is also Adjunct Professor of Physics at the University of Victoria and at Michigan State University. His research interests have ranged from H^-/proton cyclotrons to linac drivers and postaccelerators for RIB application to SRF technology. He initiated SRF research at TRIUMF and developed the ISAC-II superconducting linear accelerator commissioned in 2010.

Colin Morton is the Beam Delivery Liaison/Coordinator at TRIUMF. He received his Ph.D. in Nuclear Physics from the University of Toronto based on studies with radioactive ion beams at TISOL, TRIUMF's prototype (pre-ISAC) ISOL facility. His past research interests included nuclear beta decay and gamma ray spectroscopy at the National Superconducting Cyclotron Laboratory at Michigan State University and with the 8π and TIGRESS groups at TRIUMF. He has been a member of TRIUMF's Accelerator Division and part of the lab's beam delivery group since 2006.

Paul Schaffer is Head of the Nuclear Medicine Division at TRIUMF, Assistant Professor of Radiology at the University of British Columbia and an Adjunct Professor in the Department of Chemistry at Simon Fraser University. Dr. Schaffer received his Ph.D. in Chemistry from McMaster University in 2003 and prior to joining TRIUMF he was the Lead Scientist, Organic Radiochemistry with GE Global Research in New York, where he was responsible for the research and development of novel radiotracers for GE Healthcare's radiopharmaceutical product line.

Reviews of Accelerator Science and Technology
Vol. 6 (2013) 59–83

DOI: 10.1142/S1793626813300041

Spallation Neutron Sources and Accelerator-Driven Systems

Stuart D. Henderson
Fermi National Accelerator Laboratory,
P. O. Box 500, MS105,
Batavia, IL 60510, USA
stuarth@fnal.gov

Spallation neutron sources are the primary accelerator-driven source of intense neutrons. They require high power proton accelerators in the GeV energy range coupled to heavy metal targets for efficient neutron production. They form the basis of large scale neutron scattering facilities, and are essential elements in accelerator-driven subcritical reactors. Demanding technology has been developed which is enabling the next generation of spallation neutron sources to reach even higher neutron fluxes. This technology sets the stage for future deployment in accelerator-driven systems and neutron sources for nuclear material irradiation.

Keywords: Accelerator; spallation; neutron source; subcritical reactor.

1. Introduction

The nuclear spallation reaction [1] provides an efficient mechanism for the production of copious high energy neutrons from GeV energy protons. Neutron sources which utilize the spallation process form the basis of large scale accelerator-based neutron scattering facilities, several of which are in operation or under construction throughout the world [2]. Spallation sources are capable of delivering higher pulsed neutron fluxes than conventional reactor-based sources, making them the technology of choice for modern neutron scattering facilities [3]. Due to the high flux of fast neutrons that can be produced with spallation sources, they have attracted considerable attention for their application in accelerator-driven subcritical nuclear reactors and for material irradiation facilities.

Neutrons are produced by the interaction of a GeV energy high power proton beam with a heavy metal target. Neutron production is quite efficient; for a lead target, approximately 20 neutrons are produced per 1 GeV proton. Approximately half of the incident proton energy is absorbed in the target.

In contrast to spallation sources, fission reactors are neutron-poor. In the fission of ^{235}U, which produces ~2.5 neutrons/fission, only ~1 neutron is available for external use, 1 being required to sustain the reaction and 0.5 being lost. Since each fission reaction releases ~200 MeV of energy, the thermal energy released per neutron is about one order of magnitude higher for reactor-based sources than for spallation sources. Reactor technology, having reached the practical limits of heat removal from the core, is limited in terms of neutron flux that can be achieved.

With a spallation source it is possible to impress a time structure on the neutron flux by using a pulsed proton source, a feature which is exploited in pulsed spallation sources in which the peak flux achieved is several orders of magnitude higher than the average flux. With pulsed neutrons it is possible to use time-of-flight measurements to determine the incident neutron energy, which provides essential kinematic information for neutron scattering measurements.

The highest flux reactor-based sources reach thermal neutron fluxes of $\sim 1.3 \times 10^{15}$ neutrons/cm^2/s. Pulsed spallation sources reach peak thermal neutron fluxes of 10^{16} neutrons/cm^2/s, nearly an order of magnitude higher.

Spallation neutron sources are characterized by their time structure. Continuous neutron sources utilize continuous proton beams, which can be produced by a cyclotron or a continuous wave superconducting linear accelerator. A short pulse spallation source utilizes a long pulse high intensity proton linear accelerator, the beam from which is accumulated over many turns in a synchrotron or accumulator ring and then extracted in a single turn. The proton time structure is therefore short in duration, less than the revolution time in the ring, which is typically 1 μs. In an intermediate approach, a long beam pulse may be delivered from a high intensity linear accelerator with pulse duration of order 1 ms. The need to perform time-of-flight measurements from pulsed neutron sources requires a small beam pulse repetition rate, typically less than 60 pulses per second, in order to avoid the "frame overlap" problem, which arises when faster neutrons arrive at the experimental apparatus in time with slower neutrons from the previous pulse.

Target systems have ranged from solid metal to closed loop liquid metal systems, to rotating solid systems. For neutron scattering applications, the spallation target is surrounded by a moderator which delivers a thermalized neutron flux tailored for use in materials science measurements. The target system is part of a target core which includes the neutron moderators and massive shielding.

Spallation neutron sources operating today cover the range from ~100 kW to more than 1 MW in beam power, with systems under design for as much as 5 MW of beam power.

2. Nuclear Spallation Process

Spallation, a term coined by Glenn Seaborg in his Ph.D. thesis [1], is a process in which a light particle with energy greater than ~100 MeV strikes a heavy nucleus with the resulting emission of nucleons ("spalled" particles) through nuclear cascade, evaporation and fission processes. The process is quite efficient for the generation of neutrons. As an example, bombardment of a lead nucleus by 1 GeV protons results in the emission of ~20 neutrons per incident proton. The resulting spallation residue nucleus is of significantly smaller atomic number. The nuclear spallation process is described in Refs. 4–6.

Spallation is a complex, multistep process, in which an incident high energy particle collides with a nucleus, exciting the nucleus, which ejects nucleons and particles that subsequently interact with other nuclei in a cascade. A thorough discussion is available in Ref. 7. The process is shown schematically in

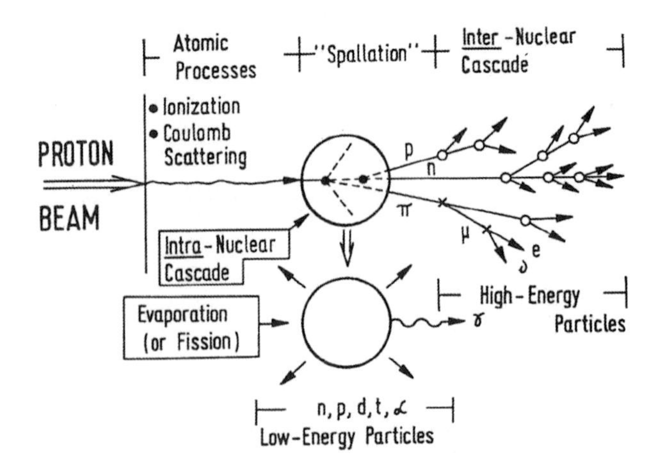

Fig. 1. Schematic diagram of the nuclear spallation process. (Reprinted from Ref. 5.)

Fig. 1. The incident proton collides with individual nucleons in the nucleus, imparting kinetic energy to individual nucleons, and occasionally ejecting nucleons which subsequently interact with surrounding nuclei. This intranuclear cascade results eventually in heating of the nucleus due to secondary and tertiary collisions. This excited nuclear state decays via nucleon evaporation and fission. The interaction of escaping nucleons with the surrounding nuclei, called the internuclear cascade, continues the process, with the net result of the emission of a high flux of neutrons and other low energy particles and fragments. The spallation process and neutron transport is modeled in several Monte Carlo codes, for example MCNPX [8].

The most important features of the spallation process from the point of view of spallation neutron sources are summarized here. A thorough discussion is available in Refs. 5 and 7. The neutron spectrum peaks in the MeV range, but extends all the way up to the primary proton energy, as shown in Fig. 2, which compares neutron spectra for two metal target materials. For low energy neutron scattering research, high energy neutrons are moderated in ambient temperature or cryogenic moderator systems.

The neutron yield per proton is an important quantity for spallation neutron source design. Figure 3 shows a comparison of measured and calculated neutron yield from a 20-cm-diameter, 60-cm-long lead target as a function of incident proton energy [10]. An often used empirical formula [5] for

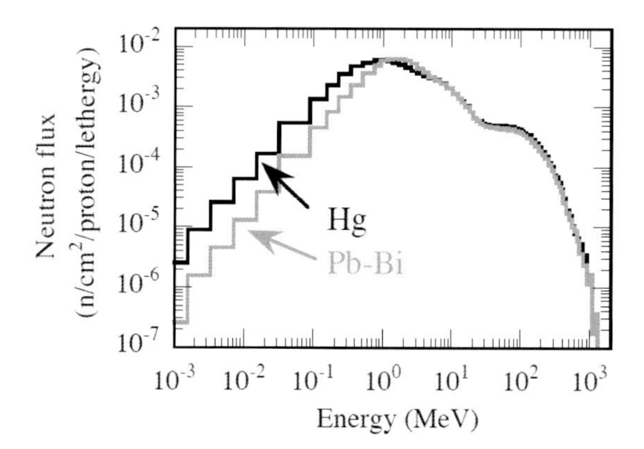

Fig. 2. Neutron spectrum for two spallation target materials, mercury and lead–bismuth eutectic, for 1.5 GeV incident protons. (Reprinted from Ref. 9.)

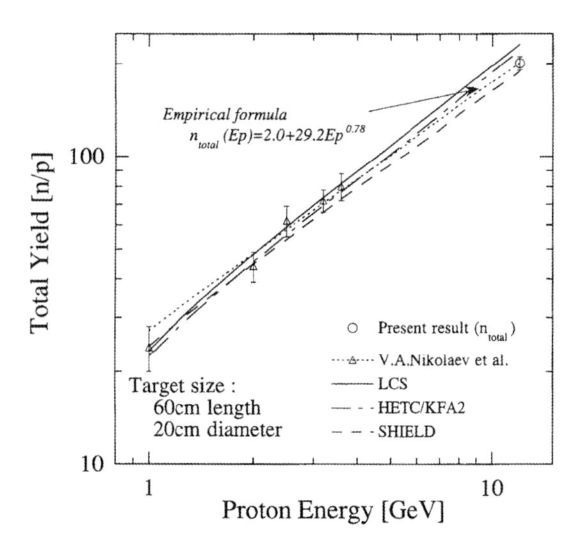

Fig. 3. Comparison of measured and calculated neutron yield for a lead target of 20 cm diameter and 60 cm length, as a function of proton energy. (Reprinted from Ref. 10.)

the approximate neutron yield, Y (neutrons per proton), of a heavy metal spallation target is given by

$$Y(E, A) = 0.1(E[\text{GeV}] - 0.12)(A + 20), \quad (1)$$

where E is the proton beam energy in GeV and A is the atomic mass of the metal target. For actinide targets the yield is approximately twice that given by Eq. (1).

The yield increases approximately linearly with proton energy in the GeV energy range. Therefore, neutron yield is proportional to beam power over a broad range of energies, from ~0.6 GeV to ~10 GeV. This fact has important consequences for the optimization of spallation neutron sources, typically favoring lower beam energies and higher beam current, as opposed to high beam energy and lower beam current.

3. Overview of Spallation Sources

Spallation sources [11] consist of an accelerator and a target system. Sources configured for neutron scattering experiments (the only type existing to date) also incorporate neutron moderator systems. Sources may be distinguished by the temporal structure of the resulting neutron flux, which itself is determined by the temporal structure of the proton beam, which in turn determines the accelerator configuration that is required.

The four main configurations are shown in Fig. 4. Short pulse sources are of two main configurations. In the first, shown in Fig. 4(a), a long beam pulse (of duration 0.1–1.0 ms) from a low energy (~100 MeV) linear accelerator is accumulated in a rapid-cycling synchrotron (RCS) and accelerated to high energy. In the second, shown in Fig. 4(b), a long beam pulse

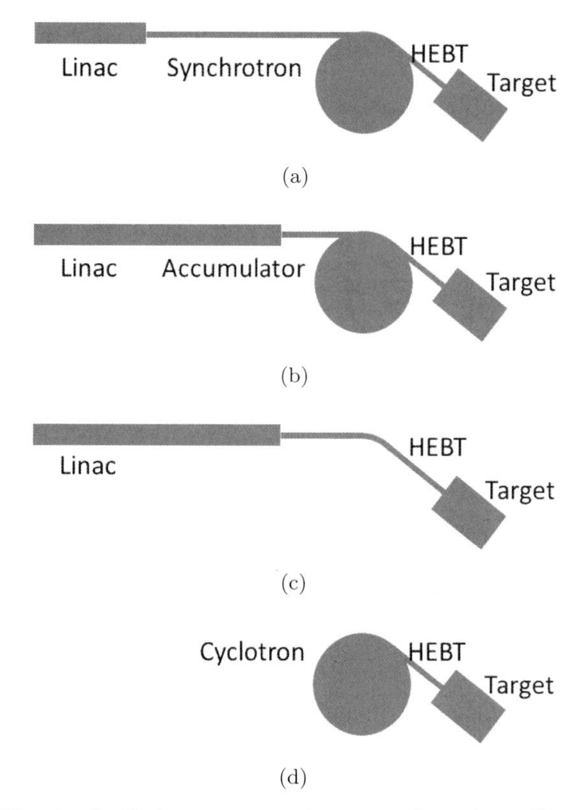

Fig. 4. Spallation source accelerator configurations. Short pulse sources use either the linac synchrotron configuration (a) or the linac accumulator configuration (b) Configuration (c) provides either a long pulse or continuous source, depending on the linac design. Configuration (d) provides a continuous source of neutrons.

from a full energy linear accelerator is accumulated in an accumulator (or compressor) ring. In each of these cases, the beam is extracted in a single turn so that the resulting proton pulse is of order 1 μs, which is a typical revolution period for such a ring. A long beam pulse of one or a few milliseconds delivered from a linear accelerator can be used without subsequent accumulation to power a "long pulse" neutron source. This configuration, shown in Fig. 4(c), can also deliver a continuous proton beam (for a continuous neutron source) provided that the linear accelerator is capable of operating in a continuous wave mode, which is one of the preferred configurations for accelerator-driven subcritical reactor systems. Finally, cyclotrons deliver a continuous beam; the configuration shown in Fig. 4(d) is utilized at the Paul Scherrer Institute (PSI) for the SINQ target.

Table 1 summarizes the main parameters of the previously operating, currently operating and planned spallation neutron sources. With the exception of SINQ at PSI, all existing spallation sources are pulsed with repetition rates ranging from 20 Hz to 60 Hz. Beam energies are in the GeV range, spanning from ~0.6 GeV to 3 GeV, a beam energy range in which the neutron flux increases linearly with beam power.

Spallation sources rely on high power proton accelerators, with the present state of the art in the megawatt range of beam power. High power can be achieved for continuous beams from cyclotrons, and recently, with the rise in capabilities of superconducting radio frequency (SRF) technology, in superconducting RF linacs. For pulsed beams both normal-conducting and superconducting linacs have been utilized, with the modern trend toward SRF systems. To date the highest beam power pulsed system, the Spallation Neutron Source (SNS) at ORNL, is based on a full energy SRF linac and accumulator ring configuration, as shown in Fig. 4(b).

The overriding accelerator design constraint in a spallation neutron source is the requirement to maintain uncontrolled beam losses less than 1 W per meter in order to limit residual activation and allow hands-on maintenance of the accelerator facility. As a point of reference, a 1 W/m loss of protons on stainless steel results in residual activation in the range of 100 mRem/h after a few hours cooldown. Such a requirement becomes extremely challenging for megawatt class accelerators, and motivates extreme attention to detail regarding beam loss mechanisms.

Target systems for pulsed application up to hundreds of kW utilize solid target technology, whereas the pulsed systems in the MW range use liquid metal technology, which is advantageous from the standpoint of heat removal from the target system.

Most sources built so far are utilized in neutron scattering facilities, and therefore incorporate moderators based on either ambient temperature water systems or cryogenic systems.

4. Historical Developments

In the postwar era, the possibility of using a high power hadron accelerator for the production of strategic nuclear materials was recognized. Under E. O. Lawrence's leadership, a project was launched in the late 1940s at the University of California Radiation Laboratory to develop the technology for producing ^{239}Pu, ^{233}U and tritium from high fluxes of accelerator-produced neutrons [12, 13]. In this

Table 1. Summary of parameters for previously operating, currently operating and planned spallation neutron sources.

	KENS	IPNS	ISIS	LANSCE	IN-6	SINQ	SNS	JSNS	CSNS	ESS
Laboratory	KEK	ANL	RAL	LANL	INR	PSI	ORNL	JPARC	IHEP	Lund
Reference	18	17	48	54	162	60	23	24	25	26
Start	1980	1981	1985	1986	1998	1997	2006	2008	2018	2019
Status	Closed	Closed	Operating	Operating	Operating	Operating	Operating	Operating	Const.	Design
Technology	Linac/ RCS	Linac/ RCS	Linac/ RCS	Linac/ Accum.	Linac	Cyclotron	Linac/ Accum.	Linac/ RCS	Linac/ RCS	Linac
Beam energy (GeV)	0.5	0.45	0.8	0.8	0.6	0.59	1.0	3.0	1.6	2.5
Beam power (MW)	0.003	0.007	0.2	0.08	0.3	0.9	1.4	1.0	0.1	5.0
Beam current (mA)	0.007	0.015	0.25	0.10	0.5	1.55	1.4	0.333	0.0625	2.0
Repetition rate (Hz)	20	30	50	20	50	CW	60	25	25	14
Target	W, U	U	Ta, W, U	W	W	Zr/Pb	Hg	Hg	W	W

program, known as the Materials Testing Accelerator (MTA) Program, one sees the birth of the field of high intensity particle accelerators.

The initial accelerator design was based on a 12 MHz drift tube linac structure, with eight 60-ft-diameter drift tubes and output energy of 15 MeV. Continuous wave acceleration of protons up to 100 mA was achieved. The next generation MTA accelerator, taking advantage of the availability of higher frequency power sources which became available in the early 1950s, included drift tube linac and quarter-wave resonator structures based on 48.6 MHz sources. That machine accelerated continuous wave beams of protons to 3.75 MeV with 75 mA current and deuterons to 7.5 MeV with 30 mA current.

The early prototype and demonstration activities focused on what was to become the front end of a very high power spallation source which remains well beyond the present day state of the art. The facility was to utilize continuous wave acceleration of deuterons to 500 MeV with 320 mA current — 160 MW of beam power. The target concept envisioned a NaK-cooled Be primary target surrounded by a depleted uranium secondary target for neutron multiplication. A water-cooled depleted uranium blanket surrounded the core for breeding plutonium.

The MTA program was abandoned in the mid-1950s with the discovery of large uranium deposits in the US. Many of the challenges encountered in that early program remain as central challenges today, which must be faced in designing and constructing modern high intensity hadron accelerators. This early work identified the importance of developing high power RF amplifiers, and developing a fundamental understanding of gradient limitations, high vacuum technology, and RF structure geometries suitable for high gradient acceleration. It further engaged the critical technical issue of accelerator–target coupling and neutronic performance.

In the mid-1960s, another ambitious high power accelerator-driven neutron source project was launched: the Intense Neutron Generator (ING) at the Chalk River Laboratory of Atomic Energy in Canada [14]. The ING concept envisioned a continuous high flux neutron source for isotope production and neutron scattering based on a 1 GeV proton accelerator providing 65 mA beam current, or 65 MW of beam power. Two designs were considered, the first a three-stage separated orbit cyclotron and the second a linear accelerator based on an Alvarez drift tube structure followed by a coupled cavity structure. The target concept consisted of flowing liquid Pb–Bi eutectic arranged to provide a 20-cm-diameter and 60-cm-long liquid target region in a vertical incident beam geometry. Surrounding the target region was a Be multiplier and heavy water moderator. The ING would have achieved a neutron production rate of 10^{19} neutrons/s and a thermal neutron flux of 10^{16} neutrons/cm^2/s. Although the project was canceled in 1968, it contributed a number of key technical developments important for the spallation neutron sources which were to be built beginning in the late 1970s, and continuing to the present.

The modern pulsed spallation neutron source concept traces its origins to the ZING project at Argonne National Laboratory in 1974 [15]. It made use of proton beams from the rapid-cycling booster synchrotron for the Zero Gradient Synchrotron (ZGS) incident on a target/moderator/beryllium reflector arrangement. The initial configuration, known as ZING-P (to indicate that it was a prototype of a larger future source), used 300 MeV protons with a power of about 0.1 kW on a lead target, and operated from 1974 to 1975. The second configuration, known as ZING-P′, used the newly constructed Booster II, which provided 500 MeV proton beams at ~3 kW beam power, and operated from 1977 to 1980. This activity laid the groundwork for the construction of the Intense Pulsed Neutron Source (IPNS), consisting of a 50 MeV linear accelerator and the 500 MeV ZGS Booster synchrotron [16, 17]. IPNS began operating in 1981 and continued until 2008.

The Weapons Neutron Research (WNR) facility at Los Alamos National Laboratory began operating in 1977, making use of ~1% of the long, ~800 μs beam pulse delivered by the LANSCE linac.

Meanwhile, the pulsed neutron source facility (KENS) was completed at KEK in 1980 [18], and it continued operating until 2006. It utilized a 500 MeV proton beam from the KEK Booster Synchrotron at 7 μA beam current for a 3.5 kW beam. KENS and IPNS were the first two spallation source facilities designed for materials research using neutron scattering and were therefore essential for establishing spallation sources as alternatives to reactor-based sources within the user community.

Reference 19 summarizes the progress in pulsed neutron sources up to 1981.

Following these first generation spallation sources, the second generation began to be built in the mid-1980s. The ISIS neutron scattering facility at RAL became operational in 1985 [20]. It is based on a high power rapid-cycling (50 Hz) 800 MeV synchrotron in which beam is accumulated through the multiturn charge exchange injection process [21]. ISIS became the world's highest power pulsed spallation source until 2008, when it was superseded by the SNS at ORNL.

At Los Alamos, a substantial increase in beam power and neutron flux was achieved with the addition of the Proton Storage Ring (PSR), a fixed energy accumulator ring, in the mid-1980s. This allowed long beam pulses to be accumulated through the multiturn charge exchange injection process, followed by single turn extraction. This facility, known as the Los Alamos Neutron Science Center (LANSCE), is one of two spallation sources in operation today in the US.

The Swiss Spallation Neutron Source (SINQ) has been operating since 1997 [22]. It is based on the 590 MeV PSI Ring Cyclotron. The beam repetition rate is 51 MHz, high enough that the neutron yield after moderation is continuous, providing characteristics similar to those of a reactor-based source. The Ring Cyclotron provides the highest power proton beam in the world, and is presently operating at 1.4 MW.

The Spallation Neutron Source (SNS) at Oak Ridge National Laboratory was constructed in the early 2000s and began operating in 2006. It incorporated a number of next generation design features (discussed further below), including the world's first superconducting high energy linear accelerator for protons, and a closed loop liquid mercury target system. The design beam power is 1.4 MW [23].

The spallation neutron source at the Japan Proton Accelerator Research Complex (J-PARC) was completed in 2008 [24]. It is based on a 180 MeV normal-conducting linear accelerator, the beam from which is accumulated and accelerated in a 3 GeV rapid-cycling synchrotron. The spallation source utilizes a closed loop liquid mercury target system. The design beam power is 1 MW.

At the present time there are two spallation neutron sources under design and construction. The China Spallation Neutron Source [25] is under construction in Dongguan. It will utilize a normal-conducting linear accelerator and rapid-cycling synchrotron. The European Spallation Source (ESS), to be built in Lund, Sweden, is under design [26]. A superconducting linear accelerator will provide a 5-MW-long beam pulse at 2.5 GeV to a solid rotating target system.

5. Applications of Spallation Neutron Sources

5.1. *Neutron scattering*

Neutrons are important probes of material structure and dynamics, owing to their fundamental properties, including zero charge, a finite magnetic dipole moment, zero — or, more precisely, so far undetected — electric dipole moment, and primary interaction with atomic nuclei through the nuclear force. As a result, they are deeply penetrating, capable of passing through thick material samples. Neutrons interact either through scattering off the atomic nuclei or from unpaired electron spins. Detection and reconstruction of the scattered neutron's energy and angle provides the basis for neutron scattering measurements and analysis [27]. These properties make neutron scattering a complementary approach to X-ray or electron scattering and diffraction.

While the scattering cross-section of X-rays from atoms increases as Z^2, the scattering cross-section of neutrons varies in accordance with the nuclear structure, and therefore displays a semirandom pattern across the periodic table. The magnetic scattering of neutrons is of the same order as the nuclear scattering, leading to the usefulness of neutrons for determining magnetic properties of materials.

One particular feature of neutron scattering which makes the technique important in the biological and polymer sciences is the large scattering cross-section for hydrogen, allowing the precise location of hydrogen atoms to be determined. Further, neutron scattering is capable of resolving the location of light atoms adjacent to heavy ones, which was crucial for the determination of the structure of high T_c superconductors.

Since the primary spallation neutrons are produced in the MeV energy range, they must be moderated to provide neutron beams of appropriate energy/wavelength for use in materials science studies. Moderator systems are incorporated around the spallation target system to reduce the neutron energies (or increase the wavelength) in order to match the elementary excitation energies (or length scales) of the material properties under study. The use of ambient temperature water (or heavy water) moderator systems produces thermal neutrons with a Maxwell–Boltzmann distribution at 25 meV; cryogenic moderator systems provide cold neutrons with energies in the few-meV range. Cold neutron beams are important for the study of systems with long length scales, such as biological molecules and polymers [28].

5.2. *Materials irradiation*

Spallation neutron sources can serve as sources of high flux fast neutrons for material irradiation studies [29]. They are in demand for the further development of fission reactor systems, particularly those that utilize fast neutron spectra, as well as future fusion devices. The development of advanced fission reactor systems requires materials capable of withstanding an order of magnitude higher radiation damage and higher temperatures. The development of nuclear fuels appropriate for such an environment is a critical developmental activity.

Future fusion systems require materials capable of withstanding even more extreme radiation damage conditions. Figure 5 gives an indication of the materials science challenge facing future next generation reactor systems and fusion devices. It shows the anticipated operating regimes for a variety of advanced reactor systems in terms of radiation damage (in displacements per atom) and temperature. The development of materials that can operate in these extreme environments requires suitable irradiation sources which can replicate these conditions.

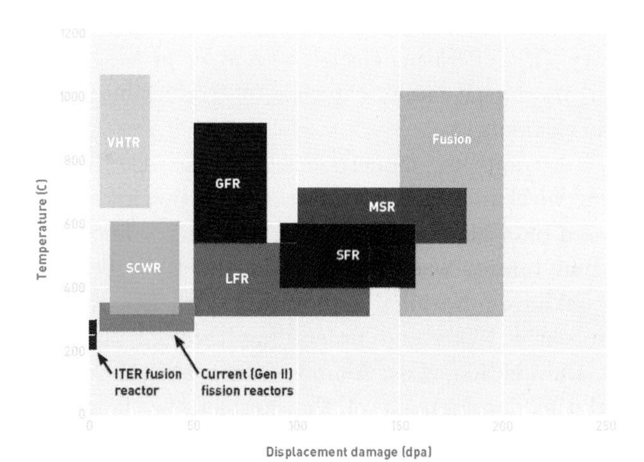

Fig. 5. Operating regions in temperature vs. displacement damage for advanced reactor systems. Fission reactors include very-high-temperature reactors (VHTR), supercritical water-cooled reactors (SCWR), gas-cooled fast reactors (GFR), lead-cooled fast reactors (LFR), sodium-cooled fast reactors (SFR) and molten salt reactors (MSR). [Reprinted from S. J. Zinkle and J. T. Busby, Structural materials for fission and fusion energy, *Mater. Today* **12**, 12–19 (2009), with permission from Elsevier.]

First-wall materials in a fusion reactor system must withstand intense fluxes of 14 MeV neutrons. Specialized material irradiation facilities optimized for fusion applications have been designed. The proposed International Fusion Materials Irradiation Facility (IFMIF) design would utilize a 40 MeV deuteron beam to produce high energy neutrons from the d-Li stripping reaction on a liquid lithium target [30].

A recent example of an accelerator-driven spallation source optimized for nuclear material irradiation, the Materials Test Station (MTS), can be found in Ref. 29. The MTS design uses the 1 MW LANSCE linac coupled to a tungsten target system driven at 100 Hz. Fast neutron fluxes of greater than $1 \times 10^{15}/\text{cm}^2/\text{s}$ can be achieved, with a spectrum very similar to that of a fast reactor. This flux, which would deliver damage rates of 17 dpa/year, is about one-half of that achieved at the world's most intense fast spectrum reactors.

An accelerator-based neutron source differs from a reactor-based source in two principal ways. First, the neutron flux is pulsed in a pulsed spallation source, and may even be intermittent, depending on the source reliability. The material response to pulsed neutron fluxes at pulse frequencies typical of spallation sources is very similar to that of a continuous source, although some difference in material response may be expected at low pulse frequencies [31]. Also, beam interruptions cause material sample cooling, which may not be representative of actual reactor operating conditions. This latter complication demands high reliability for accelerator-based irradiation sources, with an emphasis on minimizing the number of long duration unscheduled beam interruptions, as will be discussed in more detail below. Secondly, the spallation neutron spectrum contains a small fraction of high energy neutrons, up to the primary proton beam energy (see Fig. 2), which may lead to somewhat different damage response relative to a reactor-based source. Further, these high energy neutrons may be energetic enough to produce spallation products in the irradiated material. These various considerations are discussed in Refs. 32 and 33.

5.3. *Fundamental physics*

Spallation neutron sources provide neutron fluxes which allow fundamental studies in nuclear and particle physics. The electric dipole moment (EDM) of the neutron is a sensitive probe of physics beyond the standard model of particle physics. A next generation search for the EDM of the neutron with a sensitivity of less than 5×10^{-28} e-cm is in the planning stages at the SNS [34]. The experiment aims at a factor of 50–100 improvement in the present limit of $d < 2.9 \times 10^{-26}$ e-cm [35].

A spallation neutron source optimized for particle physics is envisioned at the Project-X facility at Fermilab [36]. In addition, this source is well-suited for the production of cold neutrons for neutron–antineutron oscillation experiments, for neutron EDM measurement and for the production of isotopes — ^{225}Ra, ^{223}Rn, ^{211}Fr — which are particularly sensitive to a potential finite electron EDM. Neutron–antineutron oscillation would signal baryon number violation, which, if observed, would have profound implications for the baryon asymmetry in the universe. An optimized spallation source based on the Project-X parameters would be capable of producing substantial improvements in sensitivity relative to present day techniques [36].

5.4. *Accelerator-driven subcritical reactors*

The notion of employing particle accelerators in the production of strategic nuclear materials goes back to the MTA program. The idea of using particle accelerators in a nuclear reactor system was also recognized in the early days of the nuclear energy program [37]. Nuclear systems which couple an external neutron spallation source to a subcritical reactor are referred to as accelerator-driven subcritical reactors (ADSRs) or accelerator-driven systems (ADSs).

While there are many ADS concepts and variants, they have common features. A subcritical reactor, i.e. one which has been purposely designed so that the effective neutron multiplication factor, k_{eff}, is less than 1 under all conditions, is coupled to an externally supplied source of neutrons produced by a high power proton accelerator and spallation target system embedded in the core. Such a system is shown schematically in Fig. 6. The externally produced neutrons coupled to a neutron-multiplying medium sustain the core at its design thermal power, which is extinguished once the external neutron flux is removed.

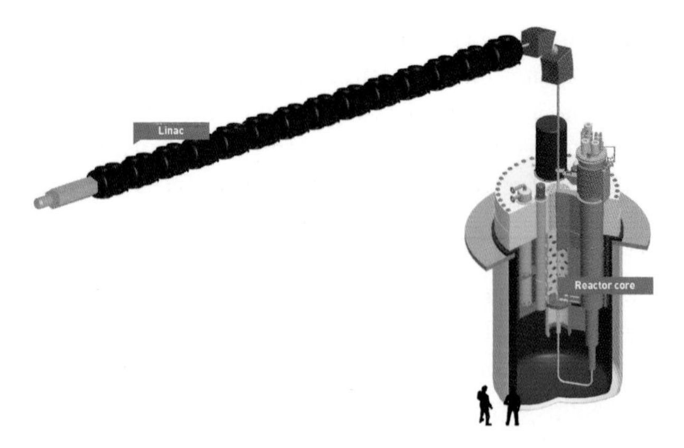

Fig. 6. Schematic diagram of an ADS. There are three main components: a high power proton accelerator, a spallation target system and a subcritical reactor assembly. (Image courtesy of SCK-CEN, Mol, Belgium.)

Such systems are under active study throughout the world, due to their potential in the development of advanced nuclear fuel cycles. ADSs may be useful for addressing three main missions: (i) transmuting problematic isotopes (e.g. minor actinides) in spent nuclear fuel to reduce the burden on a geologic repository; (ii) generating electricity and/or process heat; (iii) producing fissile materials by irradiation of fertile elements.

A complete, modern design of an ADSR system with an emphasis on transmutation of waste, based on a high power particle accelerator and spallation target, was developed in the late 1980s at Brookhaven National Laboratory [38]. Starting in the early 1990s and proceeding through the rest of the decade, Los Alamos National Laboratory led the development of the Accelerator Transmutation of Waste (ATW) design study [39, 40]. In Japan, accelerator-based transmutation R&D began in the late 1980s under the OMEGA program [41]. A roadmap for developing ATW technology was submitted to Congress [42]. In the mid-1990s Carlo Rubbia led an effort to develop an energy amplifier [43].

In the US a National Research Council study of transmutation technologies was carried out in 1995 [44]. In Europe, similar studies were carried out, comparing ADS technology and capabilities with those of fast reactors [45, 46].

The spallation target and high power accelerator technology required for ADS applications was the subject of a recent US DOE–sponsored assessment [47]. This report, discussed further below, outlines the demanding accelerator requirements and challenges which are involved in order to realize ADS applications.

6. Overview of Operating and Future Spallation Neutron Sources

6.1. *ISIS at Rutherford Appleton Laboratory, UK*

ISIS is a short pulse spallation source which began operation in 1985 [20]. The accelerator consists of an H^- ion source, a 665 MeV four-rod 202.5 MHz RFQ, a 202.5 MHz 70 MeV drift tube linac and a 163-m-circumference rapid-cycling synchrotron with a 50 Hz repetition rate. An H^- beam is accelerated and accumulated in the synchrotron via multiturn charge exchange injection utilizing a 0.25 μm alumina foil. Intensities of $\sim 2.8 \times 10^{13}$ protons are accumulated per pulse over ~130 turns. A second target station began operating in 2008; both target stations operate simultaneously, sharing the ~160–200 kW total beam power. Table 2 shows some of the key ISIS parameters. Recent descriptions of the facility and operations can be found in Refs. 48 and 49.

Reference 49 describes the challenges and primary limitations arising from beam loss. The dominant source is from the RF trapping process after injection, in which the DC injected beam is captured in two $h = 2$ buckets. Recently a dual-harmonic, $h = 4$ system was installed to increase the longitudinal bucket acceptance and to improve the bunching factor [50]. Transverse losses are dominated by emittance growth from high intensity effects, namely space charge, which generates peak transverse incoherent tune shifts of more than -0.4.

The target systems use tantalum-coated tungsten. TS1 includes two water moderators, an ~100 K liquid methane moderator and a 20 K liquid hydrogen moderator. TS2 includes a coupled hydrogen/solid methane moderator and a decoupled solid methane moderator.

Table 2. ISIS parameters.

Parameter	Value
Total beam power (MW)	0.20
Beam power TS1/TS2 (MW)	0.160/0.04
Beam energy (GeV)	0.8
RFQ/DTL frequency (MHz)	202.5
Linac macropulse duty factor (%)	1.0
Linac beam pulse length (μs)	200
Peak H^- current (mA)	22
Repetition rate (Hz)	50
Average linac current (mA)	0.25
Linac energy (MeV)	70
Synchrotron circumference (m)	163
Accumulation time (turns)	130
Ring bunch intensity	2.8×10^{13}
Harmonic number	2, 4
Pulse length on target (ns)	100

6.2. *Los Alamos Neutron Science Center, US*

The Los Alamos Neutron Science Center at Los Alamos National Laboratory is a short pulse source based on the 800 MeV LANSCE (formerly LAMPF) proton linear accelerator. LAMPF [51] began operating in 1972, and by 1977 a pulsed spallation source was in operation, based on the use of a small portion of the ~800 μs linac beam pulse.

H^- beam is accelerated in a Cockcroft–Walton generator to 750 keV, injected into a 201.25 MHz drift tube linac which accelerates the beam to 100 MeV, and further accelerated to 800 MeV in an 805 MHz side-coupled linac [51, 52]. The invention of the side-coupled structure, first demonstrated at LANSCE, provided a route to efficient high energy acceleration of proton beams. In its use for nuclear and particle physics, the LANSCE linac operated routinely at a peak beam power of 800 kW.

In 1985, a 90-m-circumference proton accumulator (called the Proton Storage Ring, PSR) was constructed to allow accumulation of the chopped ~800 μs beam pulse to provide an extracted pulse on the spallation target of 0.25 μs duration [53]. An upgrade to the PSR injection system, carried out in the mid-1990s, allowed an increase in average current to 100 μA, providing 80 kW of beam power at 20 Hz to the tungsten spallation target [54]. Two liquid hydrogen moderators and four ambient water moderators provide beams to the Lujan Neutron Science Center to support a neutron scattering user facility [55]. Table 3 provides a summary of LANSCE/PSR parameters.

Table 3. LANSCE/PSR parameters.

Beam power (MW)	0.08
Beam energy (GeV)	0.8
DTL frequency	201.25
CCL frequency	805
Beam pulse length (ms)	0.825
Peak H^- current (mA)	9
Repetition rate (Hz)	20
Average linac current (mA)	0.1
Accumulator circumference (m)	90
Ring bunch intensity	3.1×10^{13}
Pulse length on target (ns)	250

As with any spallation source, the understanding and mitigation of beam loss was a central activity. Several important high intensity limitations were encountered and studied at the PSR. An upgrade to the injection region was motivated by the limitations of beam loss arising from the H^- to proton conversion process. Beam loss mechanisms associated with stripper foil scattering, production of excited unstripped H^0 particles and handling of waste beams from the stripping process were studied in detail. The electron cloud effect, an important intensity-limiting collective effect, was identified and studied in detail in the PSR [56, 57], as was the emittance growth induced by space charge forces [58], which are described further below.

6.3. *SINQ at Paul Scherrer Institute, Switzerland*

SINQ is a continuous neutron source based on the PSI Ring Cyclotron, a high power separated sector cyclotron built in 1974 [59]. The PSI Ring Cyclotron provides the highest power proton beam in the world, recently reaching 1.4 MW [60, 61]. Routine operation at the time of this writing delivers 2.2 mA beam current (1.3 MW). Reference 61 presents an overview of high power cyclotrons.

The complex consists of a Cockcroft–Walton preaccelerator, a 72 MeV isochronous injector cyclotron and the 590 MeV Ring Cyclotron. The 72 MeV injector cyclotron was designed with high intensity in mind. The beam bunch dimensions are equal in the radial and longitudinal, which is beneficial from the standpoint of space charge effects [62].

The proton beam is continuous wave, bunched at a frequency of 50.6 MHz. It is extracted from the Ring Cyclotron via an electrostatic septum placed between the last and second-to-last turns. The extracted beam is directed to multiple targets. In the first set, a pair of rotating graphite targets is used to produce pions and muons for particle physics and a muon spin spectroscopy program. The remaining beam intensity, which is ~70% of the total intensity, is collimated and transported to the SINQ neutron production target. In routine operation of the Ring Cyclotron with 2.2 mA beam current, 1.55 mA, corresponding to 0.9 MW, is transported to the target. The target consists of a matrix of lead-filled Zircaloy tubes [63].

The relative beam loss is maintained in the few-times-10^{-4} range to limit residual activation, particularly in the extraction region. The intensity limitation in the Ring Cyclotron is due to longitudinal space charge forces which increase the energy spread, which in turn generates transverse tails between the last two turns at extraction. It has been shown that this limitation scales as the third power of the accelerating voltage [64]. Alternatively, since the number of turns in the cyclotron is inversely proportional to accelerating voltage, the maximum current increases as $1/N^3$ at fixed particle loss rates. Increases in intensity have been enabled by reducing the number of stored turns through the installation of RF systems with higher accelerating voltage per turn. Through a two-decades-long series of improvements in the RF systems, the voltage per turn has been increased from ~1.1 MV/turn to ~3 MV/turn, which has allowed a nearly tenfold increase in beam current. Figure 7 displays the achieved beam current

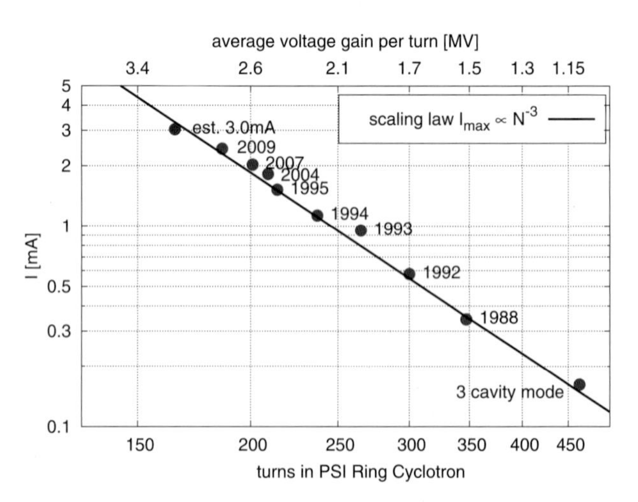

Fig. 7. Beam current in the PSI ring cyclotron as a function of ring turns. The solid line shows beam current expected from a simple space charge scaling law. The points show performance vs. time [65].

versus the number of turns, which shows remarkable agreement with the expectation from the simple scaling law described above [65].

An important R&D activity related to very high power target systems — the MEGAPIE Project — was carried out in the 2000s [66]. This project, discussed in more detail below, demonstrated for the first time the use of a Pb–Bi eutectic target loop in a spallation source. This successful demonstration is important for future use of liquid Pb–Bi in very high power applications, such as for ADSs.

6.4. *SNS at Oak Ridge National Laboratory, US*

The Spallation Neutron Source (SNS) is a short pulse neutron scattering facility at Oak Ridge National Laboratory which was constructed in the early 2000s and began operating in 2006 [23, 67]. The SNS construction project was a partnership of six US DOE national laboratories, each of which had responsibility for designing and manufacturing a portion of the facility. At 1.44 MW of proton beam power on target, the SNS was designed to operate at beam powers a factor of 8 beyond that which had been previously achieved in a pulsed source. The key SNS parameters are summarized in Table 4. The beam power achieved to date is 1.4 MW. Other parameters have been achieved individually.

The SNS accelerator complex consists of a 2.5 MeV H^- injector [68, 69], a 1 GeV linear accelerator [70–72], an accumulator ring and associated

Table 4. SNS parameters.

Parameter	Value
Beam power (design) (MW)	1.4
Beam power (achieved) (MW)	1.4
Beam energy (GeV)	1.0
RFQ and DTL frequency	402.5
CCL and SCL frequency	805
Linac macropulse duty factor (%)	6
Beam pulse length (ms)	1.0
Peak H^- current (mA)	38
Repetition rate (Hz)	60
Chopper beam-on duty factor (%)	68
Average linac current (mA)	1.6
Accumulator circumference (m)	248
Accumulation time (turns)	1060
Ring bunch intensity	1.6×10^{14}
Ring space charge tune spread	0.15
Harmonic number	1, 2
Pulse length on target (ns)	695

beam transport lines [73]. The injector (also called the Front-End System) consists of an H^- volume ion source with 50 mA peak current capability [74, 75], a radio frequency quadrupole and a medium energy beam transport (MEBT) line for chopping and matching the 2.5 MeV beam to the linac. The linear accelerator consists of a drift tube linac (DTL) with 87 MeV output energy, a coupled cavity linac (CCL) with 186 MeV output energy and a superconducting RF linac (SCL) with 1 GeV output energy [76]. At full design capability the linac produces a 1-ms-long, 38 mA peak, chopped beam pulse at 60 Hz for accumulation in the ring.

The linac beam is transported via the high energy beam transport (HEBT) line to the injection point in the accumulator ring, where the 1-ms-long pulse is compressed to $\sim$700 ns by multiturn charge exchange injection. According to the design, beam is accumulated in the ring over 1060 turns, reaching an intensity of 1.6×10^{14} protons per pulse. When accumulation is complete the extraction kicker fires during the 250 ns gap to remove the accumulated beam in a single turn and direct it into the ring to target beam transport (RTBT) line, which takes the beam to a liquid mercury target.

The integrated beam dynamics design was focused on beam loss minimization [77]. The linac beam dynamics design incorporates a standard design approach for high intensity linacs [78], including smooth longitudinal and transverse focusing and avoidance of resonances to minimize emittance and halo growth from space charge dynamics (see Ref. 79 and references therein).

The SNS linac is the first application of SRF technology for high energy acceleration of protons. The advantages of a superconducting linac include efficient transfer of energy from RF source to beam, reduced RF power capital and operating costs, a higher accelerating gradient, flexible linac operation with individually powered cavities, and low residual gas pressure to minimize losses.

Beam loss in the linac is well below the 1 W/m limit, although measurable. An experimental program was carried out to determine the source of small but measurable beam loss in the SNS linac. A comparison of losses from proton and H^- acceleration [80, 81] helped to pinpoint the dominant beam loss source in the linac as intrabeam stripping of the H^- beam [82].

The accumulator ring design [83] was likewise focused on the requirement to minimize losses. The primary loss mechanisms that informed the design were space-charge-induced emittance and halo growth during injection, handling of waste beams from the foil-stripping process, minimization of impedance to avoid collective instabilities, and mitigation of the electron cloud effect. Challenges in design of high intensity linacs and rings are discussed further below.

Beam loss in the ring remains within specifications, and is discussed in more detail in Ref. 84. An overview of beam loss challenges in such high intensity, high power accelerators is given in Ref. 85.

The liquid mercury target system [86, 87] consists of a closed loop mercury-handling system. The target lifetime is limited by the cavitation-induced erosion (discussed in more detail below), which scales as the fourth power of beam intensity [88]. The target module is designed for remote-handling maintenance by retraction into a service bay outfitted with remote manipulator systems. Neutrons are moderated in four moderators, one using ambient water and the other three utilizing supercritical hydrogen at 17–20 K.

References 89–92 review the high power operational experience at the SNS.

6.5. *Japan Spallation Neutron Source, J-PARC*

The JSNS facility [93] at J-PARC in Tokai, Japan, is the most recently completed spallation source, coming online in 2008 [24]. It is a short pulse source based on a linac and long pulse accumulation via multiturn charge exchange injection in a rapid-cycling, 3 GeV synchrotron. The JSNS parameters are summarized in Table 5.

H^- beam is produced and accelerated in a 324 MHz, 3 MeV RFQ [94]. After RF chopping and bunching in the MEBT, the beam is injected into a 50 MeV drift tube linac [95], which is followed by a separated drift tube linac (SDTL) consisting of 30 individual cavities, powered in pairs, delivering 180 MeV output energy. Beam is transported from the linac to the rapid-cycling synchrotron, where the ~500 μs pulse is accumulated. After acceleration to 3 GeV, beam is transported to a liquid mercury target system.

Table 5. J-PARC JSNS parameters.

Beam power (design) (MW)	1.0
Beam power (achieved) (MW)	0.5
Beam energy (GeV)	3.0
RFQ/DTL frequency (MHz)	324
Annular coupled structure freq. (MHZ)	972
Linac macropulse duty factor (%)	1.25
Linac beam pulse length (ms)	0.5
Peak H^- current at 181/400 MeV (mA)	30/50
Repetition rate (Hz)	25
Average linac current (mA)	0.33
Linac energy (MeV)	181
Upgrade linac energy (MeV)	400
Synchrotron circumference (m)	348
Ring bunch intensity	8.3×10^{13}
Harmonic number	2, 4
Pulse length on target (ns)	600

The rapid-cycling requirement of 0.18–3 GeV at 25 Hz puts stringent conditions on the synchrotron RF system. Fields of 23 kV/m are required, which is a factor of 2 greater than what one typically finds in ring RF systems, limited by the properties of the ferrite material. A magnetic alloy (MA)–loaded cavity was developed for this purpose [96].

Beam loss minimization is ensured in the ring through a very large aperture incorporating phase space painting injection, and a collimation system. Due to the rapidly varying magnetic field, the RCS requires a nonconducting (ceramic) beam pipe (of large aperture) with image-current-carrying conductors for RF shielding. Beam dynamics challenges for high power operation of the J-PARC RCS are discussed in Ref. 97.

At the time of this writing, construction is progressing on an annular coupled structure (ACS) linac which will increase the linac energy to 400 MeV [98].

The J-PARC target is a closed loop liquid mercury system. As in the SNS case, the target lifetime is limited by the cavitation-induced erosion [99].

The performance of the J-PARC linac is described in Refs. 100–102, and that of the J-PARC RCS in Refs. 103–105. Recent general status reports are found in Refs. 106 and 107.

6.6. *China Spallation Neutron Source (CSNS)*

Construction is underway on the China Spallation Neutron Source (CSNS) in Dongguan, China [25]. The design incorporates an H^- ion source, a 3 MeV 324 MHz RFQ and an 80 MeV 324 MHz DTL. Beam is accumulated via multiturn charge exchange injection with horizontal and vertical phase space painting in a 1.6 GeV rapid-cycling synchrotron with a 25 Hz repetition rate. The RF cavity system resonant frequency swings from 1.02 MHz to 2.44 MHz in 20 ms. Accumulated beam is extracted in a single turn. With a 62.5 μA average current, the RCS will deliver 100 kW beam power to a tungsten spallation target.

The facility is designed to incorporate a future upgrade to increase the linac output energy to 250 MeV, which would allow the RCS to deliver 500 kW beam power. The parameters of CSNS are summarized in Table 6.

6.7. *European Spallation Source (ESS)*

The European Spallation Source in Lund, Sweden, is in active design with the goal of beginning operation in 2019 [108]. ESS is envisioned as a long pulse source (~2.9 ms) at a low repetition rate (14 Hz) serving a suite of instruments optimized for long pulses [26]. The ESS parameters are summarized in Table 7.

The linac design uses an ECR proton source, since H^- is not required given the lack of subsequent accumulation in a ring. The linac architecture consists of a four-vane 3 MeV 352 MHz RFQ, a 50 MeV DTL and an SRF double spoke resonator linac with geometric $\beta = 0.57$. The remainder of the linac is based on five-cell elliptical cavities. The first portion, based on $\beta = 0.70$ structures, accelerates the beam to 606 MeV; the second portion, based on $\beta = 0.90$ cavities, accelerates the beam to 2.5 GeV. The beam is transported via an ~100 m transport line to the target, which is a solid rotating tungsten wheel cooled by helium.

Table 6. CSNS parameters.

Beam power (design) (MW)	0.1
Beam energy (GeV)	1.6
RFQ/DTL frequency (MHz)	324
Linac macropulse duty factor (%)	1.0
Linac beam pulse length (max.) (ms)	0.4
Peak H^- linac current (max) (mA)	15
Repetition rate (Hz)	25
Average linac current (mA)	0.0625
Linac energy (MeV)	80
Upgrade linac energy (MeV)	250
Synchrotron circumference (m)	228
Ring bunch intensity	1.6×10^{13}
Harmonic number	2
Pulse length on target (ns)	500

Table 7. ESS parameters.

Beam power (MW)	5.0
Beam energy (GeV)	2.5
RFQ/DTL, spoke frequency (MHz)	352.2
SRF (elliptical) frequency (MHz)	704.4
Linac macropulse duty factor (%)	4.0
Linac beam pulse length (ms)	2.86
Peak linac current (mA)	50
Repetition rate (Hz)	14
Average linac current (mA)	2.0

7. Accelerator-Driven Systems

7.1. *Introduction and motivation*

Accelerator-driven systems have been proposed to address specific missions in advanced nuclear fuel cycles and power generation. ADSs have been proposed for (i) transmutation of specific isotopes present in spent nuclear fuel ("transmutation of waste"); (ii) generation of electricity or process heat; (iii) production of fissile materials for subsequent use in critical or subcritical systems through fertile-to-fissile conversion.

There are two principal advantages that an ADS has relative to critical reactor systems. The first is that a subcritical system offers substantial flexibility with respect to fuel composition. That is, an ADS can burn fuels which would be problematic in critical reactor systems in terms of their neutronic characteristics (i.e. delayed neutron fraction and neutron lifetime). Unlike in a critical system, an ADS does not rely on delayed neutrons for reactor control. The second advantage is potentially enhanced safety of subcritical reactor systems relative to critical systems, since a properly designed subcritical core cannot reach criticality. Therefore, once the external neutron source (i.e. the accelerator) is turned off, the neutron multiplication decays.

ADSs have been proposed for power generation using the Th–^{233}U fuel cycle [109, 110]. An engineering design of such a system has been carried out by Jacobs Engineering [111]. In these energy amplifier systems [43] the spallation source provides a controlled source of fast neutrons which breeds the ^{233}U fissile material, which is consumed for power generation. Such a system is capable of operating on a pure thorium feed stream, and has much-reduced minor actinide production, which simplifies waste handling, whether the fuel cycle is used in critical reactor systems or ADSs. However, radionuclides particular to this fuel cycle have important long term radiotoxicity consequences, namely ^{231}Pa, which has a 32,500-year half-life.

7.2. *Accelerator-driven subcritical reactor fundamentals*

A short review of the fundamentals of accelerator-driven subcritical reactors is provided here. An in-depth overview is given in Ref. 112.

Consider a medium containing fissile nuclei in an infinite and homogeneous arrangement. A neutron created in that medium creates k_∞ neutrons in the second generation, ${k_\infty}^2$ in the third and so on. We assume that $k_\infty < 1$ and find the total number of neutrons in the medium following the creation of N_0 source neutrons:

$$N_t = N_0 + N_0 k_\infty + N_0 k_\infty^2 + \cdots,$$
$$N_t = N_0 \frac{1}{1-k_\infty}. \tag{2}$$

The ratio of total neutrons to source neutrons, the neutron gain, is therefore $1/(1-k_\infty)$, and the medium is neutron-multiplying. The neutron multiplication factor k_∞ is a property of the medium. A real system is finite with a specific geometry, so that some neutrons inevitably escape the multiplying medium. One also expects that the local neutron multiplication is position-dependent (for example, a variation from center to edge of the medium), which makes the situation more complex.

Therefore, to account for losses due to a finite medium we introduce an effective neutron multiplication factor k_{eff} and write the neutron gain as $1/(1-k_{\text{eff}})$. Three regimes are evident. For $k_{\text{eff}} > 1$, the number of neutrons is unbounded, leading to a criticality condition; the system is supercritical. For $k_{\text{eff}} = 1$, the system is critical. For $k_{\text{eff}} < 1$, the number of neutrons resulting from a single primary neutron is finite; the system is subcritical.

For a subcritical system the reaction is not self-sustaining. Nonetheless, large neutron gains can be achieved so that a substantial number of neutrons are produced by the introduction of external neutrons generated from a spallation source. There are two purposes for a subcritical reactor: to generate energy and to generate neutrons.

To calculate the energy gain consider that the number of secondary neutrons produced from N_0 primary neutrons is

$$N_s = N_t - N_0 = N_0 \frac{k_{\text{eff}}}{1-k_{\text{eff}}}. \tag{3}$$

Assuming that each secondary neutron is produced from a fission process in which ν neutrons are generated and a fission energy of ΔE_f is liberated, the thermal energy produced is

$$\Delta E_{\text{th}} = \frac{N_s \Delta E_f}{\nu} = \frac{\Delta E_f N_0 k_{\text{eff}}}{\nu(1-k_{\text{eff}})}. \tag{4}$$

The energy gain G is the ratio of thermal energy produced relative to input beam energy supplied:

$$G = \frac{\Delta E_{\text{th}}}{\Delta E_{\text{beam}}} = \frac{\Delta E_f N_0 k_{\text{eff}}}{N_b E_p \nu(1-k_{\text{eff}})} = G_0 \frac{k_{\text{eff}}}{1-k_{\text{eff}}}, \tag{5}$$

where N_b is the number of beam particles and E_p is their energy. The thermal power is therefore related to the beam power by

$$P_{\text{th}} = G_0 \frac{k_{\text{eff}}}{1-k_{\text{eff}}} P_{\text{beam}}. \tag{6}$$

The potential of large power gain is the basis of the energy amplifier concept [43].

For a 1 GeV proton, assuming that $N_0/N_b \approx 25$ primary neutrons per beam particle (see Fig. 3), $\Delta E_f \approx 0.2$ GeV and $\nu \approx 2.5$, we can estimate that $G_0 \approx 2$. Measured values from the CERN FEAT experiment [113] showed that $G_0 \approx 2.65$ for lead and $G_0 \approx 3.3$ for uranium for 1 GeV protons. Figure 8 displays the proton beam power required to drive a 0.8 GW subcritical core [114].

For use in transmutation, the number of excess neutrons is an important quantity. In order to create N_f fission reactions, where

$$N_f = \frac{N_0 k_{\text{eff}}}{\nu(1-k_{\text{eff}})}, \tag{7}$$

the number of neutrons required is

$$N_n = N_f \frac{\sigma_T}{\sigma_F} = \frac{N_0}{\nu} \frac{\sigma_T}{\sigma_F} \frac{k_{\text{eff}}}{1-k_{\text{eff}}}, \tag{8}$$

Where σ_F is the fission cross-section and σ_T is the total cross-section (the sum of fission and capture cross-sections). The number of neutrons required per fission is written as

$$\frac{\sigma_T}{\sigma_F} = 1 + \alpha. \tag{9}$$

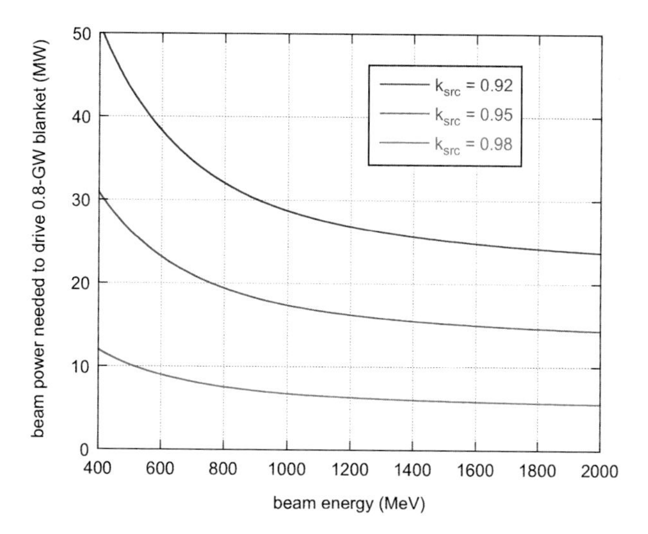

Fig. 8. Accelerator beam power required to drive a subcritical core at fixed thermal power for three different neutron multiplication factors.

Therefore, the number of neutrons available is

$$N_{\text{avail}} = N - N_n = \frac{N_0}{(1 - k_{\text{eff}})}\left(1 - k_{\text{eff}}\frac{1+\alpha}{\nu}\right). \tag{10}$$

To interpret this result, note that the number of neutrons available per fission is

$$\nu - 1 - \alpha. \tag{11}$$

Then Eq. (9) can be written as

$$N_{\text{avail}} = N_0 + N_f(\nu - 1 - \alpha), \tag{12}$$

which shows that the number of available neutrons is the sum of primary neutrons plus the number of neutrons made available after fission, a quantity which is independent of the k_{eff} value. Therefore, one concludes that the number of available neutrons (for transmutation, for example) is independent of the accelerator–reactor coupling. However, the number of neutrons available is greater by N_0 than that available in a critical fission reactor.

The major advantage of the ADS approach is that the margin to criticality becomes an engineering choice, as opposed to a fundamental property of the nuclear fuel which depends on delayed neutron fractions and neutron lifetimes. For example, in the ^{235}U fuel cycle the delayed neutron fraction, which is the margin to prompt criticality, is 0.64%. For a minor actinide burner, the delayed neutron fraction is reduced to ~0.2%, which makes the design and control of such reactors extremely challenging. In contrast, in an ADS design with $k_{\text{eff}} = 0.95$, the margin to criticality is 5%.

The evolution of the neutron multiplication factor is an important aspect of ADS design and must be carefully considered (see e.g. Ref. 115), in order to safely select the design k_{eff}.

8. Survey of Accelerator-Driven System Designs and Technical Challenges

Over the last two decades several ADS design studies have been carried out. Table 8 shows the parameters of several of these studies. The studies listed include the Los Alamos Accelerator Transmutation of Waste (ATW) design, the EUROTRANS demonstration (XT-ADS) and European Facility for Industrial Transmutation (EFIT) designs, the Multipurpose Hybrid Research Reactor for High-end Applications (MYRRHA) design, the Japan Atomic Energy Agency (JAEA) reference design, the Energy Amplifier and the Chinese Academy of Sciences ADS design (C-ADS).

All the studies summarized here are based on continuous wave proton accelerators in the GeV

Table 8. Parameters of several ADS designs.

	ATW	XT-ADS	EFIT	MYRRHA	EA	JAEA	C-ADS
Date	1999	2007	2007	2007	1995	2007	2012
Reference	40	117	117	119	43	120	121
Technology	SC linac	SC linac	SC linac	SC linac	Cyclotron	SC linac	SC linac
Number of cores	4	1	1	1	1	1	1
Beam energy (GeV)	1.0	0.6	0.8	0.6	1.0	1.5	1.5
Beam power (MW)	45	1.5	16	2.4	15	30	15
Beam current (mA)	45	2.5	20	2.5–4.0	15	20	10
Target	LBE	LBE	LBE	LBE	Pb	LBE	LBE
k_{eff} (maximum)	0.97	0.95	0.97	0.95	0.98	0.97	
Thermal power (MW)	840	57	400	85	1500	800	>1000

energy range. The underlying accelerator technology choices include linear accelerators (both normal-conducting and superconducting), cyclotrons, and advanced concepts based on fixed field alternating gradient synchrotrons.

While demonstration facilities require MW class accelerators, industrial scale facilities typically require tens of MW of beam power. These systems utilize a k_{eff} less than about 0.97 to allow for reactivity swing. Rubbia [116] has patented a concept utilizing a k_{eff} close to 1, which reduces the required beam power substantially.

There are many challenges for the accelerator systems used in ADSs. A recent summary and assessment can be found in Ref. 47. These challenges fall into four broad categories, which are discussed below.

8.1. *Acceleratortechnology*

ADSs require proton beam power that is well beyond the present state of the art. This calls for front end beam generation systems that provide continuous wave beams with high average currents. For acceleration in a linac, continuous wave SRF systems and high efficiency RF power sources are required. For a cyclotron-based system, very high power cyclotrons beyond the present state of the art would be required. Cyclotron technology may be capable of reaching beam currents as high at 10 mA [61], whereas very high power industrial scale applications will likely require superconducting linear accelerator technology. Regardless of the approach, the handling of high power beams while respecting the 1 W/m loss criterion is extremely challenging, and entails extremely careful consideration of beam loss mechanisms and collective effects, as well as the development of new beam instrumentation, diagnostics and beam-based control algorithms.

8.2. *Beam dynamics*

For accelerators in the 10 MW range, the fractional loss requirement at full energy is 0.1 parts per million per meter — an extremely demanding value. The generation and preservation of high beam quality is essential for minimizing beam loss. As the understanding of beam loss mechanisms continues to mature, the importance of high fidelity simulations, and comparison of simulations with measured performance, is essential. Due to the high reliability requirements of an ADS, fault-tolerant beam dynamics designs are needed. Finally, control and verification of the beam–target interface is another important issue which relates to the beam dynamics design of the delivery system.

8.3. *Reliability*

The most challenging design criterion is that which relates to the maximum allowable beam trip rates. Beam interruptions lead to rapid cooling of the fuel and structural components in an ADS, such as the beam window, target components and reactor vessel. Due to the potential for excessive stress and fatigue from repeated thermal cycling, beam interruptions must be minimized, and maintained at very challenging levels.

Typical requirements [117] call for keeping beam trips of duration longer than 1 s to less than about 20 per year for a low power demonstration facility, and less than 3 per year for an industrial scale facility. A recent analysis [118] considered the degradation of the structural integrity of the fuel cladding and components in the subcritical core to determine the maximum allowable beam trip rates. It revealed that the beam trip requirements are much less stringent than previously thought. Based on such recent analyses, the assessment in Ref. 47 summarized beam trip requirements for various ADS missions (shown in Table 9). While for industrial scale, power-producing plants the reliability and plant availability must be exceedingly high, systems for transmutation can accommodate more frequent interruptions. It is worth emphasizing that a frequent number of short (less than 10 s) interruptions can be accommodated in most situations, which opens up the possibility of deploying rapid fault recovery schemes to ensure the resumption of beam operation within seconds.

8.4. *Target technology*

Applications of ADSs require high power target technology beyond today's state of the art. Systems capable of handling tens of MW of beam power deposition are required. ADS target systems need to provide high efficiency for neutrons escaping the target, handle very high power deposition in the MW/liter range, operate for many months without replacement, and be capable of being replaced

Table 9. Range of parameters for various ADS missions (from Ref. 47).

	Demonstration	Industrial scale transmutation	Power production w/o energy storage	Power production w/w energy storage
Beam power (MW)	1–2	10–45	10–45	10–45
Beam trips ($t < 1$ s)	N/A	<25,000/year	<25,000/year	<25,000/year
Beam trips ($1 < t < 10$ sec)	<2500/year	<2500/year	<2500/year	<2500/year
Beam trips ($10\,\mathrm{s} < t < 5$ min)	<2500/year	<2500/year	<2500/year	<250/year
Beam trips ($t > 5$ min)	<50/year	<50/year	<50/year	< 3/year
Availability	>50%	>70%	>80%	>85%

quickly (in one week or less to maintain reasonable plant availability).

Most high power concepts call for flowing liquid metal systems which serve as both the spallation target medium and the heat removal media. Liquid Pb–Bi eutectic, with a melting point of 125°C, is the primary liquid metal under study for such applications, due to the favorable neutronic performance, capability of low temperature operation (relative to other liquid metals), low vapor pressure, high boiling point (~1950 K), and good thermal and hydraulic properties.

There are many engineering challenges which must be met in such designs, including chemistry control, development of windowless target concepts, incorporation of remote handling, and handling of the safety aspects associated with managing the radioactive inventory and accommodating off-normal and transient conditions [47].

8.5. *The MYRRHA project*

The MYRRHA Project at SCK-CEN in Mol, Belgium [122], is designing a demonstration accelerator-driven subcritical reactor system to serve multiple scientific missions, including material irradiation, isotope production, materials science, and advanced nuclear system testing. MYRRHA would be the first example of a high power proton accelerator coupled to a subcritical core. The MYRRHA design calls for a 600 MeV continuous wave superconducting linear accelerator with 4 mA maximum average current, with dual linac injector systems in order to maximize accelerator reliability [123]. The beam enters the subcritical reactor vertically from above. The target concept calls for flowing LBE in a windowless arrangement. The target produces $\sim 10^{17}$ neutrons/s.

8.6. *Other considerations*

While in this article we emphasize the accelerator and target technologies required for accelerator-driven subcritical reactors, there are many additional technological considerations which are important.

The coupling of an accelerator-driven spallation source to a reactor adds many unique considerations to the system design. As the neutron multiplication factor evolves during fuel burnup, the accelerator beam power will need to adjust in order to maintain constant system power. Therefore, accelerator beam current requires safety-class control. Likewise, the beam position and size at the target are critical parameters which require careful monitoring and control in order to avoid transient and off-normal conditions. The monitoring of very high power proton beams using nonintercepting methods represents a beam instrumentation challenge which must be addressed. Additional safety aspects require consideration, such as the change in k_{eff} due to a leak of Pb–Bi coolant into the accelerator beam pipe.

The development of transmutation systems entails advances in separations, fuel and waste disposal technologies [42]. Plant-scale separations technologies are required for the processing of spent fuel in order to separate the transuranics and long-lived fission products. Fuel forms with high minor actinide content suitable for use in an ADS must be developed. Finally, waste forms suitable for the ultimate repository strategy require development.

9. Overview of Key Technologies and R&D Challenges for Spallation Neutron Sources and Accelerator-Driven Systems

9.1. *Injector systems*

High average power injectors are required for any future high intensity applications. For proton/

H^- beam applications, state-of-the-art performance is obtained with ECR (proton) [124–127] and multi-cusp RF-driven technology (negative hydrogen ions) [128]. RFQ technology is utilized for proton/ion injector systems to deliver high quality beams at 1–6 MeV output energy. An overview of RFQs is presented in Ref. 129, and a description of their development in Ref. 130.

The state of the art in RFQ performance was demonstrated at the Low-Energy Demonstration Accelerator (LEDA) Project at Los Alamos National Laboratory [127, 131, 132]. The LEDA RFQ is a 350 MHz, 6.7 MeV output energy, 8-m-long continuous wave device. In demonstration tests, LEDA delivered 100 mA continuous wave current (670 kW beam power) for long periods of time. The LEDA RFQ meets the requirements of beam performance for ADS applications, although further testing is needed to demonstrate the high reliability required. Extensive investigations of beam halo generation in the LEDA system were performed [133].

An ECR proton source (SILHI) developed at CEA/CNRS–Saclay in the early 2000s for ADS applications demonstrated reliable performance at ∼100 mA operation [124, 125].

The performance of some recent high duty factor RFQs is presented in Refs. 134–136. Some performance limits associated with continuous wave and high duty factor operation have been observed.

9.2. *Linear accelerators*

Two examples of modern architectures for high intensity, high power linear accelerators are shown in Fig. 9. After the front end injector system, the beam is transported through a medium energy beam transport (MEBT) which incorporates beam chopping systems and provides transverse and longitudinal matching to the linac, as well as beam instrumentation for characterizing the injector. There are several possibilities for linac architecture, two of which are shown in the figure. In the first, a (pulsed) normal-conducting linac accelerates the beam to medium energies (∼200 MeV) and then transitions to an SRF linac. The frequency typically jumps at this point to capitalize on the reduced longitudinal emittance after acceleration. Multiple SRF structure geometries are required in order to achieve reasonable effective shunt impedances as the velocity increases from

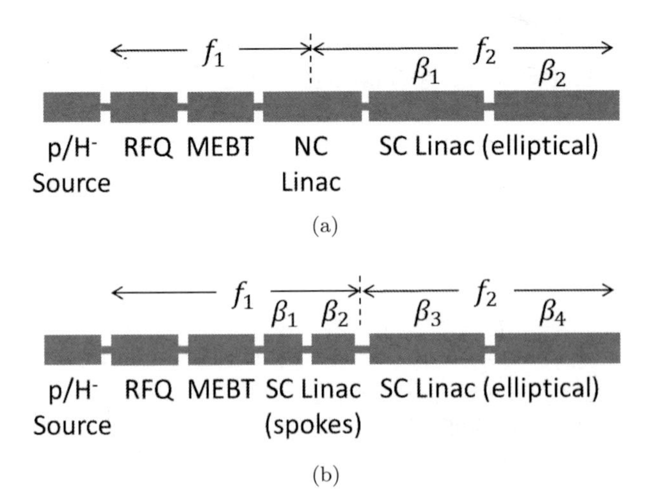

Fig. 9. Examples of high intensity, high power linear accelerator architectures: (a) utilizing a normal-conducting linac to medium energies; (b) an all-superconducting approach.

$\beta \sim 0.5$ to $\beta \sim 0.9$. This first example corresponds to the SNS linac architecture, where the linac incorporates a drift tube linac and a coupled cavity linac at twice the resonant frequency. Two SC cavity types with $\beta_g = 0.61$ and $\beta_g = 0.81$ provide large velocity acceptance.

In the second architecture shown, superconducting RF structures are utilized immediately following the RFQ. This requires the use of multiple low velocity structures, such as spoke resonators, to achieve high effective shunt impedance for low velocity particles. Eventually the linac transitions to elliptical structures for continued acceleration to high energy. This is the architecture chosen for the Project X linac [36].

SRF accelerators have become the technology of choice for high energy, high power proton/hadron applications. The advantages of SRF technology for spallation source applications include: (i) high accelerating gradients (15–20 MV/m), which allow for lower capital and operating costs; (ii) an efficient RF source to beam power transfer due to low RF structure power dissipation; (iii) a large aperture, which minimizes beam loss; (iv) a low vacuum, which minimizes beam gas scattering; (v) flexibility in adjusting the linac lattice (when using a linac architecture in which each SC cavity has independent RF phase and amplitude control); (vi) the potential for very high reliability by incorporating online spare cavities associated with rapid fault recovery schemes (for ADS applications).

The key areas of R&D include the development of high SRF quality factor processes to reduce capital cryogenic costs for continuous wave accelerators. Further development of RF structures throughout the complete velocity range from ∼0.05 to 1.0 is required. A number of RF components are challenged at high average power, including RF power couplers, windows and RF power sources.

The beam dynamics design of a high intensity proton or hadron linac is optimized to minimize emittance and halo growth, with attention paid to equipartitioning, avoiding resonances driven by space charge, and avoiding abrupt transitions in longitudinal and transverse phase advances (see e.g. Ref. 79 and references therein).

9.3. *High intensity synchrotrons and accumulator rings*

High intensity accumulator rings and synchrotrons must confront several challenges, as outlined below.

9.3.1. *Injection*

Multiturn charge exchange injection allows the accumulation of intense beam pulses in circular accelerators which are fed by relatively low current linacs. There are many aspects of the injection process that must be carefully considered (see e.g. Ref. 137), including: (i) stripping foil technology, which is capable of handling very high radiation damage and beam power deposition [138, 139]; (ii) handling of "waste beams" — the unstripped and partially stripped beam which is not accepted in the circular accelerator; (iii) careful tailoring of magnetic fields in the injection region to reduce the beam loss associated with H^0 particles which leave the stripper foil in excited (and therefore readily Lorentz-stripped) states [140]; (iv) careful handling of the stripped electrons, which are themselves a high power beam [141, 142]; and (v) engineering aspects of stripper foil handling and replacement [143–145].

9.3.2. *Collective effects*

The intensity of circular accelerators is ultimately limited by beam loss associated with collective effects. The primary collective mechanisms are: (i) the space charge force, (ii) instabilities driven by the ring impedance, and (iii) the electron cloud effect. It should be pointed out that for rapid-cycling synchrotrons and accumulator rings used for neutron spallation, the bunch lengths are quite long — tens or even hundreds of meters — due to the low harmonic number RF systems employed. This leads to some subtle differences in the dynamics relative to those of, for example, high energy synchrotrons.

The space charge effect limits the beam intensity which can be accumulated. To ameliorate some of the effects, phase space painting is employed to fill a large acceptance in a controlled manner in order to reduce emittance and halo growth from space charge forces. Several studies of space charge observations and limitations in spallation source rings have been performed [58, 146–149].

In spallation source circular accelerators, one of the dominant impedances arises from the single turn extraction kicker. An example of an instability driven by a long pulse beam is reported in Ref. 150.

The electron cloud effect (ECE) has been studied in considerable detail. Of particular importance for spallation sources are the studies performed on the ECE in long bunch proton beams at the PSR [151–155].

Long proton bunches are subject to the trailing edge multipacting mechanism, which is the primary limitation due to the ECE in accumulators. In this mechanism, seed electrons are generated through interactions of the beam with residual gas or through the interaction of lost beam particles with the vacuum chamber. These seed electrons are trapped in the electromagnetic field of the long proton bunch. Electrons gain energy during the passage of the head of the bunch as the proton beam's longitudinal density is increasing and become trapped in the beam's field. As the bunch density decreases through the passage of the beam tail, the electron oscillation amplitude increases and formerly trapped electrons strike the vacuum chamber walls, liberating secondary electrons. Some of the secondary electrons survive the beam gap and are again trapped in the proton bunch's field in the next passage. This process continues, multiplying the electron density with each successive turn. Eventually the electron density becomes high enough that it influences the proton beam motion, and the two beams undergo coupled oscillations. This mechanism leads to very rapid instability growth rates, on the order of tens of beam turns.

A number of mitigation techniques have been developed and studied [156], including specialized vacuum chamber coatings to reduce the secondary electron emission yield, clearing electrodes to attract secondary electrons, grooved vacuum chamber surfaces to reduce the effective emission yields, and active beam feedback systems [157].

9.4. *Target systems*

High power spallation target systems require challenging technologies. For MW class short pulse spallation sources liquid metal target systems are needed. An important limitation on these systems is cavitation-induced damage to the target container material, caused by the intense pressure waves generated in the liquid. Active research programs on pressure wave damage and mitigation have been carried out for the mercury target systems at both the SNS at Oak Ridge and the JSNS at J-PARC [88, 158]. A recent summary of observed pressure wave damage and mitigation approaches, including gas bubble injection and protective gas walls, is reported in Ref. 159.

The ESS long pulse source plans to use a helium cooled rotating solid tungsten wheel 2.5 m in diameter and 10 cm in height, with 0.5 Hz rotation frequency. The tungsten spallation target material is arranged in 33 arcs, each with 12 slabs, arranged in such a way as to provide helium-cooling channels between each slab. The peak tungsten temperature remains below 500°C [26].

The development of Pb–Bi eutectic spallation target systems is important for the advancement of ADSs as well as higher power spallation sources of the future. The MEGAPIE experiment at PSI [66] demonstrated for the first time the use of a Pb–Bi eutectic target loop in a spallation source. The objective of the experimental program was to demonstrate the safe operation of a complete Pb–Bi system and to assess neutronic performance. The target operated from August to December 2006 at beam powers of up to 0.8 MW. In terms of neutronic performance, the measured thermal neutron flux was a factor of ~1.8 higher than that of the SINQ solid target, in agreement with expectations [160, 161].

10. Summary

Spallation neutron sources form the basis of large scale accelerator-based neutron scattering facilities, several of which are in operation or under construction throughout the world. These facilities provide powerful capabilities for studying the structure and dynamics of materials important for energy, materials engineering, technology and health. Their potential for advancing nuclear energy has attracted worldwide attention, in terms of the potential for serving as accelerator-driven material irradiation facilities, and also as accelerator-driven subcritical reactors. Spallation sources at the state of the art rely on very challenging accelerator technologies, including high intensity linear and circular accelerators with well-controlled beam emittance and halo growth to minimize uncontrolled beam loss, and very high power liquid metal target systems for efficient neutron production and the capability of handling high pulsed and average power deposition. Spallation sources for future application in ADSs will require more than an order of magnitude greater capability relative to today's MW class sources in terms of beam power, target power deposition and accelerator reliability. This demanding technology is the subject of active R&D programs throughout the world.

Acknowledgment

The author would like to acknowledge the helpful discussions with Mike Seidel (PSI) and John Galambos (SNS).

References

[1] Glenn Seaborg, *Interaction of Fast Neutrons with Lead*, Ph.D. thesis (University of California at Berkeley, 1937).
[2] T. E. Mason, Pulsed neutron scattering for the 21st century, *Phys. Today* **59**(5), 44 (2006).
[3] J. M. Carpenter and W. B. Yelon, Neutron sources, in *Neutron Scattering* eds. K. Skold and D. L. Price (Academic, New York, 1986).
[4] G. S. Bauer, Physics and technology of spallation neutron sources, *Nucl. Instrum. Methods A* **463**, 505 (2001).
[5] N. Watanabe, Neutronics of pulsed spallation neutron sources, *Rep. Prog. Phys.* **66**, 339 (2003).
[6] M. Arai and K. Crawford, in *Neutron Imaging and Applications: A Reference for the Imaging Community*, eds. I. S. Anderson, R. L. McGreevy and H. Z. Bilheux (Springer, New York, 2009), pp. 13–30.

[7] A. Kowalczyk, Proton-induced spallation reactions in the energy range 0.1–10 GeV, Ph.D. thesis (Jagiellonian University Institute of Physics, 2008), arXiv:0801.0700v1[nucl-th].
[8] D. B. Pelowitz *et al.*, MCNPX 2.7.0 extensions, LA-UR-11-02295 (2011); http://mcnpx.lanl.gov
[9] M. Teshigawara *et al.*, JAERI-Res. 99-0101 (1999).
[10] M. Arai *et al.*, *J. Neutron Res.* **8**, 71 (1999).
[11] H. Lengeler and J. Wei, in *Handbook of Accelerator Physics and Engineering*, eds. A. W. Chao and M. Tigner (World Scientific, 2006), p. 55.
[12] P. V. Livdahl, in *Proc. 1981 Linear Accelerator Conference* (LA-9234-C), p. 5.
[13] C. M. Van Atta, A brief history of the MTA Project, UCRL-79151 (1977).
[14] G. A. Bartholomew and P. R. Tunnicliffe (eds.), *The AECL Study for an Intense Neutron Generator* (AECL-2600, 1966).
[15] C. Westfall, How Argonne's intense pulsed neutron source came to life and gained its niche: The view from an ecosystem perspective. ANL/HIST-5 (2007).
[16] C. Potts *et al.*, Performance of the intense pulsed neutron source accelerator system, *IEEE Trans. Nucl. Sci.* **NS-30**(4), 2131 (1983).
[17] C. Potts *et al.*, Recent performance of the intense pulsed neutron source accelerator system, in *Proc. 1987 Part. Accel. Conf.* (1987), p. 1651.
[18] Y. Ishikawa, Studies of condensed matter with a pulsed neutron source (KENS), *Physica B+C* **120**, 3 (1983).
[19] R. Kustom, Intense pulsed neutron sources, *IEEE Trans. Nucl. Sci.* **NS-28**(3), 3115 (1981).
[20] G. H. Rees, Status report on ISIS, in *Proc. 1987 Part. Accel. Conf.* (1987) p. 830.
[21] G. Rees, in *Handbook of Accelerator Physics and Engineering*, eds. A. W. Chao and M. Tigner (World Scientific, 2006), p. 542.
[22] W. E. Fischer, Application of spallation neutron sources, in *Proc. 2002 Eur. Part. Accel. Conf.* (2002), p. 114.
[23] S. Henderson, Status of the Spallation Neutron Source: Machine and science, in *Proc. 2007 Part. Accel. Conf.* (Albuquerque, NM, USA, 2007), p. 7.
[24] Y. Yamazaki *et al.*, Technical design report of J-PARC, KEK Report 2002–13; JAERI-Tech 2003–44.
[25] S. Fu *et al.*, Status and challenges of the China spallation neutron source, in *Proc. 2011 Int. Part. Accel. Conf.* (San Sebastian, Spain, 2011), p. 889.
[26] S. Peggs (ed.), European Spallation Source Technical Design Report, ESS Doc ESS-274-v15.
[27] I. S. Anderson, A. J. Hurd and R. McGreevy (eds.), *Neutron Scattering Applications and Techniques* (Springer).
[28] J. Fitter, T. Gutberlet and J. Katsaras (eds.), *Neutron Scattering in Biology: Techniques and Applications* (Springer, 2006).
[29] E. Pitcher, Irradiation environment of the materials test station, in *Proc. 20th International Meeting of the Collaboration on Advanced Neutron Sources (ICANS-XX)* (Argentina, 2012); LA-UR-12-22390.
[30] A. Mosnier *et al.*, The accelerator prototype of the IFMIF/EVEDA project, in *Proc. 2010 Int. Part. Accel. Conf.* (Kyoto, Japan, 2010), p. 588.
[31] N. M. Ghoniem and H. Gurol, Analytical approach to void growth in metals under intense radiation pulsing, *Radiat. Eff.* **55**, 209 (1981).
[32] P. Vladimirov and A. Moslang, Comparison of materials irradiation conditions for fusion, spallation, stripping and fission neutron sources, *J. Nucl. Mater.* 329–333, 233 (2004).
[33] B. H. Sencer *et al.*, Microstructural evolution of both as-irradiated and subsequently deformed microstructures of 316L stainless steel irradiated at 30–160°C at LANSCE, *J. Nucl. Mater.* **345**, 136 (2005).
[34] http://www.phy.ornl.gov/nedm/index.html
[35] S. K. Lamoreaux and R. Golub, Experimental searches for the neutron electric dipole moment, *J. Phys. G* **36**, 104002 (2009).
[36] S. Henderson, S. Holmes, A. Kronfeld and R. Tschirhart (eds.), *Project X Accelerator Reference Design, Physics Opportunities, Broader Impacts*, FERMILAB-TM-2557.
[37] H. Mark, Ultrahigh-current ion accelerators, *Nucl. Instrum. Methods* **28**, 131 (1964).
[38] G. J. Van Tuyle *et al.*, The Phoenix concept, BNL-52279 (1991).
[39] C. D. Bowman *et al.*, Nuclear energy generation and waste transmutation using an accelerator-driven intense thermal neutron source, *Nucl. Instrum. Methods A* **230**, 336 (1992).
[40] G. Lawrence *et al.*, A roadmap for developing ATW technology: Accelerator technology, Los Alamos report LA-UR-99-3225 (1999).
[41] T. Takizuka, T. Nishida, M. Mizumoto and H. Yoshida, JAERI R&D on accelerator-based transmutation under OMEGA program, *AIP Conf. Proc.* **346**, 64 (1995).
[42] A roadmap for developing accelerator transmutation of waste (ATW) technology: A report to Congress, DOE/RW-0519 (1999).
[43] C. Rubbia *et al.*, Conceptual design of a fast neutron operated high power energy amplifier, CERN/AT/95-44 (ET) (1995).
[44] *Nuclear Wastes: Technologies for Separations and Transmutation, National Research Council* (National Academies Press, 1996).
[45] *A European Roadmap for Developing Accelerator Driven Systems (ADS) for Nuclear Waste Incineration* (2001).
[46] *Accelerator Driven Systems (ADS) and Fast Reactors in Advanced Nuclear Fuel Cycles* (Nuclear Energy Agency, OECD, 2002).

[47] H. A. Abderrahim *et al.*, Accelerator and target technology for accelerator-driven transmutation and energy production, http://science.energy.gov/~/media/hep/pdf/files/pdfs/ADS_White_Paper_final.pdf

[48] D. J. S.Findlay, High power operational experience at ISIS, in *Proc. 2010 ICFA High-Brightness Hadron Beams Workshop (HB2010)* (Morschach, Switzerland, 2010), p. 381.

[49] D. J. Adams *et al.*, Beam loss control in the ISIS accelerator facility, in *Proc. 2012 ICFA High-Brightness Hadron Beams Workshop (HB2012)* (Beijing, China, 2012), p. 560.

[50] C. R. Prior, Studies of dual harmonic acceleration in ISIS, A-11, ICANS-X11, Report 94-025.

[51] E. Knapp, Status report on LAMPF, in *Proc. 1971 Part. Accel. Conf.* (1971) p. 508.

[52] E. Knapp *et al.*, Standing wave high energy accelerator structures, *Rev. Sci. Instrum.* **38**(7), 979 (1968).

[53] G. H. Lawrence, Performance of the Los Alamos proton storage ring, in *Proc. 1987 Part. Accel. Conf.* (1987), p. 825.

[54] D. Fitzgerald *et al.*, Commissioning of the Los Alamos PSR injection upgrade, in *Proc. 1999 Part. Accel. Conf.* (1999), p. 518.

[55] T. Wangler and P. Lisowski, The LANSCE national user facility, *Los Alamos Sci.* **28**, 138 (2003).

[56] R. J. Macek *et al.*, Electron–proton two-stream instability at the PSR, in *Proc. 2001 Part. Accel. Conf.* (Chicago, IL, USA), p. 688.

[57] R. J. Macek *et al.*, Recent observations, experiments and simulations of electron cloud buildup in drift spaces and quadrupole magnets at the Los Alamos PSR, in *Proc. 2009 Part. Accel. Conf.* (Vancouver, Canada, 2009), p. 4722.

[58] S. Cousineau *et al.*, Resonant beam behavior studies in the proton storage ring, *Phys. Rev. ST Accel. Beams* **6**, 074202 (2003).

[59] H. A. Willax, Status report on SIN, in *Proc. 7th Int. Conf. on Cyclotrons and Their Applications* (Zürich, 1975), p. 33.

[60] M. Seidel *et al.*, Towards the 2 MW cyclotron and latest developments at PSI, in *Proc. 2010 Cyclotron Conference* (Lanzhou, China, 2010), p. 275.

[61] M. Seidel, Cyclotrons for high-intensity beams. arXiv:1302.1001v1.

[62] S. Adam, Space charge effects in cyclotrons, in *Proc. 14th Int. Conf. on Cyclotrons and Their Applications* (Cape Town, South Africa, 1995).

[63] W. Wagner *et al.*, PSI experience with high-power target design and operation, in *Technology and Components of Accelerator-Driven Systems: Workshop Proceedings* (Karlsruhe, Germany, 2010) (OECD, 2011).

[64] W. Joho, High intensity problems in cyclotrons, in *Proc. 9th Int. Conf. on Cyclotrons and Their Applications* (Caen, France, 1981), p. 337.

[65] M. Seidel, private communication.

[66] C. Latge *et al.*, MEGAPIE spallation target: Irradiation of the first prototypical spallation target for future ADS, in *Technology and Components of Accelerator-Driven Systems: Workshop Proceedings* (Karlsruhe, Germany, 2010) (OECD, 2011).

[67] S. Henderson, Recent commissioning results from the spallation neutron source, in *Proc. 2006 ICFA High-Brightness, High-Intensity Hadron Beams Workshop (HB2006)* (Tsukuba, Japan, 2006), p. 6.

[68] A. Aleksandrov, Commissioning of the spallation neutron source front-end systems, in *Proc. Part. Accel. Conf.* (2003), p. 65.

[69] A. Aleksandrov *et al.*, Performance of the SNS front-end and warm linac, in *Proc. 2008 Eur. Part. Accel. Conf.* (Genoa, Axaly, 2008), p. 3530.

[70] D. Jeon, in *Proc. 2003 Part. Accel. Conf.*, p. 107.

[71] S. H. Kim, SNS superconducting linac power ramp-up status and plans, in *Proc. 2009 Part. Accel. Conf.* (Vancouver, Canada, 2009), p. 1457.

[72] S. H. Kim, SNS superconducting linac operational experience and upgrade path, in *Proc. 2008 Linear Accelerator Conference* (Vancouver, Canada, 2008), p. 11.

[73] J. Wei, Spallation neutron source ring: Status, challenges, issues, perspectives, in *Proc. 2003 Part. Accel. Conf.* (2003), p. 571.

[74] R. F. Welton *et al.*, in *Proc. 2005 Part. Accel. Conf.*, p. 472; M. Stockli, in *Proc. 2006 Linear Accelerator Conference*, p. 213.

[75] M. P. Stockli *et al.*, Highly persistent SNS H^- source fueling 1-MW beams with 7–9 kC service cycles, in *Proc. 2011 Part. Accel. Conf.* (New York, 2011), p. 1993.

[76] I. E. Campisi, in *Proc. 2005 Part. Accel. Conf.*, p. 34.

[77] N. Catalan-Lasheras *et al.*, Accelerator physics model of expected beam loss along the SNS accelerator facility during normal operation, in *Proc. 2002 Eur. Accel. Conf.* (Paris, France, 2002), p. 1013.

[78] J. Stovall *et al.*, Expected beam performance of the SNS linac, in *Proc. 2001 Part. Accel. Conf.* (Chicago, IL, USA, 2001), p. 446.

[79] S. Henderson and A. Aleksandrov, Beam dynamics in proton linacs, in *Handbook of Accelerator Physics and Engineering*, eds. A. W. Chao and M. Tigner (World Scientific, 2006), p. 565.

[80] J. Galambos *et al.*, Increased understanding of beam losses from the SNS linac proton experiment, in *Proc. 2012 Linear Accelerator Conference* (Tel-Aviv, Israel, 2012), p. 115.

[81] A. Shishlo *et al.*, H^- and proton beam loss comparison at SNS superconducting linac, in *Proc. 2012 Int. Part. Accel. Conf.* (New Orleans, USA, 2012), p. 1074.

[82] A. Shishlo *et al.*, First observation of intrabeam stripping of negative hydrogen in a superconducting linear accelerator, *Phys. Rev. Lett.* **108**, 114801 (2012).
[83] J. Wei *et al.*, Low-loss design for the high-intensity accumulator ring of the spallation neutron source, *Phys. Rev. ST Accel. Beams* **3**, 080101 (2000).
[84] M. Plum, SNS ring operational experience and ramp-up status, in *Proc. 2009 Part. Accel. Conf.* (Vancouver, Canada, 2009), p. 1952.
[85] M. Plum, Challenges facing high power proton accelerators, in *Proc. 2013 Int. Part. Accel. Conf.* (2013), p. 1.
[86] T. Gabriel *et al.*, in *Proc. 2001 Part. Accel. Conf.* p. 737.
[87] T. Gabriel *et al.*, Overview of the spallation neutron source with emphasis on target systems, *J. Nucl. Mater.* **318**, 1 (2003).
[88] J. R. Haines *et al.*, Summary of cavitation erosion investigations for the SNS mercury target, *J. Nucl. Mater.* **343**, 58 (2005).
[89] J. Galambos, Progress with MW-class operation of the spallation neutron source, in *Proc. 2009 Part. Accel. Conf.* (Vancouver, Canada, 2009), p. 1818.
[90] J. Galambos, Spallation neutron source operational experience at 1 MW, in *Proc. 2010 ICFA High-Brightness, High-Intensity Hadron Beams Workshop (HB2010)* (Morschach, Switzerland, 2010), p. 377.
[91] J. Galambos, SNS operational experience: Expectations and realities, in *Proc. 2010 ICFA High-Brightness, High-Intensity Hadron Beams Workshop (HB2010)* (Morschach, Switzerland, 2010), p. 11.
[92] S. Henderson, Spallation neutron source operation at 1 MW and beyond, in *Proc. 2010 Linear Accelerator Conference* (Tsukuba, Japan, 2010), p. 11.
[93] Y. Ikeda, *Nucl. Instrum. Methods A* **600**, 1 (2009).
[94] A. Ueno *et al.*, Beam test of a front-end system for the JAERI-KEK joint project, in *Proc. 2002 Linear Accelerator Conference* (Gyeongju, S. Korea, 2002), p. 356.
[95] H. Tanaka *et al.*, Measured RF properties of the DTL for the J-PARC, in *Proc. 2004 Linear Accelerator Conference* (Lubeck, Germany, 2004), p. 809.
[96] C. Ohmori *et al.*, High field-gradient cavities loaded with magnetic alloys for synchrotrons, in *Proc. 1999 Part. Accel. Conf.* (New York, USA, 1999), p. 413.
[97] H. Hotchi *et al.*, 1-MW beam operation scenario in the J-PARC RCS, in *Proc. 2012 ICFA High-Brightness, High-Intensity Hadron Beams Workshop (HB2012)* (Beijing, China, 2012), p. 68.
[98] H. Ao *et al.*, Status of mass production of the ACS cavity for the J-PARC linac energy upgrade, in *Proc. 2010 Int. Part. Accel. Conf.* (Kyoto, Japan, 2010), p. 618.
[99] M. Futakawa, *et al.*, *J. Nucl. Mater.* **343**, 70 (2005).
[100] K. Hasegawa *et al.*, Operating experience of the J-PARC linac, in *Proc. 2008 Linear Accelerator Conference* (Victoria, Canada, 2008), p. 55.
[101] K. Hasegawa *et al.*, Status of the J-PARC linac, in *Proc. 2010 Linear Accelerator Conference* (Tsukuba, Japan, 2012), p. 449.
[102] K. Hasegawa, Recovery of the J-PARC linac from the earthquake, in *Proc. 2012 Linear Accelerator Conference* (Tel-Aviv, Israel, 2012), p. 1069.
[103] M. Yoshii *et al.*, Acceleration of high intensity proton beams in the J-PARC synchrotrons, in *Proc. 2011 Int. Part. Accel. Conf.* (2011), p. 2502.
[104] M. Kinsho, Status of the J-PARC 3 GeV RCS, in *Proc. 2009 Part. Accel. Conf.* (Vancouver, Canada, 2009), p. 1436.
[105] M. Kinsho, Status of the J-PARC 3 GeV RCS, in *Proc. 2012 Int. Part. Accel. Conf.* (New Orleans, USA, 2012), p. 3927.
[106] Y. Yamazaki, J-PARC status, in *Proc. 2009 Part. Accel. Conf.* (Vancouver, Canada, 2009), p. 18.
[107] K. Hasegawa, Status of J-PARC accelerators, in *Proc. 2013 Int. Part. Accel. Conf.* (Shanghai, China), p. 3830.
[108] S. Peggs, The European Spallation Source, in *Proc. 2011 Int. Part. Accel. Conf.* (San Sebastian, Spain, 2011), p. 3789.
[109] R. Hargraves and R. Moir, Liquid fluoride thorium reactors, *Am. Sci.* **98**(4) 304 (July–August, 2010).
[110] S. Peggs *et al.*, Thorium energy futures, in *Proc. 2012 Int. Part. Accel. Conf.* (New Orleans, USA, 2012), p. 29.
[111] V. Ashley and R. Ashworth, The technically viable ADTR^{TM} power station, *Nucl. Future* **7**, 3 (2011).
[112] H. Nifenecker *et al.*, Basics of accelerator-driven subcritical reactors, *Nucl. Instrum. Methods A* **463**, 428 (2001).
[113] S. Andriamonge *et al.*, *Phys. Lett. B* **348**, 697 (1995).
[114] M. Todosow, private communication.
[115] K. Tsujimoto *et al.*, Neutronics design for lead–bismuth-cooled accelerator-driven system for transmutation of minor actinide, *J. Nucl. Sci. Technol.* **41**(1), 21 (2004).
[116] C. Rubbia, Accelerator driven nuclear system with control of effective neutron multiplication coefficient. PCT patent, PCT/EP2010/054132 (2010).
[117] A. C. Mueller, Prospects for transmutation of nuclear waste and associated proton-accelerator technology, *Eur. Phys. J. Spec. Top.* **176**, 179 (2009).
[118] H. Takei *et al.*, Estimation of acceptable beam-trip frequencies of accelerators for accelerator-driven

systems and comparison with existing performance data, *J. Nucl. Sci. Technol.* **49**(4), 384 (2012).
[119] D. Vandeplassche and L. M. Romao, Accelerator-driven systems, in *Proc. 2012 Int. Part. Accel. Conf.* (New Orleans, USA, 2012), p. 6.
[120] K. Tsujimoto *et al.*, Research and development program on accelerator-driven subcritical system in JAEA, *J. Nucl. Sci. Technol.* **44**, 483 (2007).
[121] Y. Chi, Progress of injector-1 and main linac of Chinese ADS proton accelerator, in *Proc. 2013 Int. Part. Accel. Conf.* (Shanghai, China, 2013), p. 3854.
[122] H. A. Abderrahim *et al.*, MYRRHA — a multipurpose fast spectrum research reactor, *Energ. Convers. Manag.* **63**, 4 (2012).
[123] D. Vandeplassche *et al.*, The MYRRHA linear accelerator, in *Proc. 2011 Int. Part. Accel. Conf.* (San Sebastian, Spain, 2011), p. 2718.
[124] R. Gobin *et al.*, High intensity ECR ion source developments at CEA/Saclay, *Rev. Sci. Instrum.* **73**(2), 922 (2002).
[125] R. Gobin *et al.*, Status of the light ion source developments at CEA/Saclay, *Rev. Sci. Instrum.* **75**(5), 1414 (2004).
[126] J. D. Sherman *et al.*, Status report on a dc 130-mA, 75-keV proton injector, *Rev. Sci. Instrum.* **69**, 1003 (1998).
[127] J. D. Schneider and R. L. Sheffield, LEDA: A high-power test bed of innovation and opportunity, in *Proc. XX Int. Linac Conf.* (Monterey, CA, USA, 2000), p. 578.
[128] M. Stockli *et al.*, Ramping up the spallation neutron source beam power with the H^- source using 0 mg Cs/day, *Rev. Sci. Instrum.* **81**, 02A729 (2010).
[129] R. H. Stokes and T. P. Wangler, Radio-frequency quadrupoles and their applications, *Annu. Rev. Nucl. Part. Sci.* **38**, 97 (1988).
[130] L. Young, 25 years of technical advances in RFQ accelerators, in *Proc. 2003 Part. Accel. Conf.*, p. 60.
[131] J. D. Schneider, Operation of the low-energy demonstration accelerator: The proton injector for APT, in *Proc. 1999 Part. Accel. Conf.*, p. 503.
[132] L. M. Young, Operations of the LEDA resonantly-coupled RFQ, in *Proc. 2001 Part. Accel. Conf.* (Chicago, IL, USA, 2001), p. 309.
[133] J. Qiang *et al.*, *Phys. Rev. ST Accel. Beams* **5**, 124201 (2002).
[134] K. Hasegawa *et al.*, Status of the J-PARC RFQ, in *Proc. 2010 Int. Part. Accel. Conf.* (Kyoto, Japan, 2010), p. 621.
[135] J. Rodnizki and Z. Horvits, RF and heat flow simulations of the SARAF RFQ 1.5 MeV/nucleon proton/deuteron accelerator, in *Proc. 2010 Linear Accelerator Conference* (Tsukuba, Japan, 2010), p. 506.
[136] S.-H. Kim *et al.*, Stabilized operation of the spallation neutron source radio frequency quadrupole, *Phys. Rev. ST Accel. Beams* **13**, 070101 (2010).
[137] J. Wei *et al.*, Injection choice for spallation neutron source ring, in *Proc. 2001 Part. Accel. Conf.* (Chicago, IL, USA, 2001) p. 2562.
[138] R. W. Shaw *et al.*, Spallation neutron source (SNS) diamond stripper foil development, in *Proc. 2007 Part. Accel. Conf.* (Albuquerque, NM, USA, 2007), p. 620.
[139] A. Takagi *et al.*, Comparative study on lifetime of stripper foil using 650 keV H^- ion beam, in *Proc. 2007 Part. Accel. Conf.* (Albuquerque, NM, USA, 2007), p. 245.
[140] M. S. Gulley *et al.*, Measurement of H^-, H^0 and H^+ yields produced by foil stripping of 800 MeV H^- ions, *Phys. Rev. A* **53**, 3201 (1996).
[141] L. Wang *et al.*, Stripped electron collection at the spallation neutron source, *Phys. Rev. ST Accel. Beams* **8**, 094201 (2005).
[142] S. Cousineau *et al.*, Dynamics of uncaught foil-stripped electrons in the Oak Ridge Spallation Neutron Source accumulator ring, *Phys. Rev. ST Accel. Beams* **14**, 064001 (2001).
[143] M. Plum *et al.*, SNS injection foil experience, in *Proc. 2010 ICFA High-Intensity, High-Brightness Hadron Beams Workshop* (Morschach, Switzerland, 2010), p. 334.
[144] M. Plum *et al.*, SNS stripper foil failure modes and their cures, in *Proc. 2010 Int. Part. Accel. Conf.* (Kyoto, Japan, 2010), p. 3969.
[145] M. Plum *et al.*, Stripper foil failure modes and cures at the Oak Ridge Spallation Neutron Source, *Phys. Rev. ST Accel. Beams* **14**, 030102 (2011).
[146] S. Cousineau *et al.*, Space charge induced resonance excitation in high intensity rings, *Phys. Rev. ST Accel. Beams* **6**, 034205 (2003).
[147] J. D. Galambos *et al.*, Comparison of simulated and observed beam profile broadening in the Proton Storage Ring and the role of space charge, *Phys. Rev. ST Accel. Beams* **3**, 034201 (2000).
[148] J. A. Holmes *et al.*, Space charge dynamics in high intensity rings, *Phys. Rev. ST Accel. Beams* **2**, 114202 (1999).
[149] H. Hotchi *et al.*, Beam loss reduction by injection painting in the 3 GeV rapid cycling synchrotron of the Japan Proton Accelerator Research Complex, *Phys. Rev. ST Accel. Beams* **15**, 040402 (2012).
[150] J. A. Holmes *et al.*, Comparison between measurements, simulations, and theoretical predictions of the extraction kicker transverse dipole instability in the spallation neutron source, *Phys. Rev. ST Accel. Beams* **14**, 074401 (2011).
[151] M. Blaskiewicz *et al.*, Electron cloud instabilities in the Proton Storage Ring and Spallation Neutron Source, *Phys. Rev. ST Accel. Beams* **6**, 014203 (2003).

[152] T.-S. Wang *et al.*, Centroid theory of transverse electron–proton two-stream instability in a long proton bunch, *Phys. Rev. ST Accel. Beams* **6**, 014204 (2003).
[153] R. J. Macek *et al.*, Electron cloud generation and trapping in a quadrupole magnet at the Los Alamos Proton Storage Ring, *Phys. Rev. ST Accel. Beams* **11**, 010101 (2008).
[154] Y. Sato *et al.*, Electron cloud simulations of a proton storage ring using cold proton bunches, *Phys. Rev. ST Accel. Beams* **11**, 024201 (2008).
[155] M. T. F. Pivi and M. A. Furman, Electron cloud development in the Proton Storage Ring and Spallation Neutron Source, *Phys. Rev. ST Accel. Beams* **6**, 034201 (2003).
[156] J. Wei *et al.*, Electron-cloud mitigation in the Spallation Neutron Source Ring, in *Proc. 2003 Part. Accel. Conf.*, p. 2598.
[157] R. J. Macek *et al.*, Active damping of the electron–proton instability at the Los Alamos Proton Storage Ring, *J. Appl. Phys.* **01**, 102 (2008).
[158] M. Futakawa *et al.*, Pitting damage by pressure waves in a mercury target, *J. Nucl. Mater.* **343**, 70 (2005).
[159] B. W. Riemer *et al.*, Status of R&D on mitigating the effects of pressure waves for the Spallation Neutron Source mercury target, *J. Nucl. Mater.* **431**, 160 (2012).
[160] W. Wagner *et al.*, MEGAPIE at SINQ — the first liquid metal target driven by a megawatt class proton beam, *J. Nucl. Mater.* **377**, 12 (2008).
[161] C. Fazio *et al.*, The MEGAPIE-TEST project: Supporting research and lessons learned in first-of-a-kind spallation target technology, *Nucl. Eng. Des.* **238**, 1471 (2008).
[162] A. A. Alekseev *et al.*, The INR RAS neutron complex — present status and new prospects, in *Proc. 20th Int. Meeting of the Collaboration on Advanced Neutron Sources (ICANS-XX)* (Argentina, 2012).

Stuart D. Henderson received his Ph.D. in Physics from Yale University. As a Research Fellow at Harvard University he worked on high intensity electron positron colliders. He then joined the research staff at Cornell University, where his work focused on performance improvements and upgrades to the Cornell Electron Storage Ring (CESR) electron–positron collider, as well as design aspects of B-factories. Subsequently, in 2001, he joined the Spallation Neutron Source Project at Oak Ridge National Laboratory, where he led the Accelerator Physics Group in the beam commissioning of the SNS accelerator complex. In 2006 he became Research Accelerator Division Head, a position in which he was responsible for the initial operation of the SNS and the ramp-up in performance to 1 MW beam power with good reliability for the neutron scattering user community. In 2010 he became Associate Laboratory Director for Accelerators at Fermilab, where he is responsible for all accelerator activities, with a particular emphasis on planning for the next intensity frontier proton accelerator for high energy physics.

Reviews of Accelerator Science and Technology
Vol. 6 (2013) 85–116

DOI: 10.1142/S1793626813300053

Accelerators for Inertial Fusion Energy Production*

R. O. Bangerter,[†] A. Faltens and P. A. Seidl

Lawrence Berkdey National Laboratory
Berkeley, CA 94720, USA
[†]*ROBangerter@lbl.gov*

Since the 1970s, high energy heavy ion accelerators have been one of the leading options for imploding and igniting targets for inertial fusion energy production. Following the energy crisis of the early 1970s, a number of people in the international accelerator community enthusiastically began working on accelerators for this application. In the last decade, there has also been significant interest in using accelerators to study high energy density physics (HEDP). Nevertheless, research on heavy ion accelerators for fusion has proceeded slowly pending demonstration of target ignition using the National Ignition Facility (NIF), a laser-based facility at Lawrence Livermore National Laboratory. A recent report of the National Research Council recommends expansion of accelerator research in the US if and when the NIF achieves ignition.

Fusion target physics and the economics of commercial energy production place constraints on the design of accelerators for fusion applications. From a scientific standpoint, phase space and space charge considerations lead to the most stringent constraints. Meeting these constraints almost certainly requires the use of multiple beams of heavy ions with kinetic energies >1 GeV. These constraints also favor the use of singly charged ions. This article discusses the constraints for both fusion and HEDP, and explains how they lead to the requirements on beam parameters.

RF and induction linacs are currently the leading contenders for fusion applications. We discuss the advantages and disadvantages of both options. We also discuss the principal issues that must yet be resolved.

Keywords: Inertial fusion energy; induction accelerators; RF accelerators; high energy density physics.

1. Introduction

The early 1970s were a transitional period in energy policy and energy research. The OPEC oil embargo led to a fourfold increase in oil prices between October 1973 and January 1974. There were long lines at US gasoline pumps. There was widespread concern about the energy crisis. By 1977 US President Jimmy Carter had declared the energy crisis the "moral equivalent of war".

The idea of using accelerators for inertial fusion energy (IFE) production was born during this period. Today it is difficult to appreciate the level of enthusiasm among accelerator experts when it appeared that accelerators — devices that they had been building for 40 or 50 years — seemed capable of solving the energy problem.

Although lasers were used for most of the initial experiments on IFE, the prospects for accelerator-based IFE appeared very promising. Nearly all reviews of the US inertial fusion program concluded that accelerators were the most promising long-term option [1–5]. With some year-to-year fluctuations, US funding for accelerator-based inertial fusion increased for two-and-a-half decades, reaching its highest level in 2000. In other countries accelerator research for fusion was mostly associated with research on accelerators being built for other purposes.

In 1994, the US Department of Energy made the decision to design and build the National Ignition Facility (NIF) at Lawrence Livermore National Laboratory [6]. The NIF is a laser-based facility designed to evaluate, in the laboratory, the feasibility of inertial fusion. It was completed in 2009 at a cost of approximately US$4 × 10^9. As the name indicates, a principal goal is to achieve thermonuclear ignition. The NIF laser does not have many of the features (such as high efficiency and high pulse

*Work supported by the US Department of Energy under contract No. DE-AC02-05CH11231.

repetition rate) needed for fusion energy production, but achievement of ignition will put to rest basic physics issues that are relevant not only to laser fusion but also to accelerator fusion.

The advent of the NIF had a profound effect on fusion accelerator research. Should research continue or should it be put on hold pending results from the NIF? In 2003, the Department of Energy decided to de-emphasize IFE research but to continue accelerator research to study high energy density physics (HEDP). Since that time funding has declined. During this interim period there has been increasing interest in HEDP not only for fusion applications but also for such fields as astrophysics and basic studies involving the equation of state for various materials [7].

Ignition at the NIF has proven to be a difficult goal. The NIF has not yet produced ignition, although a recent report by the National Research Council [8] is optimistic about eventual success. Furthermore, the report recommends the re-establishment of an aggressive IFE program, including accelerators, if and when the NIF achieves ignition.

In writing a review article, we face an important decision. If one believes that the NIF will not achieve ignition, the principal legacy of the fusion accelerator research program will lie in the science of intense beam physics, including practical details such as the effects of electrons and gas produced by a beam halo striking the walls of the chamber. The program has made major contributions to the development of computer codes to study high intensity beams. It has also made major contributions to the science of beam–plasma interactions. A review article could emphasize these topics.

If, on the other hand, one believes that there is a reasonable chance that the NIF will achieve ignition and that there will be a resurgent IFE accelerator program, then a review article could emphasize the fundamentals of accelerator-based IFE. The article could serve as a repository of issues specific to fusion and as a roadmap to help a resurgent program avoid the pitfalls and blind alleys that have been encountered and explored during the last 40 years. We have decided to take the latter approach. We hope that our references will provide a guide for those who are interested in the more basic general issues.

Accelerator IFE has always faced two fundamental issues. To ignite fusion, an accelerator must deposit a lot of energy in a small volume in a short time. This means that the beams must have an intense focus in space and time. The second issue is an issue shared by essentially all credible fusion options, namely capital cost. Before we discuss these issues, we will give an overview of IFE history, including the issues associated with ignition.

Accelerators were not the first devices to be suggested for inertial fusion. In 1972, Nuckolls *et al.* proposed producing energy by using lasers to implode (compress) and ignite small targets containing thermonuclear fuel [9]. Since the burning fuel in this scheme is not confined by magnetic fields, but by it own inertia, this type of fusion is referred to as inertial fusion, a term we have already mentioned. The beam parameters required to bring the fuel to ignition temperature (of the order of 10 keV) are formidable. Approximately 1–10 MJ of beam energy must be focused onto a target at an intensity of 10^{14}–10^{15} W/cm^2. If ignition can be achieved, detailed numerical simulations indicate that each target will produce of the order of 100 times as much energy as the 1–10 MJ supplied to the target. A typical target radius for an IFE target is ~2 mm. This radius corresponds to an area of ~0.5 cm^2 for a spherical target and consequently a total beam power usually $>10^{14}$ W. The pulse duration is typically several nanoseconds.

In some ways pulsed lasers are good tools to use for inertial fusion because they can readily produce the required high focused intensities. But there are other requirements for commercial energy production. The device (driver) used to implode and ignite the targets must be reliable, durable, and efficient. It must also be capable of producing multiple pulses per second. For example, based on the numbers given in the preceding paragraph, a typical fusion energy yield of a target might be 300 MJ. If a third of this could be converted to net electrical energy in a power plant, it would require 10 pulses per second to produce 1 GW of electrical power, the output of a typical power plant. We will discuss all these issues and requirements more fully in subsequent sections. We note parenthetically that the values of focused intensity, beam power, pulse duration, and repetition rate given above define what is meant by high intensity in the context of IFE and in the context of the title of this volume.

In 1972, there were no high power pulsed laser systems that could meet any of the requirements for commercial power production — except the requirement on focused intensity. Consequently, in addition to initiating a program to test the target physics, governmental and private groups worldwide initiated programs to develop various types of lasers that might meet the requirements. This has proven to be a difficult task. After four decades of development, the NIF, the largest laser facility in the world, does routinely deliver about 2 MJ of energy to targets, but at low repetition rates (a few shots per day) and low efficiency (<1%).

The difficulties in developing lasers for IFE led several groups and individuals to suggest alternative drivers. One early suggestion was electron and/or ion beam accelerators based on pulsed power technology [10]. Others had suggested ion accelerators based on more conventional high energy accelerator technology [11, 12]. In many ways, high energy accelerators appeared to be well suited to IFE. They had demonstrated high reliability, long life, high pulse repetition rates and, in some cases, good efficiency. In addition, large machines such as those at CERN and Fermilab had produced or stored of the order of 1 MJ of beam energy. The ISR at CERN had a DC recirculating power of approximately 1 TW. Furthermore, the cost of these large facilities was less than US$$10^9$, a cost that appeared consistent with economical energy production. Moreover, the beams from existing accelerators could be focused to the small spot sizes required by the targets. Finally, it was soon observed that accelerators have yet another important advantage. The survivability of laser optical elements in a fusion environment had emerged as a major issue for laser IFE. Since particle beams can be focused by magnetic fields rather than material optical elements, it appeared possible to design focusing elements that would survive [13–16]. In summary, accelerator technology appeared to provide solutions to most of the difficult issues facing IFE.

Despite the favorable characteristics of accelerators for IFE, there have always been some important issues. One issue is ion range. The large accelerators that produced $\geq$1 MJ of beam energy accelerated protons to $\geq$20 GeV. At these relativistic energies, the range of a proton is sufficiently large that the beams would pass through the small targets with little energy loss. This means that high energy protons or other light ions are not directly suitable for driving targets. At sufficiently low, nonrelativistic ion energies — say, $\leq$20 MeV — the range of protons or other light ions is acceptable; but it has proven difficult or impossible to generate the focused intensity needed to drive the targets. The needed currents are simply too high. One might hope to circumvent the problem of high current by employing some kind of neutralization, for example by propagating the beams in a plasma. Unfortunately, high currents also lead to high beam emittance, making it difficult to focus to the required spot sizes. Ultimately, these emittance considerations appear to be the more fundamental limitation on the use of low kinetic energy.

In about 1975, A. W. Maschke observed that one could use high energy (multi-GeV) heavy ions to minimize the required current with a range consistent with the target dimensions [17]. Figure 1 shows ion range as a function of kinetic energy for a variety of ions. For fixed range, the kinetic energy of the heaviest ions exceeds the kinetic energy of a proton by about three orders of magnitude. Because of these considerations, the entire IFE program based on high energy accelerator technology became known as the heavy ion fusion (HIF) program despite the ambiguity with nuclear physics reactions involving heavy ions. Maschke proposed using synchrotrons

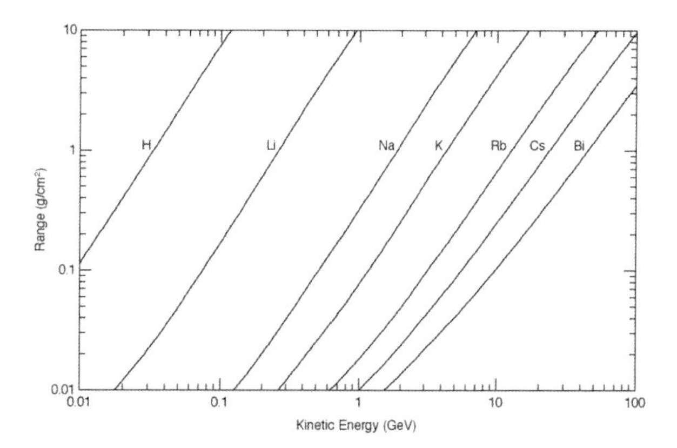

Fig. 1. Range as a function of energy for a variety of ions in low Z material (aluminum) at a temperature of 300 eV. Heavy ions rapidly strip as they strike matter, greatly increasing their stopping power. This leads to shorter range for the heavier elements. The optimal range for fusion usually lies between 0.03 and 3 g/cm^2. The chosen ions, mostly from the first column in the periodic table, are examples only. We chose bismuth rather than francium since almost no francium exists in the earth's crust.

or RF linacs followed by storage rings to produce the needed beams. Even with multi-GeV beams, the required beam current would push the tune shift in typical rings beyond 0.25, the Laslett limit. (To avoid resonances, the tune shift is usually less than this value in most machines.) Maschke demonstrated that one could transiently exceed the Laslett limit by rapidly compressing the beam in the Brookhaven AGS so that resonant conditions did not last long enough to destroy the beam [18]. Denis Keefe took another approach to HIF. He suggested the use of induction linacs from ion source to target since induction machines (relatively low impedance accelerators) are well suited to high current [19]. Section 6 discusses both RF and induction approaches to HIF.

So far our discussion has been mostly qualitative. Section 2 gives a more complete discussion on targets and their influence on beam requirements. Section 3 gives a more complete discussion on the requirements on efficiency, reliability, durability, repetition rate, economics, etc. Section 4 deals with fundamental constraints on accelerator design and beam physics, such as space charge and phase space constraints. Section 5 quantitatively discusses the choice of ion mass and charge state; and, as noted above, Sec. 6 deals with accelerators. Some important topics, such as target fabrication, target injection and tracking, and fusion chamber design, are beyond the scope of this article. Reference 8 is a good primer on most of these omitted topics.

2. Target Physics Considerations

A complete discussion of targets for HIF and their implications for accelerator design is beyond the scope of this article. There are excellent general references on IFE target physics for those readers who would like more details [20, 21]. Reference 22 is a recent reference specifically on HIF targets. For our purposes we will discuss only those aspects of target physics that are closely related to accelerator beam requirements. This section is rather long but we believe that it is necessary to understand some of the current topics in HIF, such as research on accelerators for fast ignition.

2.1. *Target scaling*

Targets for IFE and HEDP place lower limits on two fundamental quantities: the specific energy deposition ε and the beam power per unit area S. A typical IFE target requires ε greater than or of the order of 10^8 J/g; and, as mentioned above, S greater than or of the order of 10^{14} W/cm^2. The reasons for the magnitude of these quantities will be briefly explained later in this section.

As a scientific matter, the lower limits on ε and S for HEDP are much smaller than the lower limits for IFE. In fact, there is currently interest in a subfield of HEDP, namely ion-driven warm dense matter physics (WDMP) [23, 24]. WDMP is an achievable goal for near-term accelerator experiments and a variety of short pulse lasers. In the long term, lasers compete with accelerators for IFE and HEDP applications. Since lasers can already produce the values of ε and S required for fusion, it seems prudent to assume that, for planning purposes, the long-term HEDP requirements on ε and S are as stringent as those on IFE. Nevertheless, there are some differences between the requirements for the two applications. One of these is a requirement involving scale. Unlike IFE, where a certain minimum energy is needed to achieve ignition, meaningful HEDP experiments can be done at any scale as long as the appropriate values of ε and S are provided — and appropriate diagnostics are available.

The specific energies involved in IFE and HEDP are sufficiently high that the strength of the target materials is usually unimportant. The equations of ordinary hydrodynamics then provide a reasonable first approximation to the time evolution of IFE and HEDP targets. These equations are the continuity equation, Euler's equation, the energy equation, and the appropriate equations of state for the various materials in the target [25]. An examination of these equations shows that if ε and S are fixed, the various flow variables, such as density, pressure, velocity, and temperature, will depend on the spatial variables divided by the time variable, rather than on the spatial variables and time independently. In other words, if space and time are scaled by the same factor ς, the solutions to the equations are similar. This scaling is usually referred to as hydrodynamic scaling. With this type of scaling, for a fixed target design, the ion range R and the required focal spot radius r scale as ς. The beam power P scales as ς^2 and the target mass and total beam energy W scale as ς^3. This scaling explains why there is no fundamental requirement on the scale of HEDP targets. For fusion targets, however, there are important processes that do not scale hydrodynamically.

To achieve thermonuclear ignition, it is necessary to have the rate at which the fuel gains energy be greater than the rate at which it loses energy. The work done on the fuel by the implosion and the self-heating of the fuel by reaction products are energy gain mechanisms. Electron thermal conduction and radiation are energy loss mechanisms. The reaction products for deuterium–tritium fuel are 3.5 MeV alpha particles and 14.1 MeV neutrons. At a temperature of approximately 10 keV, the fusion reactions usually become the dominant heating mechanism. The alpha particles have a range of approximately $0.3\,\mathrm{g/cm^2}$ at this temperature. The nuclear reaction length of the neutrons is much larger, so the alphas do most of the heating in an IFE target. Much of the alpha energy must be trapped to achieve ignition, so the minimal acceptable value of the product of fuel density and fuel radius (a ubiquitous IFE quantity called the ρr product) is approximately $0.3\,\mathrm{g/cm^2}$.

The same considerations apply to all fusion fuels of interest for IFE. It is desirable to trap the energy of the fusion reaction products in the fuel so as to assist with ignition and burn. Since the ranges of reaction products do not scale hydrodynamically, trapping the energy of these products imposes a constraint on the allowable scale of fusion targets. Specifically, it becomes increasingly difficult to achieve ignition at a smaller scale since higher compressed fuel density is required to achieve the required ρr product at a smaller scale. Achieving higher density requires higher implosion velocity and therefore higher ε. Electron heat conduction is an additional important process that causes targets to scale poorly to smaller sizes. Finally, the fraction of the thermonuclear fuel that burns before the target explodes decreases with decreasing values of the ρr product of the compressed fuel. These (and other) considerations place a practical lower limit of roughly 1 MJ on the energy required to drive IFE targets.

2.2. *Specific energy and intensity requirements*

So far we have given the required values of ε and S without explaining the reasons for these values. Although these quantities depend weakly on scale size, as explained above, they can be quite different for different types of targets — and several types of targets have been proposed and designed. In this section we discuss various types of targets and give representative values for ε and S.

For our purposes we categorize HIF targets by their implosion (compression) mode and by their ignition mode. Direct drive and indirect drive are modes of implosion. Hot-spot ignition, shock ignition, and fast ignition are modes of ignition. Figure 2 illustrates direct drive and indirect drive for HIF targets. In the former case, the ion beams penetrate the outside layer of the fuel capsule. Owing to the rapid heating by the ions, this layer explodes or ablates, producing high pressure to implode the fuel. The fuel itself consists of a cryogenic shell of hydrogen isotopes (solid or liquid supported in a low density foam). This layer surrounds a spherical volume containing gas in equilibrium with the solid or liquid.

In the indirect case, the beams heat material inside a hohlraum (cavity) producing nearly thermal radiation that heats the ablator material on the outside of the fuel capsule. The hohlraum is made of some high Z material such as lead. High Z materials have high opacity at the wavelengths and temperatures of interest, so the radiation does not penetrate deeply into the hohlraum wall, thereby

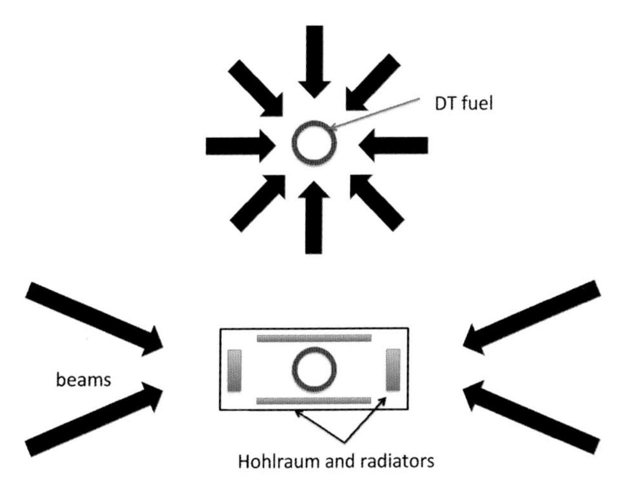

Fig. 2. Schematic diagrams of two types of targets: direct drive and indirect drive. In the first case, the ions beams impinge directly on the capsule containing the fuel. Spherical illumination is shown but other geometries are possible. In the second case, the beams heat material that fills a hohlraum (cavity) with radiation. The radiation drives the capsule. The advantages and disadvantages of these two methods of implosion are stated in the text. Either type of target can, in principle, use either hot spot ignition or fast ignition. With hot spot ignition, the implosion itself heats part of the fuel to ignition. With fast ignition, a beam pulse, usually separate from the implosion beams, heats a portion of the fuel to ignition.

minimizing energy loss. Each mode of implosion has advantages and disadvantages. The biggest advantage of direct drive is efficiency. The fraction of beam energy transferred to the imploding fuel can be ~15%, about a factor of 2 or 3 greater than the corresponding number for indirect drive. Directly driven targets are also mechanically and chemically less complex than indirectly driven targets. On the other hand, the hohlraum around the indirectly driven capsules smooths imperfections and asymmetries in the beams and allows a broader range of illumination geometries. Beam alignment tolerances are more relaxed for indirect drive (typically ~200 μm compared to ~20 μm for direct drive). Indirect drive also has an important programmatic advantage. Indirect drive with hot-spot ignition is currently the main approach at the NIF. As long as the driver beams produce nearly thermal radiation, the capsule physics is largely independent of the choice of driver (accelerator or laser). For this reason, ignition on NIF would greatly increase confidence in indirectly driven targets with hot spot ignition for HIF [26]. The connection to HIF direct drive with hot spot ignition is more tenuous, and there is little connection between the indirect drive NIF results and ion-driven fast ignition.

Note that the target examples in Fig. 2 are shown with multiple beams. Most target designs require illumination by multiple beams. More importantly, it appears extremely difficult to produce the required beam power without multiple beams.

Now consider the ignition mode. In hot spot ignition, a small quantity of fuel in the center of the capsule, such as the gas in equilibrium with the solid or liquid fuel, is heated to ignition by PdV work (and fusion reactions) as the capsule implodes. In shock ignition, a high power beam pulse is applied to the outside of the target at such a time that a strong incoming shock wave assists the heating of the central part of the target to ignition temperature. For fast ignition, the two processes — implosion and ignition — are largely decoupled and an intense beam directly heats a small portion of the fuel to ignition temperature. In all cases, the thermonuclear burn propagates from the small portion of the fuel that is ignited into the colder fuel surrounding it. The reason for requiring propagating burn is easily understood. It requires ~10^9 J/g to heat deuterium–tritium fuel to ignition but only ~10^7 J/g to compress it to ~1000 times solid density, a typical density for inertial fusion. To achieve such low compression energy, the fuel must be nearly Fermi-degenerate. Specifically, the specific energy in the fuel must not substantially exceed the Fermi specific energy (10^5 J/g for solid-density fuel). In summary, because the compression energy is low compared to the ignition energy, it is advantageous to compress most of the fuel in a nearly degenerate state and to ignite as little fuel as possible. The thermonuclear burn, under appropriate conditions, can then propagate into the colder compressed fuel. Without propagating burn, it is not possible to achieve the high energy gain (G = energy produced / beam energy) needed for IFE.

As previously noted, hot spot ignition cannot occur unless the rate of doing work on the hot spot region of the fuel exceeds the rate of energy loss out of the region. Achieving the required rate of doing work usually requires an implosion velocity $v \geq 300$ km/s. This velocity corresponds to a specific energy $\frac{1}{2}v^2 = 4.5 \times 10^7$ J/g. Since this specific energy exceeds the specific energy needed for compression alone, it is energetically advantageous to separate the ignition process from the implosion process. This consideration provides one rationale for considering shock ignition and fast ignition. It has been difficult to design credible HIF targets that use shock ignition. Shock ignition, if it is feasible for HIF, will likely have requirements between the other two cases. For this reason, in the remainder of this article we will consider only hot spot ignition and fast ignition in deriving accelerator beam requirements.

Note that, in the previous paragraph, the specific energy requirement of the inwardly moving cold fuel is 4.5×10^7 J/g. This value is close to the required value of specific energy deposition $\varepsilon \sim 10^8$ J/g given previously. The reason is the following: an implosion is basically a rocket-like process. For good efficiency, the specific energy put into the rocket exhaust must be of the same order of magnitude as the specific energy represented by the velocity of the payload. Consequently, for hot spot direct drive, the specific energy that must be deposited in the ablator by the ion beams is also of the order of 4.5×10^7 J/g. For indirect drive, the temperature of the radiators must exceed the temperature of the ablator in order to transport energy in the correct direction. Moreover, the energy in the ablator and the hohlraum wall must have been deposited in the radiator. For these reasons, the specific energy that must be deposited in the radiators by the ion beams usually exceeds 4.5×10^7 J/g by more than a factor of 2. Although this cost is substantial, there are some mitigating factors. For directly driven targets, the beam radius cannot substantially exceed the capsule radius. Otherwise part of the beam misses the target. Indirectly driven targets have the flexibility to trade off radius and ion range while keeping the mass of the radiators constant. For example, a cylindrical radiator having a radius of 2 mm and a "length" or areal density of $0.2\,\mathrm{g/cm^2}$ would have the same mass as a radiator having a radius of 4 mm and an areal density of $0.05\,\mathrm{g/cm^2}$. In a first approximation the target performance would be comparable for the two cases. In other words, the beam focal spot radius is not directly coupled to the fuel capsule radius as it is for direct drive. This flexibility in target design translates to greater flexibility in accelerator design. Furthermore, for spherical direct drive the mass that must be heated to 4.5×10^7 J/g is approximately the area of the surface times the ion range or $4\pi r^2 R$. An indirectly driven target might have as few as one or two radiators inside the hohlraum corresponding to a mass of only one or two times $\pi r^2 R$. This is an argument in favor of illuminating only one side of a hohlraum, if the capsule can be compressed uniformly. Finally, although an indirectly driven target still requires more drive energy, there is an additional mitigating factor. The ratio of required energy in the two cases — say, 2:1 for a given capsule yield — is larger than the ratio of required power. The reason is the following: the power to the capsule, in the case of indirect drive, is approximately proportional to the fourth power of the temperature in the cavity (Stefan's law). It requires a significant time, compared to the capsule implosion time, to heat the radiators and cavity to a high-enough temperature to drive the capsule effectively. This means that for indirect drive the beam energy is delivered over a time that is longer than the implosion time. In direct drive the energy is delivered in a time that is comparable to the implosion time. In other words, the energy penalty associated with indirect drive is partly paid early in the pulse when the implosion velocity is small. This is important because the constraints on accelerator design depend more strongly on power than on energy.

Although our arguments provide the correct order of magnitude for ε, they say little about the actual physical size of the targets. Why, for example, can one not make a target that consists of a very large, thin shell of fuel surrounded by an ablator? In this case, the implosion time would be long and the focal spot size would be large. Fluid instabilities are one of the important limitations on making the target shell large and thin. The implosion process is hydrodynamically unstable. Small perturbations on or in the shell grow as the shell is imploded. If they grow to sizes comparable to the shell thickness, the integrity of the shell is destroyed. Quantitatively, this and other considerations place an upper limit on the initial radius of the shell of about 2 mm for a capsule that requires of the order of 1 MJ of energy. This radius, for direct drive, is also approximately equal to the maximum usable beam radius. Knowing the radius, the required specific energy, the total energy, and the implosion velocity, one can easily calculate the other quantities of interest, such as intensity, ion range, and pulse duration. While these kinds of arguments explain the physics that is important and provide order-of-magnitude estimates for important beam quantities, it is important to emphasize that these quantities vary by significant factors, depending on the details of the target design. Large scale numerical simulations are used to obtain more accurate values [27].

There is a large difference between the beam requirements for fast ignition and the requirements for hot spot ignition. For fast ignition, the implosion velocity is not coupled to the ignition velocity, so it needs only to correspond to the specific energy of compression ($\sim 10^7$ J/g). This corresponds to a velocity of 140 km/s. Consequently, for compression, the pulse duration can be about a factor of 2 longer than for hot spot ignition and the total energy and specific energy can, in principle, be several times lower. In contrast, the beam requirements for the ignition pulse are stringent. For simplicity assume that the ignition region for fast ignition is cylindrical. The mass of this region is given by $M = \pi r^2 R$, where r is focal spot radius and R is ion range. Consider two cases: $r = 50\,\mu\mathrm{m}$, $R = 0.6\,\mathrm{g/cm^2}$ and $r = 150\,\mu\mathrm{m}$, $R = 3\,\mathrm{g/cm^2}$. The first case requires $\sim$50 kJ to heat the fuel to ignition (10^9 J/g). The second case requires $\sim$2 MJ. We can get an estimate of the pulse durations associated with these two cases by equating the time required for the ignition region to blow itself apart to the radius divided by the speed of sound in 10 keV fuel (~1000 km/s). For 50 μm we obtain 50 ps and for 150 μm we obtain 150 ps. One obviously minimizes the ignition energy by minimizing the radius. What is the practical lower limit on the beam radius? It appears that the most stringent limits arise from accelerator and beam physics considerations, as discussed in Secs. 4 and 5. The required beam power for the first case is $\sim 10^{15}$ W. For the second case it is $\sim 10^{16}$ W. For both cases the required intensity is $> 10^{19}$ W/cm^2.

There is one other important point of comparison between the two ignition modes. For fast ignition, the energy required to drive the target, and consequently the target gain, depend strongly on the assumed focal spot radius and ion range, as shown above. Targets employing hot spot ignition are significantly less sensitive to variations in ion range and focal spot radius [28].

2.3. *Published target designs*

Since the preceding parts of this section gave only estimates of beam parameters, we conclude this section by listing the requirements of four specific HIF target designs (Table 1) that have been simulated [27–31] using computer codes benchmarked to a wide variety of experiments. Three of these targets are indirectly driven target using hot spot ignition. We have chosen these targets because they are of the same type as those currently being tested on the NIF. The fourth, the X-Target, is a directly driven target using fast ignition. We will subsequently show that it is difficult to meet the beam requirements for this target, but it represents the alternative that is least like the NIF targets. It is argued that this target partially decouples the HIF program from success or failure on the NIF.

The beam energy is deposited with a time dependence of relatively low power followed by a shorter duration, higher power burst characterized by the peak intensity. Most of the energy is deposited during the short duration burst.

Direct drive with hot spot ignition is a target category that is not represented in Table 1. Many of the detailed designs in this category are dated. Also, in terms of power and energy requirements (but not beam profile or pointing accuracy), the directly driven targets are somewhat more relaxed than the indirectly driven targets. Otherwise, the requirements are similar for the two cases. The Close-Coupled target listed in the table can be used as an approximate example of direct drive.

Table 1. Examples of specific target designs and beam requirements are compared. The name of the target is the name chosen by its designers. The first three targets were designed by the same people using the same codes. The designers included some margin in the designs to cover uncertainties. The X-Target was designed using a different code. It is probably more speculative [22], although the uncertainties in all designs are large. We were unable to find a published value for the peak power of the Hybrid target. The listed value is our estimate. Different publications give somewhat different numbers for these example targets. These differences do not affect our conclusions.

Target name	Energy (MJ)	Peak power (TW)	Focal radius (mm)	Gain	Ion	Kinetic energy (GeV)
Distributed radiator	5.9	700	2.7	68	Pb	4.0
Hybrid	6.7	700	4.5	60	Bi	4.5
Close-Coupled	3.5	470	1.7	130	Pb	3.5
X-Target	5.0	3×10^4	0.26	300	Rb	20

Although the hydrodynamic scaling described in Subsec. 2.1 is not precise, we will subsequently use it to investigate the effects of changing the scale of the targets shown in Table 1.

Finally, we note that the table lists only basic target requirements. All the targets require a specified pulse shape and some require the kinetic energy to be a function of time. These details are important for detailed accelerator design studies but are not important for our purposes. There are also differences in how the various authors define "beam radius" — for example, two times rms or half width at half maximum.

3. Accelerator Considerations Relating to Target Chambers and Economics

The target considerations discussed in Sec. 2 impose stringent constraints on the design of accelerators for both fusion and HEDP. Beyond the constraints arising from target physics there are few other constraints on accelerators for HEDP. In contrast, there are a number of additional constraints for the fusion application. We discuss them in this section.

Figure 3 shows a schematic diagram of an IFE power plant. We denote the gross electrical power

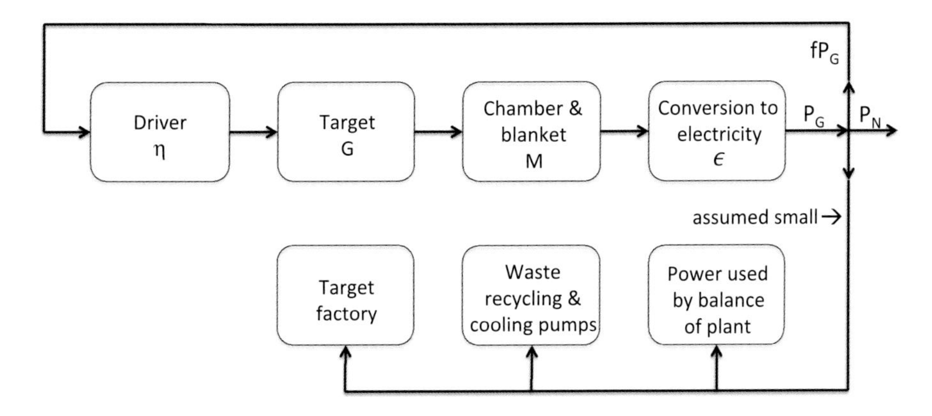

Fig. 3. Schematic diagram of the power flow in a generic inertial fusion power plant. (See text.)

output of the plant by P_G and the net power available to sell to customers by P_N. The difference between these two quantities is the power required to drive the driver plus the power required to drive all the other systems, such as controls, recycling of target debris, the target factory, pumps and lights. (We include refrigerator power needed for superconducting magnets in the power required by the driver.) A common simplification is to assume that the sum of all power requirements except the driver input power is small compared to the driver input power. One can conceive of situations where this is not a good approximation but we adopt it. Specifically, we assume that $P_N = P_G(1 - f)$. The quantity f is just the fraction of the gross power that must be recirculated to drive the driver. Let η, G, M, and ϵ respectively represent the driver efficiency, the target gain, the blanket multiplication factor, and the efficiency of converting fusion energy (say, heat) into electricity. From the figure it is clear that $P_G = fP_G\eta GM\epsilon$, leading to $f = 1/\eta GM\epsilon$. If $f > 1$, the system does not work, since it does not generate enough power to drive the driver. For economical power production, f must almost certainly be much less than 1. Otherwise the cost of the power plant, which is closely related to the gross power output, will be too high for the net power that it produces. A common assumption based on conventional costs of the nonfusion parts of the power plant is that f must be less than approximately 0.2 — but this is clearly a crude approximation. If the fusion core, including the driver, is expensive compared to the conventional equipment, the upper limit on f is smaller than 0.2 and vice versa.

Consider typical values for M and ϵ. Among possible fusion fuels, a deuterium–tritium (DT) mixture has the lowest ignition temperature. All example targets given in Table 1 assume DT fuel. It seems likely that all first generation fusion power plants will burn DT. Deuterium is readily available from ordinary water, but tritium decays with a half-life of 12.3 years, so it does not occur to any significant extent in nature. It must be created. For this reason essentially all proposed DT power plants employ a blanket of lithium or lithium compounds surrounding the fusion chamber. The neutrons from the DT reactions, or secondary neutrons from other reactions in the blanket, produce tritium via $\mathrm{Li}^6 + n \rightarrow \mathrm{H}^3 + \mathrm{He}^4$ and $\mathrm{Li}^7 + n \rightarrow \mathrm{H}^3 + \mathrm{He}^4 + n$. With proper blanket design the entire sequence of reactions involving neutrons can be exothermic, giving a blanket multiplication factor greater than unity, often in the range of $M = 1.1$–1.3 [32, 33]. Nearly all proposed fusion power plants based on DT fuel use some kind of thermal cycle with turbines and generators to convert the fusion energy to electricity. The projected efficiency of such systems is usually $\epsilon = 0.3$–0.5 [32, 33]. As a specific example, assume that $M\epsilon = 0.5$. In this typical case, we require $\eta G > 10$ to achieve $f < 0.2$. In fact, $\eta G > 10$ is commonly used as a necessary criterion for economical IFE power production, but we emphasize again that it is an approximation based on a number of typical, but not fundamental, assumptions.

If we apply $\eta G > 10$ to the target gains given in Table 1, we require the accelerator efficiency to satisfy $\eta > 15\%$ for the $G = 68$ case and $\eta > 3.3\%$ for the $G = 300$ case. Since the values of G in the table are projected numbers rather than experimental numbers, it would be desirable to have some safety factor.

In terms of cost per kWh, conventional nuclear energy competes favorably with energy based on fossil fuels [34]. (We are ignoring the interminable arguments about the safety of conventional nuclear power, greenhouse gas emissions from fossil fuels, and public acceptance.) The cost of building a typical 1 GW nuclear power plant usually lies between US$2/W and US$8/W [34]. Large variations in the regulatory environment and large variations in wages are primarily responsible for the large spread in costs. It seems unlikely that the nuclear core of an IFE power plant, excluding the driver, will cost much less than the nuclear core of a conventional fission power plant. Consequently, to be competitive, the driver cost must be small compared to the cost of a typical nuclear power plant. For illustrative purposes we assume a cost of US$4/W. Using this number, the cost of the driver must be less than approximately US$1/W, corresponding to US$$10^9$ for a 1 GW plant.

The energy yield of a target is, by definition, the product of the gain and input energy. The yield varies from approximately 400 MJ to 1500 MJ for the targets described in Table 1. Assuming that $M\epsilon = 0.5$ and $f = 0.2$, the required pulse repetition rate to obtain a net power output of 1 GW is between 1.7 and 6.3 Hz — justifying the numbers given in Sec. 1.

Power plant considerations lead to other important constraints. The competing options for energy production have very high reliability, much better than 90%, and they operate for several decades. The accelerators must do the same. Quantitatively, a pulse repetition rate of 5 Hz for 40 years at an availability of 90% corresponds to 6×10^9 pulses. The fusion chamber must be either inexpensive so that it can be replaced, or able to contain this large number of pulses without substantial damage. A large number of chamber designs have been proposed and studied. The requirements on chamber survivability, tritium breeding, pulse repetition rate, and shielding of the conductors for the final lenses typically translate into chambers having radii between 5 and 10 m [13–16, 32, 33]. The importance of this number is that it places a lower limit on the focal length of the final focusing lenses. We will show in Sec. 4 that the focal length is related to the emittance requirements of the beams.

In summary, power plant considerations place important constraints on accelerator efficiency, cost, reliability, durability, repetition rate, and the focal length of the final magnetic lenses.

4. Current Limits and Phase Space Constraints

In previous sections we have shown that both IFE and HEDP require accelerators to produce beams that can deposit large amounts of energy in a small mass in a short time. The accelerator must not only serve its usual function as an energy amplifier, but also, in conjunction with its associated focusing system, compress the beams produced by the ion source(s) by large factors in space and time. The usual function of energy amplification is straightforward. The total beam energy is the product of the ion kinetic energy, the beam current, the pulse duration, and the number of beams. The only serious limitation on making this product arbitrarily large is cost. In contrast, for fusion the beams must be compressed by unprecedented factors. Moreover, as noted in Sec. 2, the ion range cannot be arbitrarily large and still achieve the desired specific energy deposition.

The compression of an ion beam is fundamentally limited by two factors: the mutual repulsion of the charged ions and the pressure resulting from the motion of the ions in the rest frame of the beam. The first factor can be expressed in terms of normal accelerator variables such as charge density and current. The second factor can be expressed in terms of phase space density, beam brightness, or emittance.

For convenience in discussing these two fundamental factors, we divide a generic accelerator into three systems: the ion source(s) including any quasi-electrostatic acceleration or injection, the main accelerator, and the final focusing system.

4.1. *Ion sources*

For IFE the accelerator must produce a sequence of essentially identical pulses to meet the requirements of any specified target design. This means that the ion sources must be space-charge-limited rather than emission-limited. Otherwise inevitable fluctuations in emission from a spark, plasma, or a surface would create variations in source current. The space charge limit of sources is described by the Child–Langmuir formula

$$J = \left(\frac{4\epsilon_0}{9}\right)\left(\frac{2qe}{M}\right)^{1/2}\frac{V^{3/2}}{d^2},$$

where J is current density, ϵ_0 the permittivity of free space, qe ion charge, M ion mass, and V the voltage applied to the gap of width d. All quantities are in SI units. (The equation as given is for planar geometry. Similar limits apply to other geometries.) Since we must prevent variations in current, not only must the ion sources be space-charge-limited, but, the voltage must also be very stable. The quantity J/qe is the number of particles per unit area per unit time. The product of area and time is a volume in a three-dimensional space with two spatial dimensions and one time dimension. (We can calculate the velocity, so we could use three spatial dimensions instead.) In summary, space charge considerations give the maximum ion density emerging from the source. These ions are also contained is some three-dimensional volume in momentum (or momentum–energy) space. In practice several mechanisms determine the occupied volume of momentum space. One of them is the temperature of the ions. (We use the term "temperature" loosely, because the momentum distribution may not be a thermal distribution.) A second mechanism is imperfect extraction optics and a third is high frequency fluctuations in the gap voltage, or in any subsequent voltage. Assume, for example, that the ions are born with a temperature Θ expressed in energy units such as eV. Each degree of freedom has an rms velocity given by $v = (\Theta/M)^{1/2}$, where we have set the velocity of light equal to unity and expressed M in the same energy units as Θ. In calculating quantities such as emittance, we are frequently interested in "edge" quantities — dimensions that contain most of the particles. For simplicity, we assume that edge quantities are twice rms quantities so that $2v$ is the edge velocity (for the part of the transverse emittance arising from temperature). By definition the normalized transverse emittance is given by $\epsilon_N = a\theta\beta\gamma = 2av = 2a(\Theta/M)^{1/2}$, where a is the beam radius, θ is the angular spread in the beam (half width), and β and γ are the usual relativistic factors. (For heavy ions the particles are not relativistic at the source, so we have set $\gamma = 1$.) As usual, the area in two-dimensional phase space is approximately π times the normalized emittance, so we express the emittance in units of $\pi\,\mathrm{m}\cdot\mathrm{rad}$ or $\pi\,\mathrm{mm}\cdot\mathrm{mrad}$.

The situation for longitudinal emittance is slightly more complicated. We choose to work in the laboratory frame and express longitudinal emittance in terms of pulse duration τ in seconds and energy spread (half width) Δ in eV. A longitudinal rms velocity v in the beam frame corresponds to a velocity $v_0 + v$ in the laboratory frame, where v_0 is the beam velocity. Since the energy in the laboratory frame is $M(v_0 + v)^2/2 \approx M(v_0^2 + 2v_0v)/2$, the energy spread in the laboratory frame is $\Delta = 2Mv_0v$. (We multiplied by 2 to convert the rms value into the edge half width.) The longitudinal emittance ϵ_L is then given by $\epsilon_L = \tau\Delta/2$, where we have multiplied the edge energy spread by $\tau/2$ rather than τ so that the longitudinal phase space volume (area) is approximately $\pi\epsilon_L$, consistent with the transverse case. Since $v_0 = (2T/M)^{1/2}$, where T is the kinetic energy of the ions, we can also write

$$\epsilon_L = \tau(2T\Theta)^{1/2}. \quad (1)$$

Designing real ion sources involves a number of tradeoffs. For good extraction optics, the extraction gap d is usually comparable in size to the beam radius a [35]. In our experience the voltage V that one can reliably hold is approximately linear in d for $d < 1$ cm and varies approximately as the square root of d for $d > 1$ cm. If we use this scaling and the relationship between a and d, the Child–Langmuir formula shows that, for fixed beam temperature, the six-dimensional phase space density (beam brightness) increases with decreasing a and d — but so does the current. Since there will be a requirement on current, there is a limit on beam brightness. One possible way to circumvent this limit is to use multiple small aperture sources. Unfortunately, when the beamlets for the multiple sources are merged to form a beam of the desired current, the emittance increases because of space charge effects and because some unoccupied phase space is inevitably mixed with the occupied phase space. Table 2 gives the properties of some typical ion sources that might be used for HEDP or fusion [35, 36]. It is noteworthy that many elements are not suitable for good ion sources because of the difficulty of obtaining a single isotope or a single charge state.

All of the sources listed in the table are for charge state $q = 1$. In Sec. 5 we will discuss the choice of

Table 2. Properties of ion sources are compared. This table illustrates some of the considerations in the text. The brightness increases with decreasing current, and merging small beamlets to obtain adequate current reduces brightness. The brightness values for the three sources giving the higher currents are equal within about a factor of 2. The CHORDIS source is appropriate for RF machines, and the others for induction linacs.

Source	Ion	Current (A)	ϵ_N (π mm · mr)	Brightness $= I\epsilon_N^{-2}$
LBNL alumino-silicate	K^+	0.183	0.5	0.73
GSI CHORDIS	Bi^+	0.037	0.16	1.44
LBNL small aperture	Ar^+	0.0044	0.0186	12.7
Small apertures merged (simulation)	Ar^+	0.57	0.9	0.70

charge state. Since the inception of HIF there have been recurring efforts to design accelerators that use ions having $q > 1$. The rationale is that the acceleration voltage can be lower for a given kinetic energy — perhaps leading to lower cost. On the other hand, the beam power is the product of electrical current times acceleration voltage, so a higher charge state leads to higher current for fixed beam power. A higher charge state also leads to lower brightness since, according to Child–Langmuir, the current density increases as the square root of the charge state, while the number of particles for a given current decreases inversely with the charge state. A common alternative method of producing $q > 1$ is to start with singly charged ions at the source, accelerate to some modest energy, and then strip. Even if one ignores effects such as multiple scattering during stripping, this method still leads to reduced brightness since it is almost never possible to strip to a unique charge state [37, 38]. Based on inspection of various measurements, an approximate characterization of various data is that electrical current is conserved during stripping. For example, stripping to, say, +12 would be expected to reduce the brightness by a factor of ~12.

After decades of development, it appears difficult to make sources that are significantly brighter than those given in the table.

4.2. *Space charge in the main accelerator*

Since the high beam currents needed for fusion are unprecedented, one of the first tasks of the HIF program in the mid-1970s was to look carefully at the fundamental limits on transportable current. This task was originally undertaken by A. W. Maschke, leading to a theoretical expression for the maximum current that can be transported in an alternating gradient lattice. The HIF community usually refers to this expression as the Maschke formula or Maschke Limit [39]. For illustrative purposes, we will derive a very approximate version of this formula expressed in more modern form. Maschke's original idea was simple. He set the transportable current, based on some experience, so that its space charge forces were equal to one half of the average focusing forces.

We first estimate the focusing forces. Let the length of a half lattice period (the length between the centers of two quadrupoles) be L and let the length of a single quadrupole be ηL ($\eta \equiv$ occupancy). The equation of motion for a paraxial ion is given by $M\beta^2c^2\gamma x'' = \pm B_b qe\beta cx/b$ or $x'' = \pm k^2 x$, where $'$ denotes differentiation with respect to axial distance, x is the transverse displacement from the axis of the quadrupole, B_b is the magnetic field at a distance b from the axis, and qe is the charge of the ion. We have explicitly included the speed of light, c. Solving this equation gives the usual transfer matrices for the quadrupole. If we make the assumption that η is small compared to unity, it is easy to obtain $\boldsymbol{T}$, the transfer matrix for an entire period. Using the usual equation $\cos\sigma_0 = \text{trace}(\boldsymbol{T})/2 \approx 1 - \sigma_0^2/2$, we obtain $\sigma_0 \approx B_b qc\eta L^2/bp$, the phase advance per period. The particle momentum $p = M\beta\gamma$ is now expressed in eV and the other units are SI ones. The approximation (good for small η) is adequate for our purposes. It is possible to derive more accurate analytical expressions [40] but numerical methods are used for accurate work. We will use a simple smooth approximation that averages over the alternating gradient flutter. In this approximation, the beam size is given by an envelope equation [41] describing a beam with cylindrical symmetry $a'' = -\kappa^2 a + \epsilon^2/a^3 + Q/a$.

In this equation $\kappa = \sigma_0/2L$, since σ_0 is the phase advance in a full lattice period of length $2L$; and a, ϵ, and Q are respectively the beam radius, the transverse unnormalized emittance $\epsilon = \epsilon_N/\beta\gamma$, and perveance $Q = I_e qe/(2\pi\epsilon_0 Mc^3\beta^3\gamma^3)$. The quantity I_e is electrical current and the term Q/a includes both the self-electrostatic and magnetic forces acting on a particle a distance a from the axis.

We refer to beams where the emittance term in the envelope equation is larger than the perveance term as emittance-dominated beams. In the opposite case we refer to space-charge-dominated beams. Maschke originally guessed that for an equilibrium beam ($a'' = 0$), the perveance term in the envelope equation could not be larger than the emittance term. This condition sets an upper limit on the transportable current. We now know that the early guess was not entirely correct. Analytical studies first raised concerns about possible instabilities. Subsequently both numerical studies and experiments [42] showed that the perveance term could be arbitrarily large compared to the emittance term as long as σ_0 is less than about $\pi/2$. (This is in contrast to the usual case for beams that are strongly emittance-dominated, since these beams are stable for $\sigma_0 < \pi$.)

Armed with the numerical studies and experiments, we can use our simplified envelope equation to obtain an approximate limit on current. To get an upper limit we assume that the emittance term is small compared to the perveance term and write $I_e qe/(2\pi\epsilon_0 Mc^3\beta^3\gamma^3 < \sigma_0^2 a^2/4L^2$. If we use the expression for σ_0 to eliminate L^2, we obtain

$$I_e < \frac{\pi\epsilon_0 c^2\sigma_0 B_b\eta\beta^2\gamma^2 a^2}{2b}.$$

If we now set $\sigma_0 = \pi/2$ and use typical values for occupancy $\eta = 1/2$ and aperture fill factor $a/b = 1/2$, we obtain an approximate expression for the current in amperes:

$$I_e < 5 \times 10^5 B_b\beta^2\gamma^2 a. \tag{2}$$

Subsequently we will refer to Eq. (2) as the Maschke limit, although it is only approximate and it is not in the form originally written by Maschke. One can easily write similar expressions for solenoids or electrostatic quadrupoles. These focusing options may be useful at the low energy end of the accelerator; but, for simplicity, we give the expression only for magnetic quadrupoles.

4.3. *Space charge limits in final focusing*

Assume that a cylindrical beam emerging from the final lens into the target chamber has a radius r_0 and an envelope convergence angle θ_0. If we temporarily ignore the emittance term in the envelope equation, we can immediately integrate the equation to obtain $\theta_0^2 = 2Q \cdot \ln(r_0/r_t)$, where r_t is the radius at the target. We can crudely account for the possibility of neutralizing the beam in the target chamber by multiplying Q by some neutralization factor f_n. We can also rewrite the result as a current limit:

$$\begin{aligned} I_e &< \frac{\pi\epsilon_0 Mc^3\beta^3\gamma^3\theta_0^2}{f_n qe \cdot \ln(r_0/r_t)} \\ &\approx 8 \times 10^6 \frac{A\beta^3\gamma^3\theta_0^2}{f_n q \cdot \ln(r_0/r_t)} \text{amperes.} \end{aligned} \tag{3}$$

In the last expression we have expressed the ion mass in terms of the atomic mass A.

4.4. *Phase space constraints*

We have previously given examples of the phase space density or brightness available from real ion sources. We will denote the phase space density at the ion source (or at the end of a quasielectrostatic preaccelerator or injector) as B_{6i}, where the subscript 6 refers to six dimensions and i means "initial". We will now explain how the properties of targets, focusing systems, and chambers place a lower limit on the phase space density required at the end of the driver. Unless one is willing to develop non-Liouvillian beam manipulation techniques, the feasibility of any system demands that $B_{6i} \geq B_{6f}$, where B_{6f} is the minimum allowed phase space density at the end of the accelerator. (We will discuss non-Liouvillian techniques later.)

Consider the phase space volume available at the end of the final focusing system. The spatial transverse phase space volume (area) is equal to Ωf^2, where Ω is the total solid angle occupied by all the beams as they converge onto the target and f is the standoff between the end of the last lens and the target, roughly the focal length. The limit on transverse momentum space is given by $r_t p/f$, the angle subtended by the target times the beam momentum p. If we multiply the spatial area by the area in momentum space, we find that the four-dimensional transverse phase space volume is proportional to $\Omega r_t^2 p^2$.

Geometrically we know that $\Omega \leq 4\pi$. In practice it must be much smaller. One limitation is the tritium breeding ratio. For tritium self-sufficiency almost the entire solid angle surrounding the target must be occupied with the tritium breeding blanket.

This requirement usually means that Ω must be less than or of the order of 1. Target geometry requirements are often even more stringent. Most indirect drive target designs require one- or two-sided illumination where the cone surrounding all beams from one side cannot have a solid angle exceeding approximately 0.5. Only a small fraction of this solid angle can be filled with beams. There must be some standoff between the beam and the aperture of the final focusing element. Significant space is often required between the aperture and the conductor for neutron and gamma shielding, cooling, and structure. The same is true outside the conductor. If the final focusing elements are packed as tightly as possible, one usually loses another order of magnitude, so that $\Omega \sim 0.1$ is a practical upper limit.

The pulse duration required by the target sets the longitudinal time (or space) limit. Several factors limit the sixth and final dimension, the longitudinal energy or momentum spread. One factor is the target itself. If the energy spread is too large, the variations in range among the various particles cause a spreading of the ion energy deposition profile. The Bragg peak is washed out and a larger fraction of the energy is deposited near the surface of the target where it is not efficiently utilized by the target implosion or ignition. For targets where the Bragg peak plays a significant role in target performance, e.g. fast ignition targets, a 10–20% spread appears to be excessive. This limit is usually much less stringent than other limits. One of these other limits is related to longitudinal pulse compression. In all IFE and HEDP schemes it is necessary to compress the beam longitudinally as it moves toward the target. The current available from ion sources is simply much smaller than the current required by targets. The longitudinal compression (or focusing) can be achieved by applying a velocity tilt between the head of the beam and the tail of the beam. The tail of the beam is given the higher velocity so that the beam compresses as it propagates toward the target. There is a limit to the rate at which the tilt can be applied. If one attempts to apply too large a tilt (energy spread), the beam may focus longitudinally before it arrives at the target. We mention this limit because it has not been included in some systems codes developed for HIF, e.g. IBEAM [43]. IBEAM was originally developed to explore a parameter range where the limit is unimportant. Recently there have been some attempts to use IBEAM outside its original range of validity, specifically in the range appropriate for fast ignition. We found one unpublished but widely discussed example accelerator that appears to violate this limit.

Finally, and usually, the most stringent limit on energy spread is set by chromatic aberrations in the final focusing system. We examine this mechanism in some detail.

As usual, we let $x'' = k^2(s)x$ be the equation of motion in one transverse dimension for a particle with momentum p. In this equation s is the longitudinal coordinate and k^2 includes both applied focusing forces and the self-fields of the beam. For a solenoidal focusing system, k^2 is negative and varies as p^{-2}. In a magnetic alternating gradient system, it alternates in sign and varies as p^{-1}. The contribution from space charge forces is positive and varies as $(p\beta)^{-1}$. In the general case one must consider all contributions to k^2. For illustrative purposes we consider only the magnetic alternating gradient case and assume that the fractional momentum spread $\delta \ll 1$. In this case we can write a perturbed equation for an off-momentum particle as $x'' + \xi'' = k^2(1-\delta)(x+\xi)$. Using $x'' = k^2 x$ to simplify the perturbed equation and multiplying both sides by x, we obtain $x\xi'' - x''\xi = -\delta k^2 x(x+\xi)$. We can integrate both sides of the equation, leading to $x\xi' - x'\xi = -\delta \int k^2 x(x+\xi)ds$. At the beginning of the focusing system, $\xi = \xi' = 0$. At the focus x must be very small and $x' = -\theta_0$ for an edge particle having $x > 0$ upstream of the focus. In the integrand the average value of ξ is much smaller than the average value of x, so we neglect it and write

$$\xi = -\left(\frac{\delta}{\theta_0}\right) \int k^2 x^2 \, ds.$$

For a point-to-point or parallel-to-point focus we can write this in a different form by using $x'' = k^2 x$ and integrating by parts:

$$\xi = \left(\frac{\delta}{\theta_0}\right) \int x'^2 \, ds. \tag{4}$$

(For our purposes, we have ignored the difference between a small waist and a focus. Note also that for a parallel-to-point focus formed by a thin lens $\xi = \delta\theta_0 f$, a relatively obvious result.)

In an approximation where the envelope is represented by a straight line between the last lens and the focus, the integral in Eq. (4) must be at least as large as $f\theta_0^2$, so we write

$$\xi = \alpha\delta f\theta_0. \tag{5}$$

In this equation, we have introduced a parameter $\alpha > 1$. For typical waist-to-waist systems using magnetic quadrupoles, α usually lies in the range $2 < \alpha < 8$. This parameter can be quite different for the two transverse directions, because the alternating gradient flutter can be quite different in the two directions. As a consequence of this and other effects, focal spots are, in general, astigmatic and the astigmatism is time-dependent. Time-dependent astigmatism may be an important issue for directly driven targets and perhaps fast ignition targets. This issue must be addressed.

As a typical example let $\alpha = 5, f = 6\,\text{m}$, and $\theta_0 = 10\,\text{mr}$. Assume that a typical indirectly driven target can tolerate a chromatic aberration $\xi \sim 1\,\text{mm}$ and that a typical target for fast ignition can tolerate about 0.1 mm. In the former case, the final momentum spread must be limited to approximately 0.3%. In the latter case it must be limited to approximately 0.03%. These numbers are much more stringent than the other limits on momentum or energy spread; but, in our example, we assumed that $\theta_0 = 10\,\text{mr}$. Could one reduce the angle by some factor and make the limit on momentum spread less stringent? From a fundamental standpoint the answer appears to be yes — if we consider only the optics of the final focusing system. As noted above, the available transverse phase space volume is proportional to $\Omega r_t^2 p^2$. This expression depends on Ω but not explicitly on θ_0. Since $\Omega = n\pi\theta_0^2$, where n is the number of beams, we can make θ_0 small by going to an arbitrarily large number of beams. Applying similar arguments to Eq. (3) shows that the total current that can be focused also depends only on Ω. It is also true that other effects that we have not discussed, such as third order geometric aberrations, are minimized by minimizing θ_0. In summary, if one considers only the optics of final focusing, one would like to use a very large number of very small beams.

Are there considerations that limit the number and size of the beams? Some worry that the complexity and cost of dealing with a large number of beams are prohibitive. So far there has never been enough experimental information to quantify this concern. An analysis of the scaling properties of magnetic quadrupoles does, however, provide a more fundamental lower limit on beam size. Consider the integral form of Ampere's law $\oint B \cdot dl = \int \mu_0 J \cdot dA$. For a typical quadrupole, the integral on the left is linear in scale size but the integral on the right is quadratic. Thus, for a given magnetic field and current density in the conductor, a large quadrupole will devote a smaller fraction of its cross-sectional area to the conductor than a smaller quadrupole. In the limit of very small apertures, the cross-section of the quadrupole is mostly filled with conductor and there is little aperture to transport beam. At some point, decreasing the aperture and increasing the number of beams becomes economically unfeasible.

In addition to the scaling argument, there are the usual constraints on aperture size imposed by beam halo and alignment considerations. Optimization studies usually show that the optimal beam aperture for fusion does not differ greatly from the apertures used at large accelerator facilities such as the LHC [44]. The associated number of beams usually lies in the range of 4–400. Nevertheless, a complete systems analysis taking into account target scaling, final optics, magnet design, and all other considerations is currently beyond the state of the art. In any case, the optimal scale size depends on technology and economics. Improvements in superconductors, vacuum technology, alignment, and fabrication would be expected to reduce the optimal aperture.

In summary, chromatic aberrations impose the most stringent constraint on longitudinal phase space. For our purposes we will use Eq. (5) with $\xi \sim r_t$ to estimate the allowable momentum or energy spread. The allowable energy spread Δ times $\tau_f/2$, where τ_f is the pulse duration allowed by the target, is then a longitudinal admittance that is consistent with our definition of longitudinal emittance given above. This admittance times the available transverse phase space volume $\Omega r_t^2 p^2$ gives a number that is proportional to the available six-dimensional phase space volume. The number of particles needed to drive the target is E_T/T, where E_T is the total energy required by the target, so we can calculate the minimum brightness needed at the end of the machine. We noted above that feasibility requires $B_{6i} \geq B_{6f}$. This is a necessary but not sufficient condition on phase space density. Not only must the density be sufficient, but also the shape of the volume must be correct. Simply stated, the longitudinal and transverse emittance must both have acceptable values.

The needed brightness at the end of the machine $B_{6f} \propto E_T/(\Omega r_t^2 p^2 \Delta \tau_f T)$ has an important scaling property. If we assume hydrodynamic target scaling, E_T scales as $r_t^2 \tau_f$. This means that $B_{6f} \propto 1/(p^2 \Delta T)$ for fixed Ω. For chromatic aberrations, the important quantity is the fractional momentum or energy spread, so the allowable value of Δ increases with increasing T. Also, the allowable range and therefore the momentum and kinetic energy increase with increasing target energy. We conclude that larger targets, of a given type, have less stringent requirements on beam brightness. This makes it difficult to scale to small total beam energy, an unfortunate scaling if one wishes to start at a small scale to minimize buy-in costs.

Given a six-dimensional brightness from an ion source, there is significant flexibility to trade off transverse and longitudinal emittance. For example, in systems with storage rings, one can stack in transverse phase space or longitudinal space. Nevertheless, there is possibly an important constraint. Numerical simulations show that the beam temperatures in the transverse and longitudinal directions tend to equilibrate [45]. (Even if the temperatures are initially equal at the source, they can rapidly diverge if the beam is compressed differently in the different directions.) For beams that are much longer than they are wide (the common case in induction linacs), a higher temperature in the transverse direction will rapidly heat the beam in the longitudinal direction until the longitudinal temperature is some fraction of the transverse temperature. Initially this effect was not included in HIF systems codes but it can be important. Some fusion accelerator designs, which appear attractive if this effect is ignored, fail when this effect is taken into account. Quantitatively the issue is the following: the current produced by a typical ion source of a given radius is usually less than the current that can be transported in a quadrupole lattice of the same radius. The beam must be compressed transversely, and therefore heated, as it is matched into the lattice. We previously showed [Eq. (1)] that the longitudinal emittance can be expressed as $\epsilon_L = \tau(2T\Theta)^{1/2}$. We consider two examples to get a quantitative feel for the importance of this effect. From Table 2, we see that a typical transverse brightness is approximately $1\,\mathrm{A}/(\pi\mathrm{mm}\cdot\mathrm{mr})^2$.

A typical beam current at the beginning of a machine — say, at the 2 MeV point for an induction linac — is 1 A. This gives a normalized emittance of $1\pi\mathrm{mm}\cdot\mathrm{mr}$. According to the Maschke formula [Eq. (2)], the radius of a Bi beam at this energy would be about 2 cm for 5 T at the quadrupole windings. This normalized emittance and radius correspond to an effective transverse beam temperature of approximately 120 eV. According to Ref. 45, about 10% of this temperature will end up in the longitudinal direction. Other simulations find more complete equilibration, but we adopt 10%, giving a longitudinal temperature of 12 eV. At the end of a fusion accelerator, the pulse duration is of the order of 100 ns. From Eq. (1) we obtain $\epsilon_L \approx 0.03\,\mathrm{eV}\cdot\mathrm{s}$ for a 4.5 GeV beam and $\epsilon_L \approx 0.15\,\mathrm{eV}\cdot\mathrm{s}$ for a 100 GeV beam. (We have assumed that the beam radius is approximately the same size at the end of acceleration as it is at the beginning.) Other effects could make the longitudinal emittance larger than these values. There is little experimental evidence, particularly in induction accelerators, regarding transverse–longitudinal equilibration; but, to the extent that the simulations are correct, the longitudinal emittance will not likely be smaller than these values. We will subsequently discuss the consequences of these limits on emittance.

We emphasize that in the discussion in this section we have isolated various effects to estimate their importance. In any real system, these effects must be considered together. For example, phase space effects, space charge effects, and aberrations simultaneously contribute to the final focal spot size and pulse length. Effects that we have not discussed include:

(1) Fields produced by any residual gas or plasma in the target chamber; fields that are nonlinear or depend on time are particularly important, since it is difficult to correct for these effects;
(2) Nonlinear fields produced by the beam itself;
(3) First order chromatic effects;
(4) Multiple scattering;
(5) Stripping in residual gas or plasma or stripping by photons emanating from the target;
(6) Beam–plasma instabilities;
(7) Errors in the applied longitudinal bunching fields;
(8) Misalignment and beam wobble.

These topics are beyond the scope of this article. The proceedings of the heavy ion fusion workshops and symposia from 1976 to 2012 are good references on these topics [46]. Most remain topics of current interest.

4.5. *Techniques to relax constraints*

Since the space charge and phase space constraints have been found to be stringent for the high intensity beams needed for fusion, researchers have searched for techniques to relax the constraints and perhaps to lead to less expensive or more robust accelerator designs. We will give a brief description of these techniques in the remainder of this section. They fall into three main categories:

(1) Correction of aberrations;
(2) Transverse and longitudinal focusing methods that rely on beam neutralization by electrons and/or plasmas;
(3) Non-Liouvillian beam manipulations.

Two main methods of reducing chromatic aberrations have been suggested. The first — and possibly more promising — method involves pulsed focusing lenses. Since a velocity tilt must be applied to the beam to achieve longitudinal focusing, there is initially a correlation between the average momentum of the particles and their position in the bunch. This correlation and some of the velocity spread are removed as the beam compresses. The idea is to use pulsed focusing elements where there is still a strong correlation to "precorrect" for chromatic aberrations in the final lenses. Estimates show that the power requirements for the pulsed elements can, in some cases, be acceptable. Encouraged by these estimates, we performed some additional preliminary calculations to evaluate the promise of this method. Our calculations show that perfect correction is difficult, particularly for large tilts; but even imperfect correction could be important. More work is needed.

The second method is the more standard method, using dipoles and sextupoles. So far, because of the complication of high space charge, it has been difficult to develop a design that appears to work for fusion accelerators.

Historically there has been a lot of interest in beam neutralization. One recent conceptual design of a fusion power plant invoked beam neutralization [15]. Many issues remain and the feasibility of neutralization is controversial. Reference 47 is a recent review article.

Given the difficulty of meeting the phase space requirements for fast ignition, there has recently been some interest in placing small, neutralizing, self-focusing lenses very near the target [48, 49].

At least three types of non-Liouvillian techniques have been proposed for HIF. The first type dates back to about 1975 [50]. It involves stripping during injection into a storage ring. A second method uses positive and negative ions of the same mass. The two beams of opposite charge state are combined using a dipole magnet just upstream of the target [51]. A third method is telescoping. Ion bunches having different masses but the same momentum are transported to the target in the same channel. Since they differ in mass they have different velocities. If the slower bunches initially lead the faster bunches, the bunches will merge longitudinally (telescope) as they approach the target [52]. We are unaware of any detailed simulations of these methods that include all the space charge effects. A brief critique is given in Ref. 22. These methods will be discussed in somewhat more detail in Sec. 6. In any case, additional work on these methods is underway.

5. Choice of Kinetic Energy, Ion Mass, and Charge State

In this section, we will use the information developed in the previous sections to explain the considerations involved in choosing the kinetic energy, ion mass, and charge state of an accelerator for fusion applications. These considerations explain why the program is called heavy ion fusion.

First, consider the Maschke limit, $I_e < 5 \times 10^5 B_b \beta^2 \gamma^2 a$. This limit gives the largest current that can be transported in a single beam in the accelerator. It was derived assuming a matched beam so that the average beam envelope radius is constant. In a focusing system, the fields must more than satisfy the Maschke limit, since it is necessary to drive the envelope radius to a smaller value. This means that the limit is an important consideration not only in the accelerator itself, but also in the focusing system. Combining this limit with the target requirements gives considerable insight into the choice of kinetic energy, ion mass, and charge state.

First, consider only the requirements on beam power and ion range given in Table 1. The beam power is given by $P = I_p T = I_e V$, where I_p is particle current and $V = T/q$ is the total acceleration voltage. For fixed magnetic field and beam size, the power limit $P_{\max} \propto \beta^2\gamma^2 V = (T^2 + 2MT)T/qM^2$. Figure 1 shows that T is approximately proportional to $M^{4/3}$ for a fixed range in the regime of interest. In this regime, the ions are at most mildly relativistic, so we first consider the nonrelativistic limit. With a constant range we find that $P_{\max} \propto M^{5/3}/q$. To maximize the power, we maximize the mass and minimize the charge state. The price we pay is that we also maximize the acceleration voltage. We will subsequently show that, for induction linacs, this strategy also minimizes the cost of the beam transport system. The beam transport system is often a major component of the cost of the accelerator. Unfortunately, so is the system that provides the acceleration voltage — and its cost increases with increasing voltage. From these considerations alone, there is a tradeoff that leads to a minimum cost at some voltage and charge state. This is an important point, because, in our experience, newcomers to the field often immediately begin to invent schemes to use lower mass ions and/or higher charge states. They assume that high charge state and low voltage "obviously" lead to lower cost if one can satisfy the other constraints.

We emphasize again that this constraint on transportable power is ultimately an economic constraint, at least in the accelerator itself. At some cost, one can increase the aperture and/or the number of beams to make a system that works.

Now consider the effects of space charge during the propagation of the beam from the final lenses to the target. The limiting electrical current is given by the inequality (3). As a specific example, we assume 60 beams each with a convergence angle of 10 mr. We further assume a distance of 6 m from the exit of the lenses to the target. These assumptions are consistent with the considerations given in previous sections, particularly the considerations involving allowable solid angle, chamber size, and beam size in the accelerator. Since we are considering the effects of space charge, we assume that the beam is not neutralized so that $f_n = 1$. We also assume that the beam is singly charged. With all these assumptions, we can use the inequality (3) to plot maximum power as a function of kinetic energy and ion mass for any of the targets given in Table 1. We first consider the hybrid target, since it allows a large focal spot radius r_t. The results are shown in Fig. 4(a). The curves shown are for rubidium, cesium, and bismuth. Curves for lighter ions such as sodium and potassium do not lie within the limits of the plot. In each case the curve is plotted only for values of kinetic energy that give an ion range that is less than or equal to the ion range required by the target. In this particular case, it is the range of 4.5 GeV bismuth. The horizontal line at 700 TW is the required power. It is evident that the power requirement exceeds the power than can be focused. Indeed, the hybrid target was designed assuming beam neutralization. One could increase the kinetic energy with a modest increase in the total energy requirement [28]. Increasing the kinetic energy and adjusting some of the other requirements, such as the number of beams and/or the convergence angle, would allow one to meet the power requirement without neutralization. Note that the focusable power decreases as the square of the charge state, so even at 10 GeV it would be extremely difficult to focus beams having $q > 1$. HIBALL, one of the most detailed power plant studies, originally assumed $q = 2$ but finally adopted $q = 1$ because of the focusing limit [13, 14].

We emphasize that this constraint goes away if it is possible to neutralize the beams as they propagate across the chamber. Neutralization introduces a number of physics and engineering complications. The physics complications have been studied for many years [46, 47, 53, 54]. The NDCX-I and NDCX-II facilities at Berkeley Lab were specifically designed to address these issues experimentally. Unfortunately, these machines produce beams that are far from fusion requirements. Moreover, if there is a neutralizing gas or plasma in the chamber, it is difficult to prevent some of it from propagating into the beam ports. If there is too much, the beams can strip in the focusing magnets themselves, leading to beam loss. Because of these complications and the cost and difficulty of addressing neutralization at the fusion scale, some HIF researchers are reluctant to base accelerator designs on assumed neutralization. In any case, if one does not adopt neutralization, we conclude that indirectly driven targets with hot spot ignition require very heavy, singly charged ions at

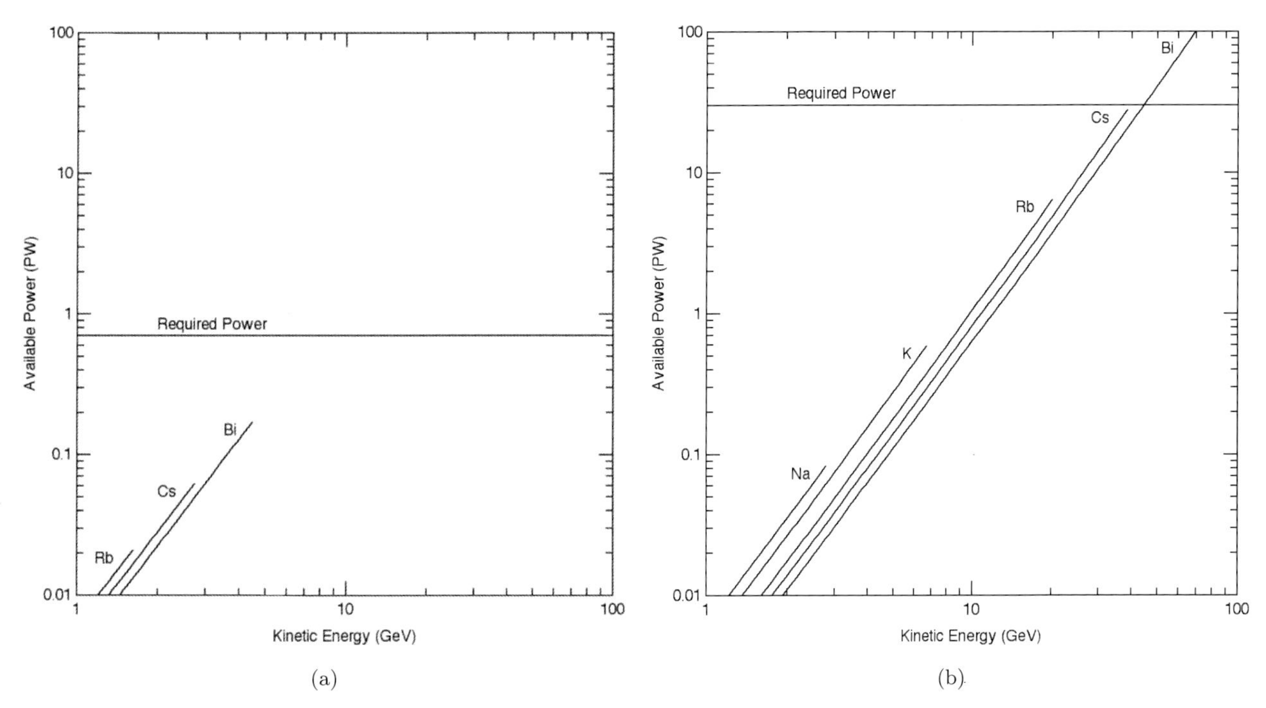

Fig. 4. (a) Power that can be focused onto the Hybrid target as a function of kinetic energy for a variety of ions. These curves assume that the beam is not neutralized as it propagates across the chamber to the target. They also assume 60 beams with an initial convergence angle of 10 mr. The horizontal line is the target's power requirement. The right end of each curve is located at the kinetic energy at which the ion range equals the range specified by the target. In this case neutralization is needed. Figure 4(b) is the same as Fig. 4(a), but it is for the X-Target.

kinetic energies of the order of 10 GeV. Otherwise, the beams cannot be focused.

The focal spot and power requirements are very different for fast ignition. Figure 4(b) shows the results for the X-Target from Table 1. In this case, results from five ions (Na, K, Rb, Cs, and Bi, from left to right) are shown. Because the range allowed by the X-Target is much larger than the range allowed by the indirectly driven targets, the upper limit on kinetic energy now extends to nearly 100 GeV. The power constraint of 30 PW can be satisfied at kinetic energies greater than about 40 GeV, again using the heaviest ions. An issue arises if one would like to employ neutralized focusing to eliminate this constraint. At 30 PW, the ion density at the focal point for a 40 GeV beam is of the order of $10^{17}\,\text{cm}^{-3}$. The stripping cross-sections [37, 55] are of the order of $10^{-16}\,\text{cm}^2$. The plasma density must be at least as high as the beam density to provide effective neutralization. These numbers imply a stripping length of roughly 1 mm, a very small number compared to the size of the chamber. Stripping would drive the required chamber pressure to even higher densities. The mix of charge states arising from stripping makes it difficult to focus the beam, because neutralization is not expected to be perfect. Of course, the beam density is much lower some distance from the target, so there might be some way to obtain some neutralization without excessively high plasma densities. This issue requires more work. While one can adopt different assumptions about things such as beam number, we conclude that fast ignition will require heavy, singly charged ions at kinetic energies of several times 10 GeV based on these focusing considerations alone.

Now consider phase space constraints. In the examples that we have considered above, the longitudinal phase space constraint appears to be the most stringent. In principle, the fundamental constraint is on six-dimensional phase space density; but, as previously explained, it may be difficult to make the tradeoff among the transverse dimensions and the longitudinal dimensions. For example, if the difficulty is in the longitudinal direction, one could, in principle, decrease the initial pulse duration, and therefore the longitudinal emittance by using more current per beam. This would lead to higher transverse emittance, but we are assuming that the transverse emittance is less constraining. Limitations on pulse duration, beam size, beam number (economics), and the fact that brightness decreases with increasing current, limit these options. To get a feel for the orders of magnitude involved, we will consider only the longitudinal constraint.

Consider first the Hybrid target using the assumptions we have previously made. These assumptions are: a longitudinal beam temperature of at least 12 eV, a pulse duration emerging from the accelerator of 100 ns, a standoff of 6 m, and a convergence angle of 10 mr. Equation (1) gives a calculated minimum longitudinal beam emittance of 0.03 eV · s at 4.5 GeV, as we have previously shown. Because of the large focal spot (4.5 mm) allowed by the Hybrid target, the allowable energy spread is also large. Equation (5) gives an allowable fractional momentum spread of 0.015. For nonrelativistic particles, the kinetic energy spread is given by $\Delta = 2T\delta$. This leads to an allowable energy spread of approximately 1.4×10^8 eV. We estimate the final pulse duration by dividing the required target energy by the peak power (Table 1). This yields a pulse duration of approximately 10 ns. If we multiply one half the pulse duration by the allowed energy spread, we obtain 0.7 eV · s, an estimate of the allowable longitudinal emittance (admittance). Since 0.7 eV · s $\gg$ 0.03 eV · s, it is evident that there is a safety factor. The factor is smaller for the other indirectly driven targets listed in Table 1. The reader can also easily show that the constraint becomes more stringent with decreasing kinetic energy.

Now consider the X-Target. With 20 GeV rubidium, the estimated minimum longitudinal emittance is approximately 0.07 eV · s. Because of the small focal spot radius (0.26 mm), the allowable fractional momentum spread is also small, leading to an allowable kinetic energy spread of approximately 3.5×10^7 eV. In this case, we estimate the pulse duration by dividing 3 MJ by the peak power, since only 3 MJ rather than the full 5 MJ is needed for ignition. We obtain 100 ps, leading to a longitudinal admittance of approximately 0.002 eV · s. The needed inequality now has the wrong sense and the constraint is not satisfied. If one used bismuth beams to drive the target, the allowable kinetic energy would be 70–80 GeV, since the range in hot fuel is about the same as the range of 20 GeV rubidium. The inequality would more nearly be satisfied.

These emittance considerations, although very approximate, reinforce our conclusions about kinetic energy, ion mass, and charge state. If we temporarily ignore the possibility of chromatic correction, non-Liouvillian techniques, and secondary focusing, most of the indirectly driven targets require heavy, singly charged ions in the GeV range. The fast ignition targets require heavy, singly charged ions at a kinetic energy of the order of 100 GeV and still appear to have trouble with the emittance constraint. The known types of accelerators that might be applicable to fusion are sufficiently expensive at 100 GeV that, barring an invention, fast ignition also appears to have an issue with economics. The use of lighter ions at lower kinetic energy for either type of target [56] seems difficult.

Are there ways to circumvent these conclusions? We have already discussed some possibilities. Chromatic correction, small secondary lenses near the target, and the use of many smaller beams have been suggested. In our opinion, more work is needed on chromatic correction. Chromatic correction has the potential to help for all classes of targets. Perhaps more importantly, our estimates on longitudinal emittance may be optimistic. But it also seems likely that continuing progress in superconductors and manufacturing techniques will gradually enable the use of more beams.

There are important issues with secondary lenses. If one tries to use these to decrease the focal spot by a large factor, one rapidly runs into constraints on transverse emittance [22]. Moreover, according to our calculations, secondary self-focusing lenses are unlikely to have the linearity needed for a good focus.

In summary, the constraints imposed by space charge and emittance are stringent. Both considerations push toward high ion mass, low charge state, and kinetic energies in the several-GeV range. Finally, we note that we have not explicitly accounted for emittance growth in the accelerator from contributions such as transverse envelope mismatch or longitudinal space charge waves. Emittance growth to some extent is inevitable, and leads to tighter constraints.

6. RF and Induction Accelerators for HEDP and Fusion

6.1. *RF accelerators*

As mentioned previously, the energy crisis of the 1970s motivated a significant effort to use accelerators to drive IFE targets. RF accelerators were a relatively mature technology. The attitude was: We know how to build them. We know how much they cost. We know that they can produce megajoules of energy; and we know that they can do it reliably for decades. Let us solve the energy problem and save the world.

Some of the constraints were also known. It appeared that if one used heavy ions, the available brightness was orders of magnitude larger than that needed by the targets. Al Maschke himself stated that he would not be willing to work on the problem if there were not initially a safety factor of a thousand or more, because that factor would inevitably decrease as we learned more. He was prescient — as we showed in the previous section. A better understanding of target physics and of issues associated with beam physics and accelerator engineering has led to a situation where the constraints are now known to be stringent. Nevertheless, the original arguments based on the success and maturity of RF technology remain strong arguments for continuing to study the RF approach. In fact, RF facilities (primarily for other applications) are currently being used to study WDMP [57]. Planned facilities will have even greater capability [58].

In the first years of the HIF program there were a number of studies of RF systems for power production. It initially appeared that one might build a suitable synchrotron for less that US$$10^8$. Interest in synchrotrons eventually waned, because it appeared difficult to meet the requirements on efficiency and pulse repetition rates. Moreover, issues such as ionization and charge exchange among the beam ions themselves led to significant particle loss for heavy ions. Perhaps more importantly, the target physicists wanted a smaller ion range (or kinetic energy) than Maschke originally assumed. Consequently, inexpensive volts, one of the chief advantages of synchrotrons, became less important. The synchrotron systems were superseded by RF linacs and storage rings. The HIBALL studies are excellent examples of early power plant studies based on the RF linac and storage ring approach [13, 14].

To minimize cost, one would like to minimize linac beam power. This situation is particularly true for an IFE target test facility or an HEDP facility, because there are no requirements on pulse repetition rate or efficiency. For example, suppose that one would be happy with a 10 GeV, 5 MJ pulse every 50 s. The linac would have to deliver 100 kW of beam power at an average current of 10 μA. Unfortunately, the inevitable beam loss from beam–beam ionization and charge exchange requires much shorter storage times and much higher current for a stored energy of this magnitude. Moreover, multiturn injection in the presence of large space charge forces can lead to unacceptable emittance growth [59]. Efficiency, beam loss, and emittance growth during injection remain important constraints on the design of RF systems. Most designs now assume a linac current of the order of 1 A.

Although the earlier designs of RF systems were excellent vehicles for elucidating the issues and providing cost estimates, they are now obsolete. Target designs have evolved and we have a better understanding of issues such as focusing and phase space requirements. There are at least three more recent studies that illustrate the state of the art and explain current thinking about RF systems. They are:

(1) The European study on Heavy Ion Driven Inertial Fusion (HIDIF) [36];
(2) Koshkarev's Charge-Symmetric Proposal [60];
(3) The Fusion Power Corporation's Single Pass RF Driver (SPRFD) [52].

The European Study Group on Heavy Ion Driven Inertial Fusion performed the HIDIF study during 1996–1998. It is the most detailed study now available. Koshkarev's charge-symmetric design uses positive and negative ions of the same mass. A magnetic dipole is used to merge the two charge states into a single channel just upstream of the target. According to recent oral presentations by the Russian HIF team, this proposal and a fast ignition target design by Basko *et al.* [61] constitute the conceptual basis of the Russian HIF program. The Fusion Power Corporation is a relatively new private company founded to develop the RF approach to IFE. The SPRFD is described in Ref. 52.

It is beyond the scope of this review to describe those three systems in detail. We will consider only some general features of each approach and list the issues that must, in our opinion, be resolved to determine the feasibility of these approaches.

In Sec. 5, we showed that the brightness requirements are stringent. RF systems typically exacerbate this problem because they depend on a number of beam manipulations, such as funnelling in linacs, multiturn injection into multiple storage rings, delays to synchronize multiple bunches, beam extraction from the rings, and longitudinal pulse compression. These manipulations usually lead to a significant loss of brightness [36, 62]. For this reason, all three studies rely on non-Liouvillian schemes to meet the brightness constraints.

There are two HIDIF designs. One is designed to deliver 3 MJ of energy; the other, 4.5 MJ. The main difference between the two designs is the number of storage rings and the number of beams. The first design, shown in Fig. 5, has 12 storage rings and 48 final beam lines. The second has 18 rings and 72 final beam lines. Both systems are based on a 10 GeV linac having a peak current of 400 mA. For the larger system, the peak power is 1100 TW and the focal spot radius is 1.7 mm. These numbers are similar to those for indirect drive given in Table 1, but they are for a specific target design given in Ref. 36. Telescoping is the non-Liouvillian technique chosen for this system. It employs three ions (^{187}Re, ^{209}Bi, and ^{232}Th) that differ by about 10% in mass. There are 48 ion sources, 16 for each isotope. Each set of 16 is merged (2-to-1) four times to provide the input to the main linac. The initial frequency at low energy is 12.5 MHz. After four mergings, the frequency of the main linac is 200 MHz. The driver is designed to have a pulse repetition rate of up to 50 Hz. At this repetition rate, the calculated efficiency of the system is 20%. The efficiency decreases with a decreasing repetition rate, so this type of system is primarily appropriate for large power plants (several GW).

At the end of the machine the three isotopes have the same momentum (so the magnetic focusing forces are equal) but different velocities. The lighter ions

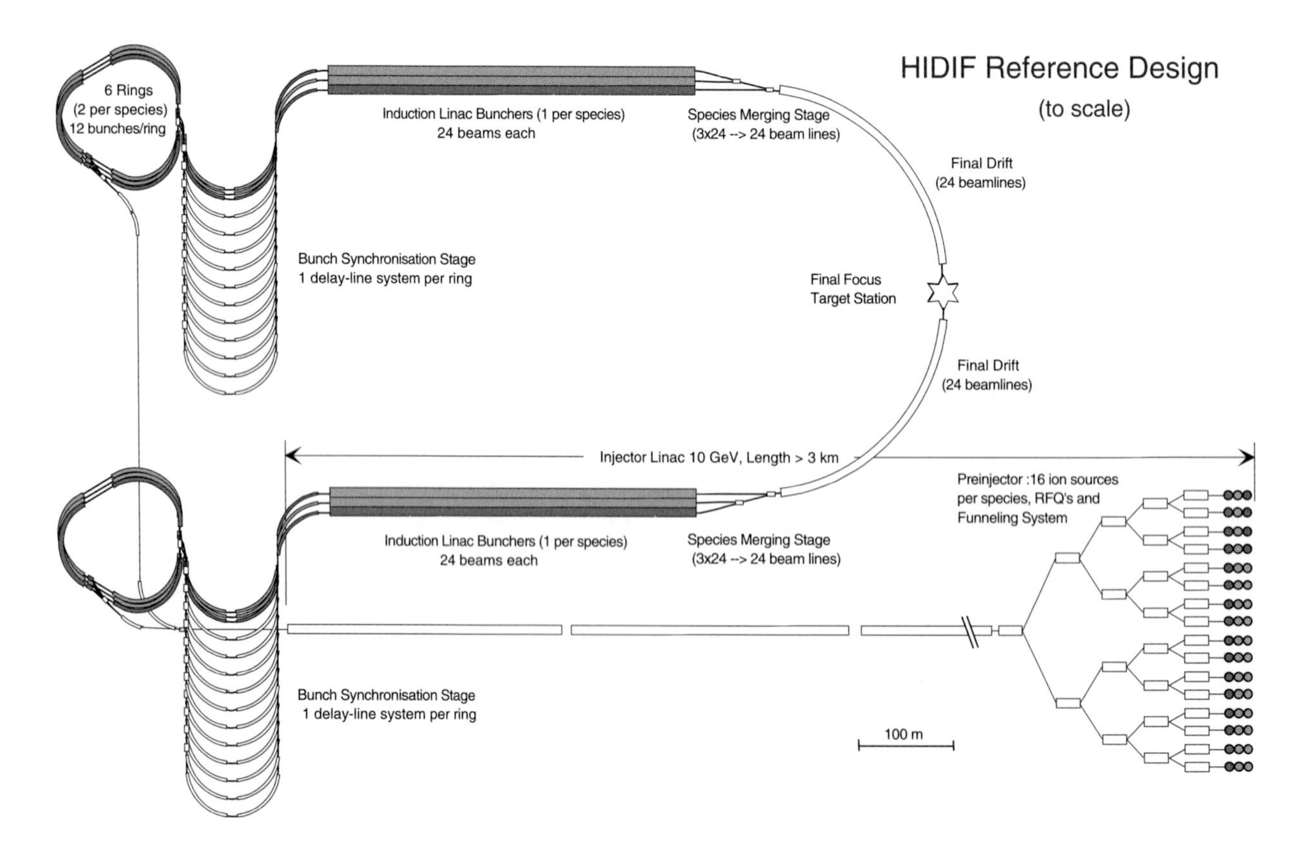

Fig. 5. The RF accelerator-based HIDIF driver design [36] begins with 48 ion sources and low frequency RF sections funnelling into a main linac. At the end of the accelerator the beam is fed to multiple storage rings, followed by an induction linac buncher.

overtake the heavier ions (telescope), and thus they all arrive simultaneously at the target.

The charge-symmetric driver is designed to provide the ignition pulse to a fast ignition target. The required focal spot radius is $50\,\mu$m and the pulse duration is 200 ps. The energy in the ignition pulse is 400 kJ, leading to a peak power of 2000 TW. The design employs both the charge-symmetric scheme mentioned above and telescoping with four different isotopes or eight types of ions (platinum 192, 194, 196, and 198, each in charge state +1 and −1). Each ion is produced by four ion sources that are merged to produce one beam. The final kinetic energy is 100 GeV. The first 25 GeV is produced with conventional RF linacs. The final 75 GeV is produced with a superconducting RF linac having a frequency of ≈1 GHz and an assumed gradient of up to ≈25 MV/m. As is the case for HIDIF, this design uses storage ring compression to achieve the required current amplification. The target assumes a single ignition beam, so the eight ion species are merged into a single telescoping beam at the end.

The fractional momentum spread at the target is given as 3×10^{-3} and the transverse emittance after merging the four beams from the ion sources is given as $\epsilon_N = 0.9\,\pi\text{mm}\cdot\text{mr}$. This corresponds to $\epsilon = 0.76\,\pi\text{mm}\cdot\text{mr}$ for 100 GeV platinum. Thus, the convergence angle to focus the beam to the required 0.05 mm must be greater than 15 mr.

If we assume our typical 6 m standoff from the target, the chromatic aberrations greatly exceed the required focal spot size. We conclude that there must be an implicit assumption of chromatic correction. There must also be an implicit assumption that the third order aberrations have been corrected [63].

This entire charge-symmetric ignition system is large. It also requires a companion system to provide several megajoules of energy for target compression. For these reasons this system also seems to be appropriate only for large power plants.

The SPRFD is, in many ways similar to the systems described above. The chief difference is that it eschews the use of storage rings in order to eliminate the brightness reduction associated with injection and extraction. This is the origin of the term "single pass." The current SPRFD design uses 16 telescoping isotopes at energies as high as 20 GeV to try to meet the brightness requirements. The proposed target uses fast ignition and has a focal spot requirement approximately the same as that of the target for the charge-symmetric driver. The proposed target has not been numerically simulated, so the uncertainties are large. While it is true that the elimination of storage rings leads to higher brightness, 20 GeV leads to higher brightness requirements than the 100 GeV used for the charge-symmetric driver. In this SPRFD case, we are unable to make the numbers consistent with the target focusing requirements without making aggressive assumptions about beam neutralization and correction of aberrations.

There are some generic issues for the RF systems described above. Unless the beams are neutralized, the beams will develop bulges owing to space charge as they begin to overlap. This means that not all parts of the beam can be matched to their proper radius as they propagate toward the target. Mismatches will undoubtedly lead to larger focal spots. We have not seen proper calculations of this effect, although simulations to address the issue are underway. This issue is one that leads to our concerns about telescoping (mentioned earlier in the article).

Finally, there is the issue of cost. Drift tube linacs are well matched to HIF applications, which are low-to-medium-β RF accelerators. The particle transit time across the RF gaps is significant at these energies, and especially at injection into the linac. Furthermore, the FODO period is short at low β, and thus space will be required for focusing elements. These considerations, which set the scale of the cavity geometry along with the cost and availability of RF power sources, strongly influence the frequency choice of ~0.2 GHz. The main options for supplying the RF power are gridded tube-based amplifiers and klystrons.

For an RF-based HIF driver, the required peak power, average power, RF pulse length, and repetition rate are important features of the overall RF requirements. Table 3 illustrates the requirements for the RF approach via the HIDIF and SPRFD

Table 3. HIDIF and SPRFD main linac parameters are compared. $\hat{P}_{\text{beam}}$ is the peak power, and $\bar{P}_{\text{beam}}$ is the average beam power.

	$\hat{P}_{\text{beam}}$ (GW)	$\bar{P}_{\text{beam}}$ (MW)	τ_{RF} (ms)	Rate (Hz)	ν (MHz)
HIDIF	2	150	1.6	50	200
SPRFD	100	200	0.2	10	200

examples. The HIDIF design of the injector and the main linac are the basis for the SPRFD design. For the RF power requirements, the two approaches differ, mainly due to the use of storage rings to accumulate charge after acceleration in the HIDIF. Thus, the RF pulse length is 1.6 ms in HIDIF and only 0.2 ms in the "single-pass" SPRFD.

The RF circuit is heavily beam-loaded, so the beam power is indicative of the supplied RF power. The tabulated values make clear that RF amplifiers in the MW (peak) output range are of interest.

For a driver with a main drift tube linac responsible for most of the acceleration, and operating around 200 MHz, the Muon Ionized Cooling Experiment (MICE) is an illustrative and relevant RF amplifier system [64]. Operating at 201 MHz, it is based on the high power Thales TH116 triode [65]. The MICE RF system will deliver 2 MW peak power with a pulse width of 1 ms. Based on Ref. 64, we extrapolate that the approximate RF amplifier costs for a similar system applied to an HIF driver is approximately US$0.1/watt (peak).

A relevant example of a klystron-powered accelerator is the Spallation Neutron Source (SNS), which utilizes 2.5 MW klystrons at 402.5 MHz to accelerate the protons up to 86.8 MeV [66, 67]. For acceleration up to the top energy of 1 GeV in the coupled cavity linac and then the superconducting RF part of the accelerator, the klystron frequency is 805 MHz. The 805 MHz klystrons are 5 MW and 0.55 MW units. High peak power klystrons at a frequency of 200 MHz are less common, but feasible in principle, and would necessarily be larger than their higher frequency counterparts.The total installed peak power at the SNS is 0.12 GW, with an average cost of about US$0.14/W.

To appreciate the importance of the cost of RF power, consider the two values of peak power listed in Table 3. Assume 25% efficiency in converting RF power to beam power. The HIDIF peak beam power of 2 GW corresponds to a peak RF power of 8 GW or US$8 $\times$ 10^8 at US$0.1/W. The SPRFD peak power of 100 GW corresponds to US$4 $\times$ 10^{10}. The SPRFD group uses lower cost numbers [52, 68] than we find for amplifier systems capable of providing $\geq$0.2 ms RF pulses near 200 MHz. Presumably, in a fusion economy, there would be significant cost reductions because of economy of scale. The difficulty seems to be the cost of a test machine.

6.2. *Induction linacs*

Nicholas Christofilos, the inventor of strong focusing for accelerators, invented the first US induction linac. It accelerated 300 ns, 400 A constant current electron pulses to 4 MeV at up to 60 times per second [69]. It is historically interesting that this early induction linac was built specifically for fusion. Its purpose was to inject electrons into the Astron fusion device, another Christofilos invention. Astron was intended to confine a fusion plasma with a rotating cylindrical sheet of electrons inside a long solenoid. But the fact that beam current was far above what an RF linac could provide was the feature that caught the attention of Denis Keefe and others regarding inertial fusion. Since the Astron days, many induction linacs have been built worldwide. The book *Induction Accelerators* [70] is a recent general reference.

Induction linacs for fusion usually differ from induction linacs for other applications in at least one major way, namely the use of multiple beams. Nevertheless, induction linacs are still conceptually simpler than the RF systems. In many designs, there is a single ion source per beam at the beginning of the accelerator (counting multiaperture sources as single sources). Each individual beam usually retains its identity all the way to the target. This simplicity means that induction machines usually have significantly fewer beam manipulations than the RF systems — leading to less brightness degradation. One could choose to merge beams or split beams at some point in the machine, but these manipulations do not appear to be necessary. In summary, induction linacs, in principle, appear to have a significant advantage over the RF systems in terms of preserving beam brightness. One principal disadvantage is that induction technology is still less developed than RF technology.

A driver layout is shown in Fig. 6. Operating at 5–15 Hz, ~100 parallel ion beams are injected into an induction accelerator. The accelerator may use electrostatic focusing quadrupoles at the front end, followed by a transition to superconducting magnetic quadrupoles for most (>90%) of the accelerator.

After acceleration, the bunch length is compressed by an order of magnitude or more and shaped as needed to meet the bunch duration required by the target design. A part of this compression section includes dipoles for each beamline to aim each beam

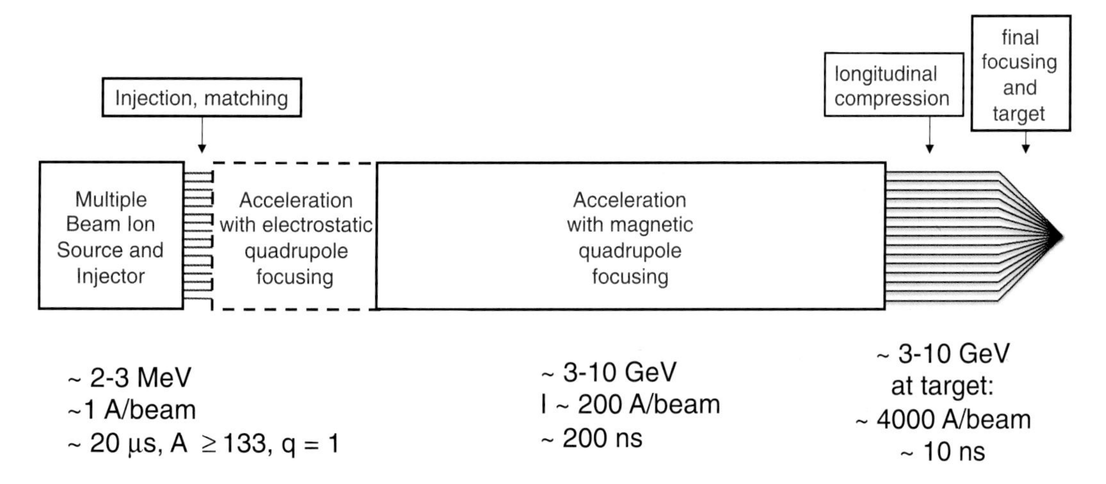

Fig. 6. Schematic of an induction accelerator driver for heavy ion fusion.

at the target according to the required illumination geometry.

Some RF linacs accelerate a short pulse, such as the original SLAC accelerator with a $1.6\,\mu s$ pulse. In such linacs it is customary to use a pulsed modulator to provide the power to a klystron, where it is converted to high frequency RF.

An induction linac uses the modulator pulse to drive the beam directly, bypassing the RF stage. With present technology, an induction cell is quite lossy, requiring hundreds to thousands of amperes to generate the acceleration fields, so it is efficient only for high current beams.

A feature of an induction linac for nonrelativistic beams is the ability to adjust the beam current during acceleration by controlling the bunch length with velocity ramps to match the beam current to the transportable current (an increasing function of energy) and by adjusting the quadrupole occupancy fraction. The typical design starts with ~90% quadrupole occupancy at the low energy and is only 5–10% at the end. These quadrupole lenses are typically less than 1 m in length and the lattice half-period increases from less than 1 m to ~10 m at the exit. Many transverse focusing options exist for the region right after injection, and a few are possible for all parts of the machine, but the magnetic quadrupole system is quite attractive.

While the ability to accelerate many kiloampere beams at durations ~20 ns, as required by the targets, was the initial motivation for considering induction linacs, the accelerator designs evolved toward designs where the bunch duration is in the 100–200 ns range within the accelerator. The final factor of ~10 (~1000 for fast ignition) shortening of the bunch and increase of the peak power comes from a ballistic compression of the bunch in the drift lines going to the target chamber. This last operation has the important property of the space charge removing the bunch tilt that causes the compression, and allowing the highest power portion of the bunch to be nearly constant energy and current.

Consider the question of multiple beams. Figure 7 shows a 60-beam array. Based on experience with common accelerators with one or two apertures, this system appears complex. Let us consider the situation quantitatively. We previously showed that quadrupole magnets do not scale well to very small apertures. If we avoid apertures that are not too small, we can derive three important results. The first follows from Eq. (2), the Maschke limit. The transportable beam current is linear in beam size (assumed proportional to magnet size in our derivation of the Maschke limit). On the other hand, the number of beams that fit into a core of a given size goes inversely as the square of the magnet size. These scaling relationships lead to the remarkable, and perhaps unexpected, result that induction core size decreases (for fixed beam current and magnetic field) with increasing beam number — as long as the magnets are not too small.

The second result is also interesting. As previously explained, Ampere's law means that, for a fixed magnetic field, the current in the conductor is

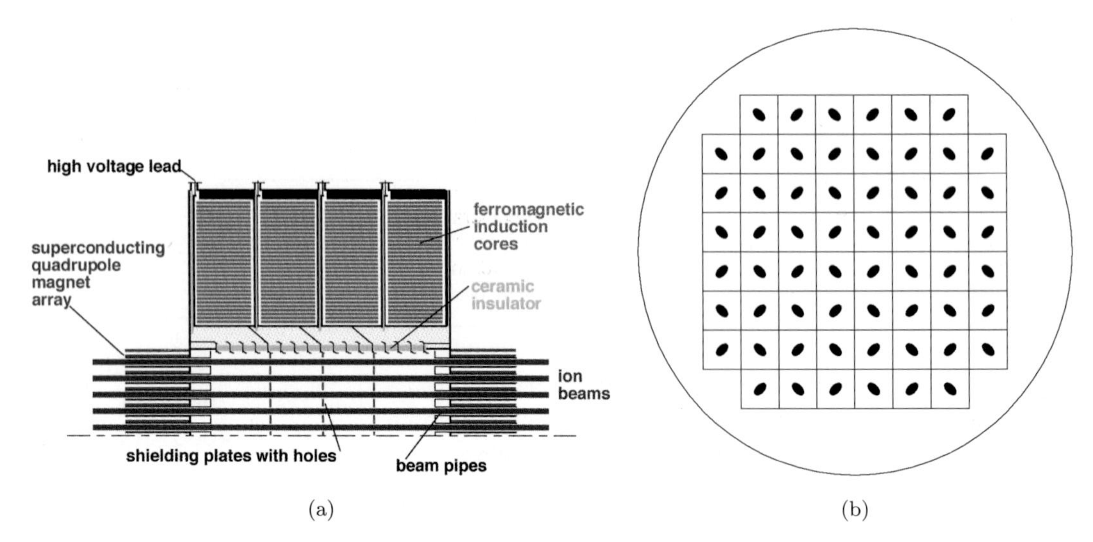

Fig. 7. (a) Schematic of one-half of a lattice period. (b) Schematic cross-section of a 60-beam quadrupole array for an induction linac. The large outer circle represents the inside of a circular induction core. Each small square represents a cell containing a quadrupole. The small filled ellipses represent the beams. Note the alternating polarity of the quadrupoles. This arrangement maximizes magnetic flux sharing among the quadrupoles and minimizes the needed conductor. It is helpful to think of the entire array as a single, multiaperture focusing element.

proportional to the magnet size — as is beam current. In this approximation, the amount of needed conductor, for fixed beam current and beam velocity, is independent of the number of beams. For these reasons, it is perhaps more reasonable to think of a quadrupole array as a single, multiaperture focusing element rather than an array of independent beams. Indeed, by placing the individual bores close together, there is significant flux sharing among the neighboring quadrupoles, leading to a reduction in needed conductor (by a factor of 2 in the limit of a large number of closely spaced beams). This proportionality between beam current and conductor current means that the challenge for these multiaperture focusing limits is to develop automated fabrication techniques so that the cost of fabrication is not large compared to the cost of conductor. (For a power plant, superconductor is necessary to get the required efficiency.)

The third result is also an important scaling law. For fixed beam power on target, the required beam current is inversely proportional to the kinetic energy. But because $\beta^2\gamma^2$ in the Maschke limit is larger at high kinetic energy, the amount of superconductor needed to deliver a given power decreases with increasing kinetic energy. This means that decreasing kinetic energy does not necessarily reduce cost.

The results that we have just discussed are approximate. The overall problem of optimization requires a systems code. LIACEP [71], the earliest systems code for HIF induction linacs, was used to address the multitude of design choices to find minimum cost solutions for HIF accelerators. Because there are numerous possible solutions for injectors and transport at the very lowest energies, the program started after the low energy region of 50 MeV or so. Similarly, there are a multitude of special solutions for delivering the beams around the chamber, and they were also omitted. LIACEP varied beam current, numbers of beams, acceleration rates, aperture sizes, clearances, focusing half period lengths, field strengths, and other similar variables. Concentrating only on the main body of the accelerator, it was found that changing the charge state or the ion mass made little difference. The cost was most closely related to the total beam energy. There was a decrease in the optimum number of beams as energy increased, but this did not include the need for a large number of beams to facilitate focusing at the end, so a choice of a number of beams of the order of 100 may be the ultimate optimum.

Although beam resonances and instabilities have been studied extensively for RF machines, there has been less work and less experience with ion induction linacs.

The beam breakup instability is a well-known instability for electron induction linacs. This instability is not a problem for HIF, because of strong focusing and high rigidity beams [72].

The longitudinal dynamics are more interesting and potentially more dangerous. To illustrate a mechanism that could increase the longitudinal phase space occupied by the beam beyond the estimates in Sec. 4, we consider the growth and decay of space charge wave perturbations. Throughout the acceleration process relativistic effects are small. High space charge intensity changes most of the familiar single particle dynamics of other accelerators to motion of space charge waves. These waves, i.e. the Hahn–Ramo space charge waves [73], have been known since 1939 and are basic to operation of microwave tubes such as the klystron. A perturbation on the beam caused by a voltage error initiates waves traveling away from the perturbation in both directions. By intentionally introducing a short voltage "error", the two waves separate, and current perturbations became evident in some of our early heavy ion beam experiments (Fig. 8). In a longer machine, these bunches would hit the ends of the beams. In an ion induction linac, the beam bunch is confined longitudinally by applying focusing voltage "ears" at the bunch ends. The perturbations reflect at the ends with the current perturbations maintaining their sign and the voltage perturbations reversing their sign. The wave going toward the front of the bunch is called a fast wave, and the wave going toward the rear is a slow wave; upon reflection the fast wave becomes a slow wave and the slow wave becomes a fast wave. (In a klystron, with a sinusoidal voltage perturbation, the current variations can drive a cavity to extract power.) In an ion accelerator, there will be random errors from each pulser and accelerator gap, and a large number of such pulses reflecting between the ends. In the early 1950s, Lan Jen Chu of MIT added to the understanding of these waves by theoretically finding that a resistive impedance along the beam tube makes the slow wave grow and the fast wave decay [74, 75]. These too reverse their roles upon reflection, so any perturbation grows for a while, reflects, attenuates for a while, and so on, without continually growing. This growth of the perturbations is dependent on the impedance seen at the accelerating gaps. Thus, is important to have low effective gap impedance at the relevant frequencies.

While over the body of the bunch, which is nominally constant amplitude current, the applied voltages are approximately flat at each gap, this

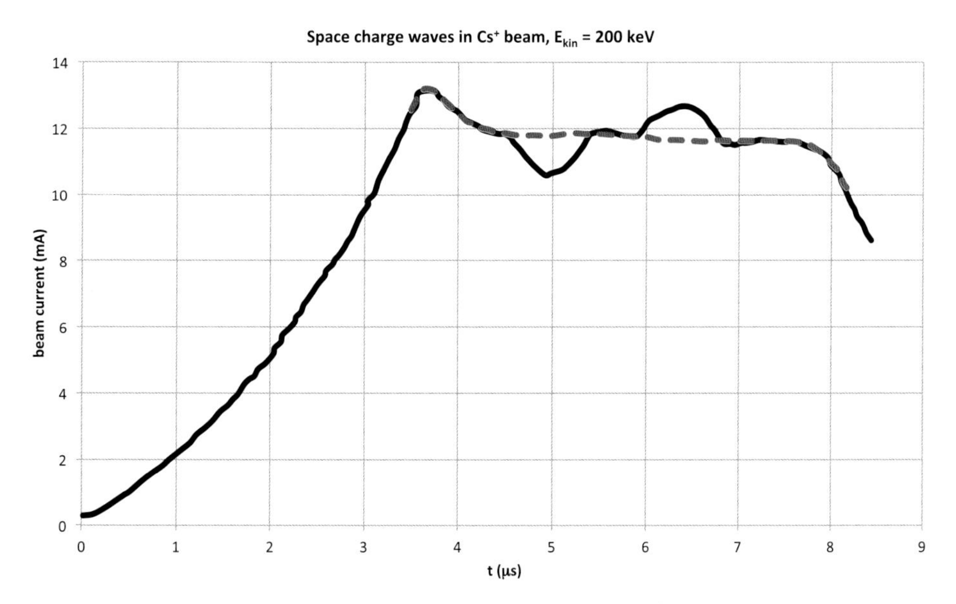

Fig. 8. Space charge waves observed in the Single Beam Transport Experiment [36] illustrate a feature of high current heavy ion beams. A short, 2 kV perturbation pulse was applied to the beam, and the resulting slow and fast space charge waves were observed 24 lattice periods downstream (7.3 m) in a Faraday cup measurement of the beam current. A Faraday cup wave form (dashed red line) in the absence of the applied perturbation is superimposed. The few-microsecond rise time of the pulse is mostly due to the longitudinal space charge of the beam, significant mostly at each end of the beam, in the absence of a confining voltage pulse.

is not true at the two ends where the longitudinal defocusing forces are located. Every few meters, ear pulsers are provided which reverse the accumulated space charge and emittance forces. This requires time variation during the pulse on the two bunch ends.

This is a somewhat noisy process, which can be affected by how closely one chooses to apply the corrections. The main acceleration modules are physically large and require tens of nanoseconds of fill time, so the end correctors are smaller modules with faster fill times. A 50 ns half sine is a reasonable approximation to an acceptable wave form, to which the many beams in the cluster would adjust their longitudinal profiles to get ~25 ns bunch ends.

Similarly, eliminating or reducing the initiating errors is important. For example, the limitations on longitudinal compression due to induction module wave form errors are studied in Ref. 76. The wave form errors were the dominant contribution, while the contribution from the injected beam was negligible. Several options to reduce effective impedance exist. They include: nonlinear stabilizing resistors, controlled switches able to use capacitive energy storage, and active amplifiers for voltage stabilization. Since the beams are slow compared to the speed of light, measured perturbations can be corrected downstream. This option is referred to as "feedforward."

For near-term progress and for the low currents of heavy ions transportable at low energies, available switches are probably adequate. In the longer term, better switches are needed. The options include: high power vacuum tubes, improved thyratrons, solid state switches, and magnetic modulators. The pulsed power and KrF laser programs have made significant progress in developing solid state switches with magnetic pulse compression, so there is some synergism with other fusion programs.

Several of the major questions for induction-accelerator-driven HIF have been settled by a few full scale and many small scale experiments. The first major question about whether suitable sources could be made was settled in the 1970s with two long-pulse 1-A Cs sources [77] and by one test on a Hibachi source for Xe at full current suitable for an induction linac and one small multiaperture Xe source suitable for an RF linac [78]. The next major experiment was an 87-electrostatic-quadrupole transport experiment [79]. A third experiment was MBE-4, which accelerated and transported four parallel beams [80], and later was used for a four-beam merging experiment [81]. A fourth significant experiment was the scaled final focusing experiment, which showed that the equations used to focus a driver beam, when scaled down correctly, led to the desired focal spot.

The newest induction linac built by the HIF group is NDCX-II at Berkeley Lab. Since it was built recently, its primary purpose is WDMP [7, 82]. Nevertheless, this machine will benefit fusion research. A discussion on NDCX-II, NDCX-I, WDMP, HEDP, and their relevance to fusion and to basic science issues is beyond the scope of this article. See Ref. 23 and references therein for details. For other important research areas for HIF, see Refs. 8 and 83.

In summary, induction linacs are attractive because they have few beam manipulations that reduce beam brightness. On the other hand, control of longitudinal dynamics and switch development are important engineering challenges. If these challenges can be met, it seems likely that induction machines will be able to satisfy the beam brightness requirements — at least for targets that do not rely on fast ignition.

6.3. *Issues common to RF and induction accelerators*

There are some issues that are shared by the RF and induction approaches. Cost reduction remains important for both approaches, but accelerators are not alone. Cost reduction is a major issue for all credible fusion options. In addition, the two approaches share a number of issues with many high intensity accelerators for other applications. These include the effects of electrons and gas produced by a beam halo striking the walls of the vacuum chamber.

The accelerator repetition rate is in the range of 10–50 Hz. Increasing the repetition rate in the reactor is limited by gas condensation times and other relaxation times in the chamber. An ongoing question for operation of the accelerator is whether the pressure inside the beam lines, which is expected to rise immediately after a beam pulse, can be lowered fast enough for the next pulse or even be damaging to the pulse itself. The low energy range (<100 MeV) is of particular interest for HIF because the cross-sections for ionization of background gas, ion stripping by background gas, and production of secondary electrons are all near their maxima or are relatively large at an ion kinetic energy of a few MeV. A thorough overview of the cross-sections is in Ref. 55.

The accumulation of electrons (electron clouds, or e-clouds) in and between the magnetic quadrupoles of the accelerator may contribute to undesirable beam parameters and emittance growth due to distortions of the space charge distribution [84]. In contrast to high energy rings, where e-clouds are also a concern [85], HIF drivers have relatively long beam bunches with a very low repetition rate. The beam potential may be a few kilovolts. At low energy, where the pulse is the longest ($\sim$10 μs in induction linacs), electrostatic quads may be used until the bunch length is only a few microseconds. The quadrupole electric fields greatly reduce the accumulation of electrons. However, for most of the accelerator, the beam bunch duration is $\sim$0.2 μs and magnetic quadrupoles are the proper choice. Strategies to mitigate the electron cloud effects for the RF linac approach may differ somewhat [86], since the bunch train is longer — 0.2 ms or 1.6 ms (Table 3). In any case, the wall-to-beam-edge clearance may be increased so that the beam is less likely to intercept the inward-moving dense cloud of halo-induced atoms, with some increase in the accelerator cost.

7. Conclusions

The NIF is expected to resolve important issues about the feasibility of inertial fusion in the near future. If these issues are resolved in the affirmative, the National Research Council recommends an expanded program in inertial fusion power production, including heavy ion fusion. For this reason, this review article is timely.

We have shown how considerations of commercial energy production, target physics, beam brightness, and space charge lead to important constraints on accelerators. The most stringent constraints are associated with focusing the beams to the small focal spots required by the targets. Chromatic aberrations and other aberrations in the final focusing system are particular concerns.

Additional research is required to better understand the mechanisms leading to emittance growth in accelerators, particularly longitudinal emittance. Additional research is also required to better understand the aberrations and all other mechanisms that determine the final focal spot size. Development of methods to minimize emittance growth and development of improved focusing systems are high priority activities. Development of targets allowing larger focal spots would also be beneficial. Finally, research aimed at cost reduction is critically important.

References

[1] National Research Council, *Review of the Department of Energy's Inertial Confinement Fusion Program* (National Academies Press, 1986).

[2] National Research Council, *Review of the Department of Energy's Inertial Confinement Fusion Program* (National Academies Press, 1990).

[3] Fusion Policy Advisory Committee report, G. Stever, Chair (1990); http://fire.pppl.gov/fpac_1990.pdf

[4] Fusion Energy Advisory Committee FEAC Panel 7 Report on Inertial Fusion Energy, R. Davidson, Chair, *J. Fusion Energ.* **13**(2/3) (1994); http://www.osti.gov/energycitations/product.biblio.jsp?osti_id=79204

[5] J. Sheffield *et al.*, Report of the FESAC Inertial Fusion Energy Review Panel, *J. Fusion Energ.* **15**(3/4) (1996); doi:10.1007/BF02266936; http://link.springer.com/content/pdf/10.1007%2FBF02266936.pdf

[6] See, for example, http://lasers.llnl.gov/

[7] *Frontiers in High Energy Density Physics: The X-Games of Contemporary Science*, Committee on High Energy Density Plasma Physics, Plasma Science Committee, National Research Council. Available from the National Academies Press at http://www.nap.edu/catalog/10544.html

[8] *An Assessment of the Prospects for Inertial Fusion Energy*, Committee on the Prospects for Inertial Confinement Fusion Energy Systems. Available from the National Academies Press at http://www.nap.edu/catalog.php?record_id=18289

[9] J. Nuckolls, L. Wood, A. Thiessen and G. Zimmerman, *Nature* **239**, 139 (1972); doi:10.1038/239139a0.

[10] G. Yonas, J. W. Poukey, K. R. Prestwich, J. R. Freeman, A. J. Toepfer and M. J. Clauser, *Nucl. Fus.* **14**, 731 (1974). The early programs were based on electron accelerators. They eventually switched to light ions.

[11] R. L. Martin, in *Proc. 1975 Particle Accelerator Conference* (Washington DC, USA; IEEE), *Trans. Nucl. Sci.* **NS-22**, 3, 1763 (1975); http://accelconf.web.cern.ch/AccelConf/p75/PDF/PAC1975_1763.PDF

[12] Private communication. J. Nuckolls, G. W. Kuswa, and others were also considering the use of accelerators for fusion in 1975 or before.

[13] B. Badger *et al.*, University of Wisconsin Report UWFDM-450 (1981); http://fti.neep.wisc.edu/pdf/fdm450.pdf

[14] B. Badger *et al.*, University of Wisconsin Report UWFDM-625 (1984); http://fti.neep.wisc.edu/pdf/fdm625.pdf

[15] S. S. Yu, W. R. Meier, R. P. Abbott, J. J. Barnard, T. Brown, D. A. Callahan, C. Debonnel, P. Heitzenroeder, J. F. Latkowski, B. G. Logan, S. J. Pemberton, P. F. Peterson, D. V. Rose, G.-L. Sabbi, W. M. Sharp and D. R. Welch, *Fusion Sci. Technol.* **44**, 266 (2003); http://escholarship.org/uc/item/6vq5x9x8

[16] J. F. Latkowski and W. R. Meier, *Fusion Sci. Technol.* **44**, 300 (2003).

[17] A. W. Maschke, in *Proc. 1975 Particle Accelerator Conference* (Washington DC, USA; IEEE), *Trans. Nucl. Sci.* **NS-22**, 3, 1825 (1975); http://accelconf.web.cern.ch/AccelConf/p75/PDF/PAC1975_1825.PDF

[18] A. W. Maschke, in *Proc. Heavy Ion Fusion Workshop* (Brookhaven National Laboratory, 17–21 October 1977), Brookhaven National Laboratory Report BNL 50769.

[19] D. Keefe, Induction linac, in *ERDA Summer Study of Heavy Ions for Inertial Fusion, Final Report* (eds. R. O. Bangerter *et al.*) (1976), p. 21; http://www.osti.gov/bridge/servlets/purl/7227551-rkrcqV/

[20] J. D. Lindl, *Inertial Confinement Fusion* (Springer-Verlag, New York, 1998).

[21] S. Atzeni and J. Meyer-ter-Vehn, *The Physics of Inertial Fusion* (Oxford, New York, 2004).

[22] R. O. Bangerter, *Proc. 19th Int. Symp. Heavy Ion Inertial Fusion* (Berkeley, CA, USA, 12–17 August 2012), *Nucl. Instrum. Methods A* **733**, 216 (2014); http://dx.doi.org/10.1016/j.nima.2013.05.065.

[23] J. J. Barnard *et al.*, *Proc. 19th Int. Symp. Heavy Ion Inertial Fusion* (Berkeley, CA, USA, 12–17 August 2012), *Nucl. Instrum. Methods A* **733**, 45 (2014); http://dx.doi.org/10.1016/j.nima.2013.05.096.

[24] J. J. Barnard *et al.*, in *Proc. 2005 Particle Accelerator Conference* (Knoxville, TN); http://accelconf.web.cern.ch/AccelConf/p05/PAPERS/RPAP039.PDF

[25] Ya. B. Zel'dovich and Yu. P. Raizer, *Physics of Shock Waves and High-Temperature Hydrodynamic Phenomena* (Dover, New York, 2002).

[26] J. D. Lindl *et al.*, *Phys. Plasma* **11**(2), 339 (2004).

[27] M. Tabak, D. Callahan-Miller, D. D.-M. Ho and G. B. Zimmerman, *Nucl. Fusion* **38**, 509 (1998).

[28] D. A. Callahan-Miller and M. Tabak, *Nucl. Fusion* **39**, 833 (1999).

[29] D. Callahan-Miller and M. Tabak, *Phys. Plasma.* **7**(5), 2083 (2000); http://dx.doi.org/10.1063/1.874031

[30] D. Callahan, M. Herrmann and M. Tabak, *Laser Part. Beams* **20**, 405 (2002); http://dx.doi.org/10.1017/S0263034602203079

[31] E. Henestroza and B. G. Logan, *Phys. Plasma.* **19**, 072706 (2012); http://dx.doi.org/10.1063/1.4737587

[32] OSIRIS and SOMBRERO Inertial Fusion Power Plant Designs Final Report, DOE/ER-54100, WJSA-92-01, March 1992; available at http://aries.ucsd.edu/LIB/REPORT/OTHER/OSIRISSOMBRERO

[33] Prometheus-L and Prometheus-H, Inertial Fusion Reactor Designs Studies, Final Report, DOE/ER-54101, MDC 92E0008, March 1992.

[34] http://www.world-nuclear.org/info/Economic-Aspects/Economics-of-Nuclear-Power/#.UdpDaDUURk. Downloaded July 2013.

[35] J. W. Kwan, *IEEE Trans. Plasma Sci.* **33**, 1901 (2005); http://ieeexplore.ieee.org/stamp/stamp.jsp?arnumber=01556676

[36] *The HIDIF Study: Report of the European Study Group on Heavy Ion Driven Inertial Fusion for the Period 1995–1998*, GSI report 98-06; http://www-alt.gsi.de/documents/DOC-2009-Feb-44.html

[37] L. Wu and G. H. Miley, *J. Phys: Conf.* **112** (2008); http://dx.doi.org/10.1088/1742-6596/112/3/032030

[38] P. A. Seidl and J.-L. Vay, in *Proc. 2011 Particle Accelerator Conference* (New York, USA); http://accelconf.web.cern.ch/AccelConf/PAC2011/papers/wep098.pdf

[39] A. W. Maschke, *Space Charge Limits for Linear Accelerators* (Brookhaven National Laboratory, Upton, NY, USA), BNL-51022 (1979); doi:10.2172/5914736; http://dx.doi.org/10.2172/5914736

[40] E. P. Lee *et al.*, *Phys. Plasma.* **9**, 4301 (2002); http://link.aip.org/link/doi/10.1063/1.1502257

[41] J. D. Lawson, *The Physics of Charged-Particle Beams* (Clarendon, Oxford, 1977).

[42] M. G. Tiefenback, *Space-Charge Limits on the Transport of Ion Beams*, LBL-22465 (1986); http://escholarship.org/uc/item/1v8770kj

[43] W. R. Meier, R. O. Bangerter and A. Faltens, *Nucl. Instrum. Methods A* **415**, 249 (1998).

[44] The LHC Design Report, Vol. 1: The LHC Main Ring (CERN-2004-003, June 2004); http://dx.doi.org/10.5170/CERN-2004-003-V-1

[45] E. A. Startsev, C. Davidson and H. Qin, *Phys. Rev. ST Accel. Beams* **8**, 124201 (2005) and references therein; http://link.aps.org/doi/10.1103/PhysRevSTAB.8.124201

[46] The collection of HIF Symposia can be found at http://ahif.lbl.gov/relevant-papers

[47] C. Olson, *Proc. 19th Int. Symp. Heavy Ion Inertial Fusion* (Berkeley, CA; 12–17 August 2012), *Nucl. Instrum. Methods A* **733**, 86 (2014); http://dx.doi.org/10.1016/j.nima.2013.05.089.

[48] P. A. Ni *et al.*, *Laser Part. Beams* **31**, 81 (2013); doi:10.1017/S0263034612001000.

[49] S. M. Lund, R. H. Cohen and P. A. Ni, *Phys. Rev. ST Accel. Beams* **16**, 044202 (2013); http://link.aps.org/doi/10.1103/PhysRevSTAB.16.044202

[50] R. L. Martin, private communication and *ERDA Summer Study of Heavy Ions for Inertial Fusion, Final Report*, eds. R. O. Bangerter *et al.* (1976), p. 13; http://www.osti.gov/bridge/servlets/purl/7227551-rkrcqV
[51] D. G. Koshkarev, *Il Nuovo Cimento* **106A**, 1567 (1993).
[52] R. Burke, *Proc. 19th Int. Symp. Heavy Ion Inertial Fusion* (Berkeley, CA, USA, 12–17 August 2012), *Nucl. Instrum. Methods A* **733**, 158 (2014); http://dx.doi.org/10.1016/j.nima.2013.05.080.
[53] C. Olson, in *Proc. Heavy Ion Fusion Workshop* (Berkeley, CA, USA, 1979), ed. W. B. Herrmannsfeldt, Stanford Linear Accelerator Center report SLAC-R-542; http://www-public.slac.stanford.edu/SciDoc/docMeta.aspx?slacPubNumber=SLAC-R-542
[54] I. D. Kaganovich, E. A. Startsev and R. C. Davidson, *Laser Part. Beams* **20**, 497 (2002); http://dx.doi.org/10.1017/S0263034602203274
[55] I. D. Kaganovich, E. Startsev and R. C. Davidson, Scaling and formulary of cross-sections for ion–atom impact ionization, *New J. Phys.* **8**, 278 (2006); http://dx.doi.org/10.1088/1367-2630/8/11/278
[56] W. R. Meier and B. G. Logan, *Nucl. Instrum. Methods A* **544**, 310 (2005); http://dx.doi.org/10.1016/j.nima.2005.01.225
[57] C. Stöckl *et al.*, *Nucl. Instrum. Methods A* **415**, 558 (1998); http://dx.doi.org/10.1016/S0168-9002(98)00370-2
[58] B. Sharkov and D. Varentsov, *Proc. 19th Int. Symp. Heavy Ion Inertial Fusion* (Berkeley, CA, 12–17 August 2012), *Nucl. Instrum. Methods A* **733**, 238 (2014); http://dx.doi.org/10.1016/j.nima.2013.05.061.
[59] C. R. Prior, in *Proc. 5th European Particle Accelerator Conference*; http://accelconf.web.cern.ch/accelconf/e96/PAPERS/MOPG/MOP026G.PDF
[60] D. G. Koshkarev, *Laser Part. Beams* **20**, 595 (2002).
[61] M. Basko *et al.*, *Plasma Phys. Control. Fusion* **45**, 12.1 A125 (2003); http://dx.doi.org/10.1088/0741-3335/45/12A/009
[62] S. M. Lund *et al.*, *Proc. 1999 Part. Accel. Conf.* (New York); http://accelconf.web.cern.ch/accelconf/p99/PAPERS/TUP127.PDF
[63] J. J. Barnard *et al.*, *Nucl. Instrum. Methods A* **544**, 243 (2005); http://dx.doi.org/10.1016/j.nima.2005.01.212
[64] A. J. Moss *et al.*, *Proc. EPAC08* (Genoa, Italy); http://accelconf.web.cern.ch/AccelConf/e08/papers/mopp099.pdf; R. A. Church, 9 April 2003, International Muon Ionization Cooling Experiment website, downloaded 28 July 2013; http://mice.iit.edu/vc/vc31/vc31_church.ppt, and R. A. Church, Mice Collaboration Meeting, 14 June 2003, *RF Power Source Design and Status*, downloaded 28 July 2013, http://www.mice.iit.edu/cm/cm6/agenda.html; D. Li (LBNL), private communication.
[65] Thales, Inc., http://www.thalesgroup.com
[66] Y. Kang *et al.*, http://accelconf.web.cern.ch/accelconf/a04/PAPERS/MOM403.PDF
[67] Rees *et al.*, 2006 CW RF Workshop (Argonne National Laboratory, 1–4 May 2006). Downloaded 1 July 2013 from http://www.aps.anl.gov/News/Conferences/2006/CWHAP06/presentations/Rees.pdf
[68] C. E. Helsley and R. J. Burke, *Proc. 19th Int. Symp. Heavy Ion Inertial Fusion* (Berkeley, CA, USA, 12–17 August 2012), *Nucl. Instrum. Methods A* **733**, 51 (2014); http://dx.doi.org/10.1016/j.nima.2013.05.095.
[69] N. Christofilos, Astron thermonuclear reactor, in 2nd *UN International Conference on Peaceful Uses of Atomic Energy*, Vol. 32, p. 279 (Geneva, Switzerland, 1958); N. Christofilos, Energy balance in the Astron device, *Nucl. Fusion Suppl. I* 159 (1962).
[70] *Induction Accelerators*, eds. K. Takayama and R. J. Briggs (Springer-Verlag, 2011); doi:10.1007/978-3-642-13917-8.
[71] A. Faltens, E. Hoyer, D. Keefe and L. J. Laslett, Design/cost study of an induction linac for heavy ions for pellet fusion, in *Proc. 1979 Part. Accel. Conf.*, *IEEE Trans. Nucl. Sci.* **NS-26**, 3, 3106 (1979); http://accelconf.web.cern.ch/accelconf/p79/PDF/PAC1979_3106.PDF
[72] S. Chattopadhyay, A. Faltens and L. Smith, Study of the beam breakup mode in linear induction accelerators for heavy ions, in *Proc. 1980 Part. Accel. Conf.*; http://accelconf.web.cern.ch/AccelConf/p81/PDF/PAC1981_2465.PDF
[73] W. C. Hahn, Small signal theory of velocity-modulated electron beams, *Gen. Elec. Rev.* **42**, 258 (1939); W. C. Hahn, Wave energy and transconductance of velocity-modulated electron beams, ibid, pp. 497–502; S. Ramo, The electronic-wave theory of velocity modulation tubes, in *Proc. IRE* **27**, 757 (1939).
[74] L. J. Chu, 1951 IRE Conference on Electron Devices (Durham, New Hampshire, USA, June 1951).
[75] C. K. Birdsall, G. R. Brewer and A. V. Haeff, in *Proc. Inst. Radio Engrs.* **41**, 865 (1953); http://dx.doi.org/10.1109/JRPROC.1953.274425
[76] S. Massidda, *et al.*, *Nucl. Instrum. Methods A* **678** (2012); http://dx.doi.org/10.1016/j.nima.2012.03.008; I. Kaganovich, *Nucl. Instrum. Methods A* **678** (2012); http://dx.doi.org/10.1016/j.nima.2012.03.007
[77] A. Faltens and D. Keefe, in *Proc. Heavy Ion Fusion Workshop* (Berkeley, USA, 1979), pp. 157–181 (Reports No. SLAC-R-542, No. LBL-10301 and No. CONF- 7910122); http://slac.stanford.edu/pubs/slacreports/reports01/slac-r-542.pdf
[78] W. Chupp *et al.*, in *Proc. Particle Accelerator Conference* (IEEE, New York, USA, 1979); http://accelconf.web.cern.ch/AccelConf/p79/PDF/PAC1979_3036.PDF

[79] T. J. Fessenden, *Nucl. Instrum. Methods Phys. Res., Sect. A* **278**, 13 (1989); http://www.sciencedirect.com/science/article/pii/0168900289911212; M. G. Tiefenback and D. Keefe, *IEEE Trans. Nucl. Sci.* **32**, 2483 (1985), http://dx.doi.org/10.1109/TNS.1985.4333954
[80] W. M. Fawley, T. Garvey, S. Eylon, E. Henestroza, A. Faltens, T. J. Fessenden, K. Hahn, L. Smith and D. P. Grote, *Phys. Plasmas* **4**, 880 (1997); http://pop.aip.org/resource/1/phpaen/v4/i3/p880_s1
[81] P. A. Seidl, C. Celata, A. Faltens, E. Henestroza and S. MacLaren, *Phys. Rev. ST Accel. Beams* **6**, 090101 (2003); http://link.aps.org/doi/10.1103/PhysRevSTAB.6.090101
[82] NDCX-II Advisory Committee, Lawrence Berkeley National Laboratory, 27 May 2009, Committee Comments submitted to DOE, June 2009.
[83] P. A. Seidl, J. J. Barnard, A. Faltens and A. Friedman, *Phys. Rev. ST Accel. Beams* **16**, 024701 (2013); http://prst-ab.aps.org/abstract/PRSTAB/v16/i2/e024701
[84] M. Kireeff-Covo *et al.*, *Phys. Rev. Lett.* **97**, 054801 (2006); http://link.aps.org/doi/10.1103/PhysRevLett.97.054801; R. Cohen *et al.*, *Phys. Plasmas* **12**, 056708 (2005), http://dx.doi.org/10.1063/1.1882292
[85] Proc. ECLOUD '12 Workshop, special edition of *Phys. Rev. ST — AB*; http://prst-ab.aps.org/speced/ECLOUD12
[86] C. Omet, H. Kollmus, H. Reich-Sprenger and P. J. Spiller, in *Proc. EPAC08* (Genoa, Italy); http://accelconf.web.cern.ch/accelconf/e08/papers/mopc099.pdf

---

Roger Bangerter received a PhD in Elementary Particle Physics from the University of California at Berkeley in 1969. During the energy crisis in the 1970s he became interested in energy issues, and joined the Inertial Confinement Fusion Program at Livermore in 1974. He served as Head of Fusion Energy Research at Lawrence Berkeley National Laboratory from 1990 to 2001. He also served as the Director of the Virtual National Laboratory for Heavy Ion Fusion, a national collaboration among people from Berkeley, Livermore, Princeton, and elsewhere. He retired in 2001 to devote more time to science. He remains actively and enthusiastically involved in accelerator and fusion research.

Andris Faltens received a BSEE in 1962 from MIT and on MSEE from Berkeley in 1964. At Raytheon (1960–63), he worked on the development of microwave tubes, including the tube that sent video from the *Apollo* moon landing. At LBNL from 1964 to 1999 and part-time after, he worked on all aspects of accelerators, including the Bevatron RF and its linac, an 800 MHz linac and multiturn injection for the 200 BeV study, and a succession of induction linacs based on the Electron Ring Accelerator injector, of which he was the primary innovator. He was an EE and physicist on HIF from 1976, and was a visiting scientist at CERN, Garching, and BNL.

Peter Seidl received a BS in Physics from McGill University in 1980 and a PhD in Nuclear Physics in 1984 from The University of Texas at Austin. He worked on numerous pion, electron, proton, and heavy ion scattering experiments at LAMPF, ANL, SLAC, and Berkeley Lab. After designing and using detectors for heavy ion collisions at the Bevalac, he joined the HIF group at Berkeley Lab in 1991. He was the lead physicist of various experiments for HIF drivers and for warm dense matter research, and supervised several graduate students. He is currently Division Deputy of the Accelerator and Fusion Research Division at Berkeley Lab.

Reviews of Accelerator Science and Technology
Vol. 6 (2013) 117–142

DOI: 10.1142/S1793626813300065

Particle Beam Radiography

Ken Peach

John Adams Institute for Accelerator Science and
Particle Therapy Cancer Research Institute,
Department of Physics, Oxford University,
Oxford, OX1 3RG, UK
Ken.Peach@adams-institute.ac.uk

Carl Ekdahl

Los Alamos National Laboratory,
Los Alamos, NM 87505, USA
cekdahl@lanl.gov

Particle beam radiography, which uses a variety of particle probes (neutrons, protons, electrons, gammas and potentially other particles) to study the structure of materials and objects noninvasively, is reviewed, largely from an accelerator perspective, although the use of cosmic rays (mainly muons but potentially also high-energy neutrinos) is briefly reviewed. Tomography is a form of radiography which uses multiple views to reconstruct a three-dimensional density map of an object. There is a very wide range of applications of radiography and tomography, from medicine to engineering and security, and advances in instrumentation, specifically the development of electronic detectors, allow rapid analysis of the resultant radiographs. Flash radiography is a diagnostic technique for large high-explosive-driven hydrodynamic experiments that is used at many laboratories. The bremsstrahlung radiation pulse from an intense relativistic electron beam incident onto a high-Z target is the source of these radiographs. The challenge is to provide radiation sources intense enough to penetrate hundreds of g/cm^2 of material, in pulses short enough to stop the motion of high-speed hydrodynamic shocks, and with source spots small enough to resolve fine details. The challenge has been met with a wide variety of accelerator technologies, including pulsed-power-driven diodes, air-core pulsed betatrons and high-current linear induction accelerators. Accelerator technology has also evolved to accommodate the experimenters' continuing quest for multiple images in time and space. Linear induction accelerators have had a major role in these advances, especially in providing multiple-time radiographs of the largest hydrodynamic experiments.

Keywords: Radiography; tomography; accelerators; neutrons; protons; electrons; photons; muons; ions; electron accelerators; linear induction accelerators; flash radiography; hydrodynamic experiments.

1. Introduction

Radiography uses incident radiations (protons, neutrons, electron, gammas and other particles such as muons) to reveal the internal structure of objects. The technique is noninvasive and largely nondestructive except for a usually small amount of radiation damage or interference.

Radiography began shortly after the discovery of X-rays by Röntgen in 1895 [1] and the first radiographic image was of the hand of Röntgen's wife Anna Bertha (see Fig. 1). Since then, X-rays have become an essential tool in medical diagnosis, with ever more sophisticated equipment, for example, Computed Tomography or CT scanning yielding 3- or 4-dimensional images of anatomical structures. Another interesting application of radiography was pursued by Alvarez [2] who used cosmic ray muons to search for hidden chambers in the 2nd Pyramid of Chephren.

This review covers particle beam radiography where the particle beam is generated from an accelerator. The general field of radiography is discussed in the rest of this section. The following sections discuss the use of neutrons, protons (for both medical and non-medical applications), ions, electrons, gammas and X-rays for radiography. There are many other accelerator-based analysis techniques (neutron scattering, ion beam analysis, SPECT, etc.) that also give detailed information about the microstructure of materials (composition, lattice)

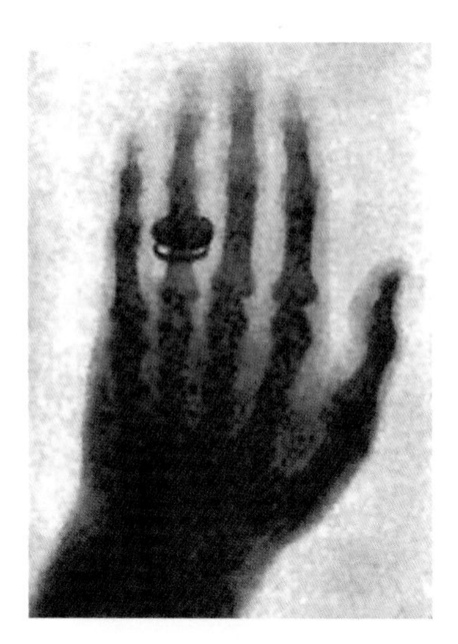

Fig. 1. The first X-ray.

which, combined with radiography, can give an extraordinarily detailed picture of an object, but which is outside the scope of this review. Similarly, while electron microscopy (particularly transmission electron microscopy) shares many features with radiography, it is excluded from this review, which concentrates on the analysis of the bulk properties of structures and systems.

1.1. *Radiography and tomography*

Radiography creates a two-dimensional projection of the object under scrutiny, with the variation in the image density proportional to the line integral of some density distribution modulated by the absorption cross-section of the radiation in the medium. Explicitly, Eq. (1) shows the lateral density distribution $I(x, y, E)$ across the image plane in a homogeneous medium for a radiation of energy E for a perfectly collimated beam:

$$I(x, y, z, E) = I_0(x, y, E) \exp\left(\int -\mu(x, y, z, E)\, dz\right), \quad (1)$$

where $\mu(x, y, z, E)$ is the attenuation coefficient as a function of the position (x, y, z) and $I_0(x, y, E)$ is the incident intensity. In writing Eq. (1), it is assumed that absorption is the dominant process; the actual situation for (say) protons, where elastic and inelastic scattering do not remove the incident particle from the beam, leads to a more complex relationship where the scattering particles contribute an inevitable background to the absorption distribution (see Ref. 3 for a discussion on the proton removal processes). Assuming a monoenergetic beam or (equivalently) that the attenuation coefficient is independent of energy, the line integral (one-dimensional) distribution for an inhomogeneous medium is given by

$$\begin{aligned} I(x, y, z) &= I_0(x, y) \lim_{\delta z \to 0} \prod_{i=0}^{\frac{z}{\delta z}} \exp(-\mu_i \delta z) \\ &= I_0(x, y) \exp\left(z \int_0^z \mu(x, y, z')\, dz'\right), \end{aligned} \quad (2)$$

where μ_i is the attenuation coefficient of the ith material at depth $i\delta z$ and $\mu(x, y, z)$ is the attenuation as a function of depth at transverse position (x, y). If $I_0(x, y)$ is constant and the medium is homogeneous, the radiograph gives a map of the integral projection of the density variation across the image. While the interpretation of the image is relatively straightforward for structure made from homogeneous materials, ambiguities can arise when complex heterogeneous structures are involved, when different combinations of the density or attenuation coefficient can produce similar intensities, or where one internal component masks another.

The advent of powerful computers and improvements in instrumentation have led to the development of computer-aided tomography (CAT, or CT), which provides a three-dimensional reconstruction of the object and its internal structure (actually the weighted density distribution multiplied by the attenuation coefficient), removing some of the ambiguities and artifacts present in radiograms. The configuration typical of a modern radiological CT scanner is shown in Fig. 2.

Briefly, a series of images $I(f, \theta)$ is accumulated from a number of line integrals given essentially by Eq. (2) as the source moves round the object. Often, the object is divided into small volume elements (voxels), and the resulting expression for the intensity distributions is

$$I(f, \theta) = \sum_i \sum_j \mu_{ij} d_{ij}(f, \theta), \quad (3)$$

where μ_{ij} is the local attenuation coefficient in voxel ij, and $d_{ij}(f, \theta)$ is the distance traversed inside voxel ij at (f, θ), with f the fan angle and θ the revolution angle. Provided that the number of images is sufficiently large, this provides an overconstrained set of equations from which μ_{ij} can, in principle, be calculated [4]. In practice, the resulting equations

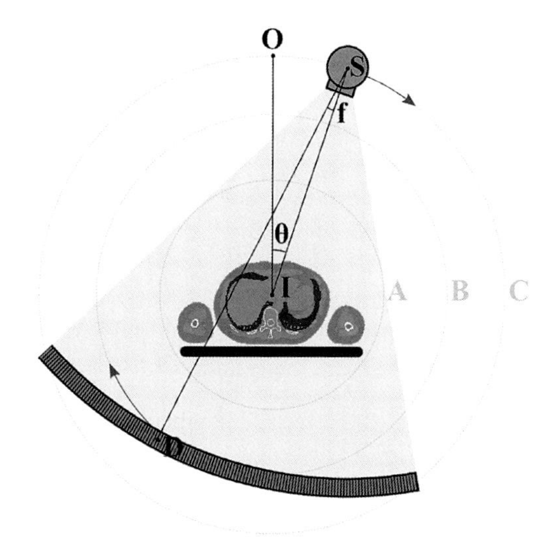

Fig. 2. The geometrical arrangement of a CT scanner, showing the source (S), the isocenter (I) detector element (D) at a fan angle (f) and the revolution angle (θ) to the origin (O). Circle A is the maximum area in which a point will contribute to a projection for all values of θ; circle B shows the path for the detector's central element for the path C traveled by the source during one revolution. (Illustration by D. Warren, University of Oxford.)

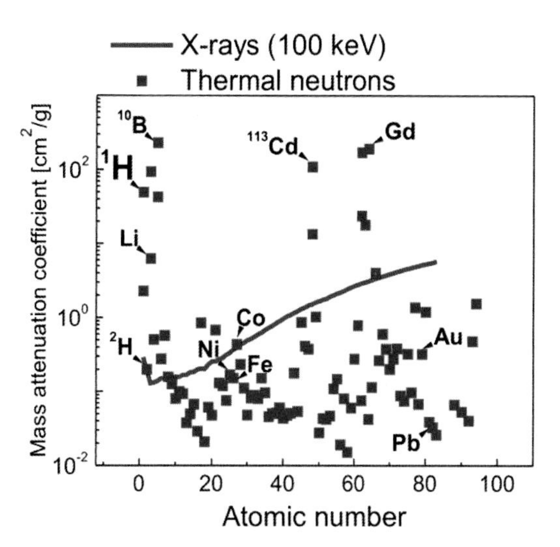

Fig. 3. The mass attenuation coefficients for thermal neutrons (points) compared with 100 kV photons (line). (Figure from Ref. 5.)

are solved using reasonable approximations to ease the computational load, but this, combined with statistical uncertainties inherent in any measurement, can result in the generation of artifacts which can degrade the quality of the reconstructed object from the sequence of images.

1.2. *Electromagnetic, hadronic and weak probes*

The most common form of radiography or tomography uses X-rays. While the technology has changed enormously since their discovery, the principles are all visible in the first radiograph. Their attenuation is essentially controlled by the local electron density, and so (for example) metals are more absorbing than the hydrogen, carbon and oxygen which account for most of the body's content, and those parts (essentially the bones) with high metal (calcium) content are revealed through their greater absorption. While for charged particles such as electrons, protons and heavier ions the processes are different, the effect is similar and the attenuation is governed largely by the local electron density, modulated by small nuclear corrections that generally increase with increasing atomic number.

Neutrons, on the other hand, have no dominating electromagnetic interaction (apart from a weak dependence through its magnetic moment) and thus their attenuation is governed by the nuclear density, and varies with the atomic number (see Fig. 3). Neutrons thus provide complementary information about the composition of the objects under study to that provided by X-rays.

Cosmic ray muons have been proposed for several radiographic purposes, from cargo screening [6] and nuclear reactor safety [7] to volcanology [8]. The advantages are their ready availability and relatively high flux ($\sim 10^4/\text{m}^2/\text{min}$), but they require large area detectors and, if directional information is required (which gives additional information), these can become very large. In principle, accelerator-created muons could also be used, but the relatively high energy required (muons lose about 1 GeV/m in iron, and therefore several GeV is required) makes this both expensive and difficult to relocate.

Cosmic ray (muon) neutrinos have been proposed as potential candidates to perform radiography on the earth (see, for example, Ref. 8). The probability $P(E_v, E_{\mu,\min})$ of observing a muon of energy greater than $E_{\mu,\min}$ from a neutrino of energy E_v is given by

$$P(E_v, E_{\mu,\min}) = N_A \int_{E_{\mu,\min}}^{E_v} dE_\mu \frac{d\sigma_v}{dE_\mu} R(E_v, E_{\mu,\min}), \quad (4)$$

where $R(E_v, E_{\mu,\min})$ is the average distance traveled by a muon with energy E_μ at its production before its energy is reduced to $E_{\mu,\min}$, the minimum energy required to trigger the detector, and N_A is Avogadro's number. The neutrino effective area $A_{\text{eff}}(\theta, E_v)$ is defined as the event rate at a given energy and angle divided by the effective flux at that energy and angle, and is given by

$$A_{\text{eff}}(\theta, E_v) = \varepsilon(\theta)A(\theta)P(E_v, E_{\mu,\min}) \times \exp[-\sigma_v(E_v)N_A X(\theta)], \quad (5)$$

where $\varepsilon(\theta)$ is the efficiency for detection of muons for a projected area $A(\theta)$ at the muon zenith angle θ, and $X(\theta)$ is the amount of matter along a chord through the earth. The neutrino effective area for a spherical 1 km^3 detector is shown in Fig. 4.

As with muons, accelerator-based sources could in theory be employed, but the high fluxes required (because of the low neutrino cross-sections) make this uneconomical. Nevertheless, the principles have been demonstrated — Fig. 5 displays the distribution of reconstructed neutrino vertices in the NOMAD [10] experiment at CERN, showing clearly the structure of the detector. The CDHS experiment [11], at CERN too, also produced an impressive tomograph of their hydrogen target using the vertices of the neutrino interactions (see Fig. 6).

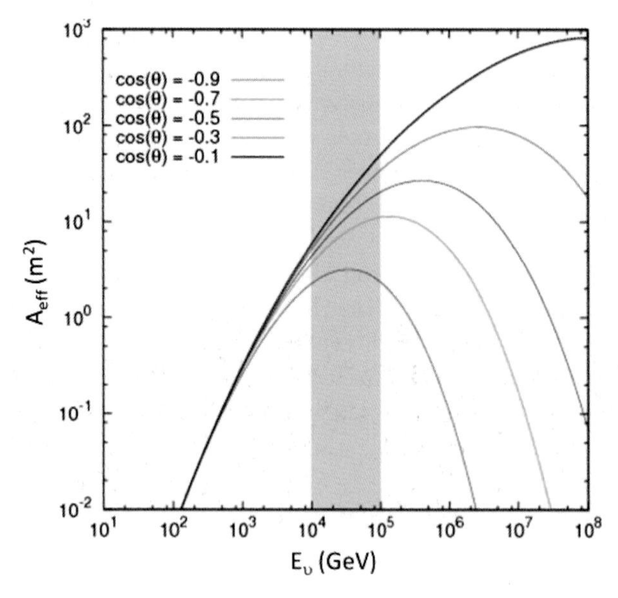

Fig. 4. Muon neutrino effective area for an idealized spherical detector with a cross-sectional area of 1 km^2 from Ref. 8.

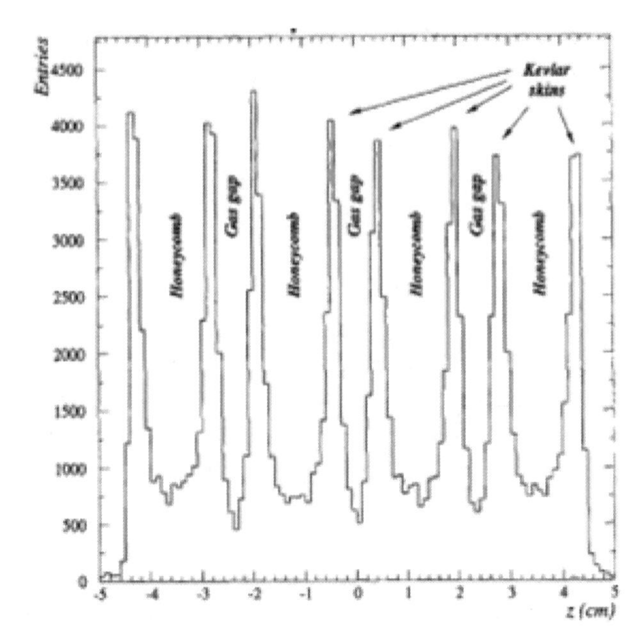

Fig. 5. The neutrino radiograph of the NOMAD detector at CERN [10].

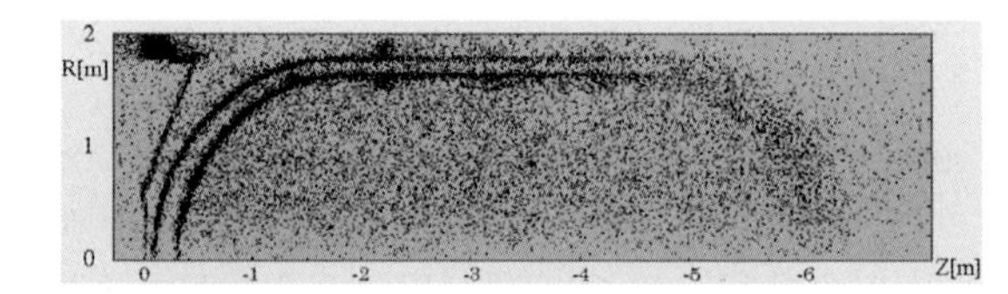

Fig. 6. A neutrino tomograph of the hydrogen target chamber in the CDHS experiment at CERN [11].

1.3. *Static, dynamic and flash radiography*

Radiography and tomography applications are usually static, or used as a sequence of static images to create a pseudo four-dimensional (dynamic) image. Flash radiography is a technique that is used to image objects moving at extremely high speed, allowing the measurement of, for example, the position, speed, shape and internal density profiles of their components. This technology has many applications, including the development of both nuclear and conventional explosions, which are used to validate computer models and to determine experimentally the properties of materials in transient phases, including high-temperature and -pressure conditions.

1.4. *Multimodal radiography*

An increased contrast can often be achieved by using two or more exposures with a different dependence of the absorption coefficients on the atomic number, such as two different X-ray energies or two different modalities (X-rays and neutrons, for example). Alternatively or additionally, further information (energy, direction) can be used to increase contrast or selectivity.

1.5. *Medical imaging*

Although medical imaging was among the first uses of radiography, there have been many developments in medical imaging technology which provide functional as well as structural information. Examples of such technologies are PET (positron emission tomography), SPECT (single photon emission computed tomography) and MRI (magnetic resonance imaging). These are outside the scope of this article, but a recent review of medical imaging techniques can be found in Ref. 12.

2. Neutron Radiography

Neutron radiography was proposed shortly after the discovery by Chadwick [13] of the neutron in 1932; a patent was granted in 1940 to Kuhn [14] for the technique. However, it was not until a high-flux thermal neutron reactor source became available in the 1950s that the techniques were developed. A comparison between an X-ray and a neutron (absorption) radiograph of a camera is shown in Fig. 7.

Although neutron radiographs were obtained in the 1930s using a weak neutron source from a DD neutron generator, the technique became more generally viable in the 1950s with the development of nuclear reactor neutron sources (see Ref. 16 for an early review). While of course fluxes and facilities vary between reactor sources, typical neutron radiography beam lines have fluxes of thermal neutrons of the order of $10^8\,\text{cm}^{-2}\text{s}^{-1}$, with the potential (for large engineering studies) of a large aperture. While there are at present many research reactors providing useful neutron fluxes, including for radiography, many of them are nearing the end of their useful life.

Since 1977 (see Table 1), accelerator-driven spallation neutron sources have provided an alternative, although there are still relatively few in operation. These provide a broad spectrum of neutrons for a variety of research purposes, including radiography, where the neutrons are thermalized using moderators surrounding the spallation target.

Such facilities can provide comparable thermal neutron fluxes to reactor-based sources, with the added option of using time-of-flight information to refine further the discriminating power of the technique for a pulsed machine (like ISIS). Spallation neutron sources can deliver useful fluxes of thermal neutrons; as an example [17], the NEUTRA (NEUtron Transmission RAdiography) beam line at PSI yields more than 5×10^6 25 meV thermal neutrons per cm^2 per second per mA of proton beam current with a maximum field of 30 cm × 30 cm.

From the accelerator perspective, the principal parameters are the beam energy and the beam power (or beam current), the target design and materials, and the moderators used to thermalize the neutrons. LANSCE [18], IN-0.6 [19], the SNS [20] and the ESS [21] are linacs (the last two are superconducting); both LANSCE and the SNS have accumulator rings to provide short pulses, whereas the ESS will be a long-pulse machine which will require beam choppers for time-of-flight measurements, except perhaps at very low energies. (LANSCE will be discussed

(a)

(b)

Fig. 7. Comparison between radiographic images of a camera taken with (a) thermal neutrons and (b) 150 keV X-rays [15].

Table 1. Spallation neutron sources (existing and planned). [(1) and (2) refer to ISIS target stations 1 and 2.]

Source	Country	Year of operation	Proton energy (MeV)	Beam power (MW)
LANSCE	US	1977	800	<1
ISIS				
(1)	UK	1984	800	0.16
(2)		2009		0.08
PSI SINQ	CH	1997	590	1.3
INS-0.6	Russia	1998	<500	0.25
SNS	USA	2006	1000	1.1
J-PARC	Japan	2008	3000	0.3
ESS	Sweden	2017	2000	5
CSNS	China	2017	1600	0.1
Phase 2				0.2

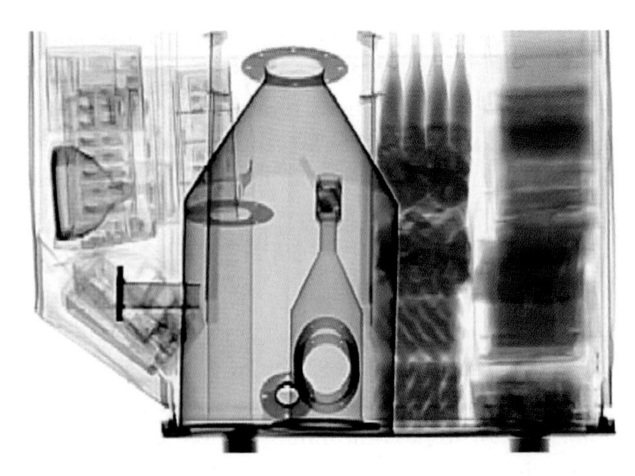

Fig. 8. A combined fast neutron and X-ray radiograph of a unit loaded device containing mixed cargo (assorted computer equipment, heavy steel industrial items, foodstuffs, and office files and papers. (Image from Ref. 30.)

below, under proton radiography.) SINQ at PSI [18] is an isochronous CW cyclotron, while ISIS [23], J-PARC [23] and the CSNS [25] are rapid-cycling (25–50 Hz) synchrotrons. Earlier spallation sources (KENS [26] at KEK — 500 MeV 7 kW — and IPNS [27] at ANL — 500 MeV and 3 kW) ceased operation in 2007 and 2008, respectively.

The other main technology for neutron radiography uses low-energy DD (DT) neutron generators to create high fluxes of neutrons at 2.4 (14.1) MeV. Such generators are relatively small, and serve a wide variety of applications [28]. Applications of such generators include security, industry, environment, art and archeology, geology and oil exploration. There are two main types of accelerator-based neutron generators, both of which are commercially available. The older type uses a Penning ion source diode generator (either DD or DT), with fluxes in the range of 10^5–10^8 (DD) or 10^7–10^{10} (DT) neutrons/s. Positively charged deuteron ions are accelerated up to 110 kV and then impinge on a metal hydride (deuterium, tritium or a mixture of the two) to produce the neutrons. Newer generators use axial radio frequency induction plasma accelerators, so far only with the DD reaction. Deuterons are first accelerated and implanted into a titanium target to form titanium hydride near the surface, and subsequently bombarded with deuterons to produce 2.4 MeV neutrons. Neutron fluxes of 10^8/s are available, and up to 10^{10} neutrons/s demonstrated. Details can be found in Ref. 28.

Developments in neutron detector technology have led to new applications for fast (14 MeV) neutron radiography. For example, Ref. 30 describes the use of fast neutrons from a DT neutron source in combination with X-rays from a 6 MeV linac in cargo scanning (see Fig. 8). There is growing interest in the development of neutron cargo scanners, especially for the detection and control of nuclear contraband and national security applications (see Refs. 31 and 32, and references therein).

Epithermal neutrons (1 eV–10 keV) have a lower scattering in hydrogenous materials, making them useful for analyzing biological materials. Recent advances with borated detectors have led to renewed interest in this technique (see, for example, Refs. 33 and 34).

3. Advanced X-Ray Sources

There have been many developments of X-ray radiography since the primitive radiographs of Röntgen, delivering sharper images with greater contrast and improved resolution, aided by improved detection technology from the photographic film to today's solid state detectors. Two of these are discussed briefly below.

3.1. *Microfocus X-ray tubes*

These are particularly used industrially for nondestructive testing and 3D microtomography, and high-resolution imaging; they are commercially available from many manufacturers. There are basically two types of microfocus X-ray tubes — those with a solid anode and metal jet anodes; the main advantage of the liquid metal jet anode is that the electron beam power can be much higher than for a solid metal target, where the beam power is limited by the need to avoid significant target damage. Typically, a metal jet anode can withstand more than an order of magnitude in power density (3–6 W/mm for gallium or tin, for example) and can thus operate with a smaller spot size (a few μm).

3.2. *Synchrotron radiation and free electron laser X-ray radiography and tomography*

Synchrotron radiation sources and free electron lasers provide high-brightness and high-brilliance beams of X-rays that can be used for radiography and tomography. The advantage over conventionally produced X-ray beams (from a linac) is not only in the photon flux but also in the ability to produce monochromatic or quasimonochromatic and coherent or partially coherent X-ray beams of a specific wavelength, and the small angular divergence and source size, leading to fine (micron and submicron) resolution and good contrast [35]. These techniques have many applications, from archeology [36] to medicine [37, 38].

4. Electron and X-Ray Flash Radiography

Flash radiography is a venerable diagnostic technique in use for large hydrodynamic experiments at many laboratories. Typically, multi-MeV radiation pulses with tens-of-ns pulse widths are used to radiograph metals driven by high explosives. The source of radiation is bremsstrahlung from high-energy electrons striking a heavy-metal target, and there has been a substantial investment in high-power electron accelerators for this purpose.

4.1. *Flash radiography*

Radiography machines for the largest experiments are based on multi-MeV electron accelerator technology, with the accelerated beam producing a bremsstrahlung radiation source from a high-Z metal target. Point projection radiography is the most common technique used to image these dynamic experiments. The requirements for this class of machines include: pulse lengths sufficiently short to stop the motion of high-velocity shocks and high-speed jets, and intensities sufficient to penetrate a few hundred g/cm^2 of metal.

4.2. *Hydrotest radiography*

Image quality is usually described in terms of contrast and resolution. Stopping the action to minimize the hydrodynamic motion blur contribution to resolution sets the permissible pulse width of the electron beam. Shock pressures in high-explosive-driven hydrodynamic experiments are multi-megabar (100 Gpa), and corresponding shock velocities can exceed 1 cm/μs, and so the bremsstrahlung radiation pulse must be $\sim$100 ns or less to achieve the desired millimeter-scale resolution. In order to produce a sufficient dose for a high-quality image through a few hundred g/cm^2 within this short pulse width, the accelerator must be capable of producing radiation at a high dose rate.

The unit of dose in common usage is the rad, which is the specific energy absorbed by a material (1 rad = 100 erg/g = 0.01 gray), so the unit is material-dependent. However, when quoted by hydrotest radiographers, usually as rads at 1 m from the converter (rad@1m), the material is seldom specified.

Detailed experimental, analytical, and Monte Carlo analyses have established that multi-kA accelerators with energies from 10 to 30 MeV are best for the large-scale hydrotests. The exception is the family of 50–70 MeV pulsed air-core betatrons at the All-Russian Institute of Experimental Physics (VNIIEF) in Sarov, Russia.

Resolution is ultimately limited by the radiation source spot size, which is measured at the different laboratories in a variety of ways [29, 40]. For this review, the spot size will be defined as the full width at half maximum (FWHM) of the intensity distribution.

4.3. *Accelerators*

There are three types of accelerator technologies now in use for radiographing hydrodynamic experiments: pulsed-power-driven diodes, air-core betatrons and linear induction accelerators (LIAs). In what follows, we will briefly describe the first two, but spend more time on LIAs, which are used to radiograph the very largest of the hydrodynamic experiments.

5. Pulsed-Power-Driven Diodes

The earliest pulsed power radiography machines were built at the Atomic Weapons Establishment (AWE), Aldermaston, UK. In this approach a short pulse of high voltage is applied directly to a large vacuum diode, and the electrons are used to produce the radiographic source spot without further acceleration.

5.1. *Pulsed power drivers*

The accelerators at AWE use a high-voltage generator to pulse-charge a fast-pulse-forming line (PFL), which is then discharged into the diode, and the resulting beam focused to produce the radiographic source spot [41]. This is a very cost-effective approach, and AWE now has many accelerators, with five of their high-explosive containment chambers capable of providing two views from separate machines. The AWE machines have a Marx-generator-charged oil-insulated Blumlein pulse-forming line, which discharges into a self-magnetically insulated transmission line (MITL) feeding the diode [42]. The parameters of the AWE accelerators are summarized in Table 2. The AWE accelerators rely on high current at relatively low energy to produce enough bremsstrahlung radiation to penetrate the experiment. Pulsed power drivers are also in use at other institutions for development of radiographic diodes [30].

One of the major engineering challenges of single PFL-driven diodes is the high-voltage vacuum insulator, which must stand off the full voltage of the diode. This difficulty can be circumvented by using an inductive voltage adder (IVA) as the driver [44]. This is a modular approach, with many lower-voltage PFLs coupled in series through inductive cells (Fig. 9) to a single-vacuum MITL that feeds the diode. In the IVA, the vacuum insulators only have to stand off the voltage of a single cell, not the entire diode voltage.

In Fig. 9 a high-voltage pulse is applied to the gap by the external pulsed power generator, usually a coaxial PFL. For short times, the high-voltage side of the gap is inductively isolated by the toroidal inductor, as in a pulse transformer. The inductance, and thus isolation time, is increased by orders of magnitude by incorporating the magnetic core. However, the magnetic field in the ferromagnetic core eventually saturates, the permeability (and inductance) drops to the vacuum value, and the isolation vanishes. In principle, integration of Faraday's law tells us that the maximum pulse width Δt that can be isolated for a drive voltage V is $V\Delta t = A\Delta B$ (volt-seconds), where A is an effective area and ΔB is the change from the initial value to saturation B_{sat}, as determined by the hysteresis loop of the material. If the core has been previously pulsed, the change ΔB is small, only from the saturation remanence to saturation. Therefore, it is common practice to reset the core by applying an opposite polarity pulse with enough volt-seconds to drive it around the hysteresis loop into saturation at $-B_{\text{sat}}$. If the material has a "square" loop, then the next pulse can be isolated for a time approximately proportional to $2B_{\text{sat}}$. Thus, the figure of merit for induction cell design is the total available flux swing ($A\Delta B$) in volt-seconds. An applied voltage pulse can only be isolated for a time shorter than that determined by the flux swing of the ferromagnetic core (volt-seconds) and the applied voltage.

Several cells are connected in series to form the IVA, which has an output voltage ideally equal to the sum of the individual drive voltages. The current in the center conductor is constant throughout the IVA, so the center conductor must be tapered or stepped down at each cell in order for the impedance of the coaxial structure to be matched

Table 2. Pulsed-power-driven diode accelerators at AWE [42]. The spot sizes are converted from AWE definition to FWHM assuming a Gaussian intensity distribution [29].

Accelerator	Energy (MeV)	Current (kA)	Pulse (ns)	Diode	Spot size FWHM (mm)	Dose (rad@1m)
Mogul E	9.5	40	70	Paraxial	4.1	400
Mogul D	8.0	30	65	Paraxial	3.4	160
Superswarf	5.0	30	65	Paraxial	3.1	62
Mini B	2.2	50	45	SMP	2.0	12
Mevex	0.8	35	60	SMP	2.0	1.2

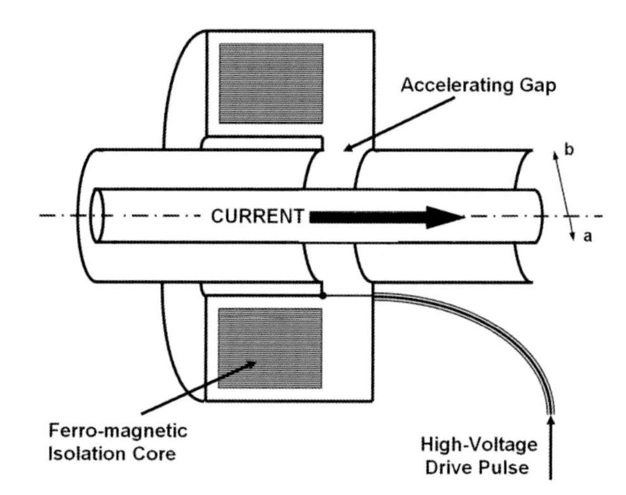

Fig. 9. Single induction cavity. The outer conductor radius of the coaxial MITL is b, the radius of the center conductor is a, and the impedance is $Z = 60\ln(b/a)$ Ω.

to the drive voltage ($Z = V/I$), which increases at each gap.

IVAs have been used for radiographic diode development at Sandia National Laboratory (SNLA) in Albuquerque, NM, USA, and at the Naval Research Laboratory (NRL) in Washington, DC, USA. At SNLA the RITS-6 IVA consists of six PFLs feeding induction cavities joined in series to a vacuum MITL. RITS-6 can deliver a 75 ns pulse to the diode with 7–11 MV peak voltage and 150–200 kA peak current [45–47]. The Mercury IVA at the NRL can deliver a 50 ns, 6 MV, 360 kA pulse to the diode [48–51].

The only IVA-based accelerator presently in use for flash radiography of hydrodynamic experiments is Cygnus, located at the Nevada National Security Site (NNSS) in Nevada, USA [52–54]. Cygnus uses two three-cell IVAs to radiograph experiments from two different directions. Each IVA delivers ~2 MV and ~60 kA in a 55 ns pulse to the radiographic diode [54].

Finally, a large IVA accelerator for radiography called Merlin is presently under construction for the AWE Hydrus facility [55]. The Merlin IVA will be able to deliver 7.5 MV to a 40 Ω diode in a 60 ns pulse.

5.2. *Radiographic diodes*

Three types of radiographic diodes are in use for flash radiography of hydrodynamic experiments: the gas-focused paraxial diode, the rod-pinch diode and the self-magnetic pinched diode (SMP). Although other types of diodes have been tested [56, 57], or proposed [58], these three are the only ones in use, or serious contenders for the future. At present, for flash radiography of hydrodynamic experiments, the paraxial and SMP diodes are in use at AWE [42], and the rod-pinch diode is in use on Cygnus [54]. The radiographic characteristics of the AWE diodes are summarized in Table 2, and the Cygnus rod-pinch diode produces more than 4 rad@1m in an ~50 ns pulse from a 0.7 mm FWHM spot [54].

6. Pulsed Air-Core Betatrons

A different tactic for radiographing hydrodynamic experiments has been pursued at VNIIEF, where they have continued development of the air-core pulsed betatron invented by A. I. Pavlovskii [59]. In this approach the dose rate required for radiography of hydrodynamic experiments results from the relatively high energy of the electrons; these betatrons accelerate the beam to energy of 65–70 MeV. They are all based on a design having an equilibrium beam orbit radius of 234 mm, giving them their name: BIM-234. The pulsed radiation output is obtained by using a fast magnetic field to change the orbit quickly, pushing the beam into a converter target located inside the toroidal vacuum tube.

There are several small BIM-234 series accelerators at different VNIIEF high-explosive sites, most with an output dose in the 2–5 rad@1m range. Some of them have been modified to produce up to three output pulses, usually 100–500 ns wide over a period of ~2 μs.

The newest and largest betatron at VNIIEF is BIM-M [60, 61]. The BIM-M injector is a cold cathode diode driven by a Marx/pulseline/MITL. This produces a 2 MeV, 10 kA, 15 ns electron beam. About 500 A is injected into the ring, and about 280 A is trapped in the magnetic well as circulating accelerated current. A 400 kJ capacitor bank is discharged into the 50 μH betatron magnet winding to accelerate the beam from the 2 MeV injection energy up to the 50–70 MeV final energy in 550 μs. As with all of the VNIIEF betatrons, the BIM-M radiation spot is elliptical (~1 mm × 4 mm). The maximum dose in a single pulse is 100–150 rad@1m. BIM-M also has three-pulse radiographic capability.

At some VNIIEF sites, multiple machines provide multiple-view radiographs of hydrodynamic experiments. One such site features three small triple-pulse machines [62], which can provide a total of nine radiographs from different directions at different times [63].

7. Linear Induction Accelerators

The ships-of-the-line for hydrotest radiography are the linear induction accelerators in the US, France and China. Invented by Christophilos in the early 1960s [64], the linear induction accelerator has found many uses [65–67], and is particularly suited to the requirements for hydrotest radiography. LIAs provide the smallest spot size with a short-pulse dose large enough to penetrate the largest hydrodynamic experiments of interest.

In the US, the flash X-ray (FXR) LIA at Livermore Lawrence National Laboratory (LLNL) in California was the first LIA designed specifically for flash radiography of large-scale hydrodynamic experiments; it began operation in 1982. Also in the US, the Dual-Axis Radiography for Hydrodynamic Testing (DARHT) facility at Los Alamos National Laboratory in New Mexico features two LIAs positioned to enable orthogonal views of the experiment; the first axis has been operational since 1999. In France, the Accelerateur a Induction de Radiographie par Imagerie X (AIRIX), until recently at the Polygon d'Experimentation de Moronvilliers (PEM), was commissioned in 1999. In China, there are three radiography LIAs at the Institute of Fluid Physics (IFP) in Sichuan. The Linear Induction Accelerator X-ray Facility (LIAXF) began operation in 1993, and more recently the two Dragon LIAs, which are orthogonal to each other, have been commissioned.

The parameters of these flash radiography LIAs are summarized in Table 3. Except for the long-pulse second axis of DARHT, these accelerators bear a close resemblance to each other (for example, use of ferrite-cored induction cells and cold cathode diode injectors), but differ significantly in design details. It is tempting to think of the LIA as simply an IVA in which the center conductor in Fig. 9 is replaced by an electron beam, and a beam-focusing solenoid is added [44]. However, LIAs cannot be extrapolated to the high currents typical of IVAs because of beam instabilities. Thus, the LIA radiography machine relies on higher electron kinetic energy to achieve the required dose, which is proportional to $IE^{2.8}$. Beam focusing in these LIAs is provided by solenoids incorporated in the cells, which generate a spatially varying but almost continuous axial magnetic field for beam transport.

Radiographic resolution is ultimately limited by the radiation source spot size [29], which is significantly affected by LIA physics. The physics influencing the spot size includes final-focus solenoid focal length and aberrations, beam emittance, energy spread and temporal variation, and beam motion [67, 68], as well as beam interactions with the bremsstrahlung target [69, 70]. The focal length is usually a design constraint, so the accelerator physics emphasis has been on beam stability, and minimizing motion and emittance.

For example, in an LIA the major contributors to beam motion are the beam breakup instability (BBU) [71] and beam "corkscrew" motion [72]. BBU is the result of beam coupling the TM_{1n0} accelerating cavity modes, and corkscrew is the result of temporal variation of the beam energy interacting

Table 3. Nominal parameters of linear induction accelerators built for flash radiography, listed in order of their commissioning dates. Final, upgraded parameters are listed where applicable; blanks indicate that the data were unpublished at press time.

Accelerator	Laboratory	Year	Energy (MeV)	Current (kA)	Number of pulses	Pulse width (ns)	Spot size FWHM* (mm)	Dose (rad@1m)
FXR	LLNL (USA)	1983	18	3	1	65	2.2	450
LIAXFU	IFP (PRC)	1992	12	2.7	1	60	4	
DARHT Axis-I	LANL (USA)	1999	19	1.8	1	60	0.7	570
AIRIX	PEM (FR)	1999	19	2	1	60	1.6	350
DARHT Axis-II	LANL (USA)	2003	17	1.7	4	20–90	< 1	80–360
Dragon-I	IFP (PRC)	2003	19	2.5	1	70	$\sim$1	
Dragon-II	IFP (PRC)	2012	20	2.5	3	60		

with random chromatic aberrations in the transport magnetic field.

Because of their high currents, BBU is especially challenging for radiography LIAs, and it is suppressed through cell design to reduce cavity Q, and by using a high magnetic focusing field. On the other hand, chromatic effects responsible for high-frequency corkscrew are inhibited by reducing the magnetic field. In principle, an optimum solution is obtained with a magnetic field that increases proportionally to the square root of the beam energy [73].

7.1. *FXR*

The LIA was invented at LLNL [64], and it is no surprise that the first LIA designed specifically for hydrotest radiography, FXR, was built there. It began operations in 1982 with beam parameters of 2.2 kA at 17 MeV [74, 75], and has since been upgraded to 3 kA at 18 MeV [76]. Although it is capable of double-pulse operation at a lower energy (8 MeV) [77, 78], FXR is operated in the higher-energy single-pulse mode to provide a sufficient dose to penetrate the hydrotest. The injector is "push–pull," with six cells energizing the cathode stalk and four more energizing the hollow anode stalk, to produce ~2.5 MV across the diode gap, creating a 65 ns, 3 kA beam pulse. The main accelerator has 44 ferrite-loaded induction cells with 300–350 kV accelerating potential per cell provided by a Blumlein PFL. The cell drive pulse is 90 ns long, which is much longer than the beam current pulse. Using only the flat top of the longer accelerating pulse minimizes the accelerating energy variation during the beam pulse. Focusing solenoids in each cell provides a high, semi-continuous axial magnetic field for beam transport and focusing.

The performance upgrade from 1991 through 1996 replaced the original 2.2 kA injector by one producing 3 kA. The upgrade also entailed replacing 62 focusing solenoids. Unlike the originals, the new magnets have bifilar windings and homogenizing rings to minimize field errors, and incorporate printed circuit steering dipoles for correction of misalignment [79]. Moreover, close attention was paid to accurate alignment of the magnets when they were installed, because corkscrew motion is due to temporal energy variation interacting with the dipoles caused by misalignment.

Mode damping ferrites were also added to the accelerating cavities during the upgrade to reduce the BBU instability, and beam position monitors were installed at 16 locations down the beam line. As a result of the upgrade, the FXR parameters were significantly improved.

Efforts to improve FXR beam quality and accelerator reliability have continued [80–87]. They include retuning the LIA to minimize emittance growth [84] and further reducing the temporal variation of the energy, thereby reducing the corkscrew [83, 86]. Many of the FXR beam optimization experiments have been made possible through implementation of advanced beam diagnostics [80–82, 84].

7.2. *AIRIX*

The AIRIX LIA [88–98] has been in operation for hydrotest radiography at PEM since 1999 [90–93]. (It is now being dismantled and moved to another site in France [98].) Unlike FXR, AIRIX uses a Blumlein PFL with coaxial step-up transformers to drive the injector diode with an exceptionally flat 60 ns, ~4 MV pulse. AIRIX uses 64 ferrite-loaded cells to accelerate the beam produced by the cold cathode diode. Each acceleration cell, including the ferrite cores, is under vacuum with an air–vacuum interface at the drive-rod feedthroughs. This eliminates the need for an oil–vacuum interface, such as required on the FXR cells, which are oil-filled. The cells are driven by Blumlein PFLs, each PFL driving two cells. Another innovation on AIRIX is the use of a hydraulic system to align the accelerator magnetic axis. This resulted in an order-of-magnitude improvement of the accelerator alignment.

A significant amount of work went into minimizing the energy variation during the flat top of the pulse. As in FXR, the accelerating pulse is longer than the beam current pulse. At 19.2 MeV the energy variation is ±0.3% over the 60 ns beam pulse. The result is that the AIRIX beam corkscrew motion is less than 2% of the beam radius at the accelerator exit.

7.3. *Accelerators at the IFP*

The Institute for Fluid Physics in China is home to three LIAs for flash radiography: LIAXF, Dragon-I and Dragon-II [99–105]. LIAXF and Dragon-I are single pulse accelerators, but Dragon-II has three pulses. In addition, Dragon-I and Dragon-II are at right angles, enabling multiview, multitime flash radiography of a hydrodynamic experiment.

7.3.1. *LIAXF*

The first flash radiography LIA built at the IFP was LIAXF. Initially commissioned at 10 MeV, it was soon upgraded to 12 MeV [99, 105]. A four-cell IVA powers the 1 MV, 2.7 kA, 60 ns diode, and 32 more cells power the main accelerator. Each ~350 kV cell is driven by a Blumlein PFL. The radiographic source spot from LIAXF is 4 mm FWHM.

7.3.2. *Dragon*-I

The next LIA built at the IFP for flash radiography was Dragon-I [100–105]. The beam is produced by a push–pull IVA injector, similar to FXR, but powered by 12 cells (7 for the cathode stalk, and 5 for the anode) to provide a higher voltage (3.6 MV) to the cold cathode diode. Each injector cell is driven by a Blumlein PFL. The rest of the accelerator consists of 72 cells, with one Blumlein driving 2 cells, as in AIRIX. As in the other single-pulse LIAs, the accelerating voltage pulse is longer than the current pulse resulting in only 2% variation in beam energy. Dragon-I produces a 1 mm class radiographic spot size (FWHM).

7.3.3. *Dragon*-II

The newest flash radiography LIA at the IFP is the three-pulse Dragon-II [104, 105], which is orthogonal to Dragon-I, enabling two-view, multitime radiography. As in Dragon-I, there is a 12-cell injector followed by 72 accelerating cells. Each injector cell is driven by three Blumlein PFLs with staggered timing to produce three 60 ns beam pulses with programmable separation. Likewise, three Blumlein PFLs with staggered timing drive two cells in the main accelerator. Thus, Dragon-II has 144 PFLs in total. The key to this scheme is a pulse adder that isolates the PFLs from each other, and also provides for the core reset. Spot size and/or dose measurements on Dragon-II have yet to be reported.

7.4. *DARHT*

The Dual-Axis Radiography for Hydrodynamic Testing (DARHT) facility at Los Alamos was built to provide the experimenter with flash radiographs from two directions [106, 107]. Two LIAs produce the radiographic source spots for orthogonal views of each test. The 1.8 kA, 20 MeV Axis-I LIA creates a single 60 ns radiography pulse. The 1.7 kA, 17 MeV Axis-II creates four radiography pulses by kicking them out of a much longer pulse that has a 1.6 μs flat top. The high-explosive-driven hydrodynamic experiments are fully contained within reusable spherical steel vessels to eliminate environmental impact.

7.4.1. *DARHT Axis*-I

The DARHT Axis-I accelerator became operational in 1999. The injector pulsed power is a glycol-insulated Blumlein PFL pulse charged by a 112 kV capacitor bank through a 1:15 transformer in ~5 μs. The output pulse from the PFL is stepped up by coaxial transmission line transformers to ~4 MV at the diode, which delivers a 60 ns 1.8 kA beam pulse from a 5 cm diameter explosive emission velvet cold cathode [108].

This beam is further accelerated by 64 ferrite-loaded induction cells [109], with each pair of two cells powered by a single Blumlein PFL delivering a 250 KV pulse. Each cell incorporates a focusing solenoid with bifilar windings, homogenizing rings and steering dipoles [110]. The beam energy variation throughout the flat top has been measured to be less than ±0.5%.

For the 1.8 kA pulse, a 570 rad@1m dose has been measured using a platinum foil calorimeter, and the 0.75 mm FWHM spot size measured with a 0.4 mm diameter pinhole and various cameras, including time-resolved measurements [70]. The dose and spot size can be tuned to match the dynamic experiment by changing the cathode to produce a different current. A range of doses from 70 rad@1m to ~1000 rad@1m is available through this method, with the spot size ranging from 0.4 mm to ~1.6 mm (FWHM).

7.4.2. *DARHT Axis*-II

The second axis of DARHT was completed in 2002 [111, 112], and was originally commissioned at 12.5 MeV in 2003 [113, 114]. Axis-II was then upgraded to 17 MeV [115–122] before radiographing its first hydrodynamic experiment in 2009. Axis-II provides radiographs from a viewpoint orthogonal to that of Axis-I. It creates and accelerates a long, few-microsecond pulse, and then chops out shorter pulses for multiple radiographs within that time window.

The Axis-II injector diode is powered by an 88-stage Marx generator configured as a pulse-forming network (PFN) that drives the diode with a 2.2 MV, ~3.5 μs FWHM pulse that has a flat top exceeding 1.7 μs. The diode produces a 1.7 kA space-charge-limited beam from a 16.5 cm diameter hot dispenser cathode, chosen to avoid the impedance collapse exhibited by the explosive emission cold cathodes used elsewhere [108].

The beam is further accelerated by 74 induction cells to 17 MeV. In order to isolate the accelerating pulse, the size of the magnetic cores in the cells has to be increased, and the material has changed from ferrite to metglass, which has a higher permeability and saturation field, thereby providing inherently larger flux swing. These ~8 tonne, super-sized cells provide ~500 mV-s of swing compared with ~30 mV-s on Axis-I. Each cell is powered by a PFN providing a 210–220 kV, ~2.3 μs FWHM pulse, which is limited by the flux swing of the cores. The beam loads down the cells, so that the actual acceleration potential is ~180 kV on the first 8 cells to ~200 kV on the remaining 66 [123]. This drive voltage is flat to within ±1.5% for ~1.6 μs to minimize chromatic effects on the beam in the accelerator and final focus.

After exiting the LIA, the long pulse is sliced into four shorter pulses by a fast kicker system in the downstream transport (DST), with the un-kicked beam diverted to an offline dump [122–130]. The durations of the kicked pulses are individually programmable, from 20 ns to >100 ns, and this feature is used to tune the individual doses to the dynamic experiment. The dose rate is ~4 rad@1m/ns, so individual pulses can be programmed to produce from ~80 rad@1m to more than 400 rad@1m with 1 mm class spots [122].

Good diagnostics, high-quality data and insightful interpretation are key to successful tuning and operation of DARHT Axis-II, and the LIA and DST are heavily instrumented with beam diagnostics [114, 131–138].

On the DARHT accelerators, BBU is mitigated through cell designs incorporating damping ferrites to reduce cell cavity Q [139, 140], and the use of high focusing field strengths. The theory of BBU growth was experimentally validated in extensive experiments on Axis-II [141], and the results used to reduce BBU at the LIA exit to less than 2% of the beam radius [142].

After correcting for non-axisymmetric injection from the diode [114, 142], low-frequency (MHz) sweep on Axis-II has been corrected using the same technique used for eliminating corkscrew [142–146].

The FWHMs of the four Axis-II radiographic source spots are consistently less than 1 mm FWHM, and they are displaced from each other by less than 0.05 mm [147]. The resulting time-resolved flash radiographs of material motion in large-scale hydrodynamic experiments have provided new insights for the experimenters.

8. Laser-Driven Electron Beam Radiography

Electron radiography was studied in the 1940s [148] using electrons from a β source. Although usable radiographs were obtained, this was effectively a "proof of principle" of the technique and was not competitive with normal X-ray radiography. Both reflection and transmission electron radiography were demonstrated employing secondary photoelectrons liberated by hard X-rays (180 keV) using emulsions [149]; again, these were relatively low-quality images (see Fig. 10), due mainly to the diffusion of the low-energy electrons.

Fig. 10. An electron radiograph of a butterfly wing by transmission (adapted from Ref. 149).

The advent of intense short-pulse electron beams from laser-excited plasmas has revived interest in this technique. Early studies [150] demonstrated the viability of the technique, and now several groups [151–153] have developed the technique further. Schumaker *et al.*, in particular [153], have recently published stunning ultrafast electron-radiographic images obtained from laser-driven electron acceleration (see Fig. 11). Furthermore, Merrill *et al.* have developed a portable electron radiography source using a converted Varian CLinac 2500 medical linac [154].

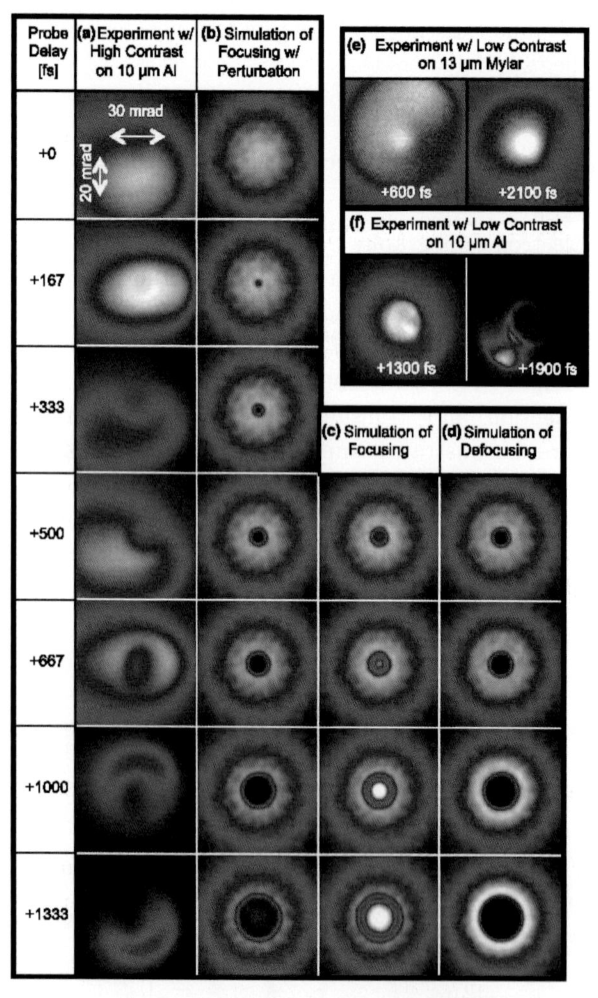

Fig. 11. (Column a) Measured radiographs taken from laser shots with high contrast and $10\,\mu$m Al at delay timings from 0 to Þ1333 fs. (Columns b–d) Simulated radiographs for each respective delay with defocusing with azimuthal perturbation (b), focusing (c) and defocusing (d). (The color and length scales are the same in each column.) Measured radiographs from low-contrast, $13\,\mu$m Mylar shots (e), and $10\,\mu$m Al shots (f) at selected delays. Note that the electron beam profile is elliptical before interacting with the target, and the simulated profiles assume radial symmetry for simplicity. (Taken from Ref. 153.)

9. Proton and Ion Beam Radiography

9.1. *Proton radiography in medicine*

Charged particle therapy (CPT) uses protons and other light ions such as carbon to treat certain localized forms of cancer (see for example, Ref. 155 for a discussion of this treatment modality). The number of CPT centers is doubling roughly every eight years, and there are now more than 40 operating or under construction. The basic accelerator requirements are, from an accelerator perspective, relatively modest, although the constraints on the dose delivery are stringent, with the dose delivered to the patient required to be within 2% of the prescribed dose. The (kinetic) energy required is at least 230 MeV (protons) and around 440 MeV/nucleon (carbon) to give access to tumors in all locations. A critical issue is the positioning of the tumor with respect to the beam. While many tumors that are treated (such as in the head and neck) are relatively immobile, others (such as in the thorax) can move significantly through the breathing cycle, and can move significantly between treatment fractions. Many installations have onboard radiography, but the number of such images taken during treatment is usually limited to avoid overexposure to radiation. There is thus interest in alternative forms of radiography which can deliver useful images with lower radiation doses.

Protons can be used to provide radiographic images [156]. For the whole body, this requires a somewhat higher energy (around 300 MeV) than is needed for therapy (~240 MeV). From an accelerator perspective, any accelerator that is suitable for therapy is, if the energy is sufficient, also adequate for proton radiography — the beam currents required for radiography are significantly less than is typical for therapy. The challenge lies in designing a suitable detection method, since both the (lateral) position and the (residual) energy of the proton must be recorded. The conceptual design of such a scanner is shown in Fig. 12. Briefly, by recording both the lateral position and the residual proton energy, the integral effective density along the proton path can be estimated which, when combined with information from a number of beam directions, allows a three-dimensional density-weighted representation of the

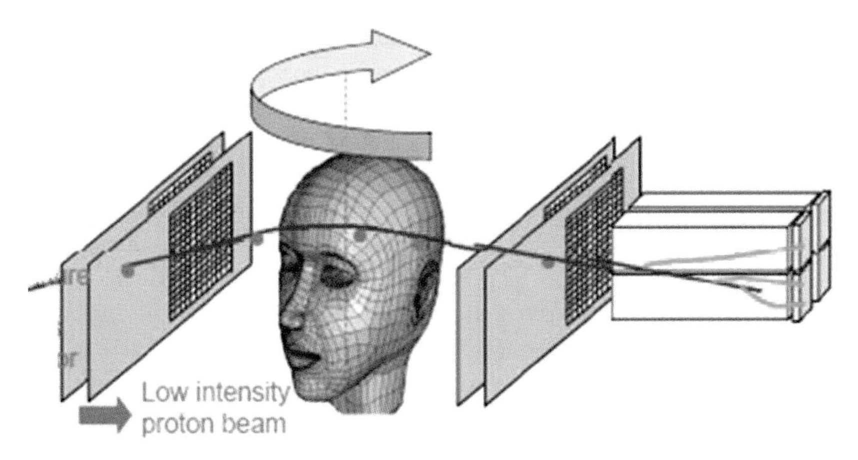

Fig. 12. A proton CT scanner concept with four (x, y) tracker planes (2 + 2) and a CsI range calorimeter. (Adapted from Ref. 157.)

anatomical structures (the integral stopping power) to be reconstructed. There is a potential advantage compared to X-ray CT scanning in the reduction of the radiation dose received by the patient (0.03–0.3 mGy for a proton CT scan of the head versus 1.4 mGy for a conventional CT scan). The time taken for a complete exposure is likely to be dominated by the time taken to rotate the gantry, but needs to be kept short (<10 min).

In principle, of course, this can also be used with heavier ions [158], where the addition mass leads to less lateral dispersion but where there is a small additional radiation dose from nuclear fragmentation and dissociation.

9.2. *Proton radiography in materials research*

Protons can be used for radiography in materials research in several ways (see Ref. 159 for an early review). The *residual range technique* (also called *marginal range radiography*) as discussed above provides an image of the integral density distribution. *Multiple scattering radiography* can be employed where there are sharp lateral density changes, making use of the feature that the multiple scattering within a homogeneous medium about any point is symmetric, but along a sharp boundary between materials of different densities, more particles are scattered into the light medium than are scattered out of it, producing clear "edges." The difference between these two techniques is well illustrated in Fig. 13, which shows radiographs of a leaf obtained from marginal range radiography and multiple scattering radiography, using 12 MeV protons.

If only the position is recorded in the image plane, the multiple scattering can produce a blurred image, but if both position and angle are measured in the image plane, the object can be reasonably well reconstructed if projected back to the object plane. The contrast can be further improved if the momentum is also measured, allowing elastically scattered events to be preferentially selected. For inelastic events, the distribution of the vertices in the object will image the (nuclear) density distribution with a precision determined by the vertex resolution of the reconstruction algorithm. There were several demonstrations of the potential for charged particle radiography (using pions and kaons as well as protons) in the 1960s and 1970s, where most particle physics experiments could produce radiographs of their apparatus, similar to those shown in Figs. 5 and 6. However, this process is very inefficient (in energy terms) and offers no additional advantage when compared with, for example, protons, and is not pursued.

A third method, related to marginal-range radiography, measures the energy loss of individual protons, and also delivers a measurement of the integral density distribution, but requires protons of significantly higher energy and more extensive particle-tracking detectors and possible magnetic spectrometry.

As discussed in Sec. 3, there is a large field of application of radiography for studying the evolution of dense materials under extreme conditions.

It can be shown [144] that the optimal interaction length $\lambda_{\rm opt}$ (defined as $\lambda = A/\sigma$, where A is the atomic weight in gm and σ is the cross-section in $\rm cm^2$) is given by $\rho_A/2$, where ρ_A is the areal density $\int_0^L \rho(x, y, z)dz$. As noted in Subsec. 3.1, there is a requirement to image large heavy metal objects with high areal densities of several hundred $\rm g/cm^2$, which is much greater than the typical interaction length for X-rays of the order of $25\,\rm g/cm^2$ or less. Protons, however, have much greater interaction lengths (typically $200\,\rm g/cm^2$ in uranium, for example), and (at a sufficiently high energy where inelastic scattering dominates) can give sharp radiographs. The technique has been developed since 1996, when lens-focused proton radiography was proposed [160]. A comprehensive recent review of charge particle flash radiography can be found in Ref. 3.

For hydrotest proton radiography, high energy is an advantage since it reduces considerably blurring due to multiple scattering; the mean multiple scattering angle θ_m for a particle of momentum p (in MeV/c), velocity βc, charge z is given by $\theta_m = (13.6/\beta cp)z\sqrt{x/X_0}\lfloor 1 + 0.038\ln(x/X_0)\rfloor$, where x is the distance traversed and X_0 is the radiation length. In addition, a series of short proton pulses (bunch trains) allows the dynamical development of the system under observation to be followed.

The relatively simple proton radiography described in Ref. 159 has a number of limitations which lead to image degradation and blurring. In particular, multiple scattering leads to blurring of images, especially when the distance from the object to the image plane is large. Many of these can be mitigated by using quadrupole focusing [161], taking advantage of the position–angle correlation, making it possible to improve contrast in the image plane by selecting the appropriate trajectories in the Fourier plane, as illustrated in Fig. 14. The introduction of magnetic elements also introduces chromatic aberration. For the simple four-quadrupole scheme shown in Fig. 14, the dispersion in the first pair of quadrupoles can be canceled, at least to first order, in the second pair. The initial tests [161] were performed by experiment E912 using the 24 GeV AGS at Brookhaven with a secondary 10 GeV proton beam, and further tests were performed at LANSCE at Los Alamos. The introduction of the magnetic lens improves the resolution by up to two orders of magnitude [162]. With a unit magnifying lens, a resolution of $\sim 180\,\mu$m has been obtained across a 120 mm field of view at LANSCE. The 1 ms proton pulse length can be rapidly chopped by a deflector to generate a series of time-resolved radiographic images.

More recently, it has been shown [162] that the resolution can be improved by a further factor of about 3 by removing also the second order chromatic aberrations in the focusing system, while at the same time increasing the magnification. To see how this works, consider the usual beam transport equation (6) relating two points [in this case, the object (o) and image (i) planes] in a magnet system, modified to take account of the second order chromatic effects:

$$[X]_i = [R][X]_o + [T][X]_o[X]_o, \qquad (6)$$

where $[X]_j = [x, x', y, y', l, \delta p/p]_j$, with x, x', y, y' the position and direction of the particle at point j, and $l, \delta p/p$ are the deviation of the particle from the ideal trajectory of the integral path length and

(a)

(b)

Fig. 13. A proton radiograph of a leaf using (a) marginal range radiography and (b) multiple scattering radiography (from Ref. 159).

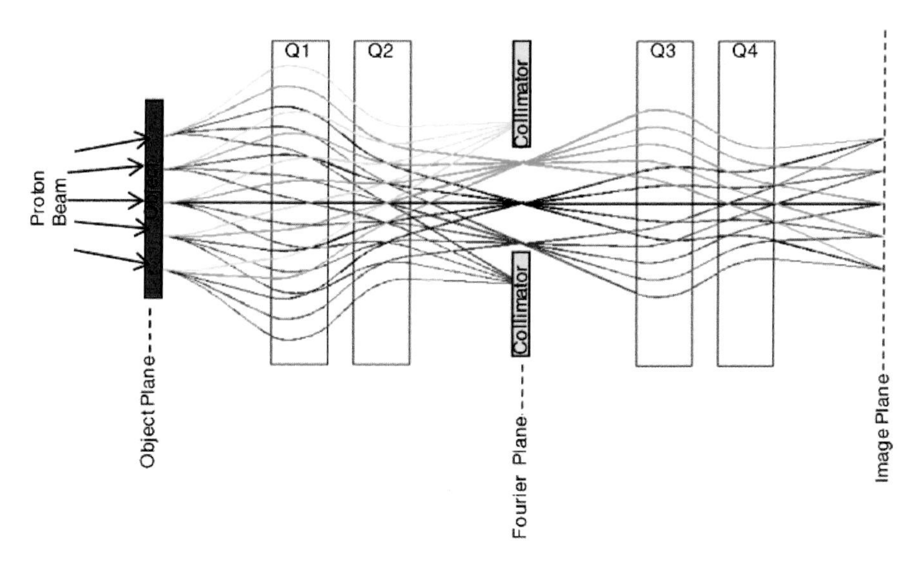

Fig. 14. A plot of trajectories through a four-quadrupole lens, Q1, Q2, Q3 and Q4. The incoming trajectories are shown by the black arrows, and the trajectories for scattering angles of 5.0, 2.5, 0.0, −2.5 and −5.0 mrad are shown in blue, red, black, green and yellow, respectively. The vertical scale has been magnified. Because of the incident position angle correlation, the position at the Fourier plane depends only on scattering in the object plane. (Taken from Ref. 3.)

the momentum respectively at that point. Here $[R]$ is the usual Twiss matrix, with the elements R_{11} and R_{33} essentially giving the magnifications in the horizontal and vertical planes. An achromatic lens system requires that the coupling between the first four components (x, x', y, y') and the momentum deviation vanish in both R and T. What is shown in Ref. 162 is that a simple six-lens system consisting of a singlet, two closely coupled doublets and a final singlet quadrupole (see Fig. 15) can achieve a significant improvement in image quality. With this arrangement, overall magnification of 2.65 has been achieved. The chromatic aberrations arising from the couplings in the horizontal (vertical) planes are

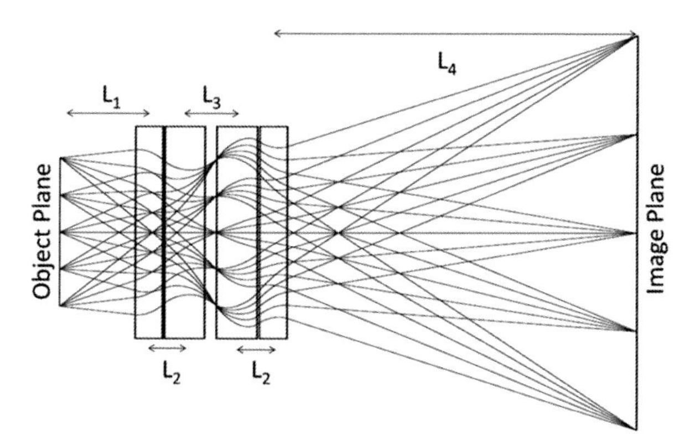

Fig. 15. Horizontal trajectories of 800 MeV protons traveling through the magnifier system. The middle two lenses have two quadrupoles closely coupled. (Taken from Ref. 162.)

reduced by the magnification, with values of the chromatic length of the system T_{126}/M (T_{346}/M) equal to 4.55 m (3.54 m). With this system, object plane resolutions of 53 μm (63 μm) were obtained.

There have been significant improvements in the technique, through better focusing and more advanced detection systems. Advanced proton radiography facilities have been established not only at Los Alamos and Brookhaven, but also at the Terrawatt Accumulator of the Institute for Theoretical and Experimental Physics in Moscow [164], where the 800 MeV proton facility provides four 70 ns pulses of up to 10^{11} protons/pulse and has a permanent magnet magnifier lens with four quadrupoles. This achieved a position resolution of 50 μm over 2 cm in the object plane. There is also a proton radiography facility [165] at the U-70 proton accelerator at IHEP in Protvino that has been operating since 2004 [166–168]; this is currently the highest-energy proton radiographic facility available. The proton energy can be selected up to a maximum of 70 GeV, with a high bunch intensity capability (up to 1.5×10^{13} protons per cycle, with a multiframe capability of up to 29 frames separated by 169 ns (see Fig. 16). The proton pulse length is between 10 and 30 ns across an aperture of 60 mm, although there are plans to increase the aperture to around 250 mm.

There are plans for a 20 GeV proton accelerator with a radiographic capability in China [169] and

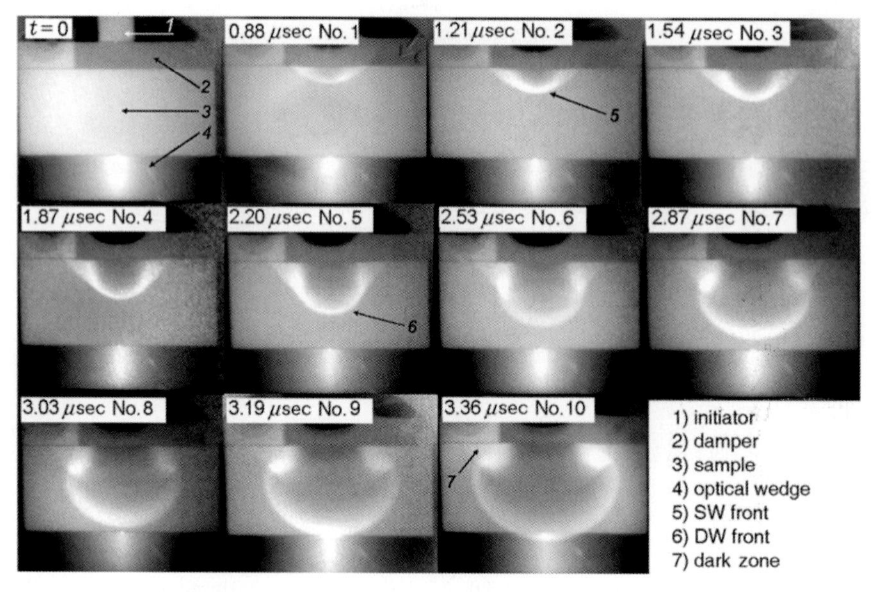

Fig. 16. Images of the transition from initial shock to detonation driven shock in an explosive. These images, taken with the U-70 proton radiographic facility at IHEP in Protvino, show the overtaking of the initial shock wave (SW) by the detonation driven shock wave (DW). The features indicated by the numbered black arrows are (1) initiator, (2) damper, (3) sample, (4) optical wedge, (5) initial shock wave (SW) front, (6) detonation driven shock wave (DW) front, (7) dark zone. (Image from Ref. 166.)

ideas for a proton radiography facility (PRIOR — PRoton mIcroscOpe at FAIR) [170].

9.3. *Laser-driven proton radiography*

The next major development in particle beam radiography uses plasma-driven particle beams, which have several potential advantages over conventional accelerator-based systems — shorter bunch lengths, higher bunch intensity, smaller emittance and an approximately point source, all of which lead to improved resolution and reduced blurring. Cobble *et al.* [171] showed that 2–3 μm object plane resolution could be achieved with protons generated from thin foils (gold or aluminum) with energies up to a few MeV, driven by short pulses (∼0.6 ps) from the LANL Trident laser with about 1 J on the target. (The laser power and irradiance were estimated to be 1.5 ± 0.3 TW and $\sim 3 \times 10^{19}$ W/cm^2, respectively.) This is about two orders of magnitude better resolution than was obtained (see above) with an 800 MeV conventional accelerator source, and with an experiment setup which is considerably simpler.

The proton spectrum from these experiments shows a broad, approximately exponential spectrum typical of the Target Normal Sheath Acceleration (TNSA), with the maximum proton energy being related to the total laser power. Although this might seem to be a disadvantage when compared with the quasi-monoenergetic protons from a conventional accelerator, this can be overcome by using either fast timing at the image plane to select the proton energy, or a range chamber to provide a sequence of images similar to the residual range technique.

Choi *et al.* [172] have compared the radiographic images of several phantoms using 1.8 MeV protons from a tandem van de Graaff accelerator and TNSA-generated protons with a maximum energy of 1.8 MeV, and have shown that the laser-driven protons give much sharper images (see Fig. 17).

Laser-produced proton beams have been used [173] to provide self-radiographic images of the evolution of the thin metal target and its glass holder during irradiation, and [174] to image the plasma expansion following the interaction of an intense laser pulse with the inner surface of a gold hohlraum. Finally, high-intensity-laser-generated protons with energies of several tens of MeV have been used [175] to image the electromagnetic field configuration around an imploded capsule 860 μm in diameter filled with helium 1t 18 atm subjected to a 17 kJ shock from all 60 beams of the OMEGA Nd:glass laser.

As the above examples show, the broad spectrum typical of TNSA-generated protons can

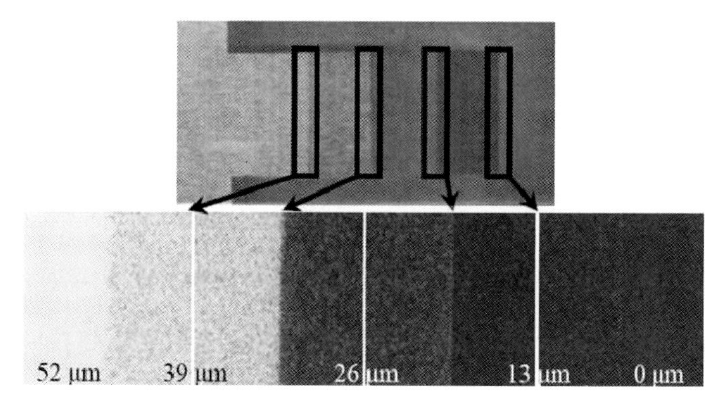

Fig. 17. *Top*: The image of a step-wedge phantom take with femtosecond laser-driven protons. *Middle*: The enlarged images of each boundary from the upper image. *Bottom*: The enlarged images of the same step-wedge phantom obtained with 1.8 MeV protons from the van de Graaff. (Data adapted from Ref. 172.)

be accommodated, and might even confer some advantage. However, particularly for thicker or more complex structures, intense, short-pulse high-energy (several hundred MeV and above) would be better. Alternative acceleration mechanisms, for example radiation pressure acceleration (RPA) [176], might permit the generation of higher-energy, tunable, narrowband proton beams, and even higher energies might be accessible from circularly polarized laser pulses [177].

9.4. *Other potential charged particle radiography techniques*

Several other noninvasive techniques can be employed to yield structural information about objects using high-energy protons or other ions. For example, the nuclear excitations along the proton path give a gamma ray map of the density distribution along the proton trajectory. With heavier ions such as carbon, other imaging techniques may be possible, for example in-beam Positron Emission Tomography from the ^{11}C which is produced from fragmentation along the ion path [178]. Ion beam analysis (IBA) covers a number of techniques to resolve material structures and compositions near the surface of objects; these techniques include elastic and inelastic scattering, X-ray emission and channeling (see for example the reviews in Refs. 179 and 180). These techniques are important industrially as well as for academic research. Finally, ions heavier than protons can also be used for charged particle radiography, where the larger ballistic effect means that multiple scattering is reduced when compared with protons, leading to less blurring and a smaller penumbra.

10. Conclusions

Radiography has been an essential tool in many disciplines, from medicine to engineering, over the past century or so. With the development of accelerators, new radiographic methods have been developed, using particle beams (neutrons, protons, electrons and potentially other particles such as ions, muons, pions or neutrinos) to explore different domains. Naturally occurring particle beams (cosmic rays) can also be used for radiography. As well as serving several scientific disciplines, particle beam radiography is an essential tool in medicine, industry and security.

Flash radiography as a means of diagnosing the motion of dense materials by freezing it with a short pulse of penetrating high-energy photons has been under development since early in the last century. The last decade has seen significant advances in the capability to obtain high-quality flash radiographs of large hydrodynamic experiments with up to hundreds of g/cm^2 of areal mass. The LIA-based machines have achieved spot sizes less than 1 mm FWHM, with a corresponding dose of over 500 rad@1m, and have made high-resolution multitime flash radiography a reality. The development of proton beam flash radiography adds new diagnostic capability to larger and more optically dense conditions.

In the longer term, multimodal radiography (X-ray or electron in combination with proton) is likely to develop still further as the technique is extended into the submicron spatial and subns temporal regime using laser and FEL-based particle beam sources.

Acknowledgments

We would like to thank Chris Morris, Frank Merrill and Martin Schulze at the Los Alamos National Laboratory, USA; Winfried Kockelmann at the STFC Rutherford Appleton Laboratory, UK; John Banhart of the Institute of Applied Materials in Berlin, Germany; and Roger Webb of the Surrey Ion Beam Centre, University of Surrey, UK, for valuable input.

Part of this research (C. E.) was supported by the US Department of Energy contract DE-AC52-06NA25396.

References

[1] W. C. Röntgen, Über eine neue Art von Strahlen, *Sitzb. Wurzb. Phys.-med. Gesells.* **9**, 132–141, Wurzburg: Stahl (1895).
[2] L. W. Alvarez *et al.*, Search for hidden chambers in the pyramids using cosmic rays, *Science* **167**, 832–839 (1970).
[3] C. L. Morris *et al.*, Charge particle radiography, *Rep. Prog. Phys.* **76**, 046301 (2013).
[4] For a general discussion on the reconstruction of CT images, see for example A. C. Kak and M. Slaney, *Principles of Computerized Tomographic Imaging* (Society for Industrial and Applied Mathematics, Philadelphia, PA, USA, 2001).
[5] Image from J. Banhart (ed.), *Advanced Tomographic Methods in Materials Research and Engineering* (Oxford University Press, New York, 2008).
[6] See for example M. Hohlmann *et al.*, Design and construction of a first prototype muon tomography system with GEM detectors for the detection of nuclear contraband, IEE NSS/MIC 971–975 (2009).
[7] See for example H. Fujii *et al.*, Performance of a remotely located muon radiography system to identify the inner structure of a nuclear plant, *Prog. Theor. Exp. Phys.* 073C01 (2012).
[8] See for example K. M. Hiroyuki *et al.*, Detecting a mass change inside a volcano by cosmic-ray muon radiography (muography); first measurement at the Asama volcano, Japan, *Geophys. Res. Lett.* **36**, L17302 (2009).
[9] T. K. Gaisser, Atmospheric neutrinos in the context of muon and neutrino radiography, *Earth Planets Space* **62**, 195–199 (2010).
[10] J. Altegoer *et al.*, The NOMAD experiment at the CERN SPS, *Nucl. Instrum. Methods Phys. Res. A* **404**, 96–128 (1998).
[11] The CDHS detector is described in: M. Holder *et al.*, A detector for high-energy neutrino interactions, *Nucl. Instrum. Methods Phys. Res.* **148**, 235–249 (1978); the plot is available on the website http://knobloch.home.cern.ch/knobloch/cdhs/h2.html
[12] M. J. Darby, D. A. Barron and R. E. Hyland, *Oxford Handbook of Medical Imaging* (Oxford University Press, 2012).
[13] J. Chadwick, Possible existence of a neutron, *Nature* **129**, 312 (1932).
[14] E. Kuhn, US patent number 2186 757 (1940).
[15] Image from J. Banhart, *Advanced Tomographic Methods in Materials Research and Engineering* (Oxford University Press, New York, 2008).
[16] J. Thewliss, Neutron radiography, *Br. J. Appl. Phys.* **7**, 345 (1956).
[17] E. Lehmann *et al.*, Properties of the radiography facility NEUTRA at SINQ and its potential for use as a European reference facility, *Nondestr. Test Eval.* **16**, 191–202 (2001); doi: 10.1080/10589750108953075.
[18] See for example K. W. Jones and K. F. Schoenberg, Operational plans and status of the Los Alamos Neutron Science Center (LANSCE), in *Proc. LINAC08* (Victoria, BC, Canada, 2008), pp. 88–90.
[19] See for example E. A. Koptelov *et al.*, Spallation neutrons at INR RAS — a Facility Status Report, ICANS-XVII, 17th *Meeting of the International Collaboration on Advanced Neutron Sources* (Santa Fe, New Mexico, USA, 2005).
[20] S. Henderson, Spallation neutron source progress, challenges and upgrade options, in *Proc. EPAC 2008* (Genoa, 2008), pp. 2892–2896.
[21] See for example H. Danared, The design of the ESS accelerator, in *Proc. Int. Particle Accelerator Conference* (New Orleans, USA, 2012), pp. 3904–3906.
[22] See for example M. Seifel *et al.*, Production of a 1.3 MW proton beam at PSI, in *Proc. IPAC 2010* (Kyoto, 2010), pp. 1309–1313.
[23] See for example D. J. S. Findlay, ISIS – Pulsed neutron and muon source, in *Particle Accelerator Conference* (Albuquerque, New Mexico, USA, 2007), pp. 695.
[24] See for example K. Hasegauva, M. Kinsho and H. Oguri, Status of J-PARC accelerators, in *Proc. IPAC 2013* (Shanghai, 2013), to be published in *Phys. Rev. ST Accel. Beams.*
[25] S. Fu *et al.*, Status of CNSN Project, in *Proc. IPAC 2013* (Shanghai, 2013), to be published in *Phys. Rev. ST Accel. Beams.*
[26] See for example Y. Fujii, Basic to industrial research on neutron platform in Japan, *Pramana J. Phys.* **71**, 617–622 (2008).
[27] See for example G. E. McMichael, Accelerator research at the Rapid Cycling Synchrotron at

IPNS, in *Proc. European Accelerator Conference 2006* (Edinburgh, 2006), pp. 339–341.

[28] See for example Neutron generators for analytical purposes, in *IAEA Radiation Technology Report* #1 (2012).

[29] C. Ekdahl, Characterizing flash-radiography source spots, *J. Opt. Soc. Am. A* **28**, 2501 (2011).

[30] B. D. Sowerby *et al.*, Recent developments in fast neutron radiography for the interrogation of air cargo containers. IAEA International Topical Meeting on Nuclear Research Applications and Utilization of Accelerators (Vienna, 4–8 May 2009), Paper SM/EN-01.

[31] S. Pesente *et al.*, Progress in tagged neutron beams for cargo inspections, *Nucl. Instrum. Methods Phys. Res. B* **261**, 268–271 (2007).

[32] D. A. Strellis *et al.*, Explosives (and other threats) detection using pulsed neutron interrogation and optimized detectors, in *Proc. SPIE* 8017, Detection and Sensing of Mines, Explosive Objects, and Obscured Targets XVI, 801717 (2011).

[33] M. Balaskó *et al.*, A novel type epithermal neutron radiography detecting and imaging system, *Nucl. Instrum. Methods Phys. Res. A* **424**, 263–269 (2009).

[34] H. Tomita *et al.*, Development of epithermal neutron camera based on resonance-energy-filtered imaging with GEM, *JINST* **7**, C05010 (2011).

[35] See for example A. Rack *et al.*, High resolution synchrotron-based radiography and tomography using hard X-rays at the BAM *line* (BESSY II), *Nucl. Instrum. Methods Phys. Res. A* **586**, 327–344 (2008).

[36] See for example I. Reche *et al.*, Synchrotron radiation and laboratory micro-X-ray computed tomography — useful tools for the material identification of prehistoric objects made of ivory, bone or antler, *J. Anal. At. Spectrom.* **26**, 1802–1812 (2011).

[37] See for example R. T. Lopes *et al.*, X-ray transmission microtomography using synchrotron radiation, *Nucl. Instrum. Methods Phys. Res. A* **595** 604–607 (2003).

[38] M. Marinescu *et al.*, Synchrotron radiation X-ray phase micro-computed tomography as a new method to detect iron oxide nanoparticles in the Brain, *Mol. Imag. Biol.*, doi: 10.1007/s11307-013-063906 (2013) in press.

[39] G. Barnea, Penumbral imaging made easy, *Rev. Sci. Instrum.* **65**, 1949 (1994).

[40] B. T. McCuistian *et al.*, Temporal spot size evolution of the DARHT First Axis Radiographic Source, in *Proc. European Part. Accel. Conf.* (2008), pp. 1206–1208.

[41] *J. C. Martin on Pulsed Power*, eds. T. H. Martin, A. H. Guenther and M. Kristiansen (Plenum, New York, 1996).

[42] M. Sinclair, Current radiographic pulsed power machines at AWE, in *Proc. 15th IEEE Int. Pulsed Power Conf.* (2005), pp. 124–129.

[43] J. R. Boller, J. K. Burton and J. D. Shipman, Jr., Status of the upgraded version of the NRL Gamble II Pulse Power Generator, in *Proc. 2nd IEEE Int. Pulsed Power Conf.* (1979), pp. 205–208.

[44] I. D. Smith, Induction voltage adders and the induction accelerator family, *Phys. Rev. ST Accel. Beams* **7**, 064801 (2004).

[45] N. Bruner *et al.*, Modeling particle emission and power flow in pulsed-power driven, non-uniform transmission lines, *Phys. Rev. ST Accel. Beams* **11**, 040401 (2008).

[46] K. D. Hahn *et al.*, Measurement and simulations of plasma evolution in the A-K gap of the self-magnetic pinch diode fielded on the RITS-6 accelerator, *IEEE Trans. Plasma Sci.* **38**, 2652 (2010).

[47] D. B. Seidel, T. D. Pointon and B. V. Oliver, Controlling feed electron flow in MITL-driven radiographic diodes, in *Proc. 18th IEEE Int. Pulsed Power Conf.* (2011), pp. 875–880.

[48] R. J. Commisso *et al.*, Status of the Mercury Pulsed-Power Generator, a 6-MV, 360-kA, magnetically-insulated inductive voltage adder, in *Proc. 14th IEEE Int. Pulsed Power Conf.* (2003), pp. 383–386.

[49] R. J. Allen *et al.*, Electrical modeling of Mercury for optimal machine design and performance, in *Proc. 14th IEEE Int. Pulsed Power Conf.* (2003), pp. 887–890.

[50] R. J. Allen *et al.*, Initialization and operation of Mercury, a 6-MV MIVA, in *Proc. 15th IEEE Int. Pulsed Power Conf.* (2005), pp. 318–332.

[51] T. A. Holt *et al.*, Analysis of switch performance on the Mercury pulsed power generator, in *Proc. 15th IEEE Int. Pulsed Power Conf.* (2005), pp. 128–131.

[52] D. Weidenheimer *et al.*, Design of a driver for the Cygnus X-ray source, in *Proc. 13th IEEE Int. Pulsed Power Conf.* (2001), pp. 591–595.

[53] V. Carboni *et al.*, Pulse power performance of the Cygnus 1 and 2 radiographic sources, in *Proc. 14th IEEE Int. Pulsed Power Conf.* (2003), pp. 905–908.

[54] J. Smith *et al.*, Cygnus dual beam radiographic source, in *Proc. 15th IEEE Int. Pulsed Power Conf.* (2005), pp. 334–337.

[55] K. Thomas *et al.*, Status of the AWE Hydrus IVA fabrication, in *Proc. 18th IEEE Int. Pulsed Power Conf.* (2011), pp. 1042–1047.

[56] J. E. Maenchen *et al.*, Intense electron beam sources for flash radiography, in *Proc. 14th Int. Conf. on High-Power Particle Beams, AIP Conf. Proc.* **650**, 117–122 (2002).

[57] S. Portillo *et al.*, Time-resolved spot size measurements from various radiographic diodes on the RITS-3 accelerator, *IEEE Trans. Plasma Sci.* **34**, 1908 (2006).

[58] C. Ekdahl and S. Humphries, Grid-focused diodes for radiography, in *Proc. 15th IEEE Int. Pulsed Power Conf.* (2005), pp. 868–871.

[59] A. I. Pavlovskii *et al.*, High-current ironless betatrons, *Sov. Phys. Dokl.* **10**, 30 (1965).

[60] Yu. P. Kuropatkin *et al.*, Characteristics of the installation for flash radiography based on the uncored betatron BIM-M, in *Proc. 11th IEEE Int. Pulsed Power Conf.* (1997), pp. 1663–1668.

[61] Yu. P. Kuropatkin *et al.*, Uncored betatron BIM-M a source of bremsstrahlung for flash radiography, in *Proc. 11th IEEE Int. Pulsed Power Conf.* (1997), pp. 1669–1673.

[62] N. I. Egorov *et al.*, Use of pulsed radiography for investigation of equation of state of substances at megabar pressures, *Contrib. Plasma Phys.* **51**, 333 (2011).

[63] M. A. Mochalov *et al.*, Measurement of quasi-isentropic compressibility of helium and deuterium at pressures of 1500–2000 GPa, *J. Exp. Theor. Phys.* **115**, 614 (2012).

[64] N. C. Christophilos *et al.*, High current linear induction accelerator for electrons, *Rev. Sci. Instrum.* **35**, 886 (1964).

[65] S. Humphries, Jr., *Charged Particle Beams* (Wiley, New York, 1990), pp. 283–325.

[66] R. J. Briggs and G. Westenskow, *Induction Accelerators*, eds. K. Takayama and R. J. Briggs (Springer, New York, 2011), pp. 2–22.

[67] G. Westenskow and Y.-J. Chen, *Induction Accelerators*, eds. K. Takayama and R. J. Briggs (Springer, New York, 2011), pp. 165–184.

[68] N. Pichoff *et al.*, Contributors to AIRIX focal spot size, in *Proc. 2006 European Part. Accel. Conf.* (2006), pp. 2164–2166.

[69] D. R. Welch and T. P. Hughes, Effect of target-emitted ions on the local spot of an intense electron beam, *Lasers Charged Part. Beams* **16**, 285 (1998).

[70] B. T. McCuistian *et al.*, Temporal spot size evolution of the DARHT first axis radiographic source, in *Proc. 2008 European Part. Accel. Conf.* (2008), pp. 1206–1208.

[71] V. K. Neil, L. S. Hall and R. K. Cooper, Further theoretical studies of the beam breakup instability, *Part. Accel.* **9**, 213 (1979).

[72] Y. J. Chen, Corkscrew modes in linear accelerators, *Nucl. Instrum. Methods Phys. Res. A* **292**, 455 (1990).

[73] E. Merle *et al.*, Transport optimization and characterization of the 2kA AIRIX beam, in *Proc. 20th Int. Linac Conf.* (2000), pp. 494–496.

[74] B. Kulke *et al.*, Initial performance parameters on FXR, in *Proc. 15th IEEE Power Modulator Symp.* (1982), pp. 307–311.

[75] B. Kulke and R. Kihara, Recent performance improvements on FXR, *IEEE Trans. Nucl. Sci.* **30**, 3030 (1983).

[76] R. D. Scarpetti *et al.*, Upgrades to the LLNL flash X-ray induction linear accelerator (FXR), in *Proc. 11th IEEE Int. Pulsed Power Conf.* (1997), pp. 597–602.

[77] M. Ong, C. Avalle, R. Richardson and J. Zentler, *LLNL Flash X-Ray Radiography Machine (FXR) Double-Pulse Upgrade Diagnostics* in *Proc. 11th IEEE Int. Pulsed Power Conf.* (1997), pp. 430–435.

[78] L. G. Multhauf *et al.*, The LLNL Flash X-ray Induction Linear Accelerator (FXR), *25th Int. Conf. on High-Speed Photography and Photonics, Proc. SPIE* **4948**, 622 (2003).

[79] J. M. Zentler and R. D. Van Maren, Improved focus solenoid design for linear induction accelerators, in *Proc. 16th Int. Linac Conf.* (1992), pp. 459–461.

[80] M. Ong *et al.*, FXR fast beam imaging diagnostics, in *Proc. 12th IEEE Int. Pulsed Power Conf.* (1999), pp. 636–639.

[81] W. E. Nexsen, R. D. Scarpetti and J. Zentler, Reconstruction of FXR beam conditions, in *Proc. 2001 Part. Accel. Conf.* (2001), pp. 2383–2385.

[82] M. M. Ong and G. E. Vogtlin, Flash X-ray (FXR) accelerator optimization beam-induced voltage simulation and TDR measurements, in *Proc. 15th IEEE Int. Pulsed Power Conf.* (2005), pp. 54–57.

[83] M. Ong, Flash X-ray (FXR) accelerator optimization injector voltage-variation compensation via beam-induced gap voltage, in *Proc. 15th IEEE Int. Pulsed Power Conf.* (2005), pp. 112–115.

[84] T. Houck *et al.*, Tuning the magnetic transport of an induction linac using emittance, in *Proc. Int. Linac Conf.* (2006), pp. 444–446.

[85] M. Ong *et al.*, Estimating the reliability of Lawrence Livermore National Laboratory (LLNL) flash X-ray (FXR) machine, in *Proc. 16th IEEE Int. Pulsed Power Conf.* (2007), pp. 1078–1081.

[86] W. J. DeHope *et al.*, "Real-life" pulse flattening on the LLNL flash X-ray (FXR) machine, in *Proc. 16th IEEE Int. Pulsed Power Conf.* (2007), pp. 1261–1263.

[87] B. R. Kreitzer, T. L. Houck and O. C. Luchterhand, Suppressing thermal energy drift in the LLNL flash X-ray accelerator using linear disk resistor stacks, in *Proc. 18th IEEE Int. Pulsed Power Conf.* (2011), pp. 1408–1412.

[88] C. Cavailler, AIRIX, an induction accelerator facility developed in CEA for flash radiography in detonics, in *Proc. 23rd Int. Congress on High-Speed Photography and Photonics, Proc. SPIE* **3516**, 25 (1999).

[89] F. Bombardier *et al.*, AIRIX induction accelerator performances at CEA, in *Proc. 13th IEEE Int. Pulsed Power Conf.* (2001), pp. 1693–1695.

[90] M. Mouillet *et al.*, First results of the AIRIX induction accelerator, in *Proc. 20th Int. Linac Conf.* (2000), pp. 491–493.

[91] E. Merle *et al.*, Transport optimization and characterization of the 2kA AIRIX electron beam, in *Proc. 20th Int. Linac Conf.* (2000), pp. 494–496.

[92] E. Merle *et al.*, Transport of the 1.92–3.1 kA AIRIX electron beam, in *Proc. 2001 Part. Accel. Conf.* (2001), pp. 3481–3483.

[93] E. Merle *et al.*, The first years with the AIRIX flash X-ray radiographic facility, in *Proc. 25th Int. Congress on High-Speed Photography and Photonics*, *Proc. SPIE* **4948**, 652 (2002).

[94] M. Caron *et al.*, High intensity high energy e-beam interacting with a thin solid state target: First results at AIRIX, in *Proc. 2005 Part. Accel. Conf.* (2005), pp. 1982–1984.

[95] O. Mouton *et al.*, Computer-assisted beam characterization at AIRIX facility, in *Proc. 2007 Part. Accel. Conf.* (2007), pp. 3250–3252.

[96] H. Dzitko *et al.*, Reliability study of the AIRIX accelerator over a functioning period of ten years (2000–2010), in *Proc. 2011 Part. Accel. Conf.* (2011), pp. 1882–1884.

[97] F. Poulet *et al.*, AIRIX measurement chain optimization for electron beam dynamic and dimensional characteristics analysis, in *Proc. 2012 Rus. Part. Accel. Conf.* (2012), pp. 677–679.

[98] H. Dzitko *et al.*, Operational efficiency of the AIRIX accelerator since its commissioning, in *Proc. Int. Part. Accel. Conf.* (2012), pp. 4017–4019.

[99] J. Deng *et al.*, Upgrading of linear induction accelerator X-ray facility (LIAXF), in *Proc. 19th Int. Linear Accel. Conf.* (1998), pp. 389–390.

[100] J. Deng *et al.*, Design of the DRAGON-I linear induction accelerator, in *Proc. 21st Int. Linear Accel. Conf.* (2002), pp. 40–42.

[101] J. Deng *et al.*, DRAGON-I linear induction accelerator, in *Proc. 23rd Int. Linear Accel. Conf.* (2006), pp. 49–51.

[102] W. Zhang *et al.*, Beam instability and correction for "DRAGON-I," in *Proc. 2007 Part. Accel. Conf.* (2007), pp. 4114–4116.

[103] J. Liu *et al.*, Decreasing the scatter effect in density reconstruction in high-energy X-ray radiography, *Nucl. Instrum. Methods Phys. Res. A* **716**, 86–89 (2013).

[104] S. Chen *et al.*, Design and characterization of a high-power induction module at megahertz repetition rate burst mode, *Nucl. Instrum. Methods Phys. Res. A* **579**, 941–950 (2007).

[105] J. J. Deng *et al.*, R&D status of high-current accelerators at IFP, *J. Kor. Phys. Soc.* **59**, 3619 (2011).

[106] M. J. Burns *et al.*, Status of the dual-axis radiographic hydrotest facility, in *Proc. 18th Int. Linear Accel. Conf.* (1996), pp. 875–877.

[107] M. J. Burns *et al.*, DARHT accelerators update and plans for initial operation, in *Proc. Part. Accel. Conf.* (1999), pp. 617–621.

[108] R. B. Miller, Mechanism of explosive electron emission for dielectric fiber (velvet) cathodes, *J. Appl. Phys.* **84**, 3880 (1998).

[109] M. J. Burns *et al.*, Cell design for the DARHT linear induction accelerator, in *Proc. 1991 Part. Accel. Conf.* (1991), pp. 2958–2960.

[110] M. J. Burns *et al.*, Magnet design for the DARHT linear induction accelerators, in *Proc. 1991 Part. Accel. Conf.* (1991), pp. 2110–2112.

[111] M. J. Burns *et al.*, Status of the DARHT phase 2 long-pulse accelerator, in *Proc. 2001 Part. Accel. Conf.* (2001), pp. 325–329.

[112] M. V. Zumbro *et al.*, DARHT second axis — status and plans, in *Proc. 21st Int. Linear Accel. Conf.* (2002), pp. 314–315.

[113] C. Ekdahl *et al.*, First beam at DARHT-II, in *Proc. 2003 Part. Accel. Conf.* (2003), pp. 558–562.

[114] C. Ekdahl *et al.*, Initial electron-beam results from the DARHT-II linear induction accelerator, *IEEE Trans. Plasma Sci.* **33**, 892–900 (2005).

[115] R. D. Scarpetti *et al.*, Status of the DARHT 2nd Axis at Los Alamos National Laboratory, in *Proc. 15th IEEE Int. Pulsed Power Conf.* (2005), pp. 37–42.

[116] K. E. Nielsen *et al.*, Upgrades to the DARHT second axis induction cells, in *Proc. 15th IEEE Int. Pulsed Power Conf.* (2005), pp. 43–46.

[117] T. P. Hughes *et al.*, Numerical model of the DARHT accelerating cell, in *Proc. 15th IEEE Int. Pulsed Power Conf.* (2005), pp. 143–146.

[118] J. Barraza *et al.*, Mechanical engineering upgrades to the DARHT-II induction cells, in *Proc. 15th IEEE Int. Pulsed Power Conf.* (2005), pp. 402–406.

[119] W. L. Waldron *et al.*, Reliability and lifetime of the DARHT second axis induction cells, in *Proc. 15th IEEE Int. Pulsed Power Conf.* (2005), pp. 47–49.

[120] B. A. Prichard *et al.*, Technological improvements in the DARHT-II accelerator cells, in *Proc. 2005 Part. Accel. Conf.* (2005), pp. 169–173.

[121] C. Ekdahl *et al.*, Electron-beam dynamics in the DARHT-II linear-induction accelerator, in *Proc. 25th Int. Linear Accel. Conf.* (2008), pp. 311–313.

[122] M. Schulze *et al.*, Commissioning the DARHT-II accelerator downstream transport and target, in *Proc. 25th Int. Linear Accel. Conf.* (2008), pp. 434–436.

[123] K. Nielsen, Design and performance of the DARHT second axis accelerator, in *Proc. 18th IEEE Int. Pulsed Power Conf.* (2011), pp. 1048–1051.

[124] A. C. Paul *et al.*, The beamline for the second axis of the Dual Axis Radiographic Test Facility, in *Proc. 1999 Part. Accel. Conf.* (1999), pp. 3254–3256.

[125] G. A. Westenskow *et al.*, The DARHT-II downstream transport beamline, in *Proc. 2001 Part. Accel. Conf.* (2001), pp. 3487–3489.

[126] F. W. Chambers *et al.*, Parallel measurement and modeling of transport in the DARHT II beamline on ETA II, in *Proc. 15th IEEE Int. Pulsed Power Conf.* (2005), pp. 139–142.
[127] K. C. D. Chan *et al.*, Ion effects in the DARHT-II downstream transport, in *Proc. 2005 Part. Accel. Conf.* (2005), pp. 375–377.
[128] J. T. Weir *et al.*, DARHT II scaled accelerator tests on the ETA II accelerator, in *Proc. 15th IEEE Int. Pulsed Power Conf.* (2005), pp. 135–138.
[129] Y.-J. Chen *et al.*, Scaled accelerator test for the DARHT-II downstream transport system, in *Proc. 15th IEEE Int. Pulsed Power Conf.* (2005), pp. 789–792.
[130] M. Schulze *et al.*, Commissioning the DARHT-II scaled accelerator downstream transport, in *Proc. 2007 Part. Accel. Conf.* (2007), pp. 2627–2629.
[131] C. A. Ekdahl, *Rev. Sci. Instrum.* **55**, 1221 (1984).
[132] J. Johnson, C. Ekdahl and W. Broste, DARHT Axis-II beam position monitors, in *Proc. 11th Beam Instrum. Workshop, AIP Conf. Proc.* **732**, 317 (2004).
[133] C. Ekdahl, *Rev. Sci. Instrum.* **76**, 095108 (2005).
[134] J. B. Johnson, C. A. Ekdahl and W. B. Broste, B-dot detector signal recording at the DAHRT-II accelerator, in *Proc. 16th IEEE Int. Pulsed Power Conf.* (2007), pp. 490–492.
[135] H. Bender *et al.*, Quasianamorphic optical imaging system with tomographic reconstruction for electron beam imaging, *Rev. Sci. Instrum.* **78**, 013301 (2007).
[136] D. Frayer *et al.*, Fielding a time-resolved tomographic diagnostic, in *Proc. 11th Conf. Novel Opt. Sys. Design Optimization, Proc. SPIE* **7061**, 70610Y-1–10 (2008).
[137] D. Frayer *et al.*, Data and analysis from a time-resolved optical beam diagnostic, in *Proc. 2010 Beam Instrum. Workshop* (2010), pp. 122–126.
[138] C. Ekdahl *et al.*, Electron beam dynamics in the long-pulse, high-current DARHT-II linear induction accelerator, in *Proc. 2008 European Part. Accel. Conf.* (2008), pp. 968–970.
[139] L. Walling *et al.*, Transverse impedance measurements of prototype cavities for a dual-axis radiographic hydrotest (DARHT) facility, in *Proc. 1991 Part. Accel. Conf.* (1991), pp. 2961–2963.
[140] R. Briggs *et al.*, Transverse impedance measurements of the DARHT-2 accelerator cell, in *Proc. 2001 Part. Accel. Conf.* (2001), pp. 1850–1852.
[141] C. Ekdahl *et al.*, Long-pulse beam stability experiments on the DAHRT-II linear induction accelerator, *IEEE Trans. Plasma Sci.* **34**, 460–466 (2006).
[142] C. Ekdahl *et al.*, Suppressing beam-centroid motions in a long-pulse linear induction accelerator, *Phys. Rev. Spec. Top. Accel. Beams* **14**, 120401 (2011).
[143] C. Ekdahl *et al.*, Beam dynamics in a long-pulse linear induction accelerator, *J. Kor. Phys. Soc.* **59**, 3448 (2011).
[144] Y.-J. Chen, Control of transverse motion caused by chromatic aberration and misalignments in linear induction accelerators, *Nucl. Instrum. Methods Phys. Res. A* **398**, 139–146 (1997).
[145] C. Ekdahl, Tuning the DARHT long-pulse linear induction accelerator, *IEEE Trans. Plasma Sci.* **41** (2013), in press.
[146] C. R. Rose, C. Ekdahl and M. Schulze, Beam-energy-spread minimization using cell-timing optimization, *Phys. Rev. Spec. Top. Accel. Beams* **15**, 040403 (2012).
[147] M. Schulze and S. Balzer, Personal communication (2013).
[148] H. S. Tasker and S. W. Towers, "Electron radiography" using secondary β-radiation from lead intensifying screens, *Nature* **156**, 50–51 (1945).
[149] J.-J. Trillat, Electron radiography and microradiology, *J. Appl. Phys.* **19**, 844–852 (1948).
[150] S. P. D. Mangles *et al.*, Table-top laser–plasma accelerator as an electron radiography source, *Laser Part. Beams* **24**, 185–190 (2006).
[151] V. Ramanathan *et al.*, Submillimeter-resolution radiography of shielded structures with laser–accelerated electron beams, *Phys. Rev. ST Accel. Beams* **13**, 104701 (2010).
[152] G. C. Bussolino *et al.*, Electron radiography using a table-top laser-cluster plasma accelerator, *J. Phys. D* **46**, 245501 (2013).
[153] W. Schumaker *et al.*, Ultrafast electron radiography of magnetic fields in high-intensity laser–solid interactions, *Phys. Rev. Lett.* **110**, 015003 (2013).
[154] F. Merrill *et al.*, Electron radiography, *Nucl. Instrum. Methods Phys. Res. B* **261**, 382–386 (2007).
[155] T. F. Delaney and H. M. Kooy (eds.), *Proton and Charged Particle Radiotherapy* (Lippincott Williams and Wilkins, Philadelphia, 2008).
[156] A. Cormack and A. Kohler, Quantitative proton tomography: Preliminary experiments, *Phys. Med. Biol.* **21**, 560–569 (1976).
[157] G. Coutrakon *et al.*, Design and construction of the 1st proton CT scanner, *AIP Conf. Proc.* **1525**, 327 (2013).
[158] J. Telsemeyer, O. Jäkel and M. Martišiková, Quantitative carbon ion beam radiography and tomography with a flat-panel detector, *Phys. Med. Biol.* **57**, 7957–7971 (2012).
[159] J. A. Cookson, Radiography with protons, *Naturwissenschaften* **61**, 184–191 (1974).
[160] A. Gavron *et al.*, Proton radiography, *LANL* LA-UR-96-420 (1996).
[161] C. T. Mottershead and J. D. Zumbro, Magnetic optics for proton radiography, in *Proc. 17th Particle Accelerator Conference* (Vancouver, Canada, 1997), pp. 1397–1399.

[162] F. J. Merrill *et al.*, Magnifying lens for 800 MeV proton radiography, *Rev. Sci. Instrum.* **82**, 103709 (2011).

[163] K. L. Brown, A second-order magnetic optical achromat, *IEEE Trans. Nucl. Sci.* **NS-26**, 3490–3492 (1979).

[164] S. A. Kolesnikov *et al.*, Application of charged particle beams of TWAC-ITEP for diagnostics of high dynamic pressure processes, *High Pres. Res.* **30**, 83–87 (2010).

[165] Y. M. Antipov *et al.*, A radiographic facility for the 70 GeV proton accelerator of the Institute of High Energy Physics, *Instrum. Exp. Tech.* **53**, 319–326 (2010).

[166] V. V. Burtsev *et al.*, Initiation of detonation in explosives on a U-70 proton accelerator, *Combust. Explos. Shock Waves* **47**, 350–356 (2011).

[167] V. V. Burtsev *et al.*, Use of multiframe proton radiography to investigate fast hydrodynamic processes, *Combust. Explos. Shock Waves* **47**, 627–638 (2011).

[168] V. D. Selemir *et al.*, Investigation of the foil destruction dynamics in an explosively formed fuse opening switch by proton radiography, *Zh. Tekh. Fiz.* **82**, 95–100 (2012).

[169] T. Wei *et al.*, A lattice scenario for a proton radiography accelerator, *Chin. Phys. C* **34**, 1754–1756 (2010).

[170] F. E. Merrill *et al.*, Proton microscopy at FAIR, *AIP Conf. Proc.* **1195**, 667–670 (2009).

[171] J. A. Cobble *et al.*, High resolution laser-driven proton radiography, *J. Appl. Phys.* **92**, 1775–1779 (2012).

[172] C. I. Choi *et al.*, Comparison between proton radiography images using a high-power femtosecond laser and a tandem van de Graaff accelerator, *JKPS* **50**, 721–725 (2011).

[173] Y. Paudel *et al.*, Self-proton/ion radiography of laser-produced proton/ion beam from thin foil targets, *Phys. Plasmas* **19**, 123101 (2012).

[174] G. Sarri *et al.*, The application of laser-driven proton beam to the radiography of intense laser-hohlraum interactions, *New J. Phys.* **12**, 045006 (2010).

[175] A. B. Zylstra *et al.*, Using high-intensity laser-generated energetic protons to radiograph directly driven implosions, *Rev. Sci. Instrum.* **83**, 013511 (2012).

[176] T. Ezirkepov *et al.*, Highly efficient relativistic ion generation in the laser-piston regime, *Phys. Rev. Lett.* **92**, 175003 (2004).

[177] A. P. L. Robinson *et al.*, Radiation pressure acceleration of thin foils with circularly polarized laser pulses, *New J. Phys.* **10**, 013021 (2008).

[178] W. Enghardt *et al.*, Positron emission tomography for quality assurance of cancer therapy with light ion beams, *Nucl. Phys. A* **654**, 1047–1050 (1999).

[179] C. Jeynes and R. Webb, Ion beam analysis: A century of exploiting the electronic and nuclear structure of the atom for materials characterisation, *RAST* **4**, 41–62 (2011).

[180] C. Jeynes *et al.*, "Total IBA" — where are we?, *Nucl. Instrum. Methods Phys. Res. B* **271**, 107–118 (2012).

Ken Peach obtained his B.Sc. and Ph.D. degrees in physics from the University of Edinburgh. He is a particle physicist who spent 25 years at Edinburgh working in neutral kaon physics, and was part of the NA31 and NA48 collaborations at CERN that discovered direct CP-violation, for which he shared the 2005 EPS HEP Prize. In 1996 he was appointed Deputy Leader of the Experiments Division at CERN, and in 1998 was appointed Director of Particle Physics at the Rutherford Appleton Laboratory, where he was responsible also for managing the UK particle physics programme. While at RAL he restarted accelerator R&D for particle physics applications, and was awarded the 2006 Institute of Physics Rutherford Prize and Medal for his work on kaons, and for the accelerator initiative. After leaving RAL he moved to the University of Oxford, where he was appointed Director of the newly-created John Adams Institute for Accelerator Science. There he created the Particle Therapy Cancer Research Institute. He is a Fellow of the Institute of Physics and the Royal Society of Edinburgh.

Carl Ekdahl obtained his B.A. degree in physics from San Diego State College, San Diego, and his M.S. and Ph.D. degrees in physics from the University of California at San Diego, La Jolla. He has practiced experimental and theoretical physics at Smyth Research Associates and the Scripps Institution of Oceanography in San Diego, at Mission Research Corporation and Sandia National Laboratories, in Albuquerque, New Mexico and at the Los Alamos National Laboratory. His research interests have included accelerators and intense relativistic electron beams, pulsed-power, high-power microwaves, high energy density plasmas and hydrodynamics, warm dense matter, and atmospheric CO_2. He is currently engaged in further improving beam quality on the long-pulse linear induction accelerator at the Dual-Axis Radiography for Hydrodynamic Testing (DARHT) facility at the Los Alamos National Laboratory. Dr. Ekdahl is a life member of the American Physical Society, and a Senior Member of the IEEE. He serves on the International Advisory Committee for the High Power Particle Beam Conferences, and he has served on several IEEE award and conference organizing committees. He is currently the Senior Editor of the *IEEE Transactions on Plasma Science for Charged Particle Beams and Sources.*

Reviews of Accelerator Science and Technology
Vol. 6 (2013) 143–169

DOI: 10.1142/S1793626813300077

Rapid Cycling Synchrotrons and Accumulator Rings for High-Intensity Hadron Beams

Jingyu Tang
Institute of High Energy Physics,
Chinese Academy of Sciences,
Yuquan Road 19, Beijing 100049, P. R. China
tangjy@ihep.ac.cn

Boosted by the needs in high-energy physics and nuclear physics and also multidisciplinary applications, high-intensity proton synchrotrons and accumulator rings have been developed quickly around the world over the last 30 years. New projects and plans are proposed with even higher beam power. The proton beam power has increased from less than 10 kW in the 1970s to about 1 MW level today, and the required beam power in the coming decade is a few MW. This article reviews the achievements in designing and constructing rapid cycling synchrotrons (RCSs) and accumulator rings (ARs) and the future development trends, principally on proton beams but also including heavy ion beams. It presents the evolution of RCS and AR machines, today's design philosophy, relevant accelerator physics, and also state-of-the-art accelerator technology.

Keywords: High-intensity; rapid cycling synchrotron; accumulator ring; space charge effects; beam loss.

1. Introduction

High-intensity and high-repetition-rate proton synchrotrons include rapid cycling synchrotrons and accumulator rings. Different from other hadron synchrotrons, they provide very high beam power, usually for generating secondary beams such as neutrons, muons or neutrinos, and also act as medium-energy boosters for high-energy proton synchrotrons or colliders for high-energy physics research. The product of beam energy and average beam current represents the beam power, ranging from tens of kilowatts to a few megawatts. The synchrotrons with beams of high energy and low average current are not placed in this category, such SPS at CERN, the Main Injector at FNAL and AGS at BNL, which have much lower cycling rates and can be considered as high-energy accelerators.

In the 1970s, energy boosters were proposed to boost the injector energy for several proton synchrotrons, such as ANL/ZGS, CERN/PS and KEK/PS, for a competitive race to increase the intensities. Increasing the output energy of the linac injector was considered very expensive at the time, so an energy booster became a cheaper choice. With the appearance of H-minus stripping injection [1, 2], the RCS booster proved to be successful in increasing the number of particles per pulse.

The first proton RCS in history is the 20 Hz, 3 GeV ring built in 1957–1963 at Princeton University. The next is the booster at Fermilab, which was commissioned in 1971. With the final energy of 8 GeV, it acted as the bridge between the 200 MeV linac and the 200 GeV Main Ring in the early years [3]. The other two RCSs built in the 1970s are the PS booster [4] at KEK and the ZGS booster [5] at ANL, also for high-energy physics, and both were shut down in the 2000s. After the shutdown of ZGS in 1979, the RCS booster became the proton driver for the first and pioneering spallation neutron source IPNS for matter structure studies, operated from 1978 to 2006. With a beam energy of 450 MeV and a beam power of only 7.5 kW, many neutron scattering techniques were developed at IPNS. The pioneering work at IPNS established the unique scientific value of this new type of neutron source. IPNS was followed by a similar facility — KENS at KEK — and higher-intensity spallation neutron sources, such as ISIS at RAL [6], LANSCE [7] at LANL, SNS [8] at ORNL, and J-PARC/JSNS [9] at JAEA. There are also other spallation neutron sources: SINQ is a spallation neutron source using the CW beam from a high-power cyclotron complex; CSNS at IHEP is under construction [10] and ESS [11] at Lund is preparing for construction.

The ISIS/RCS at RAL, commissioned in 1985, was the first of such accelerators to reach a beam power above 100 kW. It has achieved great success in exploiting the advantages of spallation neutron sources in matter structure studies, and stimulated the studies and construction of MW-class spallation neutron sources in the last two decades. In the late 1980s, BNL/AGS increased its proton intensity greatly by adding an RCS-type booster between the linac and AGS [12, 13].

The RCS is also considered one of the major candidates for proton drivers of a few MW for neutrino factories and superbeams. Different schemes for this purpose have been studied.

On the other hand, accumulator rings (ARs) came up to compress a long-pulse beam from the linac into a short-pulse beam for the short-pulse spallation neutron source and neutrino factory, e.g. PSR in LANSCE, AR in SNS, AR in the early ESS design [14], and AR for the SPL superbeam [15]. In these cases, full-energy linacs are used to alleviate space charge effect in the rings and avoid the complicated accelerator technologies for RCSs. The other possible use is to accumulate several batches from the upstream linac or RCS to match the low repetition rate of the next-stage synchrotron [16, 17].

The beam power of proton RCSs and ARs has been increased steadily in the last three decades, from a few kW in the 1970s to about 1 MW today; this can be seen in Fig. 1. In the next decade, we can expect to see a machine of a few MW in operation.

Due to the difficulty in obtaining high-current heavy ion beams from ion sources, a low acceleration rate due to a lower charge–mass ratio and even stronger space charge effects in rings, high-intensity heavy ion synchrotrons have progressed very slowly, until the appearance of the FAIR project at GSI, which is under construction [18].

One very important technical innovation in boosting beam intensity in proton synchrotrons is the H-minus stripping injection. The H-minus ion source and the stripping injection method were first developed at BINP in the 1960s [1]. It was first applied in ZGS and later to the RCS booster [5]. In Sec. 3,

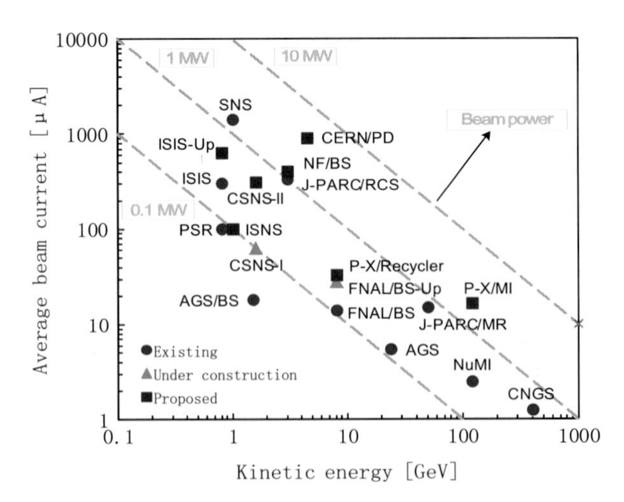

Fig. 1. Beam energy and average current for high-power proton synchrotrons. The three dashed lines stand for beam power levels. BS denotes the booster, Up is for "planned upgrading," and NF for "neutrino factory" [19].

details will be given to explain how it can increase the beam intensity in proton rings.

Along with the development of proton RCSs and ARs, the proton linac, or more precisely the H-minus linac, has always served as the injector for an RCS or AR and has also undergone much development, from the invention of RFQ to high-brightness H-minus ion sources to superconducting structures.

This article will give an overview of high-intensity proton RCSs and ARs, with a repetition rate larger than a few hertz. It includes the related accelerator physics, design methods and the special techniques in hardware development for these types of accelerators. Table 1 shows the main parameters for high-power rapid cycling synchrotrons and accumulators which are either in operation or under construction or planned projects.

2. General Beam Dynamics

2.1. *Modern design philosophy*

2.1.1. *Low beam loss requirement for maintenance*

Unlike ordinary weak-intensity synchrotrons, where more-than-10% beam loss can be tolerated, beam loss in high-intensity synchrotrons should be controlled much more strictly [20, 21]. Large beam loss can lead to a high radiation dose rate hindering hands-on maintenance, shortening device lifetime due to irradiation, heating up the irradiated devices,

Table 1. The main parameters for high-intensity rapid cycling synchrotrons and accumulator rings.

	Type	Energy (GeV/u)	Inj. energy (GeV/u)	f_rep (Hz)	Accu. part. (10^13)	I_ave (mA)	P_ave (MW)	Status
ISIS	RCS	0.8	0.07	50	3.75	0.3	0.24 (0.2)	Operating
ISIS-Upgrade	RCS	0.8	0.18	50	7.8	0.625	0.5	?
LANSCE/PSR	AR	0.8	0.8	20	3.1	0.1	0.08	Operating
SNS	AR	1.0	1.0	60	14.6	1.4	1.4 (1.0)	Operating
J-PARC/RCS	RCS	3.0	0.4	25	8.3	0.333	1.0 (0.3)	Operating
FNAL/Booster	RCS	8	0.4	7.5	0.45	0.014	0.043	Operating
FNAL/Booster-U	RCS	8	0.4	15	0.9	0.028	0.086	2016
AGS/Booster	RCS	1.5	0.2	7.5	1.5	0.018	0.027	Operating
CSNS-I	RCS	1.6	0.08	25	1.56	0.0625	0.1	2018
CSNS-II	RCS	1.6	0.25	25	7.8	0.313	0.5	?
ISNS	RCS	1.0	0.1	25	2.4	0.1	0.1	?
CERN/PD	AR	4.5	4.5	50	15	0.889	4	Fréjus?
FNAL/Recycler	AR	8	8	0.83	16	0.033	0.267	Project-X?
NF/Booster	RCS	3	0.2	50	5	0.4	1.2	?
FAIR-SIS100	RCS	1.5	0.2	0.65	0.051	0.0015	0.019	2018

Note: (1) The numbers in brackets are the beam powers achieved in the present day; (2) for FAIR-SIS100, the beam is U28+.

etc. Although the ALARA (as low as reasonably achievable) principle [22] is practised in most accelerator facilities, a more explicit requirement for the design and operation of high-power proton accelerators is to have 1 mSv/h at 30 cm from the location for hands-on manipulation after a 4 h beam shutdown, which corresponds to an uncontrolled beam loss rate of 1 W/m for energy larger than tens of MeV [23, 24]. The counterpart for heavy ion beams can be somewhat relaxed, e.g. 5 W/m for uranium beam [25]. Controlled beam loss at specific locations such as collimators and beam dumps, where heavy shielding is provided, can be much higher.

Therefore, it is very important to identify the beam loss locations and the mechanisms which cause the losses. Major beam loss mechanisms in high-power proton synchrotrons include: various losses at the H-minus stripping injection, longitudinal beam loss during the RF trapping process, extraction loss caused by kickers' misfiring, and beam loss due to gradual emittance growth by single-particle dynamics by space charge and nonlinear magnetic fields and by collective beam instabilities. In addition, the residual-gas scattering may produce nonnegligible beam losses and vacuum instability. There are also beam losses due to intrabeam scattering, but they are negligible compared with other loss mechanisms in high-repetition-rate rings.

Different measures, such as alleviating space charge effects, high-efficiency stripping, redundancy in extraction kickers, or avoiding collective instabilities, should be applied in high-power rings to minimize beam losses, as will be presented later in the article in detail. When there are inevitable beam losses, it is preferable to have the losses at lower energy, to minimize the lost power for the same loss rate. On the other hand, the localization of beam losses is very important, and this means a good beam collimation system. In the regions where the radiation dose rate is high enough to hinder hands-on maintenance, the remote handling method should be applied. High reliability in a hard radiation environment is required for many accelerator components.

2.1.2. *Easy upgrading in the future*

Large projects are often planned to be built in multiple phases. The compromises should be made between the investment in the first phase and the upgrading phases. For RCSs and ARs, the upgrading is usually done on the beam power. As the RCS extraction energy is not easy to increase due to the top field limitation in magnets as discussed in Subsec. 6.2, the upgrading can be pursued by increasing either the injection energy or the repetition rate. The increase in injection energy is more straightforward than the other method as the linac output energy is easy to increase by adding more acceleration sections if the space is reserved. To increase the repetition rate of an RCS is considered more difficult in terms of building many more RF stations and changing the power converter network. Many studies on the injection line and the RCS should be carried out even in the designing of the first-phase facility, to avoid difficulties in the future. For example, the injection region with higher injection energy, the space for adding secondary harmonic RF cavities, the collimators with higher beam energy, etc., should be studied from the very beginning. Nevertheless, some price must be paid for keeping the upgrading potential. Collective beam instabilities should be studied based on the ultimate phase if possible.

Power upgrading in ARs is easier, as the linac beam may be upgraded in beam energy and duty factor, and possibly in beam current.

2.1.3. *RCS or AR?*

As mentioned in Sec. 1, for different beam powers and applications it is better to adopt different acceleration schemes. For applications of short-pulse beam, a ring, either RCS or AR, should be used with the linac. For applications of hundreds-of-kW beam, the combination of a low-energy linac and an RCS has a significant advantage over the combination of a full-energy linac and an AR, due to lower cost and good upgradability for the former; however, for applications of MW-class beam, the situation is different. There have been many discussions on the pros and cons of RCS versus AR [26, 27], with the former associated with a medium-energy linac and the latter associated with a higher-energy linac. The SNS studies show that the RCS solution is even more expensive than the AR solution [28]. SNS has also found that the intrascattering effect in the H-minus linac is the leading source of beam losses [29], which is more serious in higher-energy linacs. Table 2 lists the pros and cons for the two difference schemes.

The choice for the accelerator scheme is often influenced by other beam utilizations also included in the project. For example, J-PARC adopted the combination of a 400 MeV linac and a 3 GeV RCS due to the fact that it should also provide 3 GeV beam for the injection into the 50 GeV MR for high-energy physics. A full-energy linac of 3 GeV was considered too expensive for the project. However, with the advancement in superconducting RF technology, a multiple-GeV linac also becomes feasible and this changes the above discussions to some extent.

2.1.4. *Mature technology combined with new technical achievements*

As a rule of thumb, if a project is to be launched without time to pay for R&D studies, only mature technology can be used. In the case where the R&D already shows that the new technology has evident advantages over the conventional one, it can be integrated into the engineering construction, but sometimes the fallback solutions with conventional design should also be prepared. For some novel but key technologies which are necessitated by the new accelerator project, a long-time R&D phase before starting the project construction is needed for the technical development.

2.2. *Lattice designs*

For a high-intensity RCS, the lattice design is confined by many imposed conditions, though different

Table 2. Pros and cons for RCS and AR schemes for an MW-class and short-pulse beam.

	RCS	AR
Extraction beam	Higher energy, lower current	Lower energy, higher current
Injector linac	Modest cost	Expensive
	Medium energy	Full energy
	Lower current	Higher current
	Lower loss level	Higher loss, intrabeam scattering
Total cost (linac + ring)	More expensive	Expensive
Upgradability	Better	Modest
Ring technology	Difficult	Easy
Availability	Better with a short linac	Worse (less critical with SRF)
Commissioning and operation	More difficult	Easier
Injection	Easier with lower injection energy	More difficult
Acceleration	Fast acceleration	No acceleration
Ext. beam emittance (trans. and long.)	Small due to acceleration	Large
Beam loss ratio	More tolerated with lower injection energy	Less tolerated

designs have been studied at the laboratories around the world and compared within the same project. But, for ARs, the imposed conditions are much more relaxed or more flexibility is available for designing a lattice. Modern lattice designs have the common properties explained as follows [20, 30].

2.2.1. *Uninterrupted long straight section*

To design a lattice for an RCS, among many functions to be incorporated in the global design, injection, extraction, RF acceleration and collimation are most important. As explained in greater detail in the following subsections, these functions also occupy large space in a ring. In most cases, it is preferred that each of these functions can be realized in uninterrupted long straight sections or without an arc region in between. This is helpful for cabling, and equipment installation in service building and maintenance.

Naturally, a fourfold lattice is well matched to this, with each function occupying a long straight section [21, 31]. Figure 2 shows the CSNS/RCS functional layout. A threefold lattice has also been chosen at J-PARC/RCS [9], where the major reasoning is the geometry layout and ESS/AR [14], where the injection was put in one of the arc regions.

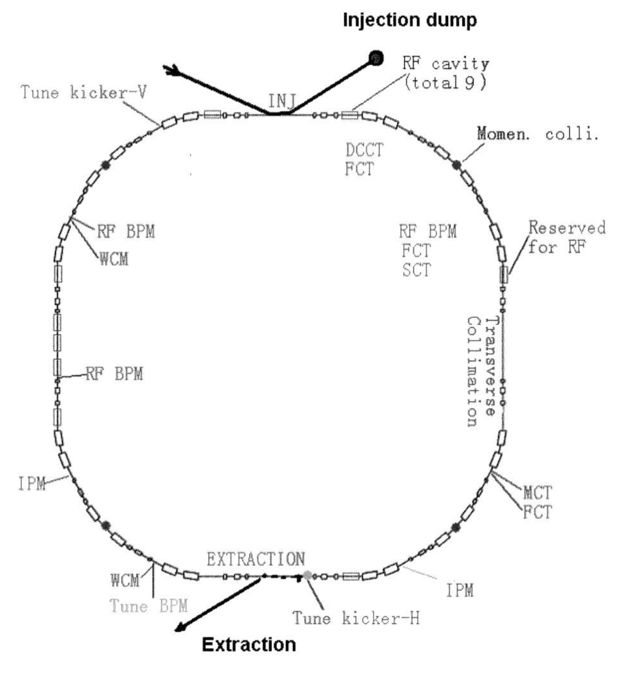

Fig. 2. Separate functions of the long straight sections in CSNS/RCS.

2.2.2. *Betatron function optimization*

Betatron function optimization is required to reduce magnet apertures in all kinds of synchrotrons, but in high-intensity rings the apertures for the magnets are even more important, as they influence the construction cost of the magnets and the relevant power supplies critically due to very large physical acceptance here. The lattice optimization should emphasize the betatron functions at the magnets, especially the dipole magnets, which are usually longer than those in electron rings [14, 31, 32]. A less modulated betatron function is good for reducing the largest apertures in the quadrupoles. Small betatron functions are also very helpful in reducing the apertures of injection magnets, extraction magnets and RF cavities.

2.2.3. *Focusing structures*

Different focusing structures based on FODO cells, doublet cells, triplet cells and hybrid focusing cells are used in the lattice design of synchrotrons. The FODO lattice has the advantages of uniform quadrupole apertures, smaller quadrupole strength and much stability against field variation; the doublet lattice has the advantages of making long drifts with small betas; the triplet lattice is an enhanced doublet lattice, with a possibility of making similar betatron functions or double waists in long drifts. It is considered that a uniform focusing lattice is useful in beam matching when space charge is very important [33].

In order to obtain dispersion-free long straight sections, for different focusing structures, one has to adopt different methods for dispersion suppression at arcs. For FODO-based arcs, depending on the number of periodic cells, one can use the half-field method [34] or the full 2π phase advance method; for triplet-based arcs, dispersion suppression by a symmetrical lattice can be used.

For example, the FODO lattice is used at J-PARC/RCS, the triplet lattice is used at CSNS [31], ESS [14] and AUSTRON [32], and the hybrid lattice is used at SNS [35]. At CSNS, the triplet lattice was found very attractive in obtaining long drifts with small betatron functions, which benefits the design and cost reduction of the injection, extraction, collimation, dipole magnets and RF cavities. Figure 3 shows the lattice functions for CSNS/RCS, SNS/AR, ESS/AR and J-PARC/RCS.

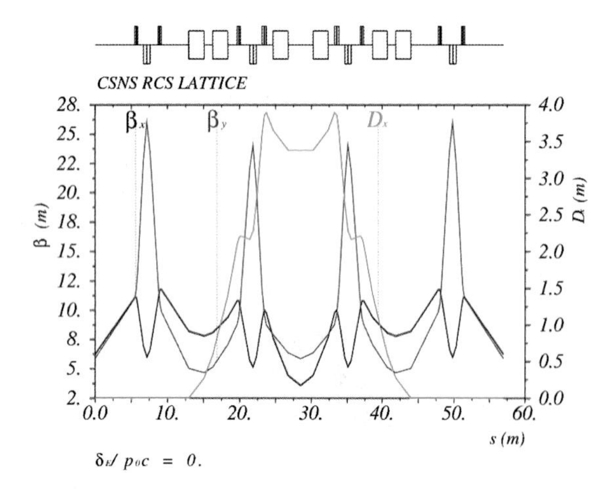

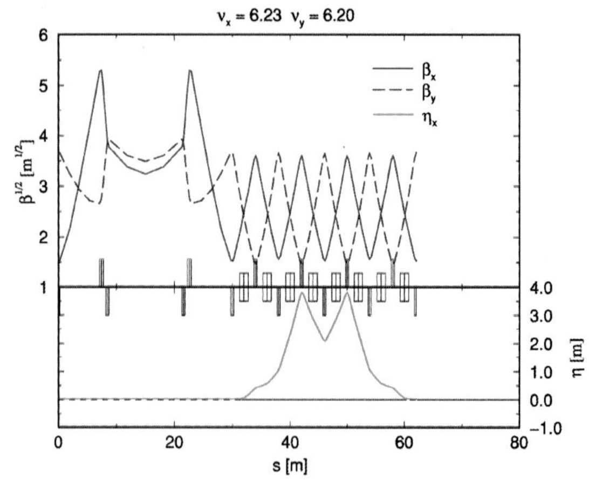

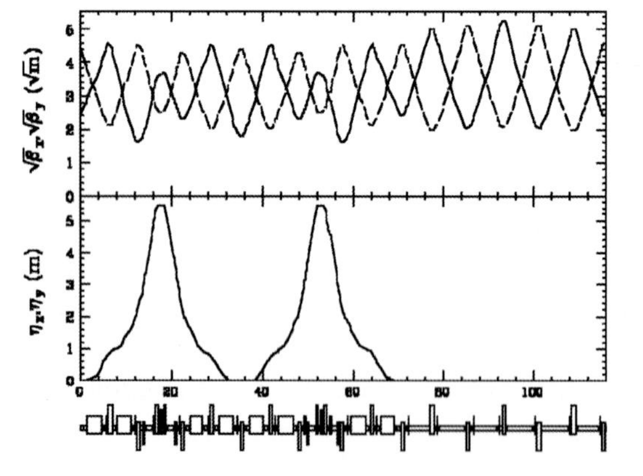

Fig. 3. Different lattice designs for high-intensity rings. From upper to lower: (1) fourfold triplet lattice for CSNS/RCS; (2) fourfold hybrid lattice for SNS/AR; (3) threefold FODO lattice for J-PARC/RCS.

No matter what kind of focusing structure is used, it is better to keep the phase advance per period less than 90°, and this is even more important in high-intensity synchrotrons, where space charge is very important [36].

2.2.4. *Shorter circumference for reducing RF voltage*

A shorter circumference is useful for reducing the costs for both the tunnel and the RF stations. The total RF voltage needed for fast acceleration in an RCS is strongly dependent on the circumference. For a given ramping rate and RF synchronous phase pattern, the required RF voltage is proportional to the ring circumference, as indicated by the following formula:

$$V_{\mathrm{RF}} \sin \phi_s = 2\pi R \rho \dot{B}, \tag{1}$$

where $V_{\mathrm{RF}}, \phi_s, R, \rho$ and $\dot{B}$ are the RF voltage, synchronous phase, average ring radius, curvature radius in dipoles and field ramping rate. In addition, more RF stations increase the ring circumference further. Therefore, the lattice should be optimized to reduce the circumference. For the same reason, the use of magnetic-alloy-loaded RF cavities instead of ferrite-loaded cavities can save much space. The choice of cavity types will be discussed in greater detail in Subsecs. 4.2 and 6.5.

2.2.5. *No-gamma-transition design*

Although mature techniques for gamma transition in proton synchrotrons have been developed, beam losses are almost unavoidable during the crossing of transition energy due to chromatic nonlinearity, self-field mismatch and instabilities [37–39]. This becomes even more unacceptable in high-intensity synchrotrons, as a larger beam loss rate due to stronger space charge at the crossing of transition energy and the total beam power is much higher. Thus, no gamma transition lattice design is required. There are two methods for designing a gamma-transition-free lattice: one is to design the transition gamma beyond the energy range, either higher than the extraction energy or lower than the injection energy; the other is to use the imaginary gamma design, which can be realized by making a proper dispersion function to have a negative momentum compaction [9, 40–42] and has the property always below transition energy. For the maximum energy below 5 GeV, the first method can be easily applied, while for higher energy the imaginary gamma design can be applied, such as J-PARC/MR [9] and FAIR/HESR [43, 44].

What is more, it is preferred to have the lattice design with the transition gamma significantly larger than the extraction energy, in order to guarantee sufficient longitudinal acceptance [32].

2.2.6. *Superperiod symmetry and structure resonances*

Quite often, the supersymmetry of the lattice is determined by the geometry of the facility site, though supersymmetry has important impact on the dynamic performance of the ring. Larger supersymmetry has the advantage of sparser structure resonance lines in the tune diagram, which is good for the selection of the working point; however, it is not good for attaining dispersion-free long straight sections. Therefore, a threefold ($S = 3$) or fourfold ($S = 4$) lattice is usually used for high-intensity synchrotrons as they provide a good compromise, with preference for $S = 4$.

Structure resonances are those resonances driven by magnetic field errors having the same supersymmetry as the lattice, and can be expressed by

$$mQ_x \pm nQ_y = pS \quad (m, n, p = 1, 2, 3 \ldots), \qquad k = m + n, \qquad (2)$$

where k is the resonance order, $+$ is for the sum resonance and $-$ is for the difference resonance.

For example, the old-fashioned design of ISIS/RCS has a lattice supersymmetry $S = 10$, so there are no dispersion-free long straight sections [6]; SNS, CSNS and the UK proton driver RCS for the neutrino factory have a lattice supersymmetry $S = 4$; ESS/AR, AUSTRON/RCS and J-PARC/RCS have a lattice supersymmetry $S = 3$.

2.2.7. *Working point selection*

In high-intensity rings, the strong space charge effects produce large tune shift and tune spread. Thus, the crossing of some low-order nonlinear resonances of some particles is often unavoidable. However, the lattice design should avoid the crossing of low-order structure resonances as described by Eq. (2). It is important to place the tune footprint in a region where there are no low-order structure resonances. Figure 4 shows the simulated tune footprint at SNS.

If split tunes or closed tunes should be selected, people will have different opinions. On the one hand,

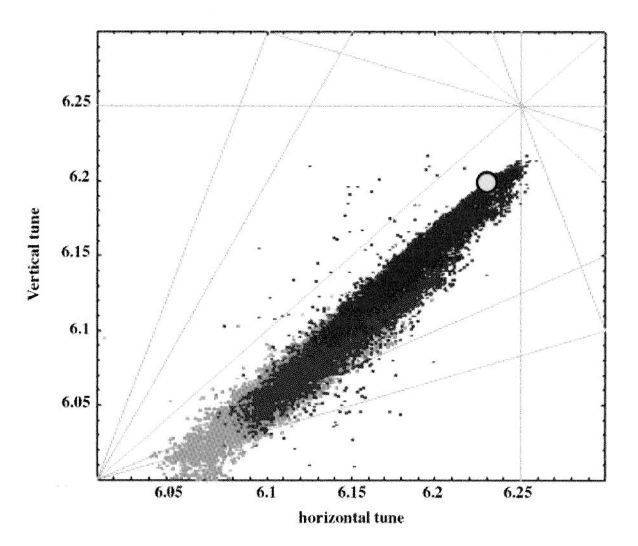

Fig. 4. Simulated tune footprint of the beam at SNS/AR [45]. The yellow spot is the nominal working point; the red is the tune spread at 263 turns, the blue is at 526 turns, and the green is at 1052 turns of accumulation.

some argue that split tunes are good for suppressing the space charge coupling resonance between the two transverse phase planes. On the other hand, closed tunes are considered to give a large clear area in the tune diagram without important resonances, and the coupling resonance driven by space charge is in the fourth-order difference resonance, which plays a role not so important in driving emittance growth, especially when the two emittances are similar.

As the tune shifts change with acceleration and some types of beam collective instabilities are relevant to the bold tunes, it is often required to vary the bold tunes during the beam cycle by adjusting trim quadrupole magnets. Figure 5 shows the variation of the bold tunes during the 10 ms cycle in the ISIS synchrotron [46]. In SNS/AR and CSNS/RCS, drift spaces are reserved for future rapidly changing trim quadrupoles. In J-PARC/RCS, adding such trim quadrupoles is under study.

2.2.8. *Large physical acceptance*

Very large physical acceptance for the beam ducts in high-intensity RCSs and ARs is needed to accommodate beam of very large size and momentum spread and provide good collimation efficiency, such as a few hundred πmm-mrad instead of a few ten πmm-mrad for low-intensity synchrotrons. As the Laslett tune shift expression shows (see Subsec. 2.3), the transverse space charge tune shift can be reduced largely

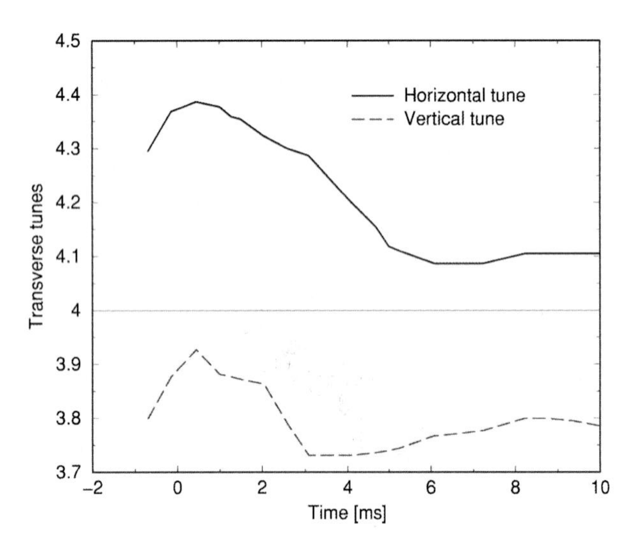

Fig. 5. Variation of the transverse tunes during the beam cycle in the ISIS synchrotron when the beam is accelerated from 70 to 800 MeV in about 10 ms. The maximum incoherent tune depression occurs at about 1 ms [46].

by painting the injection beam into a large emittance. Almost identical acceptance for the horizontal and vertical planes is preferred, in contrast to larger horizontal acceptance for low-intensity synchrotrons.

As will be explained in Subsec. 2.5, larger diameters of beam ducts are also useful for weakening space charge impedance. However, large apertures have important impact on the cost of magnets and power supplies, and also cause field interference between magnets [47]. Table 3 shows the physical acceptance and collimation acceptance for the newly designed machines and the early machines.

2.2.9. *Chromaticity correction or not?*

The chromaticity has two major effects in high-intensity rings: one is the tune spread from its combined effect with momentum spread, and the other is its role in controlling the collective instability (see Subsec. 2.5). Due to quite large momentum spread that comes from fast acceleration and longitudinal phase painting, for example about $\pm 1\%$ at the injection energy, the tune spread by a natural chromaticity can be about ± 0.06, but it is still significantly smaller than the tune shift/spread by space charge, which is often between -0.1 and -0.4. The necessity for chromaticity correction depends on how well one wishes to control the total tune shift/spread. Apparently, it is more worthwhile to make the correction in megawatt rings where the tune shift is required to be larger than -0.2; whereas in fast rapid synchrotrons with lower beam power it might be unnecessary, as the tune shift by space charge is dominant.

For chromaticity correction, one just needs to arrange some families of sextupole magnets in dispersive locations. At least two families are needed for the correction of both horizontal and longitudinal chromaticities [30]. When it is required to correct the second-order chromaticity, four families are needed. The use of chromatic sextupoles may shrink the dynamic apertures of the ring, if they are not well arranged by proper phase advances or balanced by additional resonance correction sextupoles.

At ISIS/RCS and LANSCE/PSR, no chromaticity correction has been made although the spaces for correction sextupole magnets were reserved in the construction phase. At SNS [48] and J-PARC/RCS [49], the chromaticity corrections by four-family and three-family sextupole magnets were used, respectively; for the latter they are ramping together with the main magnets. At CSNS, only two-family fixed field sextupole magnets are used, mainly for chromaticity correction at low energy where momentum spread is large. At ESS (ESS-2002), the tune spread by natural chromaticity was considered helpful in controlling beam instabilities, and thus no sextupole magnets were planned for this purpose.

Table 3. Physical acceptance and collimation acceptance for the newly designed machines and the early machines.

	Physical acceptance (πmm-mrad)	Collimation acceptance (πmm-mrad)
Fermilab/Booster	24.3	15.0
AGS/Booster	250(H)/125(H)	185(H)/90(V)
ISIS/RCS	540(H)/430(V)	300
SNS/AR	310	240
J-PARC/RCS	486	324
CSNS/RCS	540	350
ESS/AR (2002 version)	480	270

2.3. *Space charge effects*

Space charge plays a key role in designing and operating high-intensity proton synchrotrons [35, 50, 51]. Different space charge effects often drive halo production and emittance growth, which will probably result in beam losses.

2.3.1. *Incoherent and coherent space charge*

Free space charge produces defocusing force in all the three dimensions, but there is also image charge from the conducting vacuum wall, which is focusing in one transverse plane and defocusing in the other plane. For a circular vacuum duct, there is no focusing or defocusing effect. For the image-current-producing magnetic field, the boundary is often defined by high-permeability material surface such as magnet poles. The image charge has two different effects: the incoherent effect and the coherent effect. For the incoherent effect, the image force acts on individual particles, and can be added to the direct space charge force, but the former's strength is significantly weaker, for example less than one fifth of the latter's. For the coherent effect, the force acts on the whole beam bunch, and can be seen when the beam is not aligned in the symmetry axis of the vacuum chamber.

2.3.2. *Transverse tune shift*

The defocusing effect of space charge has an important impact on the transverse motion of beams by moving the working point downward. The change in the betatron tune is called a tune shift. If the distribution is not a KV distribution, the tune shift of a beam spreads out to become a tune footprint in the Q_x–Q_y diagram. The space charge tune shift in the transverse planes decreases with increasing beam energy drastically, due to the self-cancelation of the Lorentz force by the electric field and magnetic field. This means that the space charge tune shift is much more important at low energy, such as the injection energy for an RCS.

The Laslett tune shift [52] is often used to describe the averaged tune shift by the space charge in a ring, and it can be expressed by

$$\Delta Q_{x,y} = -\frac{Nr_c}{2\pi\varepsilon_{x,y}\beta^2\gamma^3 B_f}, \tag{3}$$

where $N = h_{\mathrm{RF}}n_{\mathrm{bunch}}$ is the total particles in the ring with all buckets filled, $r_c = q^2/4\pi\varepsilon_0 m_0 c^2$ is the particle classical radius (for the proton, 1.53×10^{-18} m), q and m_0 are the charge and mass of the particle, β and γ are for the Lorentz velocity and energy factors, ε is the unnormalized emittance and B_f is the bunching factor.

For more accurate calculations, the image charge should be added, as explained above. For a round beam pipe, the electric field of the image charge has a self-canceled effect on the incoherent transverse tune but an identical effect on the coherent tunes for the two transverse planes [50, 53].

For a bunched beam with nonuniform longitudinal distribution, the local tune shift of a long bunch can be modified from Eq. (3):

$$\Delta Q_{x,y}(s) = -\frac{r_c I(s) R}{2\pi\varepsilon_{x,y}\beta^2\gamma^3}, \tag{4}$$

where $I(s)$ is the local circulating beam current. The longitudinal distribution usually contributes to the largest part of the tune spread.

As space charge in high-intensity rings introduces large tune shift and tune spread, it may cause partial particles to cross low-order resonances, with the most dangerous half-integer structure resonance [54].

To control the beam loss rate level for rings of different beam power, the requirement on the Laslett tune shift is also different, for example about −0.3 or even smaller for 100–200 kW at ISIS [55] and CSNS-I [56], and about −0.2 for 0.5–1 MW at J-PARC/RCS [9] and CSNS-II [57]. The tune shift in ARs should be controlled more strictly than in RCSs, due to higher beam power at the injection energy, such as −0.14 for SNS at 1.4 MW [11, 45] and −0.07 for ESS at 2.5 MW [14].

2.3.3. *Longitudinal tune shift*

The longitudinal space charge behaves quite differently from the transverse one. Below the transition energy, it has a defocusing effect as the transverse space charge does; above the transition energy, it has a focusing effect. Almost all high-intensity synchrotrons with a strong focusing lattice work below the transition energy without crossing it during acceleration. The longitudinal motion including space charge can be expressed as [36]

$$z'' + k_{z0}^2 z - \frac{K_L}{z_m^3} z = 0,$$
$$k_{z0} = \sqrt{\frac{qV_{\mathrm{RF}} f_{\mathrm{RF}} \eta \cos\phi_s}{\bar{R}\beta^3\gamma m_0 c^3}}, \quad K_L = -\frac{3}{2}\frac{gNr_c}{\beta^2\gamma^3}\eta, \tag{5}$$

where f_{RF} is the RF frequency, η the phase slipping factor, and g the geometry factor. For a small tune shift, it can be expressed as

$$\Delta Q_z = \frac{3}{4}\frac{gNr_c R^2\eta}{\beta^2\gamma^3 z_m^2 Q_{z0}}. \tag{6}$$

For long bunches in proton synchrotrons, the longitudinal space charge effect is much weaker than the transverse counterpart.

2.3.4. *Space-charge-driven resonances*

Not only does the space charge produce large tune shift/spread and cause particles to cross the resonances driven by magnetic field errors or higher-order field components, but also it plays the role of a driving source for some resonances. Typically, the fourth-order resonance driven by the transverse space charge may occur [58].

$$2Q_x - 2Q_y \approx 0. \tag{7}$$

With this resonance (also called the Montague resonance), one can observe significant emittance change between the two transverse phase planes [59, 60].

2.4. *Field errors and correction schemes*

There are different field error types in synchrotrons, and some of them are specially related to RCSs. They come from magnetic field imperfection, installation errors, fringe field interference and errors in power supplies. There are also synchronization errors, because the saturation effect during magnet ramping may not be compensated for when resonant magnet power supplies are used. Among the orbit effects induced by errors, the first type is the closed orbit errors, which produce orbit distortions. The conventional closed orbit correction scheme by BPMs and correctors can be applied. However, due to very high ramping speed in RCSs, the correction should be performed very quickly and special magnets and power supplies should be built. Usually one needs to perform 10–20 times for a cycle.

The second type of errors is from quadrupoles. For the same reason of uncorrected saturation effect as for dipoles, quadrupole field errors in RCSs can be corrected by a few sets of trim quadrupoles. These trim quadrupoles are also quite useful for changing the bold working point during ramping to alleviate space charge effect and avoid some collective instabilities. It is also considered useful to maintain the ideal lattice functions with linear space charge included [32], and this can be performed by trim quadrupoles. The method for moving the working point during the cycle has been practised quite well at ISIS. CSNS has reserved space for installing these trim quadrupoles in the future. J-PARC is also planning to add trim quadrupoles in its RCS.

The third type of errors is from the higher-order field components in dipoles and quadrupoles. Since the beam circulating time in a cycle is short in an RCS or AR and the working point is chosen to avoid lower-order structure resonances, halo formation due to the resonance crossing is usually tolerable. The fourth type of errors is from the RF system. The errors in RF frequency, voltage and phase can produce nonsynchronization between the magnet system and the RF system. For RF setting errors, the nonsynchronization between magnets and RF can be corrected by incorporating the BPMs at the dispersive regions into the LLRF feedback control loops.

2.5. *Collective beam instabilities*

2.5.1. *Coupling impedance in rings for high-intensity hadron beams*

When a beam circulates in the vacuum beam duct, its accompanying electromagnetic field is affected by the boundary of the metallic vacuum duct and other electromagnetic devices such as RF cavities and magnetic cores. There are two kinds of expressions for the interaction between the beam and the environment [50, 61]: one is the wake field function, which describes the interaction in the time-position domain; the other is the coupling impedance, which describes the interaction in the frequency domain [62]. In synchrotrons, the coupling impedance is usually employed. There are four types of coupling impedance: space charge impedance, resistive wall impedance, wideband impedance and narrowband impedance. They come from different accelerator devices: space charge impedance is from the image charge, resistive wall impedance from the nonperfect conductivity of the vacuum wall, wideband impedance from the nonsmooth transition along the vacuum duct such as bellows, valves, pump ports, collimators or kickers, and narrowband impedance from RF cavities.

The relation between the longitudinal and the transverse impedances is given by the Panofski–Wenzel theorem:

$$Z_m^{//}(\omega) = \frac{\omega}{c} Z_m^{\perp}(\omega). \tag{8}$$

The impact of the coupling impedance on the beam is usually a many-turn process, and therefore the sum of all the impedances by different devices should be considered as the entirety. The total impedance dictated by the collective instabilities to be explained below requires impedance budget studies to distribute the impedance to different sources.

2.5.2. *Typical instabilities in high-intensity proton rings*

Although the circulating beam current is very high in high-intensity RCSs and ARs, for example in the tens of amperes, a relatively short circulation time helps the elimination of many collective beam instabilities which take a longer time to develop. Many types of instabilities are related to strong space charge effect, as the space-charge-induced incoherent tune shift lowers the threshold currents [63]. Relatively large momentum spread by fast acceleration in RCSs and by longitudinal painting in ARs also facilitates the Landau damping, which is a critical mechanism in suppressing most instabilities [64–66]. Besides controlling the impedance budget and operating the machine with a proper setting, one can use active feedback dampers to suppress certain instabilities [67, 68]. The mostly concerned instabilities include longitudinal microwave instability, electron–proton instability, resistive wall instability, head–tail instability, and longitudinal and transverse coupled bunch instabilities. Some more detailed descriptions are as follows:

Transverse head–tail instability. There are different kinds of transverse head–tail instabilities, slow or fast depending on the strength of the instability. Generally, it is due to the real part of the broadband impedance and resistive wall impedance, which contributes to the coherent motion of the beam. As the tune spread in the betatron motion is much larger than in the synchrotron motion in high-intensity rings, one can assume that the Landau damping usually takes effect. This instability has been observed in the CERN PS Booster, as shown in Fig. 6. Another kind of head–tail instability is mainly driven by the resistive wall impedance, which shows the coherent bunch oscillation and was observed at ISIS [69]. Its development is inhibited by lowering the tune rapidly, away from the integer tune.

Longitudinal microwave instability. This is called microwave instability due to the fact that the impedance at the high-frequency end (in the GHz range) provokes the instability. The bunch will break into many small bunches, and it is a fast process, as the growth rate is much faster than synchrotron frequency. It is caused by the integrated longitudinal impedance, namely both the real and imaginary parts contribute to the effect. In the presence of momentum spread, the instability may be Landau-damped. There is a so-called Keil–Schnell criterion defining the stability condition [71–73], which can be described as

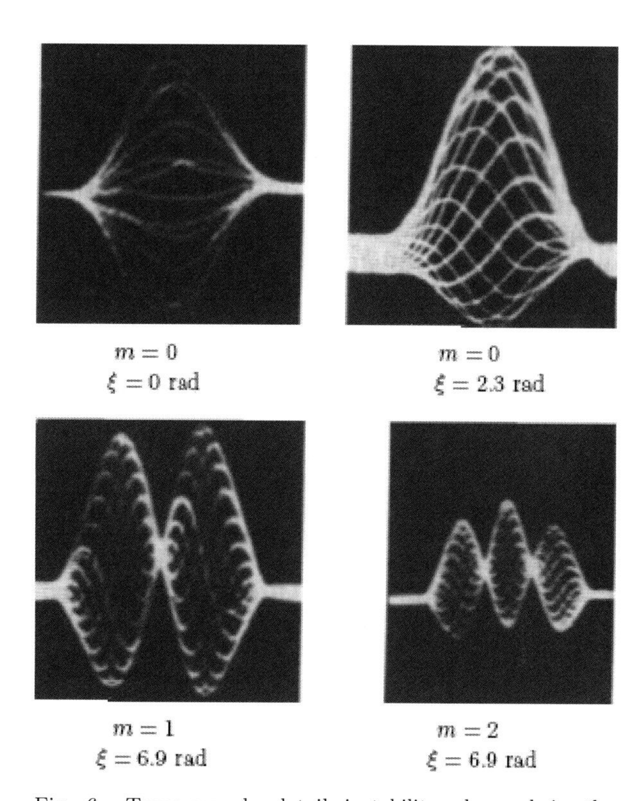

Fig. 6. Transverse head–tail instability observed in the CERN PS Booster [70]. The signal is from a wideband pickup for a single bunch: vertical axis — difference signal; horizontal axis — time (50 ns per division). The azimuthal mode number and head–tail phase shift in each plot are labeled.

$$\left|\frac{Z_{//}}{n}\right| \leq \frac{2\pi\beta^2 E\sigma_\delta^2|\eta|F}{eI_0}. \tag{9}$$

However, the threshold defined by the criterion can be easily surpassed in low-to-medium-energy machines where the space charge impedance is dominant and the beam distribution is Gaussian-like [74, 75], as observed in some machines, such as ISIS, CERN/PS and the BNL/AGS Booster. The experiments at PSR also show that the total impedance dominated by space charge can be effectively lowered by adding an inductive insert [76].

Longitudinal Robinson instability. This is caused by the beam loading effect on the narrowband impedance of the accelerating cavities of the ring. These are detuned to compensate for the reactive component of the beam loading, positively when below transition and negatively when above. Unless remedial action is taken, the instability occurs when the beam loading power equals the generator power (which would have to be high for heavy beam loading). The threshold level may be increased by about an order of magnitude by the additional use of beam feedforward or RF feedback techniques, or both.

Electron–proton instability. This is the only collective beam instability which cannot be described by coupling impedance. It is due to the interaction between the circulating proton beam and the electron cloud along the vacuum duct mainly formed by the proton beam itself [77]. It becomes more important with large numbers of stored protons and the neutralization rate, together with a short bunch gap and proton leakage into the bunch gap. Thus, it is more important in ARs than in RCSs. There are several measures to reduce the electron cloud density, including a lower vacuum pressure, TiN coating for the surfaces of in-vacuum devices to reduce the secondary electron emission, collection of electrons, etc. Large tune spread also helps to damp the collective oscillation by the Landau damping mechanism. There have been many studies about the electron cloud effect and the electron–proton instability in major high-intensity RCSs and ARs [06, 78–82].

2.6. *Multiparticle simulations and tools*

2.6.1. *Importance of multiparticle simulations*

After the optics design which defines the parameters for the lattice, injection, extraction and acceleration, it is also important to carry out multiparticle simulations for the whole acceleration cycle. Certainly, multiparticle simulations are now performed in almost all the machines. However, in high-intensity accelerators multiparticle simulation plays an even more crucial role in the design and beam dynamics studies. There are three reasons: the first is that the beam loss level is so important that one often needs to study the dynamic behavior of a very sparse halo of the order of 1 part in 10^6 or even 10^8; the second is that the space charge calculation is strongly dependent on the number of macroparticles and their distribution; the third is that the beam collective effect becomes more important here. Quite often, one needs to reoptimize the optics design based on the simulation results.

2.6.2. *Methods used in major simulation codes*

With the advancement of computation techniques and computer capability, very powerful simulation codes for beam dynamics in high-intensity synchrotrons have been developed. These codes must incorporate all or most of the following important features: lattice definition, transverse and longitudinal space charge, tracking with a large number of macroparticles, injection painting in the phase spaces, collimation, nonlinear resonances, collective effect with impedance, etc.

Different techniques have been employed in the widely used codes to deal with the above features. However, some techniques are shared within most of the codes. For example, parallel processing is now widely used in those codes tackling heavy computation; PIC (particle-in-cell) and FFT (fast Fourier transform) are used for calculating internal space charge forces. Tracking for RCS or AR rings is typically needed only over 10^4 turns, so it is not a critical issue. Different methods, from simple thin lenses, matrices to Lie algebra to differential algebra, are used to track through elements in different codes. A simple method is helpful in saving the simulation time when a large number of macroparticles are used.

2.6.3. *Benchmark efforts between codes and with experiments*

During the development of a large-scale simulation code, there are always bugs in the code. Checking with standard models, comparison among different codes and comparison with experimental results are all important in finding the bugs [83, 84]. There is an international collaboration on collaborating and benchmarking the codes [85].

3. Injection and Extraction

3.1. *H-stripping injection*

Single-turn injection is used only for the beam transfer from one ring to another. It is seldom used in RCSs and ARs. It has been proposed for use in the case of an RCSs proton driver chain for a neutrino factory or muon collider [19].

The multiturn injection method by proton beam has been almost abandoned for injection from a linac to an RCS, due to its poor injection efficiency [86]. H-minus stripping injection is a must to obtain highly accumulated protons in rings by hundreds of turns [2, 87–89].

Foil lifetime was considered critical in megawatt-level machines [90], but SNS operation proves that it is not so pessimistic [91, 92]. Now a good quality of stripper foils can be obtained from commercial companies. A novel stripping method using laser was also proposed and is under development [93, 94].

Careful treatments of nonstripped H-minus and partially stripped H-zero are important in designing the injection system, as they may produce large beam losses in the injection area. It is required to send either both types of particles or at least H-zero to a beam dump [95–97]. For the stripped electrons of the H-minus beam, a good collection system is important for removing the heat load and reducing the electron cloud. At SNS, the foil bracket was damaged by the stripped electrons in a specially designed but not well-studied magnetic field [92].

Traversal of circulating protons in the stripper foil poses serious beam loss concern due to nuclear scattering and multiple scattering, and it also accelerates the degrading process of the foil. The traversal per proton can be as large as more than 10 [98, 99]. It should be studied carefully to reduce the average traversal per proton by designing an appropriate program for the local orbit bump.

The Lorentz stripping of H-minus beam and the stripping of H-zero in the excited Stark states in the magnetic field are also important causes of beam losses [100, 101]. Careful selection should be made for the magnetic fields of the magnets in the injection line and injection bump magnets [95, 96].

The H-minus injection can be performed either in a dispersion-free long straight section or in the middle of a highly dispersive arc. Different layout schemes have been studied and used, for example: the ESS design uses a special arc dipole with the stripper foil in the middle [14]; SNS uses a chicane and two sets of painting bump magnets [102]; J-PARC/RCS uses a decay chicane and two sets of painting bump magnets [9]; CSNS uses a very long drift hosting a DC chicane and two sets of painting bump magnets [103]. Figure 7 shows the injection layouts for CSNS/RCS.

3.2. *Transverse phase space painting*

As mentioned in Subsec. 2.3, the space charge plays a key role in producing emittance growth and beam loss. One important measure to alleviate the space charge effects is the phase space painting during

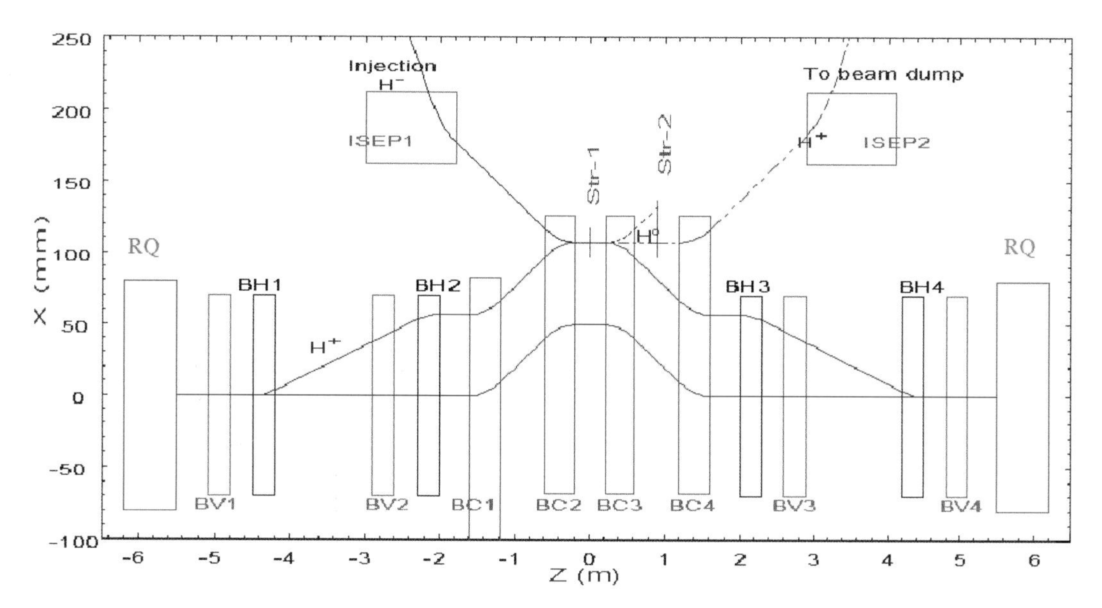

Fig. 7. CSNS/RCS injection layout. BC1–BC4 — DC orbit bump magnets; BH1–BH4 — horizontal painting bumpers; BV1–BV4 — vertical painting bumpers, ISEP1, 2 — septa; RQ — ring quadrupoles; Stri-1, 2 — primary and auxiliary strippers.

injection. This is to reduce the peak charge density, the average tune shift and the maximum tune shift in both transverse and longitudinal phase spaces. Thanks to the property that the multiturn injection process by H-minus stripping disobeys Liouville's theorem, the injected linac beam can be overlapped in the transverse phase spaces so that more-or-less uniform beam distribution can be obtained by properly moving the local orbit bumps — the so-called phase space painting.

The painting can be performed either by moving the local orbit bump or by sweeping the angle of the injection beam at the stripper foil. For the latter, the sweeping magnets are in the injection line, instead of being in the ring. Different combinations of the local orbit bump moving and injection angle sweeping can be used in a realistic design, as shown in Table 4. Due to the emittance growth by space charge during the painting, it is preferred to paint from the inner part to the outer part in the acceptance, so that the lately injected beam can paint over the halo. When the painting is executed from the inner part to the outer part in the ring acceptance for both of the transverse phase planes, it is called the correlated painting method; when the paintings are done from inner to outer in one plane and reverse in the other plane, it is called the anticorrelated painting method. There are also other painting methods proposed, such as the one using combined local orbit bump moving and angle sweeping for both of the transverse planes to obtain self-consistent distributions [104].

Table 4. Comparison of the usual transverse painting schemes for injection in a dispersion-free region. "Bump" means a dynamic local orbit bump, and "sweeping" means injection angle sweeping.

Type	Advantage	Disadvantage
Correlated/ bump–bump	Paint over halo	Singular density Coupling emittance growth Rectangular profile
Anticorrelated/ bump–bump	Immune to coupling Elliptical profile	Halo growth Extra aperture
Correlated/ bump–sweeping	Paint over halo Save injection space	Foil support difficult Susceptible to operation error
Anticorrelated/ bump–sweeping	Immune to coupling Elliptical profile Save injection space	Foil support difficult Susceptible to operation error Rise time of sweeping magnet

As some examples, SNS adopted the correlated painting method by moving local orbit bumps in the two transverse planes; J-PARC/RCS adopted the anticorrelated painting method by local orbit bump moving for the horizontal plane and angle sweeping for the vertical plane; CSNS adopted the anticorrelated painting method by moving local orbit bumps in the two transverse planes. Figure 8 shows the painting scheme at CSNS.

A well-prepared beam in the injection beam line is helpful in controlling the beam losses during injection and obtaining good painted beam quality [9, 14, 105–107]. This includes halo collimation, and reduction in energy jittering and spread.

3.3. *Fast extraction*

In contract to conventional low-intensity proton rings, fast extraction in high-power rings requires very careful treatment of beam loss during the extraction, although it benefits from the straightforward single-turn extraction that produces a naturally low beam loss rate compared with that for a resonant slow extraction scheme. The main beam loss mechanisms are the transverse halo particles lost in the extraction channel due to the limitation of acceptance, insufficient kicking due to accidental misfiring of kicker units, wobbling of triggering times and

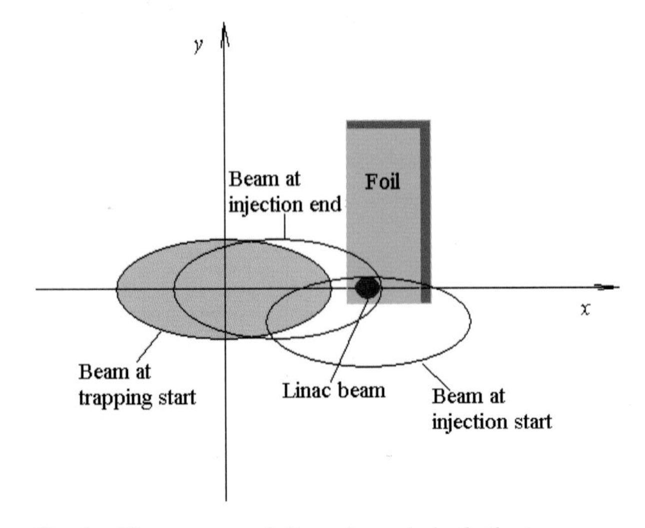

Fig. 8. Phase space painting scheme in both the transverse planes by moving local orbit bumps at CSNS.

nonflatness of the deflection field for the kickers, a distorted closed orbit due to a strong stray field of the septum magnets, and the particles in the bunch gap lost during the rise time of the kickers in the case of an accumulator ring.

Sufficient time gap between the circulating bunches is needed for the risetime of kickers, which is not always easy to be reduced to about 200 ns due to large kicker sizes and small PFN impedance. The longitudinal beam dynamics design should take care of this issue by producing a short bunch before the extraction.

Occasional kicker failures due to PFN misfiring are important in the extraction beam loss. Two measures are used to reduce this type of beam loss: one is to use derated operation parameters for the kickers, such as a modest PFN voltage less than 40 kV to obtain higher reliability; the other is to use a large number of kicker units to reduce the deflection angle per unit so that one kicker failure can be almost tolerated. For example, SNS was designed to have 14 kicker units for the extraction [108], as shown in Fig. 9.

A stray field in the extraction septum magnets is important due to its large aperture, very high field and thin septum. The situation becomes worse in the case where a Lamberston magnet is used. Careful shielding design is very important. For example, SNS met with very serious beam distortion due to the stray field of the Lambertson magnet [109].

In low-intensity proton synchrotrons, a slow orbit bump at the extraction can be used to ease the requirements for the kickers by taking the benefit

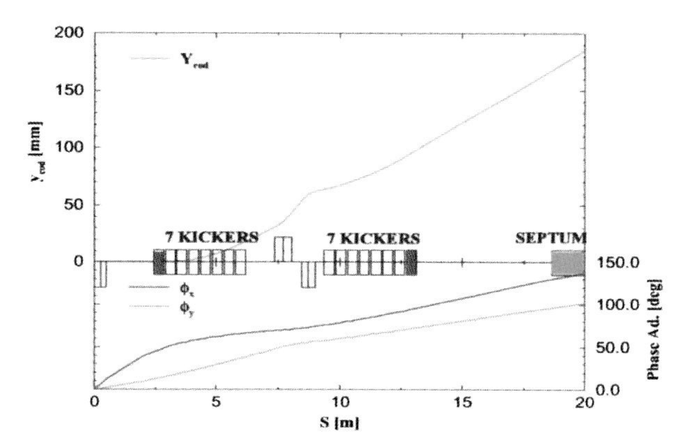

Fig. 9. SNS/AR extraction layout. The single-turn extraction is by 14 kickers and one Lambertson magnet in two dispersion-free long drifts. The orbit deviation by kickers and phase advances are also shown.

that the beam emittance shrinks during acceleration [110]. However, it should be more prudent to use the method for high-intensity beams, because a beam halo may be generated during acceleration. At ISIS, a slow orbit bump by the ring COD correctors has been used to assist the fast extraction, with the observation of no additional beam loss due to the manipulation. One new method uses simultaneous orbit bumps at both the extraction and the transverse collimators so that the beam halo can be collimated at low energy [111]. Besides reducing the requirement of the extraction kickers, the collimation-assisted extraction method is helpful in reducing the acceptance of the extraction channel, the downstream beam transport line and the next-stage synchrotron.

4. Fast Acceleration and RF Gymnastics

4.1. *Adiabatic RF capture*

Conventionally, the beam is injected to form coasting beam in rings. Adiabatic RF capture is then applied to compress the beam into bunches with high capture efficiency [30]. However, in RCSs, there is insufficient time to make adiabatic RF capture, due to fast ramping in the magnetic field. The dilution of longitudinal emittance becomes too important, so that some particles will be lost during the initial RF capture and the shrinking of the RF bucket in the period of fast acceleration [112]. For example, at ISIS there was about 7% beam loss due to the RF capture when single harmonic RF was used [113]. A chopped beam which can be produced in the LEBT or MEBT of the injector linac is preferred to reduce the beam loss during the RF capture. In ARs, chopped beam is a must to obtain clean bunch gaps for single-turn extraction, as there is no acceleration process to shorten the beam bunch as in an RCS.

4.2. *Longitudinal phase painting*

As mentioned in Subsec. 2.3, the longitudinal beam distribution along with a bunch has a very important impact on both the transverse and the longitudinal space charge effects. Therefore, it is important to inject the beam, preferably uniformly, into the RF bucket. At the same time, a large momentum spread is required for the Landau damping of collective beam instabilities. Therefore, it is required to paint the large RF bucket with a chopped linac beam which has a very small momentum spread. Depending on the injection layout design, the longitudinal motion during injection is correlated or not with the horizontal motion, as mentioned in Subsec. 3.1. The following longitudinal painting schemes have been studied to obtain the desired distribution in the longitudinal phase plane: (1) with off-momentum injection, without longitudinal space charge the injected beam will form a ribbon ring; (2) with momentum sweeping by using an "energy spreader" cavity in the injection beam line which works in the phase-scanning mode, the injected beam will fill the RF bucket several times; (3) with large momentum spread produced by a debuncher in the injection beam line, the injected beam repeatedly paints over the same area in the RF bucket. The first one is usually adopted by RCSs [9, 57, 114, 115]. The second one is usually adopted by ARs [14, 102]. However, the longitudinal space charge strongly changes the painting process and the final beam distribution. Detailed multi-particle simulations including space charge should be carried out to search for the best painting scheme.

In RCSs, the changing RF bucket during injection has a significant influence on the longitudinal painting. One can consider desynchronizing the RF system with the ramping magnetic field which oscillates in the DC-biased cosine form to keep the RF bucket constant during the injection painting period [57].

4.3. *Fast acceleration*

Fast acceleration in RCSs means both high RF voltage and large synchronous phase. High RF voltage means many RF stations which are very costly and require more precious longitudinal space in rings; a large synchronous phase means the shrinking of the bucket area, which is not good to keep the beam of large longitudinal emittance within the bucket. This situation becomes even more critical when the repetition rate is larger than several Hz so that resonant magnets should be used. With resonant magnets excited by DC-biased AC power supplies, the field ramping rate is defined by the following form:

$$\dot{B} = 2\pi f_{\rm rep} B_0 \sin\Big(2\pi f_{\rm rep} t + \frac{\pi}{4}\Big), \qquad (10)$$

where $f_{\rm rep}$ and B_0 are the repetition rate and AC field amplitude, respectively. Thus, the acceleration rate follows the ramping rate of the magnetic field, which is sinusoidal, starting from zero. The maximum acceleration rate occurs just in the middle of the acceleration period or a quarter of the whole cycling period, which is significantly larger than a linear field ramping. As the bucket area decreases dramatically with the synchronous phase [30], the maximum excursion of the synchronous phase is usually controlled not to exceed 45°.

To provide RF voltage of hundreds of kV, many RF stations are required. For example, a two-gap ferrite-loaded cavity about 3 m in total length can provide about 24 kV at the maximum [116]. Although the new type of MA (magnetic alloy)-loaded RF cavity can provide 20–40 kV/m, it has not been widely used in high-power proton synchrotrons except at J-PARC [117].

In order to reduce the maximum acceleration rate in RCSs, there were proposals to use dual harmonic magnetic fields to prolong the ramping time with a given cycling period [118–120], but the method is technically difficult and has never been put into engineering construction.

4.4. *Dual harmonic RF and RF barrier*

In order to increase the bunching factor that is important in the average tune shift as expressed by the Laslett tune shift and improve the uniformity of the longitudinal distribution that defines the tune spread, dual harmonic RF is considered very helpful. Use of the second harmonic with fundamental RF cavities can increase the bucket area and flatten the bucket height. Dual harmonic RF systems have been successfully applied in several RCS and AR machines [121–126], and used in the designs [14, 31, 57]. For example, during the ISIS upgrading, four $H = 4$ (harmonic number) cavities were added to the six $H = 2$ cavities, in order to increase the beam power by 50%; see Fig. 10.

For ARs, a more powerful method — the so-called RF barrier — to produce long and uniform bunches has been studied. Taking the advantage of the possibility of injecting different harmonic RF powers into a single MA-loaded cavity with a very low Q value, a nearly perfect potential barrier and a long field-free space can be composed to confine the beam [29, 127–130]. Another method for forming

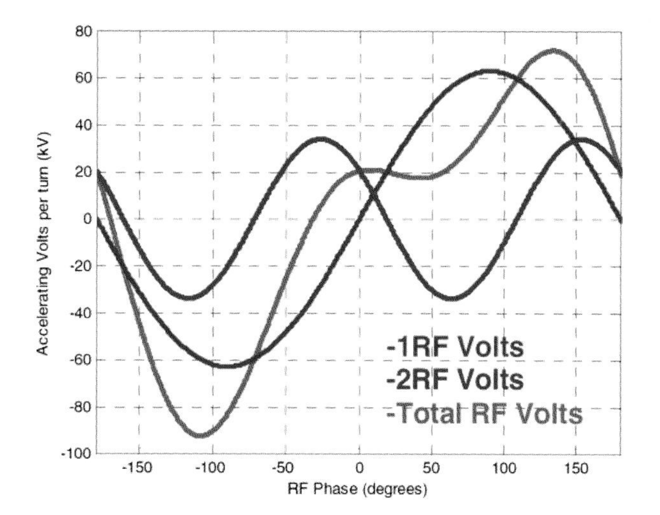

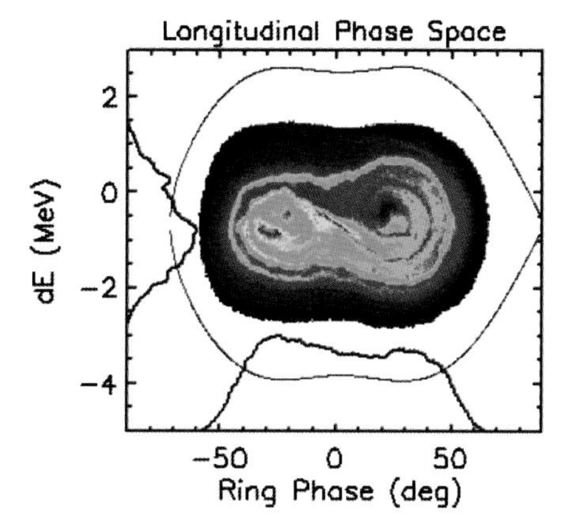

Fig. 10. Dual harmonic RF acceleration at ISIS. Upper: Wave form of the second harmonic and the fundamental RF components. Lower: Beam distributions in $(\varphi, \Delta E)$ at the end of injection [133, 134].

an RF barrier is by an induction cavity [131]. A less perfect RF barrier with a few harmonic components [132] can also be considered.

4.5. *Bunch rotation*

Bunch compression is required for very short bunch extraction which is required for secondary beam acceleration such as muon acceleration in the neutrino factory [135–138], for white neutron applications for better time resolution in the time-of-flight method [139], and for secondary radioactive beam production such as at FAIR/SIS100 [140]. It can be practised in both RCSs and ARs.

Compared with simple adiabatic RF voltage programming, the bunch rotation method is more complex. It employs an adiabatic process to lengthen the bunch by decreasing the RF voltage slowly, and then a nonadiabatic process to make the bunch rotation by increasing the RF voltage rapidly, so that it can produce a very short bunch for fast extraction. Other methods to obtain a very short bunch are: a lattice with the extraction energy just below the transition energy and adding a higher harmonic RF system before the extraction [135].

There are some concerns about making bunch rotation in high-intensity rings. First, in order to make bunch rotation at extraction, it is important to make a small longitudinal emittance at injection, which is contradictory to the general design for high-intensity rings. Thus, one must study the beam collective behaviors carefully [137]. Second, during the bunch rotation process, there will be a very strong space charge effect when the beam becomes very short [136, 137]. Third, when very low RF voltages are applied at the injection and bunch lengthening, the beam loading effect is very strong, so it complicates the LLRF control for the RF system.

5. Collimation Method

5.1. *General considerations about collimation methods in rings*

The localization of beam losses is very important in reducing noncontrolled beam losses which hinder hands-on maintenance, as explained in Subsec. 2.1. Collimation in rings is quite different from that in beam transport lines, where the impact parameter is usually large. A large impact parameter is critical in obtaining high collimation efficiency. As the beam halo usually grows slowly, the impact parameter is usually very small in rings, typically a few μm. Thus, the two-stage collimation method is developed [141–145].

With the two-stage collimation method, a thin blade, primary collimator or scraper scatters halo particles, and those with large divergence angles hit the thick absorber, secondary collimators with large impact parameters, and so are absorbed efficiently. As shown in Fig. 11, one needs two secondary collimators to collimate the particles with both positive and negative divergence in two different locations, respectively. It is also required to have appropriate phase advances between the collimators, e.g. about $\pi/6$ and $2\pi/3$. Therefore, it is good to place the whole collimation system in an uninterrupted long straight section. To make a complete collimation, it will take

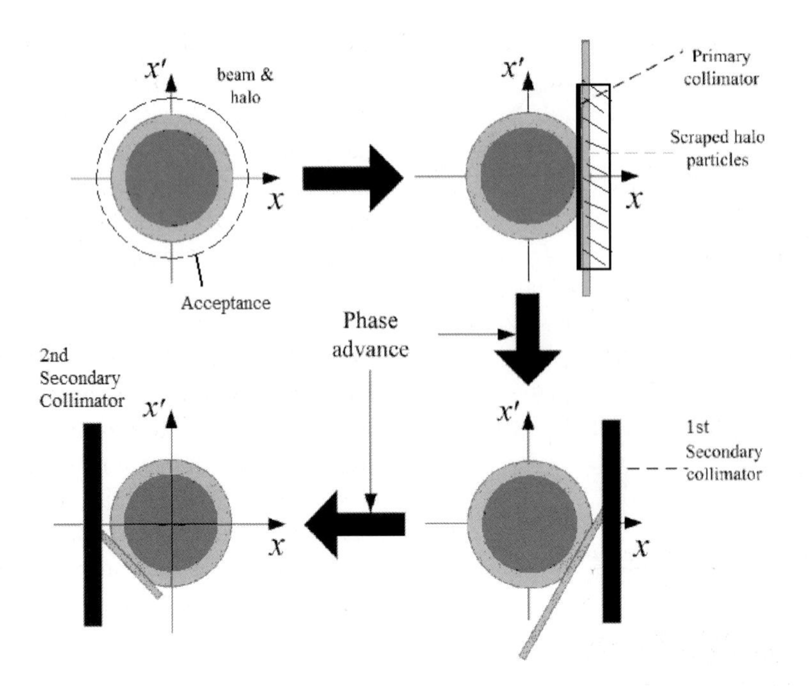

Fig. 11. Schematic for the two-stage collimation method.

many turns to scrape all the halo particles around the beam core and make repeated scattering for those particles with small divergent angles which will miss the secondary collimators. The collimation efficiency is defined by the ratio between the total absorbed particles in the secondary collimators and the total scattered particles. The particles lost in other locations in the ring are considered uncontrolled beam loss. The typical collimation efficiency for a two-stage collimation system is about 90%. To obtain high collimation efficiency, it is important to have a reasonable gap between the collimation acceptance defined by the primary collimator and the physical acceptance.

Certainly, one can design the collimation for both transverse planes in the same locations. Due to different betatron functions, the collimation effects are somewhat different from each other. It should be pointed out that the collimation effect is also dependent on the beam correlation between the two transverse phase planes, which comes from the injection painting scheme [146].

For a simple collimation system like the one used at ISIS/RCS, a combined transverse and longitudinal collimation method can be employed. However, new designs prefer to have a more or less independent collimation system for the transverse and longitudinal phase planes [146, 147].

In early days, for halo collimation at low energy, e.g. 70 MeV at ISIS/RCS, simple one-stage collimation is still efficient if carefully designed. ISIS uses two massive collimators with quite a large phase advance in between. Collimation efficiency of about 90% can be obtained [113]. For collimation at high energy, one needs to design even multiple-stage collimation instead of two-stage collimation. The main reason is that the required impact parameter for good collimation efficiency increases with the stopping range of the particle in collimators. Like LHC, five-stage collimation was designed [148].

5.2. *Transverse collimation*

Usually, the collimation for the two transverse phase planes is performed simultaneously in an uninterrupted long straight section, preferably in a dispersion-free region. As the emittance growth mainly happens at low energy where the space charge effects are the strongest, for example during the injection period and the early acceleration period, one needs only to design a transverse collimation system for a small energy range, which helps a lot in defining the thickness for the collimators. Usually, the primary collimators are made movable to have flexible collimation acceptance, and the secondary collimators can be either fixed [105, 144] or movable [147], which also has important impact on the mechanical and shielding design of the collimators. Furthermore, the supposed power deposition in the collimators has impact on the structure design.

If the collimation energy range is a few hundreds of MeV or higher, it is better to have the tertiary collimators just following each of the secondary collimators to collect the outscattered particles from there [147], but they are often not distinguished from the secondary collimators.

5.3. *Longitudinal collimation*

In RCSs and ARs, a longitudinal beam halo has different meanings and comes from different mechanisms: first, it means the particles with a large momentum deviation which miss the RF capture; second, it means the particles which lose momentum in the stripper foil and the collimators; third, it means the particles between bunches due to the EP instability.

Momentum collimation should involve collimators in dispersive regions [105, 142, 146, 147]. When there is sufficient space in arcs like in high-energy synchrotrons [148, 149], momentum collimation can be performed completely in one or two arcs. In most cases where the long drifts in arcs are lacking, one can place the primary collimator at a location with high normalized dispersion and employ the transverse collimation system as the secondary collimators. In order to improve the collimation efficiency, a combined momentum collimation method using both the longitudinal and the transverse secondary collimators can be employed to deal with the complicated longitudinal–transverse correlations [146].

For the cleaning of the beam halo in bunch gaps in ARs, a so-called beam-in-gap kicker (BIG) is used to blow up the transverse emittance for those particles and the transverse collimation system will collimate them as normal transverse halo particles [150]. However, the operation at SNS has not observed important beam loss without BIG.

6. Key Accelerator Technologies

6.1. *Special features with the accelerator components in high-power rings*

Owing to fast ramping and high beam power, there are some special features in high-intensity RCS and AR rings, including eddy current effect in magnets and vacuum chambers, high RF voltage and beam loading for RF systems, lifetime of injection strippers, fast detection of beam loss and wide dynamic range for the beam diagnostics, complicated collimation, etc. Thus, RCS and AR rings often require different accelerator technologies from ordinary synchrotrons, and some of them are introduced as follows.

6.2. *Large-size and fast-ramping magnets*

For RCSs, room temperature "conventional" magnets are still needed, since superconducting magnets cannot afford a very high field ramping rate (>4 T/s), although efforts have been made in NUKTRON and FAIR [151–153].

Large apertures and fast ramping make the magnets very difficult to design and fabricate, so they are not ordinary conventional magnets. The critical issues include eddy current in coils and magnet cores, high-voltage insulation and mechanical vibration. Some special techniques have been developed to tackle the problems, such as coils based on stranded conductors [154–156] or braided conductors [156, 157] to alleviate eddy current in coils, thin laminated sheets and low saturation to weaken eddy current effect and hysteresis loss in magnet cores, cutting slots in end cores and supporting plates to decrease the temperature which is also caused by eddy current [158], design optimization to have the maximum induction voltage in coils and cables below 10 kV, unremovable but machined shimming parts and a specially designed supporting system to control the mechanical vibration, a maximum magnetic field of about 1 T to assert linear inductance during field ramping, etc. Stranded conductor made of aluminum wires and stainless steel cooling pipe, as shown in Fig. 12, is considered a good technique for magnets working in tens Hertz.

Due to the higher radiation dose rate exposed to the magnets, it is required to have somewhat radiation-hard designs, such as epoxy resin with a special radiation-resistant formula or polyimide resin [155]. No other organic insulation materials can be used here.

6.3. *Resonant power converters for RCS main magnets*

For an RCS with a repetition rate less than 10 Hz, conventional power supplies by bridge rectifiers or

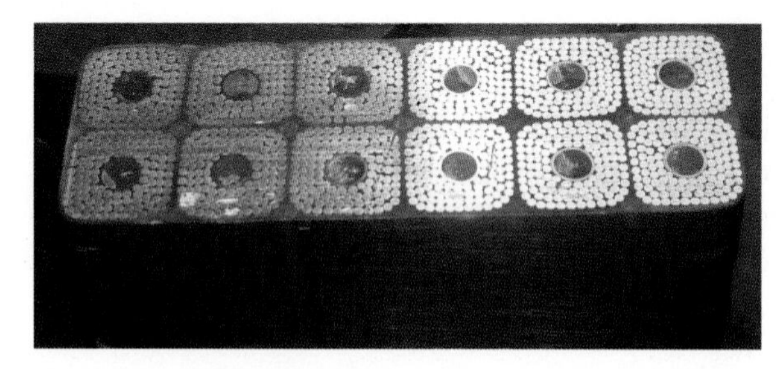

Fig. 12. Stranded conductor made of aluminum wires and a stainless steel cooling pipe developed for CSNS/RCS magnets.

bridge rectifiers with local energy storage directly connected to the power grid can be used, such as in the AGS Booster. However, for a higher repetition rate, large no-work energy transfer to and from the magnets during fast ramping up and down is not acceptable for conventional power converters using a trapezoidal wave form, as it produces enormous active/reactive power peaks and amplitude variations on the power distribution network. The resonant magnet excitation technique should be used, which is a resonant circuit (known as a "White Circuit") [9, 159–162], with DC-biased AC excitation. For each magnet family, the power converter, magnets, choke transformers and capacitor banks altogether form a parallel or series network divided into cells. The cell size is limited by the maximum applicable induction voltage of 10 kV. The network including choke transformers and capacitors will increase the cost and complexity of the power supply system to a large extent.

Due to the saturation effect in the magnetic field, the inductance of the magnet and perhaps also the choke transformer is not perfectly constant during field ramping, so there is a dynamic inductance which deforms the field oscillation curve. It is important to have good field synchronization among the magnet families including dipoles and quadrupoles for the whole cycling range. A possible remedy is to inject proper multiple harmonic components from the power supply into the circuit to restore the sinusoidal wave form for all the magnetic fields [9, 163].

6.4. *Large ceramic vacuum chambers*

For the vacuum chambers used in RCS main magnets, the heat deposition and magnetic field induced by eddy current become very important. Different solutions have been studied and developed, such as a ceramic vacuum chamber, ribbed thin metallic vacuum chamber [120], or bellows-type vacuum chamber [164]. However, for the vacuum chambers with large apertures, only ceramic vacuum chambers can meet the requirement [165].

It is very difficult to develop ceramic vacuum chambers of large length and aperture, especially those with special cross-sections other than cylindrical ones. Short ceramic tubes should be joined together to form long ceramic chambers, especially the curved ones used for dipole magnets. The first of such ceramic chambers was studied for an RCS booster at LBNL [166]. Since then, two different techniques have been developed to make long chambers: the glass-glazing joint technique [167] and the metal-brazing technique [168].

RF shielding, which is required to reduce the coupling impedance, is another technical challenge to the fabrication of vacuum chambers. Different solutions have been developed, such as metallic cages inside the vacuum chamber at ISIS [6], electroformed thin copper stripes on the external surface of ceramic tubes at J-PARC, and wrapped copper stripes on the external surface of ceramic tubes at CSNS. The conducting passage is segmented by special capacitors, so it looks like an open circuit by the slowly changing eddy current. Figure 13 shows the ceramic vacuum chamber used at J-PARC.

TiN coating for all the ceramic chambers is usually required in most RCSs and ARs to reduce the secondary electron emission and avoid the EP instability, as discussed in Subsec. 2.5.

6.5. *Ferrite or MA-loaded RF cavities*

As mentioned in Subsec. 4.3, for RCSs fast acceleration demands a large number of RF cavities, as each cavity loaded with magnetic material can provide RF voltage of only tens of kV, such as about 10 kV/m

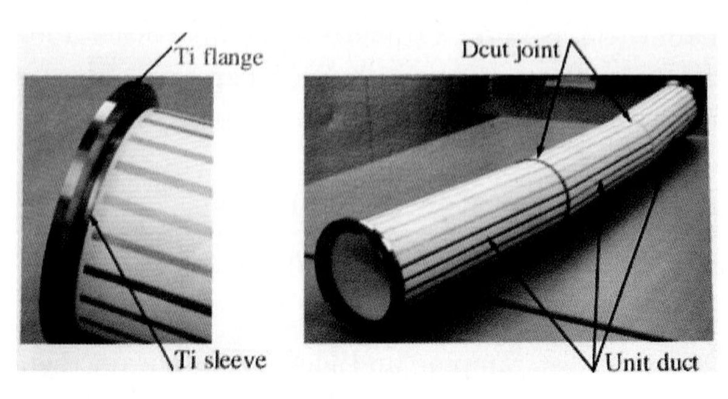

Fig. 13. A long and curved ceramic vacuum duct developed for J-PARC/RCS using the metal-brazing technique [168].

for ferrite-loaded RF cavities and 20–40 kV/m for MA-loaded RF cavities. Fast acceleration also means fast frequency sweeping, and this poses technical difficulties in ferrite-loaded cavities, where bias current power supplies are used for the frequency tuning. The bandwidth of the tuning system should be large enough to follow the fast ramping and deal with the fast change in beam loading during beam injection. The nonlinear characteristics of ferrite cores make the situation more complicated. Big change in impedance during the frequency sweeping requires a flexible RF amplifier.

Although the MA-loaded cavity is very successfully applied in synchrotrons with low RF power such as synchrotrons for hadron therapy with the benefit of there being no need for cavity tuning, its use in high-intensity RCSs demanding high-power RF is still in the development stage. In particular, the application of the cavities using the cut core configuration to raise the quality factor was not so successful at J-PARC [169], and cavity degrading appeared gradually. Nevertheless, its successful application and continuous improvement at J-PARC took a big step forward.

Heavy beam loading effect due to very high circulating beam current should be compensated for to keep the RF system stable, and this is especially important for ferrite-loaded cavities. Quality factor decrease by liquid resistor is used at ISIS to deal with the beam loading at the injection energy when the RF voltage is very low [170]. For MA-loaded cavities, the concern about beam loading is that different harmonic components of the beam can be excited in the cavity due to its very low quality factor.

A sophisticated low-level RF control system of as many as six or seven control loops for each RF station is usually needed, such as control loops on the RF phase, RF voltage, frequency tuning, feedbackward, feedforward, direct feed and local feedback [171]. Digitalized systems are showing significant advantages over traditional analog systems, and have been widely adopted [171–173]. When a dual harmonic RF system is adopted, the LLRF control system and second harmonic cavity beam loading become even more complicated [174].

6.6. *Collimators and stripper foils*

The material selection and structure design for collimators are based on the energy and power of beam, prompt and residual dose rates, cooling effect and fabrication techniques. The material also has an impact on collimation efficiency, because different materials have different effects in scattering beam and reducing the average momentum [146]. For thin blade primary collimators, independently controlled jaws are used for changing the collimation acceptance if necessary. As the heat deposition and radiation dose rate are relatively low here, indirect cooling and simple shielding can be designed. For secondary collimators, intercepting modest beam power, movable jaws together with indirect cooling or forced air-cooling by fans can be considered [9], whereas for those intercepting beam powers higher than a few kW, a more complicated design used at SNS can be considered [175]. As this is a high-radiation region, the components and accessories such as motors, cables or vacuum tubes should have radiation-hard design, and local shielding with remote handling is needed.

The material, supporting structure and mechanical driving system for stripper foils depend on the beam power level in RCSs and ARs. For a primary stripper, the foil suffers heavy bombardment of the injection beam and the circulating beam, and raises the concern about lifetime. On one hand, recent global efforts succeeded in developing high-quality carbon foils to meet the requirement for megawatt beams, such as corrugated nanocrystalline diamond stripper foils [176] and hybrid boron mixed carbon (HBC) stripper foils [177]. On the other hand, many foils can be installed with remote control for quick online replacement of the failed foils [92, 178]. Different foil-supporting structures have also been developed.

6.7. *Magnets and power supplies for injection and extraction*

Fast bump magnets are used for multiturn injection. In order to obtain good beam distribution by phase space painting, programmed field curves for the bump magnets are required, either for the falling field or for the rising field, depending on the painting scheme. Only a few coil windings can be employed here to fit with the fast-changing field, and thus very high excitation current has to be used, such as about 30,000 A at J-PARC/RCS [179] and 18,000 A at CSNS [180]. Large magnet apertures and close installation lead to important field interference, and particle trackings with the integrated field distribution are required [181, 182]. Special coil clamping can be helpful in reducing the field inference [183].

For extraction kickers, both the transmission-line-type and lumped-type magnet cores are applicable. The former has the advantage of producing a magnetic field of shorter rise time for a given magnet size and the latter has the advantage of being more robust, which is important when working at a high repetition rate. The pulse-forming network (PFN) is always used to trigger the very high excitation current required for the magnet. Different PFN schemes, such as PFL using transmission cables, lumped PFN, the Blumlein circuit to reduce the charging voltage, either with magnets short-ended or with magnets matched to earth, are used to meet different requirements. A good flat top field is important for limiting the emittance growth through the extraction channel; for example, J-PARC uses the combined field by shifting the triggering times to different kicker modules [184]. Derated parameters should be used to increase the kickers' reliability; for example, SNS limits the maximum PFN voltage to 35 kV [185]. Other important issues include the following: low impedance is required by the total impedance budget, and this can be helped by adding a parallel resistor to the PFN [185, 186]; many units in a long vacuum chamber poses problems for installation and *in situ* baking.

The septum magnets (especially the ones for extraction) used for high-intensity synchrotrons should have large apertures and be radiation-resistant to some extent. There are two types of septum magnets: the conventional coil-blade magnet and the Lambertson magnet. The former is more straightforward in beam optics, but has high power consumption due to a large aperture and needs special insulation to be radiation-resistant. The latter has the advantages of saving power consumption and being radiation-resistant, but has disadvantages in the complicated optics for the extracted beam and severe stray field which needs to be compensated for carefully [109, 187].

6.8. *Beam instrumentation*

Beam diagnostics in accelerators has been advancing steadily in the last few decades, and it helps a lot in modeling accelerators in operation and controlling accelerators more precisely. There is no exemption in high-intensity proton RCSs and ARs [188].

Special requirements and conditions for the beam diagnostic system used in high-intensity synchrotrons include: fast detection of beam loss at the injection and extraction by beam loss monitors, a wide dynamic range for most of the beam diagnostic devices, very large apertures for the probes, avoiding interceptive monitors as it is possible, electronics outside the tunnel due to a relatively high radiation level in the tunnel, etc. There have been new developments in this domain, such as the IPM (ionization profile monitor; see Fig. 14) for noninterceptive profile measurement [189, 190], the turn-by-turn BPM, and fast tune measurement. Because of very high beam power, the MPS (machine protection system) and PPS (personnel protection system) also rely on the fast and reliable beam diagnostics.

On the other hand, through combination with the accelerator physics application programs and the control system, the beam monitoring system can help build a virtual accelerator which is very useful for modeling the accelerator and finding problems during the beam commissioning and operation. The accelerator physics software package XAL, which was initially developed for SNS, has now become a common platform, with several laboratories collaborating to develop its new framework — Open XAL [191]. Not only does it show the beam status, but it also can play back the hardware parameters and beam status in looking back at the abnormalities which occurred in the past.

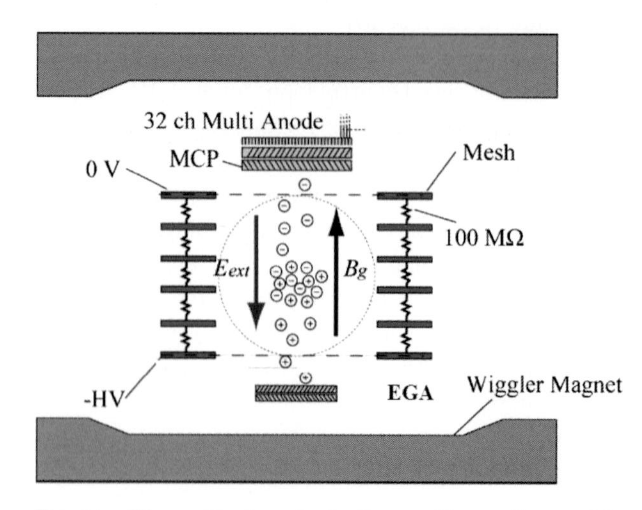

Fig. 14. The residual gas ionization profile monitor used in J-PARC/RCS [190].

6.9. *Future development trends*

The demand for beams of a few megawatts and short pulse is foreseen in the next decade, for example as in the proton drivers of neutrino superbeams and the neutrino factory (or the muon collider as the Higgs Factory) [192, 193]. This includes new machines such as SPL neutrino beam, ESS neutrino beam, CSNS neutrino beam, and also the upgrading of the existing machines, such as the Recycler and Main Injector for Project-X [16, 194], or PS2 for LHC intensity upgrading [195]. Even higher beam power has not been proposed.

Spallation neutron sources will include more realistic projects, such the RCS at PEFP [196] and ISNS. CSNS is being planned to be upgraded to 0.5 MW, and ISIS is also seeking the upgrading of its RCS power by building a new linac injector. The 2 MW upgrading of SNS was planned long ago. J-PARC/RCS is still on the way to achieving its design goal of 1 MW from the present about 300 kW, with the linac extension from 181 MeV to 400 MeV.

With the rapid advancement in the superconducting proton linac in the last decade, it has become not so difficult to build high-energy and high-duty-factor linacs. The injection energy for RCSs and ARs can be higher, so the peak linac current is not so demanding, which is difficult for H-minus acceleration. This also explains the defeat of the linac–RCS combination by a larger linac in the competition to replace the current linac–booster as the proton driver at Fermilab. The new ESS design also increased the linac energy from 1.334 GeV with the 2002 version to 2.5 GeV (with proton beam) when keeping the maximum beam power of 5 MW.

In technology, fast ramping superconducting magnets may become available in the coming years, so they may be employed in high-energy synchrotrons of a few Hz. This will decrease the circumference and cost significantly. MA-loaded RF cavities are expected to be widely used in the synchrotrons where very high RF voltage is needed, especially in the cases of a low RF duty factor. With more sophisticated beam diagnostics, one can control the beam more precisely.

6.10. *Discussion and conclusions*

In high-power proton accelerators, an uncontrolled beam loss rate is a key issue. With higher beam power, the total loss rate should be controlled to a lower level. On the other hand, a more sophisticated collimation should be designed to reduce the uncontrolled beam loss rate. In order to achieve a lower loss rate, one has to investigate all the mechanisms which potentially result in beam loss. Much more powerful multiparticle simulations are needed to understand the development of the beam halo. Collective beam instabilities also become more important when the circulating current increases. To operate machines of very high beam power, fast and reliable machine protection systems are imperative. Advanced control systems with imbedded physics application programs are needed to diagnose the problems during commissioning and operation. A long process usually lasts a few years to ramp up the beam power to the design goal and beyond.

With regard to the technology evolution, MA-loaded RF cavities have already been applied at J-PARC, though there are still some problems in high-power applications, which are expected to be solved in the next decade. With big international efforts in the 2000s, it looks like now specially prepared stripper foils which have already been commercialized can meet the requirement for the beam power of a few MW.

With the successful construction and operation of SNS/AR and J-PARC/RCS, the knowledge gained will strengthen our confidence in successfully building RCSs and ARs of a few megawatts in the coming years. Several of them have been designed or studied in the last decade or even earlier.

Acknowledgments

Many thanks to the students of mine, Jinfang Chen, Ye Zou, Zhen Guo, Yingpeng Song, Biao Sun and Jilei Sun, who helped me in preparing some manuscript material. Warm thanks also go to Wu-Tsung Weng of BNL, who kindly read through the draft and gave valuable suggestions. Partial support from the National Natural Science Foundation of China (Projects 11235012 and 10875099) and the project "China Spallation Neutron Source" is acknowledged.

References

[1] G. I. Budker and G. I. Dimov, in *Proc. Int. Conf. on High-Energy Accelerators* (Dubna, 1963), p. 1372.

[2] C. W. Potts, *IEEE Trans. Nucl. Sci.* **NS-24**(3), 1385 (1977).
[3] National Accelerator Laboratory, Design Report, Second Printing (Fermilab, 1968).
[4] I. Yamane, E. Takasaki, S. Hiramatsu *et al.*, in *Proc. PAC 1989* (Chicago, 1989), p. 301.
[5] High Energy Physics Division, Argonne National Laboratory, ANL-HEP-TR-06-44 (2006).
[6] B. Boardman (ed.), RAL Report, RL 82-006 (1982).
[7] P. W. Lisowski and K. F. Schoenberg, *Nucl. Instrum. Methods A* **562**, 910 (2006).
[8] T. E. Mason *et al.*, in *Proc. LINAC 2000* (Monterey, 2000).
[9] Accelerator Technical Design Report for J-PARC, J-PARC 03-01 (Mar. 2003).
[10] J. Wei, H. S. Chen, Y. W. Chen *et al.*, *Nucl. Instrum. Methods A* **600**, 10 (2009).
[11] S. Peggs, for ESS in *Proc. PAC 2011* (New York, 2011), p. 2549.
[12] AGS Booster Project, *Booster Design Manual*, ed. W. T. Weng, BNL Report (1988).
[13] W. T. Weng, *Part. Accel.* **27**, 13 (1990).
[14] European Spallation Neutron Source Project, Vol. III, Technical Report.
[15] A. Alekou, M. Aiba, I. Efthymiopoulos *et al.*, in *Proc. IPAC 2013* (Shanghai, China, 2013).
[16] S. Nagaitsev, *Proc. PAC 2011* (New York, 2011), p. 2566.
[17] W. Yang and J.-Y. Tang, *Chin. Phys. C* **37**(9), 097002 (2013).
[18] http://www.fair-center.eu.; FAIR Baseline Technical Report, Vol. 2, Mar. 2006.
[19] ISS Accelerator Working Group *et al.*, *J. Instrum.* **4**, P07001 (2009).
[20] J. Wei, *Rev. Mod. Phys.* **75**(4) (2003).
[21] J. Wei, D. T. Abell, J. Beebe-Wang *et al.*, *Phys. Rev. ST Accel. Beams* **3**, 080101 (2000).
[22] http://hps.org/publicinformation/ate/q435.html
[23] T. P. Wangler, *Principles of RF Linear Accelerators* (Wiley, New York, 1998).
[24] N. V. Mokhov and W. Chou (eds.), in *Workshop on Beam Halo and Scraping* (Lake Como, 1999).
[25] I. Strask, E. Mustafin and M. Pavlovic, *Phys. Rev. ST Accel. Beams* **13**, 071004 (2010).
[26] W. Chou, in *Proc. HB 2002*, (2002), p. 29.
[27] Y. Yamazaki, in *Proc. HB 2012* (Beijing, China, 2012), p. 15.
[28] J. Wei, J. Beebe-Wang, M. Blaskiewicz *et al.*, in *Proc. EPAC 2000* (Vienna, Austria, 2000), p. 981.
[29] M. A. Plum, in *Proc. HB 2012* (Beijing, China, 2012), p. 36.
[30] S. Y. Lee, *Accelerator Physics* (World Scientific, 1999).
[31] S. Wang, Y.-W. An, N. Huang *et al.*, in *Proc. IPAC 2010* (Kyoto, Japan, 2010), pp. 630–632.
[32] P. Bryant, M. Regler and M. Schuster (eds.), CERN/PS 95-48 (DI).
[33] J. A. Holmes, J. D. Galambos, J. H. Whealton *et al.*, *AIP Conf. Proc.* **448**, 344 (1998).
[34] Y. Cho *et al.*, in *Proc. EPAC 1996* (Sitges, Spain, 1996), p. 521.
[35] J. Wei, C. Gardner, Y. Y. Lee and N. Tsoupas, in *Proc. EPAC 2000* (Vienna, Austria, 2000), p. 978.
[36] M. Reiser, *Theory and Design of Charged Particle Beams* (John Wiley & Sons, 1994).
[37] A. Sørenssen, *Part. Accel.* **6**, 141 (1975).
[38] J. Wei, Ph.D. dissertation, 1990 (revised 1994), SUNY, Stony Brook.
[39] J. Wei, in *Handbook of Accel. Phys. and Eng.*, ed. A. Chao and M. Tigner (World Scientific, 1998), p. 281.
[40] L. C. Teng, *Part. Accel.* **4**, 81 (1972).
[41] R. C. Gupta, J. I. M. Botman and M. K. Craddock, *IEEE Trans. Nucl. Sci.* **NS-32**, 2308 (1985).
[42] D. Trbojevic, D. Finley, R. Gerig *et al.*, in *Proc. EPAC 1990* (Gif-sur-Yvette, France, 1990), p. 1536.
[43] Yu. Senichev, S. An, K. Bongardt *et al.*, in *Proc. EPAC 2004* (Lucerne, Switzerland, 2004), p. 653.
[44] Yu. Senichev, *Proc. PAC09* (Vancouver, Canada, 2009), p. 677.
[45] S. M. Cousineau, *Nucl. Instrum. Methods Phys. Res. A* **561**, 297 (2006).
[46] C. M. Warsop, in *AIP Conf. Proc. No. 448* (Melville, NY, 1998), p. 104.
[47] Y. Papaphilippou, Y. Y. Lee and W. Meng, in *Proc. PAC 2001* (Chicago, 2001), p. 1667.
[48] N. Tsoupas, C. Gardner, Y. Y. Lee *et al.*, in *Proc. EPAC 2000* (Vienna, Austria, 2000), p. 1581.
[49] Y. Watanabe, N. Tani, T. Adachi *et al.*, in *Proc. IPAC 2011* (San Sebastián, Spain, 2011), p. 3338.
[50] A. W. Chao, *Physics of Collective Beam Instabilities in High Energy Accelerators* (Wiley, New York, 1993).
[51] K.-Y. Ng, Physics of collective beam instabilities. FERMILAB-Conf-00/142-T, June 2000.
[52] L. J. Laslett, Brookhaven National Laboratory Report 7534 (1963), p. 324.
[53] A. Hofmann, CERN 94-04 (1994).
[54] A. V. Fedotov and I. Hofmann, *Phys. Rev. ST Accel. Beams* **5**, 024202 (2002).
[55] C. M. Warsop, D. J. Adams, B. Jones *et al.*, in *Proc. HB 2008* (Nashville, Tennessee, 2008), p. 143.
[56] S. Y. Xu, S. X. Fang and S. Wang, in *Proc. HB 2010* (Morschach, Switzerland, 2010), p. 420.
[57] J.-F. Chen, J.-Y. Tang and X.-Y. Zhang, in *Proc. HB 2012* (Beijing, China, 2012).
[58] B. W. Montague, CERN-Report No. 68-38 (CERN, 1968).
[59] A. V. Fedotov, N. Malitsky, Y. Papaphilippou *et al.*, in *Proc. PAC 2001* (Chicago, 2001), p. 2848.
[60] I. Hofmann, G. Franchetti and S. Y. Lee, in *Proc. HB 2006* (Tsukuba, Japan, 2006), p. 268.

[61] B. Zotter and S. Kheifets, *Impedances and Wakes in High-Energy Particle Accelerators* (World Scientific, 1998).
[62] A. Sessler and V. Vaccaro, CERN Report ISR-RF/67-2 (1967).
[63] O. Boine-Frankenheim, I. Hofmann and V. Kornilov, in *Proc. EPAC06* (Edinburgh, Scotland, 2006), p. 1882.
[64] V. K. Neil and A. M. Sessler, *Rev. Sci. Instrum.* **6**, 429 (1965).
[65] S. Koscielniak, *AIP Conf. Proc. No. 496*, eds. T. Roser and S. Y. Zhang (Melville, NY, 1999), p. 391.
[66] V. Kornilov and O. Boine-Frankenheim, *Phys. Rev. ST Accel. Beams* **13**, 114201 (2010).
[67] V. Danilov *et al.*, *Phys. Rev. ST Accel. Beams* **4**, 120101 (2001).
[68] R. J. Macek, R. C. McCrady and S. B. Walbridge, in *Proc. HB 2006* (Tsukuba, Japan, 2006), p. 94.
[69] G. H. Rees, *Part. Accel.* **39**, 159 (1992).
[70] J. Gareyte and F. Sacherer, in *Proc. IXth Int. Conf. on High Energy Accel.* (1974), p. 341.
[71] C. E. Nielson *et al.*, in *Proc. Int. Conf. on High-Energy Accel. and Instrum.* (1959), p. 239.
[72] E. Keil and W. Schnell, CERN Report TH-RF/69-48 (1969).
[73] D. Boussard, CERN Report Lab II/RF/Int./75-2 (1975).
[74] I. Hofmann, *Laser Part. Beams* **3**, 1 (1985).
[75] K. Woody, J. A. Holmes, V. Danilov *et al.*, in *Proc. PAC 2001* (Chicago, 2001), p. 2057.
[76] K. Y. Ng, D. Wildman, M. Popovic *et al.*, in *Proc. PAC 2001* (Chicago, 2001), p. 2890.
[77] A. Groeber, in *Proc. 1977 Int. Conf. on High Energy Accel.* (Protvino, 1977), p. 277.
[78] D. Neuffer, *Nucl. Instrum. Methods Phys. Res. A* **321**, 1 (1992).
[79] R. Macek *et al.*, in *Proc. PAC 2001* (Chicago, Piscataway, NJ, 2001), p. 688.
[80] M. Blaskiewicz, M. A. Furman, M. Pivi *et al.*, *Rev. ST Accel. Beams* **6**, 081002 (2003).
[81] K. Ohmi, in *AIP Conf. Proc. 642* (2006), p. 360.
[82] G. Franchetti, I. Hofmann, W. Fischer *et al.*, *Phys. Rev. ST Accel. Beams* **12**, 124401 (2009).
[83] G. Franchetti, I. Hofmann, S. Machida *et al.*, *Proc. HB 2006* (Tsukuba, Japan, 2006), p. 344.
[84] G. Franchetti, W. B. Bayer, F. Becker *et al.*, *Proc. PAC09* (Vancouver, Canada, 2009), p. 3239.
[85] https://oraweb.cern.ch/pls/hhh/code_website.disp_allcat
[86] G. H. Rees, CERN Accelerator School: General Accelerator Physics, CERN 94-04 (1994).
[87] V. C. Kempson, C. W. Panner and V. T. Pugh, *IEEE Trans. Nucl. Sci.* **28**(3), 3085 (1981).
[88] G. Rees, *Proc. EPAC 1994* (London, UK, 1994), p. 241.
[89] G. Rees, in *Handbook of Accel. Phys. and Eng.* eds. A. Chao and M. Tigner (1998), p. 497.
[90] W. Chou, J. Lackey, Z. Tang *et al.*, *Proc. PAC07* (Albuquerque, 2007), p. 1679.
[91] S. G. Lebedev, *Nucl. Instrum. Methods A* **613**, 442 (2010).
[92] M. A. Plum, S. M. Cousineau, J. Galambos *et al.*, *Phys. Rev. ST Accel. Beams* **14**, 030102 (2011).
[93] V. V. Danilov, in *Proc. HB 2008* (Nashville, Tennessee), p. 284.
[94] T. V. Gorlov, A. V. Aleksandrov and V. V. Danilov, in *Proc. PAC 2011* (New York, 2011), p. 2035.
[95] D. T. Abell, Y. Y. Lee and W. Meng, in *Proc. EPAC 2000* (Vienna, Austria, 2000), p. 2107.
[96] I. Sakai *et al.*, in *Proc. PAC 2003* (Portland, Oregon, 2003), p. 1512.
[97] J. Y. Tang *et al.*, in *Proc. EPAC 2006* (Edinburgh, Scotland, 2006), p. 1783.
[98] J. Beebe-Wang, Y. Y. Lee, D. Raparia and J. Wei, in *Proc. PAC 2001* (Chicago, 2001), p. 1508.
[99] D. Raparia, in *Proc. PAC 2011* (New York, 2011), p. 1975.
[100] M. Furman, in *Handbook of Accel. Phys. and Eng.*, eds. A. Chao and M. Tigner (1998), p. 438.
[101] R. J. Damburg and V. V. Kolosov, in *Rydberg States of Atoms and Molecules*, eds. R. F. Stebbings and F. B. Dunning (Cambridge Univ., 1983), pp. 31–71.
[102] J. Wei, J. Beebe-Wang, M. Blaskiewicz *et al.*, in *Proc. PAC 2001* (Chicago, 2001), p. 2560.
[103] J. Qiu, J. Y. Tang, N. Huang and S. Wang, in *Proc. IPAC 2011* (San Sebastian, Spain, 2011), p. 2739.
[104] V. Danilov *et al.*, in *Proc. EPAC 2004* (2004), p. 2227.
[105] N. Catalan-Lasheras and D. Raparia, in *Proc. PAC 2001* (Chicago, 2001), p. 3263.
[106] J. Y. Tang, G. H. Wei and C. Zhang, *Nucl. Instrum. Methods Phys. Res. A* **572**, 601 (2007).
[107] J. Tang, L. Liu, J. Qiu *et al.*, in *Proc. HB 2008* (Nashville, Tennessee, 2008), p. 320.
[108] N. Tsoupas, M. Blaskiewicz, Y. Y. Lee *et al.*, in *Proc. EPAC 2000* (Vienna, Austria, 2000), p. 2270.
[109] J. G. Wang, *Phys. Rev. ST Accel. Beams* **12**, 042402 (2009).
[110] M. Tanaka *et al.*, in *Proc. PAC 1995* (Dallas, 1995), pp. 1930–1932.
[111] J. Y. Tang, *Nucl. Instrum. Methods Phys. Res. A* **575**, 328 (2007).
[112] P. J. Bryant and K. Johnsen, *The Principles of Circular Accelerators and Storage Rings* (Cambridge University Press, 1993).
[113] C. M. Warsop, in *Proc. EPAC 2004* (Lucerne, Switzerland, 2004), p. 1464.
[114] M. Yamamoto, M. Yoshii *et al.*, *Phys. Rev. ST Accel. Beams* **12**, 041001 (2009).
[115] L. Lin, J.-Y. Tang, Q. Jing and W. Tao, *Chin. Phys. C* **33** (Suppl. II), 4 (2009).
[116] H. Sun, X. Li, F. C. Zhao *et al.*, in *Proc. IPAC 2013* (Shanghai, China, 2013).

[117] M. Yoshii, E. Ezura, K. Hara *et al.*, in *Proc. IPAC 2010* (Kyoto, Japan, 2010), p. 615.
[118] E. Griesmayer, CERN/PS/CA/Note 98-18 (1998).
[119] IPNS Upgrade, Argonne National Laboratory, 1995.
[120] Proton Driver Design Study, Fermilab-TM-2136 (2000).
[121] J. Baillod, L. Magnani, G. Nassibian *et al.*, *IEEE Trans. Nucl. Sci.* **30**(4), 3499 (1983).
[122] W. T. Weng, in *Proc. APAC 1998* (Tsukuba, Japan, 1998), p. 329.
[123] M. E. Middendorf, F. R. Brumwell, J. C. Dooling *et al.*, in *Proc. PAC07* (Albuquerque, New Mexico, 2007), p. 2233.
[124] A. Seville, I. Gardner, J. Thomason *et al.*, in *Proc. EPAC 2008* (Genoa, Italy, 2008), p. 349.
[125] J. A. Holmes, S. M. Cousineau, V. V. Danilov *et al.*, in *Proc. HB 2006* (Tsukuba, Japan, 2006), p. 298.
[126] F. Tamura, M. Yamamoto, M. Yoshii *et al.*, *Phys. Rev. ST Accel. Beams* **12**, 041001 (2009).
[127] J. Griffin, J. MacLachlan, A. G. Ruggiero *et al.*, *IEEE Trans. Nucl. Sci.* **NS-30**, 2630 (1983).
[128] J. M. Brennan *et al.*, 1994, in *Proc. EPAC 1994* (1994), p. 1897.
[129] R. Garoby, in *Handbook of Accel. Phys. and Eng.*, eds. A. Chao and M. Tigner (World Scientific, 1998), p. 284.
[130] O. Boine-Frankenheim, *Phys. Rev. ST Accel. Beams* **13**, 034202 (2010).
[131] K. Takayama and J. Kishiro, *Nucl. Instrum. Methods Phys. Res. A* **451**, 304 (2000).
[132] S. Z. An, Ph.D. dissertation (China Institute of Atomic Energy, July 2010).
[133] A. Seville, D. Adams, C. Appelbee *et al.*, in *Proc. PAC 2007* (New Mexico, 2007), p. 1649.
[134] C. M. Warsop, D. J. Adams and D. J. S. Findlay, in *Proc. IPAC 2011* (San Sebastián, Spain, 2011), p. 2760.
[135] C. R. Prior, in *Proc. EPAC 2000* (Vienna, Austria, 2000), p. 963.
[136] R. Cappi, J. Gareyte, E. Métral and D. Möhl, *Nucl. Instrum. Methods A* **472**, 475 (2001).
[137] K. Y. Ng, FERMILAB-FN-0702 (Aug. 2001).
[138] J. Pasternak, M. Aslaninejad, K. Long *et al.*, in *Proc. PAC09* (Vancouver, Canada, 2009), p. 1433.
[139] Y. Zou, J. Y. Tang and J. F. Chen, in *Proc. NA-PAC13* (Pasadena, CA, 2013).
[140] P. Spiller, J. Ahrens, K. Blasche and M. Emmerling, in *Proc. PAC 2001* (Chicago, 2001), p. 3278.
[141] L. C. Teng, FNAL Report, FN-196/0400 (1969).
[142] P. J. Bryant and E. Klein, CERN Report, CERN SL/92-40 (AP) (1992).
[143] L. C. Trenkler, CERN Report, SL/92-50(EA) (1992).
[144] N. Catalan-Lasheras, Y. Y. Lee, H. Ludewig *et al.*, *Phys. Rev. ST Accel. Beams* **4**, 010101 (2001).
[145] T. Wei and Q. Qin, *Nucl. Instrum. Methods Phys. Res. A* **566**, 212 (2006).
[146] J.-Y. Tang, J.-F. Chen and Y. Zou, *Phys. Rev. ST Accel. Beams* **14**, 050103 (2011).
[147] K. Yamamoto, in *Proc. HB 2008* (Nashville, Tennessee, 2008), p. 304.
[148] R. W. Assmann, in *Proc. HB 2012* (Morschach, Switzerland, 2010), p. 21.
[149] J. E. Johnson, in *Proc. PAC 2007* (2007), p. 1967.
[150] S. Cousineau *et al.*, in *Proc. PAC 2001* (Chicago, 2001), p. 1723.
[151] A. D. Kovalenko *et al.*, *Atom. Energ.* **112**(2), (2012).
[152] G. Moritz, in *Proc. EPAC 2006* (Edinburgh, Scotland, 2006), p. 2577.
[153] E. Fischer, A. Mierau, P. Schnizer *et al.*, in *Proc. PAC09* (Vancouver, Canada), pp. 277–279.
[154] H. Sasaki, H. Someya, T. Adachi *et al.*, KEK 86-101 (1986).
[155] N. Tani *et al.*, *IEEE Trans. Appl. Supercond.* **14**(2), 409 (2004).
[156] W. Kang *et al.*, *IEEE Trans. Appl. Supercond.* **22**(3), 4001204 (2012).
[157] R. T. Elliot, J. A. Lidbury and M. R. Harold, *IEEE Trans. Nucl. Sci.* **NS-26**(3), 3922 (1979).
[158] X.-J. Sun, C.-D. Deng and K. Wen, *Chin. Phys. C* **36**, 443 (2012).
[159] M. G. White, F. C. Shoemaker and G. K. O'Neill, *CERN Symposium* (1956), p. 525.
[160] W. Bothe, in *Proc. Power Converters for Particle Accelerators* (Montreux, Switzerland, 1990), p. 271.
[161] J. W. Gray and W. A. Morris, in *Proc. EPAC 2000* (Vienna, Austria, 2000), p. 2202.
[162] J. Zhang, X. Qi, F.-L. Long *et al.*, in *Proc. PAC 2009* (Vancouver, Canada, 2009), p. 1541.
[163] X. Qi, 3rd Workshop on Power Converters for Part. Accel. (Hamburg, May 2012).
[164] G. Horikoshi *et al.*, *IEEE Trans. Nucl. Sci.* **NS-24**(3), 1290 (1977).
[165] J. R. J. Bennett, R. J. Elsey and A. J. Dossett, *Vacuum* **28**, 507 (1978).
[166] P. T. Clee and H. P. Hernandez, in *Proc. PAC 1967* (Washington), pp. 809–814.
[167] J. R. J. Bennett and R. J. Elsey, *IEEE Trans. Nucl. Sci.* **NS-28**(3), 3336 (1981).
[168] M. Kinsho, Y. Saito, Z. Kabeya *et al.*, *Vacuum* **73**, 187 (2004).
[169] C. Ohmori, O. Araoka, E. Ezura *et al.*, in *Proc. IPAC 2011* (San Sebastian, Spain, 2011), p. 2885.
[170] P. Barrat *et al.*, in *Proc. EPAC 1990* (Nice, France, 1990), pp. 949–951.
[171] X. Li, in *Proc. IPAC 2013* (Shanghai, China, 2013).
[172] S. Peng, L. Hoff and K. S. Smith, in *Proc. PAC 2005* (Knoxville, Tennessee, 2005), p. 3697.
[173] F. Tamura, A. Schnase, M. Nomura *et al.*, in *Proc. EPAC 2008* (Genova, Italy, 2008), p. 364.

[174] A. Seville, D. B. Allen, N. E. Farthing *et al.*, in *Proc. IPAC 2011* (San Sebastian, Spain, 2011), p. 466.
[175] H. Ludewig, N. Simos, J. Walker *et al.*, in *Proc. PAC 1999* (New York, 1999), pp. 548–550.
[176] R. W. Shaw, M. A. Plum, L. L. Wilson *et al.*, in *Proc. PAC 2007* (Albuquerque, 2007), p. 620.
[177] I. Sugai *et al.*, in *Proc. EPAC 2006* (Edinburgh, Scotland, 2006), p. 1753.
[178] Y. Takeda, M. Kinsho, J. Kamiya *et al.*, *Nucl. Instrum. Methods A* **590**, 213 (2008).
[179] T. Takayanagi *et al.*, *IEEE Trans. Appl. Supercond.* **16**(2), 1358 (2006).
[180] L. Shen, Y. L. Chi and C. Huang, in *Proc. PAC 2007* (Albuquerque, 2007), p. 2140.
[181] J. G. Wang, in *Proc. HB 2008* (Nashville, Tennessee, 2008), p. 265.
[182] M. J. Shirakata, H. Fujimori, Y. Irie *et al.*, *Phys. Rev. ST Accel. Beams* **11**, 064201 (2008).
[183] Y. Chen and J. Y. Tang, *Phys. Rev. ST Accel. Beams* **11**, 032401 (2008).
[184] J. Kamiya, T. Takayanagi and M. Watanabe, *Phys. Rev. ST Accel. Beams* **12**, 072401 (2009).
[185] W. Zhang, J. Sandberg, R. Lambiase *et al.*, in *Proc. PAC 2003* (Portland, Oregon, 2003), p. 550.
[186] M. Watanabe, N. Hayashi, Y. Shobuda *et al.*, in *Proc. IPAC 2012* (New Orleans, 2012), p. 3629.
[187] Y. Chen, CSNS Internal Note (2012).
[188] K. Wittenburg, http://arxiv.org/abs/1303.6767
[189] S. A. Whitehead, P. G. Barnes, G. M. Cross *et al.*, in *Proc. BIW10* (Santa Fe, 2010), p. 106.
[190] K. Satou, S Lee, T. Toyama *et al.*, in *Proc. HB 2010* (Morschach, Switzerland, 2010), p. 506.
[191] http://xaldev.sourceforge.net/
[192] J. W. G. Thomason, R. Garoby, S. Gilardoni *et al.*, *Phys. Rev. ST Accel. Beams* **16**, 054801 (2013).
[193] Y. Alexahin and D. Neuffer, in *Proc. IPAC 2012* (New Orleans, Louisiana, 2012), p. 1260.
[194] S. Holmes (ed.), Fermilab, Project X-doc-776-v5 (2013).
[195] M. Benedikt and B. Goddard, in *Proc. PAC 2009* (Vancouver, Canada, 2009), p. 1828.
[196] J. H. Jang, H. J. Kwon, H. S. Kim *et al.*, in *Proc. IPAC 2010* (Kyoto, Japan, 2010), p. 678.

Jingyu Tang is a professor at the Institute of High Energy Physics, Chinese Academy of Sciences (CAS). His main interests are accelerator physics and technology for high-intensity hadron accelerators and beam applications. He played key roles in projects including the China Spallation Neutron Source, the China Accelerator-Driven System and a future neutrino superbeam facility. He teaches a graduate course, "Proton Accelerators," at the University of Chinese Academy of Sciences. From 1984 to 1990, he did his MS and PhD studies at the Institute of Modern Physics, CAS. He has worked in several European laboratories, such as GANIL in Caen, FZ-Juelich, and MEDICYC in Nice. Before 2001, he had worked many years on cyclotrons for heavy ion nuclear physics and hadron therapy.

Reviews of Accelerator Science and Technology
Vol. 6 (2013) 171–196
© World Scientific Publishing Company
DOI: 10.1142/S1793626813300089

Superconducting Hadron Linacs

Peter Ostroumov
Physics Division, Argonne National Laboratory,
S. Cass Ave., IL 60439, USA
ostroumov@anl.gov

Frank Gerigk
BE-RF-LRF, CERN,
CH-1211 Geneva 23, Switzerland
frank.gerigk@cern.ch

This article discusses the main building blocks of a superconducting (SC) linac, the choice of SC resonators, their frequencies, accelerating gradients and apertures, focusing structures, practical aspects of cryomodule design, and concepts to minimize the heat load into the cryogenic system. It starts with an overview of design concepts for all types of hadron linacs differentiated by duty cycle (pulsed or continuous wave) or by the type of ion species (protons, H^-, and ions) being accelerated. Design concepts are detailed for SC linacs in application to both light ion (proton, deuteron) and heavy ion linacs. The physics design of SC linacs, including transverse and longitudinal lattice designs, matching between different accelerating–focusing lattices, and transition from NC to SC sections, is detailed. Design of high-intensity SC linacs for light ions, methods for the reduction of beam losses, preventing beam halo formation, and the effect of HOMs and errors on beam quality are discussed. Examples are taken from existing designs of continuous wave (CW) heavy ion linacs and high-intensity pulsed or CW proton linacs. Finally, we review ongoing R&D work toward high-power SC linacs for various applications.

Keywords: Linac; superconducting; ion; design; beam dynamics.

1. Introduction

In the past 20–30 years, superconducting RF technology has found wide application for ion beam acceleration. Several SC linacs for the acceleration of CW ion beams have been developed and are operational since the early 1980s. The main requirement for an ion SC linac is that it should provide acceleration of any ion species from hydrogen to uranium. This can be accomplished by using independently phased short cavities which are capable of covering a wide range of particle velocities. The Argonne Tandem Linear Accelerator System (ATLAS) is a good example of a CW ion linac [1]. Following ATLAS, several laboratories and universities worldwide have developed and built SC linacs, primarily for the acceleration of low-intensity stable or radioactive ions, as discussed in Sec. 7 of this article. Currently, no high-power CW linac exists. Substantial research-and-development work has been performed in the past decade for several high-power CW linac projects worldwide. The most advanced projects are the 200 kW light ion linac at Grand Accélérateur National d'Ions Lourds (GANIL) [2] for nuclear physics research, the deuteron linac for the International Fusion Materials Irradiation Facility (IFMIF) [3], and the 400 kW ion linac for the Facility for Rare Isotope Beams (FRIB) [4]. Capability to accelerate several milliamps of CW proton beam by SC cavities has been demonstrated in the prototype cryomodule at SARAF [5].

SC technology can be efficiently used in pulsed high-power linacs too, as was recently demonstrated at SNS [6]. SNS has begun a new era for the application of SRF technology to high-power pulsed proton accelerators. There were several advantages of SC technology in the SNS linac compared to the normal conducting (NC) option: (a) lower construction and operation cost, (b) higher accelerating gradients, (c) greater feasibility for energy upgrades and (d) a larger bore diameter.

Due to the thermal handling problems, there are no demonstrated NC accelerating structures suitable for acceleration of CW hadron beams with velocities above ~0.1 c (c is the speed of light). Even if we assume that such NC accelerating structures can be developed, the SRF technology is superior to NC technology for CW accelerators due to its significantly reduced operational cost. Below, we estimate wall plug power for an ~1 GeV, 10 MW, CW proton linac based on NC technology. The total wall plug power P_w (not including beam power) for a CW linac based on NC structure can be estimated as

$$P_w = \frac{V_{\text{eff}}^2}{\eta L R_{\text{sh}}}, \tag{1}$$

where V_{eff} is the total effective voltage required to obtain 1 GeV protons, L is the total length of the accelerating structures, R_{sh} is the shunt impedance (linac definition), and $\eta \approx 0.3$ is the efficiency of the transformation of the wall plug power into the RF power dissipated on the accelerating structures including efficiency of the cooling systems. For a $W = 1.0\,\text{GeV}$ linac, if we assume that $E_{\text{acc}} = 3\,\text{MV/m}$, $\varphi_s = -25°$, and $R_{\text{sh}} = 30\,\text{M}\Omega\text{/m}$, then $P_w = 369\,\text{MW}$. If we perform similar estimates for SC structures, the efficiency of the conversion of wall plug power into liquid helium should be taken into account. Overall, the efficiency of the CW SC linac with similar parameters is better by a factor of 20.

For many applications in pulsed SC linacs, SC accelerating structures can work much more power efficient than NC linacs and can therefore significantly lower the operational cost and the cost of RF power sources. This is simply due to the fact that in NC standing wave linacs a high fraction of the RF power is dissipated in the cavity walls (typically between 50% and 95%), while in SC cavities almost 100% of the power, which is fed into the cavities, is transferred to the beam. Depending on the beam duty factor, the beam current and the final beam energy of the linac there can be large differences in power consumption between NC and SC machines. However, one should not forget that for certain parameters NC traveling wave cavities in electron linacs have reached efficiencies of 95% (e.g. the CLIC drive beam accelerator [7]). Another argument to consider is the real-estate accelerating gradient, which can be achieved and which impacts on linac length and initial cost. For each application one should carefully consider the implications for power consumption, real-estate length, and initial capital investment before deciding whether to use NC or SC technology.

There is a significant difference between the acceleration of electrons and ions due to the particle velocity. While the acceleration of charged particles with velocities above ~0.8c can be effectively performed with $\beta_G = 1$ (β_G is the so-called geometrical beta) elliptical cavities, ion linacs require a suite of SC accelerating structures to cover the complete velocity range, from the lowest radio frequency quadrupole (RFQ) velocities to ~0.8c. The number of different cavity types has to be low, in order to reduce the capital cost of the linac. Typically in modern designs, four or five types of SC structures are sufficient to achieve beam velocities above ~0.8c. The acceleration of low charge-to-mass ratio (q/A) heavy ions in the front end requires a factor of A/q higher voltages as compared to a proton linac. Therefore, the velocity of ion beams exiting an RFQ is lower than in a proton RFQ and it can result in an additional type of low-velocity accelerating structure in the front end of an SC heavy ion linac.

The current status of the SC technology can be effectively applied for the construction of very-high-power linacs largely up to 100 MW; it can be achieved in SC linacs at a reasonable cost.

2. Design of SC Linacs

2.1. *Overview of design concepts*

Independent of ion species, the proposed designs of high-intensity SC linacs comprise three well-distinguished sections: (1) a front end, (2) a low-energy section, and (3) a high-energy section. Depending on the final beam energy, sections (2) and (3) are combined in some linacs. The front end consists of an ion source, low-energy beam transport (LEBT) system, an RFQ accelerator, and a medium-energy beam transport (MEBT) system. In some older-generation linacs, the function of the RFQ is performed by a DC high-voltage accelerator and RF bunching system. The transition from the front end to the low-energy section in CW linacs coincides with the transition from NC to SC structures, while the transition from low- to high-energy sections is usually accompanied by a frequency jump in the accelerating structures. In superconducting ion linacs, several frequency transitions may take place and the transition from low to high energies is related to the location of an ion stripping station.

The front end of all modern hadron linac designs includes an RFQ. The application of RFQs for the acceleration of high-current beams has resulted in two main breakthroughs: (1) the capability to form a compact longitudinal emittance with a substantially lower halo compared to a front end based on external bunchers and drift tube linacs (DTLs), and (2) the transverse emittances remain unchanged in spite of the bunched structure of the beam exiting the RFQ.

The primary focus of high-power linac designs is to deliver high-power beams over hundreds of kilowatts with minimal beam losses to avoid excessive radioactivation of accelerator components. Extensive study of high-intensity beam physics in the late 1980s and the 1990s resulted in the very important conclusion that a higher fundamental frequency, which is usually defined by the RFQ, results in a lower number of particles in a bunch for a given peak or average current. This approach allows better control of the beam emittance and the halo of high-intensity light-ion beams. As an example, a CW 352 MHz RFQ was successfully operated, and accelerated proton beams up to 100 mA [8]. In a heavy ion linac, the RFQ frequency can be much lower, ~60–100 MHz, since space charge effects are less significant and low-frequency SC structures downstream of the RFQ are required.

2.2. *Frequency choice*

Besides the advantage of higher shunt impedance for SC cavities versus NC structures, there is an appreciable difference in the achievable accelerating gradients. Due to the long filling time of SC structures, they are generally used for long-pulse or CW operation. In this regime the maximum peak surface field does not depend on the RF but rather on the maximum magnetic surface field. In NC structures one has to differentiate between (i) structures that are used with pulse lengths in the range of the filling time and (ii) longer pulse lengths. For the first case, recent research [9] has shown that the maximum accelerating gradients are limited by peak values of the modified Poynting vector (a weighted combination of its real and imaginary parts) on the surface [9]. In the long-pulse case, the only available criterion dates from experiments in 1957 by W. D. Kilpatrick [10], and it predicts the level of the electric fields at which an RF breakdown occurs as a function of frequency. His experimental data can be approximated by [11]

$$f(\text{MHz}) = 1.64{E_K}^2 e^{-8.5/E_K}, \tag{2}$$

and predict higher breakdown field limits, E_k, for higher frequencies. Today's RF designers usually surpass this limit by a factor between 1 and 2.

In practice, as with NC structures, the frequency of SC structures is defined by the distance $\beta\lambda$, where β is the relative beam velocity and λ is the RF wavelength in free space. Almost all known SC cavities operate in π mode; therefore the distance between adjacent accelerating gaps is equal to $\beta\lambda/2$. The design of the cavity must provide sufficient space for a cryogenic liquid; therefore the cavity frequency has to be below 100 MHz for very low particle velocities ($\beta < 0.1$). One can select even lower frequencies to provide higher accelerating voltages per cavity. This approach is used to optimize the overall cost of ion accelerators. Modern SC cavities in the low-β range can provide real-estate accelerating gradients up to ~4 MV/m in CW mode [12, 13].

Overall, the frequencies of SC structures being used or proposed for proton and ion accelerators are in the range of 48–1300 MHz. A 48 MHz quarter wave resonator (QWR) is used in the first cryomodule of ATLAS, accelerating ion beams from 300 keV/u. The 1300 MHz elliptical structures being well developed under ILC R&D [14] are proposed for the pulsed SC linac in Project X in the energy range of 3–8 GeV [15].

In high-power, high-energy linacs, it is preferable to minimize jumps of operating frequencies along the linac. As the efficiency and accelerating gradients of high-frequency SC cavities in the high-energy section increase, a higher-frequency front end is preferable. For this particular reason, the front end of the 700 MHz, 100 MW SC APT linac [16] was designed to operate at 350 MHz fundamental frequency.

3. Accelerating Structures and Focusing Lattice

The physics design of the linac is an important step in overall cost minimization and must be well coordinated with the engineering design of all accelerator systems. The primary focus of the physics design is to deliver the linac lattice for the following engineering analysis and the development of fabrication drawings. The linac design must satisfy the major requirement of high-power accelerators — the avoidance of excessive uncontrolled beam losses. This can be achieved through minimization of the emittance growth, avoidance of beam halo formation, and choice of apertures. The accelerator lattice must be integrated with other systems, such as the beam chopper, beam collimation, diagnostics system, beam corrective steering, vacuum system, or cooling system. The physics design and engineering development of the linac must be iterated several times to assure final linac performance. High-power accelerators must tolerate "hands-on maintenance" of the accelerator components. The latter implies a radioactivation limit of 20 mrem/h at a distance of 1 m from the component surface after extended operation of the linac (~100 days) and 1 h of downtime [17]. This typically requires losses of less than 1 W/m, especially above 100 MeV/u.

3.1. *Accelerating structures*

Two main classes of SC cavities are being used or are under consideration for hadron linacs. The first class is based on singly or multiply loaded structures where each loading element supports a TEM mode. Included in this class are the types of cavities that are used extensively in the low-β region [13, 18]. Some cavity designs have not survived real-life testing for problems such as microphonics and low performance. The first successful type of cavities developed and used in the world's first SC linac, ATLAS, were split ring cavities [19]. Afterward, QWRs found application worldwide in low-energy (below ~25 MeV/u) CW SC linacs. Numerous types of QWR cavities over a frequency range of 50–240 MHz have been built or are proposed for a variety of applications. Other well-known TEM-class accelerating structures are half-wave resonators (HWRs), singly and multiply loaded spoke resonators being developed in the frequency range of 160–400 MHz for proton or ion beam acceleration in the energy range of 2–600 MeV/u.

The other class is based on a group of several cylindrical cells operating in the cylindrically symmetrical TM_{010} mode and is known as elliptical cavities. This class is topologically identical to the one used at $\beta = 1$, but with a reduced cell length. We will refer to these two classes of cavity geometry as TM and TEM geometries [20], respectively. TEM and TM cavity geometries differ in three major respects: the transverse dimensions, the cell-to-cell coupling, and the localization of the electromagnetic field. The transverse dimension of TM structures is of the order of 0.9λ, while for TEM structures it is of the order of 0.5λ. Thus, at the same frequency, TEM structures have about half the transverse size of TM structures. Alternatively, at the same transverse size, they will operate at about half the frequency.

The majority of SC cavities developed for the acceleration of low-velocity particles ($\beta < 0.4$) are TEM-class cavities. In the intermediate-velocity range, $0.4 \leq \beta \leq 0.7$, both TEM-class and TM-class SC cavities are applicable. The frequency–velocity chart for SC cavities in ion linac applications is well described in Ref. 13.

There are several reasons why SC structures are short and designed for fixed velocities: (i) fabrication cost is significantly reduced; (ii) focusing elements are required between SC cavities; (iii) the SC accelerating structure cannot be very long, due to the necessity of limiting the stored RF field energy; (iv) the RF surface processing and cavity handling in a clean room limit cavity dimensions.

While SC technology is the technology of choice for CW linacs, pulsed linacs can be composed from NC accelerating structures. However, successful operation of the SNS linac proves that SC technology is a strong candidate for future high-power pulsed linacs. SC structures offer much larger apertures, which is especially critical for high-intensity linacs. In order to avoid or to minimize the beam intercepting the walls of the accelerating structures, the ratio of the physical aperture to the beam size must be kept large. As we will discuss in the following sections, the beam bore size must be ≥12 times the rms beam size. This condition can be easily satisfied in SC linacs. Due to the strong dependence of the shunt impedance on the aperture, large-bore NC accelerating structures can drive the cost of the linac to unacceptable levels.

The first step of the iterative procedure in the design of an SC linac is to define a preliminary number and type of SC cavities and transition energies. For this analysis we have derived the following formulas for the transit time factor, T, of cavities operating in π mode of electromagnetic oscillations with an approximation of the harmonic spatial accelerating field distribution [21]:

$$T(N,\beta,\beta_G) = \left(\frac{\beta}{\beta_G}\right)^2 \cos\left(\frac{\pi N \beta_G}{2\beta}\right) \times \frac{(-1)^{\frac{N-1}{2}}}{N((\frac{\beta}{\beta_G})^2-1)} \quad (3)$$

for an odd number of cells N in the cavity,

$$T(N,\beta,\beta_G) = \left(\frac{\beta}{\beta_G}\right)^2 \sin\left(\frac{\pi N \beta_G}{2\beta}\right) \times \frac{(-1)^{\frac{N+2}{2}}}{N((\frac{\beta}{\beta_G})^2-1)} \quad (4)$$

for an even number of cells N in the cavity, and

$$T = \frac{\pi}{4}, \quad \text{if } \beta = \beta_G; \quad (5)$$

here N is the number of accelerating cells in a cavity. These formulas provide a good approximation of multicell cavities. Figure 1 shows the transit time factor, T, as a function of velocity for different numbers of cells. As can be seen from the figure, a lower number of cells provides wider "velocity acceptance." The maximum acceleration is achieved for the velocity slightly larger than β_G, primarily due to the fringe field in the end cells. The velocity where the maximum T is achieved, β_{opt}, is called the optimal beta. In low- and medium-energy regions, velocity changes fast and therefore it is cost-efficient to use a lower number of accelerating cells in a cavity. For TEM-class cavities, the transit time factor is a complicated function of both the field distribution and the particle velocity, which may change appreciably during the passage through the multiple-gap cavity. For this reason, T is most conveniently calculated numerically.

By the use of SC cavity parameters, such as the designed accelerating field and transit time factor, the energy gain in the linac composed of several types of cavities can be calculated. The goal of this exercise is to minimize the total number of cavities. Transition energies have been selected to equalize the voltage gain in the transition region. Typical voltage gains in the earlier design of the Project X 8 GeV linac [21] are shown in Fig. 2. The geometrical betas of spoke-loaded SC cavities are selected to cover a velocity range from $0.145c$ to $0.724c$ with the lowest number of cavities. In this velocity range, it is not reasonable to reduce the number of cavity types to two, due to an appreciable increase in the number of cavities which is caused by the large variation of the cavity's T.

The geometry of TM-class elliptical SC cavities is axially symmetric and the accelerating field does not contain multipole components. For most-practical applications, the dipole electric field created by the coupling antenna can be neglected. In contrast, QWRs, HWR, and spoke-loaded cavities can produce multipole components of the electromagnetic field distribution in the accelerating gaps. Recent studies have revealed an important drawback of the QWR: the presence of beam steering fields in the aperture [22]. Beam steering is induced by

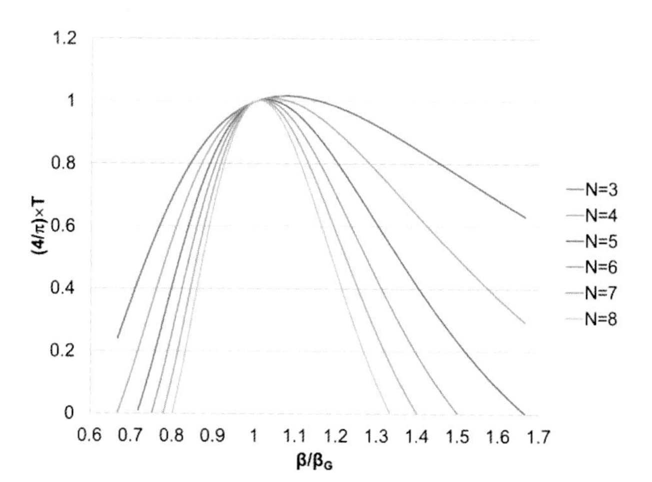

Fig. 1. Transit time factor as a function of normalized beta for a different number of accelerating cells.

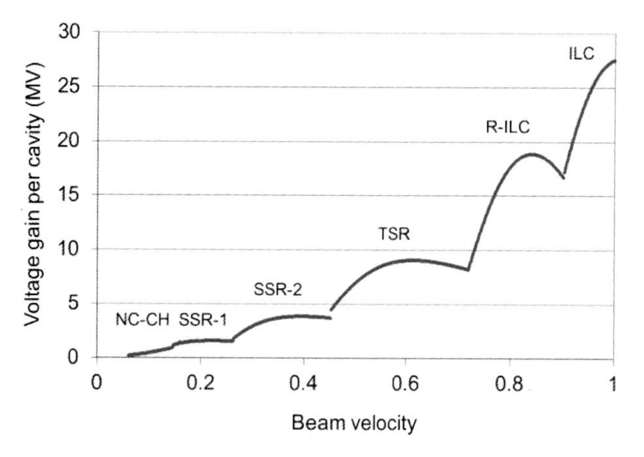

Fig. 2. Voltage gain per cavity for Project X as a function of beam velocity. RT-CH is the NC cavities; SSR-1 and SSR-2 are the spoke-loaded cavities, TSR is the triple-spoke resonator, S-ILC the squeezed ILC cavity, and ILC the International Linear Collider (ILC) cavity.

dipole components of the field and is a strong function of the RF phase, which couples the longitudinal and transverse motion. This coupling can cause not only beam deflection but also a level of transverse emittance growth which is unacceptable for many applications. Also, steering effects will be most pronounced for light ions and beams of large longitudinal emittance. Such emittance growth cannot be compensated for by static fields and can be a particularly serious problem in applications for high-intensity light-ion beams. As was shown in Ref. 23, the steering can be largely compensated for by two different methods. Simply offsetting the cavity beam axis by 1–2 mm can often provide adequate compensation. In this method, the available range of steering is limited by the reduction of the useful aperture. Offsetting can be effectively applied for low-intensity heavy ion accelerators dealing with $q/A < 1/3$ in a velocity range of $\sim 0.01c$–$0.15c$. This method has been applied, for example, in the ISAC-II project [24]. More generally, steering can be largely eliminated over the entire useful velocity range by shaping the drift tube and cavity wall faces adjacent to the beam axis to provide appropriate corrective vertical electric field components, as was proposed in Ref. 23 and implemented at ATLAS [12]. In some cases steering is sufficiently small that the QWR can be used without any correction.

There is, however, another problem in QWR, HWR, and spoke-loaded cavities: the drift tube design can introduce quadrupole terms in the transverse Lorentz force which can cause appreciable emittance growth when the linac lattice includes transverse focusing by SC solenoids. Solenoidal focusing provides a compact lattice and maximizes transverse acceptance while maintaining low longitudinal emittance. Recently, several TEM-class cavity geometries have been proposed [25–27] to eliminate both the dipole and the higher-order components in the equations of motion in the transverse planes, while keeping the ratio of surface to accelerating field low. The resulting QWR, HWR, and spoke-loaded resonator designs minimize emittance growth, which is critical in most applications.

3.2. *Focusing structures*

Focusing of charged particles in an SC linac is trivial in the medium- and high-energy sections, where long focusing periods, many times exceeding $\beta\lambda$, are easily tolerated by the beam dynamics. Depending on the properties of the particular linac, the focusing element can be placed inside or outside of the cryomodule. The cryomodule can be long enough to reduce the linac cost by minimizing the number of cold–warm transitions. If the focusing by electromagnetic quadrupoles is preferable, as in the case of light ion beams, quadrupoles can be placed between cryomodules to provide access for alignment. This option has been implemented in the SC section of the SNS linac. In heavy ion linacs with multiple-charge-state acceleration capability, focusing by SC solenoids is more beneficial for maintaining smaller transverse emittance.

In the low-energy section, the focusing elements can be placed between the SC cavities. These elements are either SC solenoids or SC quadrupoles which can be placed in a common cryostat with SC cavities. In high-intensity light ion linacs, it is mandatory to have short focusing periods so as to suppress the space charge effects. Using SC solenoids for transverse beam focusing together with SC resonators offers several advantages. The solenoids can be placed close to the cavities inside the cryostats, and a short focusing period can be achieved. A lattice with solenoidal focusing is compact, and maximizes both transverse and longitudinal acceptance. SC solenoids are perfectly suitable for SC environments with SRF; they are easily retunable to adjust to the accelerating gradient variation from cavity to cavity and they can also be supplemented with dipole coils for corrective steering of the beam centroid. SC solenoids have been used at the ATLAS facility for several decades [1] and are being developed for the application in a high-power SC linac for the IFMIF [3]. The idea of using SC solenoids in the front end of high-intensity SC proton linacs was discussed conceptually in Ref. 28. In high-intensity SC linacs, solenoids make much higher real-estate accelerating gradients possible with better control of space charge by reducing the length of the focusing periods. Furthermore, lattices using SC solenoids are less sensitive to misalignments, errors, and beam mismatches, as discussed in Ref. 29.

Quadrupole focusing for H^- beams is recommended above $\sim 160\,\mathrm{MeV}$, due to stripping of negative ions in the fringe fields of solenoids. There were proposals to develop SC structures for low-velocity beams ($\beta < 0.2$) which can combine RF electric field focusing [30]. This technology is not yet well developed and SC RF focusing structures are not available.

There are CW SC ion accelerator designs where acceleration is provided by single-cavity cryomodules and focusing is achieved by using warm quadrupole doublets located between the cryomodules. This lattice is acceptable in the front end of low-intensity accelerators. Due to the long drift spaces (many $\beta\lambda$s) between the accelerating cavities, the longitudinal dynamics becomes susceptible to the phase and amplitude errors in the cavities. As a result this lattice can produce a large longitudinal halo in the machine with errors. Also, from a practical point of view, short cryomodule systems are expensive due to the increased number of cold–warm transitions, valve boxes in the helium distribution system, and the static heat load into the cryogenic system.

In the design of periodic focusing lattices of SC linacs, several important issues should be taken into account. Standard criteria such as stability of the transverse motion and the maximum possible acceptance certainly should be applied. In SC linacs, due to the availability of high accelerating gradients and the relatively long focusing periods, strong coupling between transverse and longitudinal motion may occur. Particularly, the transverse–longitudinal coupling can excite parametric resonances (see also Subsec. 6.2.1 for transverse oscillations [31]). The condition for an nth-order parametric resonance of transverse motion for a zero-current beam is

$$\sigma_{0T} = \frac{n}{2}\sigma_{0L}, \tag{6}$$

where σ_{0T} and σ_{0L} are the phase advances of the transverse and longitudinal oscillations per focusing period. In the smooth approximation of the equation of motion [17], the longitudinal phase advance can be written as

$$\sigma_{0L} = \sqrt{2\pi \frac{q}{A}\frac{1}{(\beta_s\gamma_s)^3}\frac{S_f^2}{\lambda}\frac{eE_m \sin\varphi_s}{m_u c^2}}, \tag{7}$$

where γ is the relativistic factor, S_f is the length of the focusing period, E_m is the amplitude of the equivalent traveling wave of the accelerating field, φ_s is the synchronous phase, m_u is the atomic mass unit, and subindex s indicates the parameters for a synchronous particle. Usually, it is sufficient to avoid the strongest first-order resonance, $n = 1$. For the given transverse phase advance, the expressions (6) and (7) result in a limiting longitudinal accelerating field, E_m. Note that the voltage gain in a SC cavity is

$$V_0 = \frac{1}{N_C}\frac{1}{T}S_f E_m, \tag{8}$$

where N_C is the number of cavities per focusing period.

In the high-intensity linac, long focusing periods can easily shift the transverse beam dynamics to the space-charge-dominated regime, according to the expression for the tune depression [17]:

$$\left(\frac{\sigma_T}{\sigma_{0T}}\right)^2 = 1 - \frac{q}{A}\frac{3I\lambda(1-f_B)}{I_0\beta_s^2\gamma_s^3(r_x+r_y)r_x r_z}\left(\frac{S_f}{\sigma_{0T}}\right)^2, \tag{9}$$

where I is the beam current, $I_0 \approx 3.13 \cdot 10^7$ A is the characteristic current for protons [32], $r_{x,y,z}$ are the bunch semiaxes for a uniformly charged ellipsoid in the corresponding directions, and f_B is the bunch shape form factor defined in Ref. 17.

3.3. *Accelerating gradient and cryogenic load*

The preliminary design of an SC accelerator starts with the selection of the accelerating gradient in all types of cavities required to cover the full velocity range. The quantity of each type of cavity depends directly on the available accelerating gradient. In SC cavities, the highest achievable accelerating field is limited by a peak magnetic field which can quench the cavity. The theoretical limit for the quench magnetic field is ~200 mT at 2 K for niobium cavities. In practice, the quench magnetic field is limited by the quality of cavity fabrication and RF surface processing. Each type of SC cavity must undergo an extensive electromagnetic optimization procedure to reduce the ratio of the peak electric, E_{peak}, and magnetic fields, B_{peak}, to the accelerating field, E_{acc}. The latter is defined as a maximum electric field for optimal beta. Another limitation on the accelerating field is the excessive X-ray radiation due to field emission. The best practice of RF surface processing and cleaning usually results in $E_{\text{peak}} \cong 70$ MV/m at an acceptable level of radiation. Two more parameters, the ratio of shunt impedance, R_{sh}, to the intrinsic quality factor, Q_0, and geometry factor, G, should be maximized during the electromagnetic design of the cavity. Indeed, the dynamic heat load, P, into the cryogenic system due to the dissipated RF power is expressed as

$$P = \frac{V_0^2 R_S}{\frac{R_{\rm sh}}{Q_0} G}, \tag{10}$$

where $V_0 = E_{\rm acc} L_{\rm eff}$ is the voltage gain in an SC cavity at optimal beta $\beta_{\rm opt}$, $L_{\rm eff}$ is the effective length of the cavity, and R_S is the cavity surface resistance. While the shunt impedance and geometry factor can be maximized by EM design of the cavity, the surface resistance entirely depends on SC properties of the cavity material and the quality of RF surface processing.

The cryogenic load is an important parameter in SC linacs and directly impacts both the capital and operation costs. Higher accelerating gradients can reduce the number of accelerating cavities, cryomodules, and the length of the tunnel. However, the dynamic heat load in the LHe system is proportional to the square of the accelerating voltage, as is seen from (10). Therefore, in a CW linac there is an optimum accelerating gradient that can minimize the combined capital and operation costs. The optimum value of the accelerating gradient can be further increased by providing a lower surface resistance for the cavity. Table 1 lists the achieved and planned real-estate accelerating gradients in several accelerator facilities.

Table 1. Real-estate accelerating gradient, $E_{\rm RE}$, in several operational and planned facilities.

Facility	Cavity type	β_G	$E_{\rm RE}$ (MV/m)
ATLAS	QWR	0.077*	3.4†
	QWR	0.145*	3.4†
SNS‡	Ellipt.	0.61	3.61†
	Ellipt.	0.81	7.42†
FRIB	HWR	0.285*	3.6
	HWR	0.53*	4.7
Project X	SSR1	0.186	1.7
	SSR2	0.431	3.6
	Ellipt.	0.61	5.6
	Ellipt.	0.9	11.7
SPL‡	Ellipt.	0.65	6.2
	Ellipt.	1.0	15.1
ESS‡	DSR	0.50	2.6
	Ellipt.	0.67	6.7
	Ellipt.	0.86	8.8

*Optimal beta; †achieved in operation; ‡pulsed.

The real-estate accelerating gradient is calculated as a total voltage provided by the cryomodule at optimal β and divided by the length of the cryomodule and focusing elements if they are located outside the cryomodule.

As in electron linacs, there is a different set of requirements for pulsed SC cavities. Specifically, the accelerating gradient can be higher than in CW linacs because the dynamic cryogenic load is lower. In pulsed accelerators, SC cavities require fast frequency tuning to compensate for the Lorentz detuning.

Another feature of pulsed operation is the power overhead needed for filling the cavity and the increased cryogenic load associated with the filling and "emptying" process. This can be understood by looking at the voltage profile in a pulsed SC cavity, as shown in Fig. 3. Define the filling time constant of a loaded cavity as

$$\tau_l = \frac{Q_l}{\omega_0} = \frac{Q_0}{\omega_0(1+\beta_C)}, \tag{11}$$

with β_C being the waveguide-cavity coupling factor. The loaded Q can be calculated from

$$Q_l \approx \frac{V_{\rm acc}}{(R_{\rm sh}/Q) I \cos\phi_s}. \tag{12}$$

Integrating over the curves in Fig. 3, one can derive [33] the duty cycles for the beam ($D_{\rm beam}$), the RF generator ($D_{\rm gen}$), and the cryogenic system ($D_{\rm cryo}$):

$$\begin{aligned} D_{\rm beam} &= f_{\rm rep} t_{\rm pulse}, \\ D_{\rm gen} &\approx f_{\rm rep}(1.39\tau_l + t_{\rm pulse}), \\ D_{\rm cryo} &\approx f_{\rm rep}(1.55\tau_l + t_{\rm pulse}). \end{aligned} \tag{13}$$

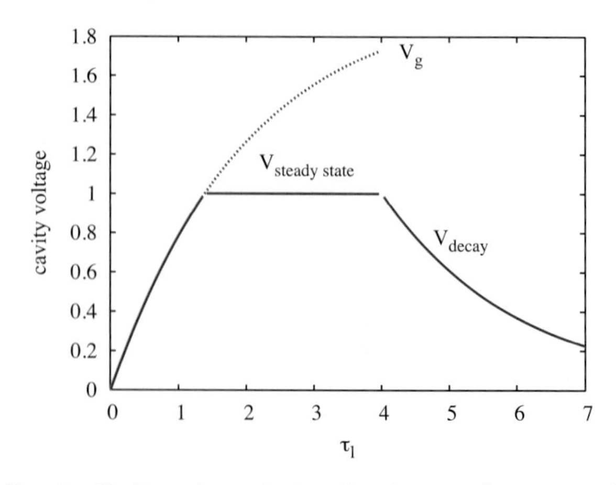

Fig. 3. Cavity voltage during charging, steady state, and discharging.

Due to the high quality factors, the filling time can easily be as long as the beam pulse itself, which means that the use of SC cavities is not automatically more efficient than using NC cavities. In addition, one has to add the power needed to evacuate the heat load at cryogenic temperature. In a large cryogenic system (>10 kW), one needs approximately 990 W to cool 1 W lost at 2 K (or 210 W to cool 1 W at 4.5 K) [34], and in smaller systems the efficiency gets even worse.

3.4. *Cryomodule design*

3.4.1. *Low-energy designs*

As mentioned in Subsec. 3.2, there are two types of cryomodules for ion accelerators: (i) with and (ii) without focusing elements inside the cryomodule. In the first case, the cryomodule length should be as long as possible, but usually it is limited by the available infrastructure. In the second case, the cryomodule length is limited by a tolerable length of the focusing period. A compact cryomodule design with a high packing factor of SC cavities and focusing elements is essential for preserving the quality of the ion beam at low and medium energies. The cryomodule-to-cryomodule transition introduces irregularity into the beam focusing and accelerating lattice. This should be mitigated by short transitions and beam matching in 6D phase space by adjusting parameters of cavities and focusing elements located near the transitions. Below, we present the designs of two compact cryomodules developed for the low-energy end of (i) a heavy ion accelerator and (ii) a high-intensity proton accelerator. A new cryomodule has been developed for the ATLAS intensity upgrade. It consists of seven $\beta_{\text{opt}} = 0.077$ QWRs operating at 72.75 MHz SC cavities and four 9 T SC solenoids [35]. New cryomodule features include the separation of the cavity and the cryogenic vacuum systems, and top loading of the cleaned and sealed cavity string subassembly. The new cryomodule was designed to create a total voltage of 17.5 MV for the acceleration of any ion with charge-to-mass ratio $q/A \geq 1/7$. Offline cavity testing demonstrated very high performance, and therefore the expected total voltage exceeds 17.5 MV and most likely will be limited by the available cooling capacity of liquid helium at ATLAS. The side cross-section of the cryomodule is shown in Fig. 4. The focusing period includes two SC cavities and a solenoid. The last focusing period includes a solenoid, cavity, and intercryomodule drift space.

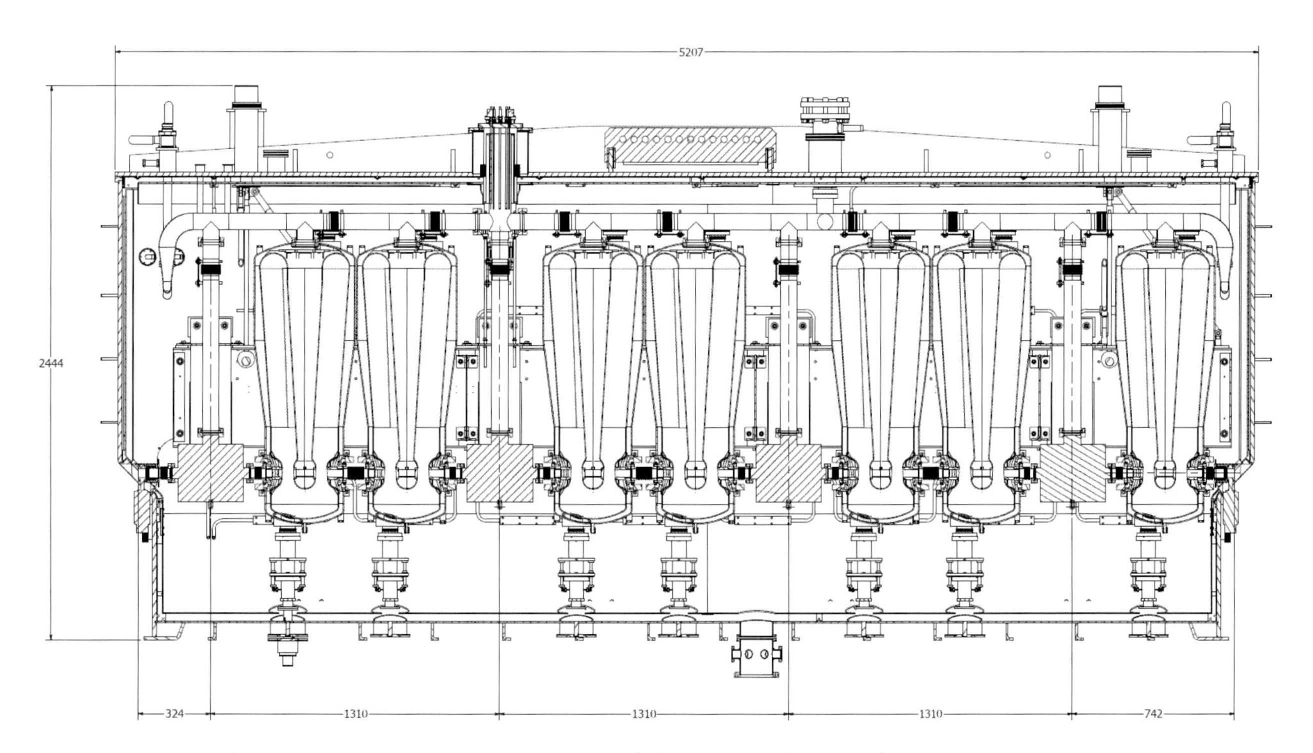

Fig. 4. The side cross-section of the new ATLAS cryomodule. The dimensions are in mm.

The first cryomodule for Project X at FNAL [36] is designed to accelerate the H^- beam from 2.1 to 11 MeV. To maintain a high beam quality, an adiabatic ramp of the real-estate accelerating gradient is necessary. The cryomodule has eight accelerating–focusing periods, with each period based on a 162.5 MHz SC HWR and an SC solenoid for focusing. The solenoid has integrated x–y steering coils and a beam position monitor. Reduced RF defocusing due to both the low frequency and the small synchronous phase angle results in a much faster energy gain without emittance growth. The beam dynamics optimization determined that a cavity beta of $\beta_{\text{opt}} = 0.112$ is optimal. The side cross-sectional view of the cryomodule is shown in Fig. 5.

Beam dynamics requires the solenoids to be aligned to a better than ± 0.5 mm peak transversely with $\pm 0.1°$ for all of the rotation angles, with similar constraints on the cavities. The experience with the ATLAS cryomodules shows that alignment of all components in the cryomodule can be performed with an accuracy that is a factor of 2 better than the above values.

3.4.2. *High-energy designs*

The principal considerations for designing a cryomodule for high-energy acceleration of protons are similar to the ones for the low-energy case of ions, which is explained above. One of the main choices is whether to use long continuous modules with cold magnets or separated modules with cold–warm transitions and warm magnets in between. So far the latter choice has been preferred for most hadron projects since it allows a quicker exchange of faulty modules and, more importantly, it allows the use of fast vacuum valves, which can be employed to isolate modules in case of a leak. Without fast isolation of faulty modules a potential vacuum leak may contaminate a long section of cavities, reducing their maximum accelerating gradients. This aspect is particularly important for high-gradient machines like SPL [37] or ESS [38].

Figure 6 presents the design of a high-energy module with $\beta = 1$ cavities. Shown is the so-called SPL short module, a half-length prototype, which has been developed for the SPL at CERN. The "full-size" module is foreseen to house eight cavities in order to minimize cold–warm transitions. However, one advantage of the short module is that there are many clean rooms and RF test facilities, which can accommodate this size of module, while for a full-size module very few places worldwide are big enough. For this reason ESS has recently decided to have only four cavity modules so as to facilitate and accelerate prototyping and to ease the transition from prototyping to series production. A special feature of the CERN development is the support of the SC cavities via the power couplers, which simplifies the structural elements on the inside but puts more constraints on the mechanical design of the coupler.

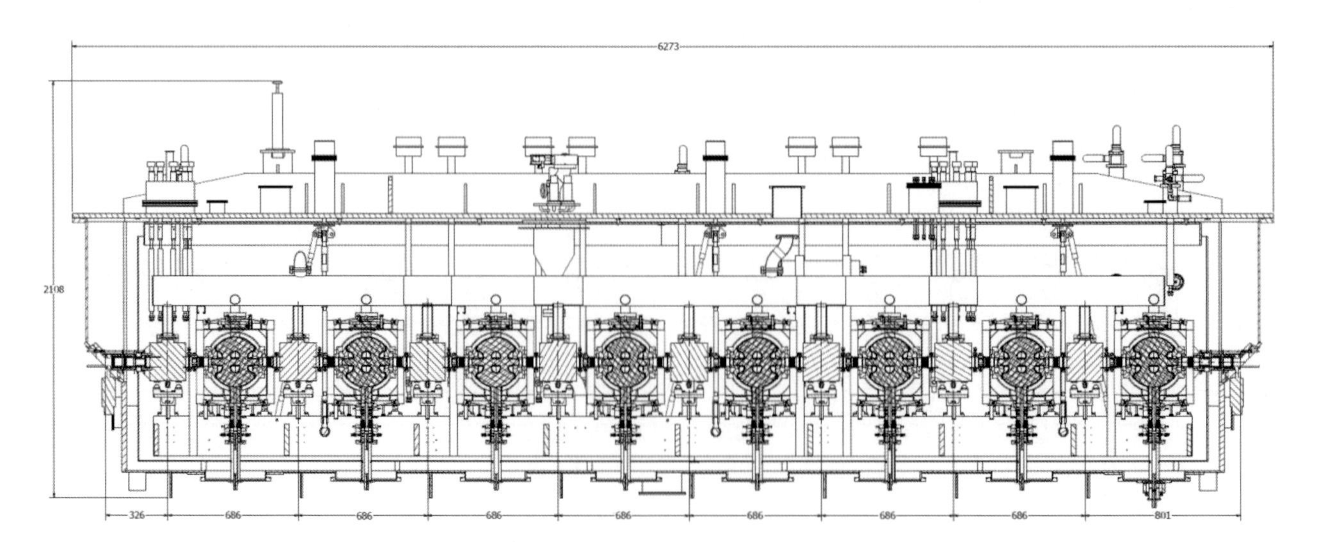

Fig. 5. The side cross-section of the Project X HWR cryomodule. The dimensions are in mm.

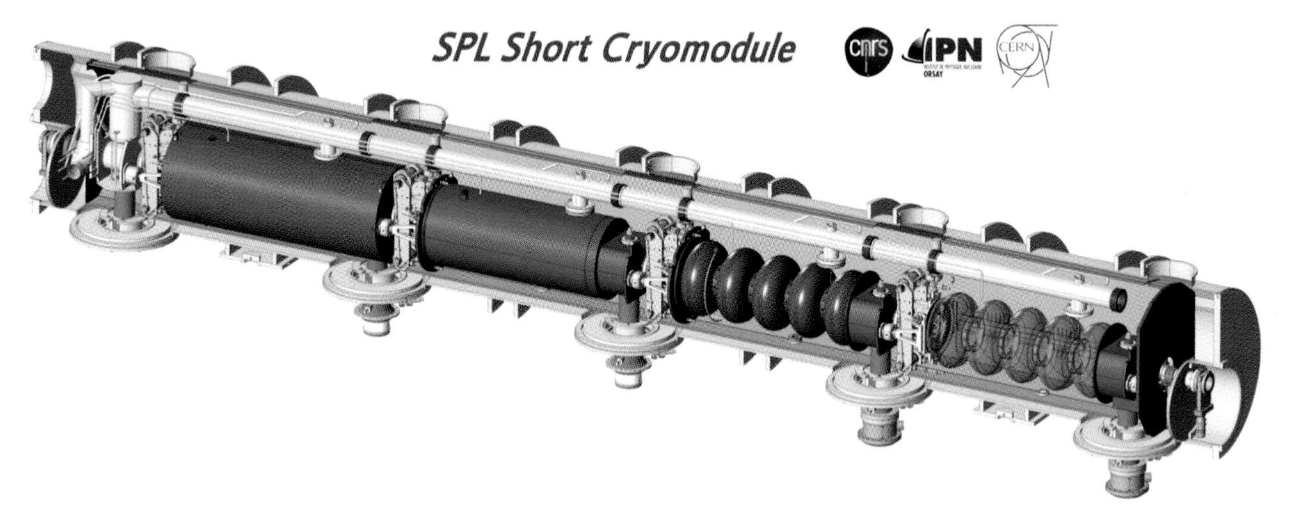

Fig. 6. The SPL short cryomodule.

While cryomodules are not the main cost drivers of an SC linac, they are very likely to be one of the schedule drivers. Due to the highly complex integration of cooling systems, RF equipment (power couplers, HOM couplers, pickups, etc.), cavity-tuning mechanisms, vacuum, thermal insulation, and mechanical alignment, the design and prototyping phase for a single module easily spreads over several years. When time is an issue it is advisable to follow existing design approaches, such as the space frames used at SNS [39] or the XFEL/ILC [40] concept, where the helium lines inside of the module provide mechanical rigidity.

4. Front Ends of SC Linacs

The front end of an SC linac includes an ion source, LEBT, RFQ, and MEBT, as in any NC high-intensity linac. In CW SC linacs, transition from the NC to the SC sections often takes place immediately after the RFQ with the exceptions discussed below. Since the invention of the RFQ in 1968 [41], it has found wide application as an accelerator immediately following the ion source. There are several dozen RFQs worldwide operating in pulsed mode, and the design and fabrication of these devices are widely covered in the literature. The first RFQ accelerator, based on a 148.5 MHz double H-type resonator, was commissioned in 1975 at the Institute of High-Energy Physics, Protvino [42]. The maximum accelerated proton beam current was 140 mA at 1.97 MeV, with a normalized emittance less than $2.5\pi \times \text{mm} \times \text{mrad}$ (for 85% of the particles). The first SC RFQ was developed and built at Legnaro, Italy for the acceleration of heavy ion beams extracted from an ECR source [43]. An 80 MHz SC RFQ provides vane voltages about a factor of 2 higher than in NC RFQs. However, CW RFQs for heavy ions can be built to operate at room temperature, as was demonstrated at TRIUMF and Argonne [44, 45].

A 6.7 MeV proton RFQ operating at 352 MHz in CW mode was successfully demonstrated at LANL [8]. This RFQ enables transition to an SC linac at moderately high velocity and mitigates space charge effects in the SC section. The LEDA RFQ and the following SC section were designed for acceleration of 100 mA proton beam.

The transition energy from NC to SC structures is one of the optimization parameters for a new generation of pulsed high-power proton and H^- linacs. In proton or H^- pulsed high-intensity linacs, an NC classic DTL remains a cost-efficient option. Besides, the DTL has a proven record of acceleration of high-quality, high-intensity beams. There is a proposal [46] for an SC ion linac capable of accelerating any ions from hydrogen to uranium to ~100 MeV/u for injection into the synchrotron. In this linac, the NC–SC transition energy is ~5 MeV/u and an additional NC structure is used between the RFQ and the SC linac.

Normally, an RFQ is followed by an MEBT. The main function of the MEBT is to provide a space for a fast chopper, beam instrumentation, and matching of the beam into the following linac lattice. In an

SC linac, the main function of the MEBT is a direct matching of the RFQ beam to the following SC linac lattice due to the significant difference of the lattice parameters. For example, an RFQ has very short focusing periods and a lower accelerating gradient than the following accelerating structure; therefore an adiabatic ramping of the accelerating gradient in the following linac section is required.

5. Beam Dynamics in Heavy Ion Accelerators

SRF technology creates opportunities for a CW multibeam driver accelerator which can deliver megawatt-level power to targets for the production of rare isotopes. The layout of an SC driver linac capable of providing 400 MeV/u for the heaviest uranium ions is shown in Fig. 7 [47]. Basing on the accelerator layout given in Ref. 47, below we apply the demonstrated operational performance of SC cavities to update the design of the heavy ion driver linac. Initial acceleration up to 500 keV/u in this driver linac is provided by an NC RFQ operating at 81.25 MHz. Two strippers are required to increase the charge state of the heaviest ions and reduce the required total voltage of the driver to ~1.5 GV. An SC linac is the technology of choice for several reasons. The ability to operate economically in a CW mode minimizes heating problems in the production targets and also makes the best use of existing ion source technology and performance. The independent phasing of an SC cavity array allows the velocity profile to be varied, permitting higher energies for the lighter ions, for example 800 MeV for protons. Also, the lower peak current in CW operation reduces space charge effects. This keeps both longitudinal and transverse beam emittance small, and also reduces the beam halo. Finally, the short, high-gradient SC cavities provide large transverse and longitudinal acceptance. The latter enables the incorporation of a novel feature of the linac, which is the acceleration of beams containing more than one charge state through portions of the linac to maximize beam current for the heavier ions. State-of-the-art ECR ion sources [48] are able to produce up to ~12 pμA uranium beam at charge state 34+. By simultaneously accelerating several charge states, the efficiency of charge stripping is greatly enhanced, since a much higher portion of the stripped beam can be utilized. Multi-charge-state operation not only provides a substantial increase in the available beam current, typically a factor of 4, but also enables the use of multiple strippers, which reduce the size of the linac required for 400 MeV/u beams. Taking uranium as an example, two charge states can be transported and accelerated simultaneously to the first stripper (12 MeV/u). Between the first stripper and the second stripper (85 MeV/u), the beam has an average charge state $q_0 = 75$; in this region we can accelerate five charge states, which encompasses 80% of the incident beam.

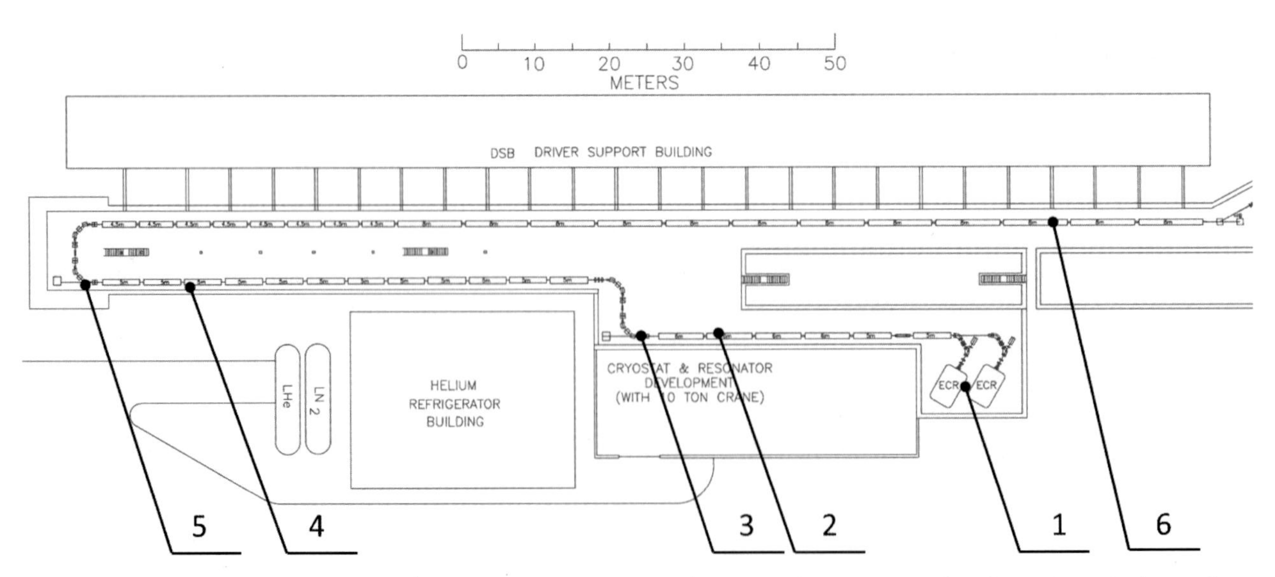

Fig. 7. Schematic layout of a 400 MeV/u uranium driver linac. 1- injector, 2 – low-energy section, 3 – stripper 1, 4 – medium energy section, 5 – stripper 2, 6 –high-energy section.

After the second stripper, 99% of the beam is in four charge states neighboring $q_0 = 90$, all of which can be accelerated to the end of the linac. Numerical simulations show such operation to be straightforward, with the consequent increase in longitudinal and transverse emittance well within the linac acceptance. By accelerating multiple-charge-state beams of the heavier ions, the linac described above would be capable of producing 1 MW beams of virtually any stable ion. The beam intensity for heavy ions is limited by the ion source.

The cost of the driver linac is largely defined by the number of SC cavities and the cryogenic load. The accelerating field in SC cavities is fundamentally limited by the peak surface magnetic field, B_{peak}. In practice, the limit is lower and defined by the field emission and excessive X-ray radiation. Another limiting factor is the dynamic cryogenic load due to RF power losses on cavity RF surfaces.

In the design stage the cavity geometries are optimized using modern 3D simulation codes with the goal of reducing the ratios $B_{\text{peak}}/E_{\text{acc}}$ and $E_{\text{peak}}/E_{\text{acc}}$, while increasing the shunt impedance and geometry factor. Advances in RF surface processing and clean room techniques initially developed at DESY, KEK, and JLAB in the early 1990s for $\beta = 1$ elliptical cavities are now systematically adapted to low-β TEM-class cavities. For example, quarter-wave, half-wave, and spoke cavities developed at ANL and FNAL achieve simultaneously high surface fields and very low residual resistance. Based on demonstrated technology, the 1.5 GeV driver linac can be composed from five types of SC cavities, as detailed in Table 2. In these calculations, the design peak field is $E_{\text{peak}} = 40\,\text{MV/m}$ while the peak magnetic field remains below 74 mT for all types of cavities. These are rather conservative numbers compared to the demonstrated values. For example, in recently built and tested QWRs at ANL, an average E_{peak} value of 73 MV/m and $B_{\text{peak}} = 110\,\text{mT}$ have been achieved. In addition, the residual resistance at $B_{\text{peak}} = 80\,\text{mT}$ is below 3 nΩ. The voltage gain per cavity for a uranium beam along the proposed driver linac is shown in Fig. 8. The beam dynamics studies on a driver linac with similar parameters as shown in this section have been detailed in recent publications [49, 50]. A new tracking code, TRACK [22, 51], has been developed to include up-to-date multicomponent heavy ion beam physics in all types of focusing and accelerating elements being used in linacs. For reliable beam loss studies in the presence of field errors and element misalignments, the code has been parallelized to run up to a million particles in hundreds of randomly seeded accelerators.

Table 2. The main parameters of the 400 MeV/u uranium driver linac.

Cavity type	f (MHz)	β_{opt}	L_{eff} (m)	V_0 at β_{opt} (MV)	N_{cav}	N_{cryo}
QWR	81.25	0.045	0.17	1.4	7	1
QWR	81.25	0.1	0.37	3.2	28	4
HWR	162.5	0.25	0.46	4.1	72	12
HWR	325	0.38	0.35	3.5	48	8
Elliptical	650	0.62	0.71	12.7	72	12
Total					227	37

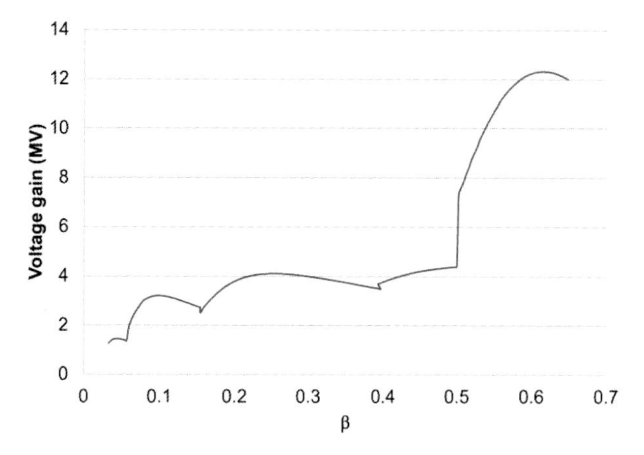

Fig. 8. Voltage gain per cavity as a function of the uranium beam velocity.

5.1. *Concept of multiple-charge-state beam acceleration in a heavy ion linac*

A heavy ion linac is usually designed for the acceleration of many ion species. In an SC linac the cavities, fed by individual RF power sources, can be independently phased. The phase setting can be changed to vary the velocity profile for synchronous motion along the linac. For a given, fixed phase setting, the synchronous velocity profile is fixed, and the transit time factor, $T(\beta, \beta_G)$, is constant. This is similar to the case of a DTL, in which the phase is fixed, and the velocity profile is determined by the DTL geometry. In this case, to accelerate ions with a charge-to-mass ratio $(q/A)_i$ different from the design value, the following relation must be satisfied:

$$\left(\frac{q}{A}\right)_i E_i = \left(\frac{q}{A}\right)_0 E_0. \tag{14}$$

That is to say, the phase setting of individual cavities is kept the same as for the design beam, and only the amplitude of the RF field is changed in order to fulfill the condition (14). In this way, a beam of lower charge-to-mass ratio can be accelerated by using higher electric fields. The velocity and the accelerated beam energy per nucleon do not depend on the ion species.

In an independently phased cavity array such as an SRF ion linac, beams of different charge-to-mass ratios can be accommodated by changing either the phase or the amplitude of the electric field, or both. Allowing both parameters to change enables the velocity profile to vary. This can provide higher energies per nucleon for ions with a higher charge-to-mass ratio.

The driver linac will accelerate uranium ions at charge state $q_0 = 75$ after the first stripper and at $q_0 = 90$ after the second stripper. The simultaneous acceleration of neighboring charge states becomes possible because the high charge-to-mass ratio makes the required phase offsets small. We note that different charge states of equal mass will have the same synchronous velocity profile along the linac if the condition

$$\left(\frac{q}{A}\right)_i \cos\varphi_{s,q} = \frac{q_0}{A}\cos\varphi_{s,0} \tag{15}$$

is fulfilled. The simultaneous acceleration of ions with different charge states requires injection of the beam with each charge state q at a synchronous phase which is determined from (15):

$$\varphi_{s,q} = -\mathrm{Arc}\cos\left[\frac{q_0}{q}\cos\varphi_{s,q_0}\right]. \tag{16}$$

Figure 9 shows the synchronous phase as a function of the charge state calculated for uranium ions in the medium-energy section of the driver linac. This particular example shows that if the linac phase is set for charge state $q_0 = 75$, it can accelerate a wide range of charge states. In fact, 15 charge states of uranium beam could be accepted and accelerated. For q ranging from 70 to 85, even for the worst case charge state of 70, only a small change in the synchronous phase is required, from 30° to 23°. At this phase, the longitudinal focusing is sufficient to accept the anticipated beam emittance.

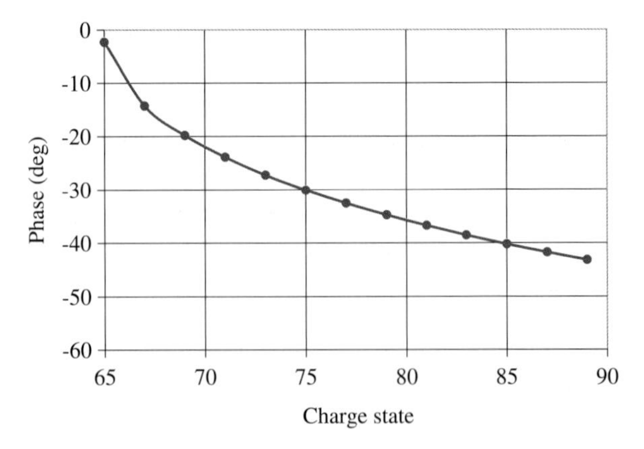

Fig. 9. The synchronous phase as a function of the uranium ion charge state. The designed synchronous phase is −30° for $q_0 = 75$.

The phase trajectories of the linear synchrotron oscillations are given by elliptical trajectories

$$\left(\frac{g}{g_m}\right)^2 + \left(\frac{\psi}{\psi_m}\right)^2 = 1, \tag{17}$$

where $\psi(q) = \varphi - \varphi_{s,q}$ and $\psi_m(q)$ is the amplitude of the phase oscillations. Each particle with a different charge state q oscillates around its own synchronous phase with a different amplitude. The amplitude of the relative momentum oscillations is

$$g_m(q) = \sqrt{\frac{q}{A}\frac{\gamma_s eE_m \sin|\varphi_{s,q}|}{2\pi\beta_s m_e c^2}}\,\psi_m(q). \tag{18}$$

It would be entirely feasible to eliminate the relative oscillations. If the linac has been tuned for the acceleration of some charge state q_0, then the particle bunches of different, neighboring charge states could be injected into the linac at different, neighboring RF phases in order for each charge state to be matched precisely to its own phase trajectory. The higher the charge state is, the sooner it must arrive at a given point to be matched. One possible method of adjusting the phase of multiple charge states would be a magnetic system, such as a chicane, designed with appropriately varying path lengths for the various charge states.

For the present application, however, such a system is not necessary since the acceleration of a multiple-charge-state beam is possible even without matching of different charge states to the proper synchronous phase. If all charge states are injected at the same time (at the same RF phase), then, as described above, each charge state bunch will perform coherent oscillations with respect to the tuned charge state q_0. One can view this as an increase in the total (effective) longitudinal emittance of the multiple-charge-state beam relative to the (partial) longitudinal emittance of the individual-charge-state bunches. For the heavy ion SRF linac being considered, the longitudinal rms emittance is determined by the injector RFQ, and can be made as small as $\sim 0.15\,\text{keV/u} \times \text{ns}$ for a single-charge-state beam. The SC linac provides ample headroom for the effective emittance growth introduced by the acceleration of multiple charge states.

It should be noted that if no phase-matching is done for different charge states, additional emittance growth will occur at frequency transitions in the linac. Heavy ion linacs may have several such transitions to permit efficient operation over the large velocity range required.

In the transverse phase planes, the main difference between single- and multiple-charge-state beams is that a mismatch of a single-charge-state beam is generally correctable, and does not lead to transverse emittance growth. For multiple charge states, correction is more difficult, and will generally induce growth in transverse emittance. For highly charged heavy ion beams, the difference in Twiss parameters for five charge states is sufficiently small that all the charge states can be injected into the linac with the same transverse parameters. If these parameters are chosen to match, for example charge state $q_0 = 75$ for uranium after the stripping at 12 MeV/u, the other charge states will be only slightly mismatched.

Multiple-charge-state beams are also more severely affected by misalignment errors. Misalignments coherently deflect the beam. For a single-charge-state beam, misalignment causes lateral displacement of the beam, but no emittance growth as long as the beam remains in the linear region of the focusing elements. With a beam containing multiple charge states, the differing betatron periods, as well as the differing displacements, cause growth in the transverse emittance. Even for multiple-charge-state beams, however, emittance growth can be substantially reduced by simple corrective steering procedures. The detailed simulation of transverse dynamics of multiple-charge-state beam in the presence of misalignment and other errors in a driver linac is presented in Ref. 49.

6. High-Intensity/High-Power Beam Dynamics in SC Linacs

High-intensity/high-power beams are usually produced with protons, H^-, or deuterons. The applications of high-power linacs are very diverse and comprise drivers for: (i) radioactive ion beam facilities (RIB), (ii) neutrino facilities (superbeam, beta beam, and neutrino factories), (iii) spallation neutron source, and (iv) accelerator-driven systems (ADSs) for the transformation of nuclear waste into varieties with shorter half-lives. The parameters of the application determine the parameter choices for the accelerator. In a CW machine or in high-duty-cycle machines, one will operate at a lower cavity gradient to limit the cryogenic load; while in pulsed machines operating in the medium-current range (10s of mA) one is interested in very high gradients to keep the length of the machine short.

If a high-power linac is injecting into a synchrotron or an accumulator ring, then it is almost mandatory to use H^- instead of protons in order to enable charge exchange injection into the ring. The use of H^-; however, complicates the design and operation of the linac, for the following reasons: (i) H^- sources for high duty cycles and/or high currents are notoriously difficult; (ii) the outer electron of the H^- particles can be stripped (see Subsec. 6.4) by various mechanisms, making loss control much more challenging; (iii) when injecting into a ring with an RF system, one usually tries to adapt the length of the linac bunch trains to the length of the RF bucket in the ring in order to avoid beam loss by uncaptured particles. This is done by employing a low-energy beam chopper after the RFQ. At this energy the beam is usually space-charge-dominated and the chopper line introduces a strong disturbance, which can easily yield halo development in the subsequent parts of the linac.

6.1. *Lattice design*

For beams without space charge, Smith and Gluckstern [52] analyzed the matrix solution to the transverse single-particle equations of FODO channels and found that particle lattice resonances are triggered for transverse phase advances per period of 180°. For beams with space charge, one needs to consider the envelope equations, and in doing so it is found [53] that one needs to keep the zero-current phase advance per period below 90° to avoid envelope lattice resonances. More recent work [54, 55] suggests that a fourth-order resonance ($4\sigma = 360°$) has a much stronger effect on 3D beams, and yields pronounced emittance growth and halo formation for zero-current phase advances around 90°. All three resonance types (particle lattice, envelope lattice, and fourth-order resonance) are avoided by keeping the zero-current phase advance per period in all three planes below 90°, which is why this design rule is so fundamental.

A second "golden rule" is to keep the phase advances per meter smooth across lattice transitions. This not only ensures a stable matching of transitions but also results in a design which is matched for a wide range of beam currents. Together with keeping the phase advance smooth, all lattice transitions have to be matched for 6D beams including space charge.

A third rule can be derived from the core–core resonances, which are explained in Subsec. 6.2.2. It is basically imposed to keep the ratio between longitudinal and transverse phase advances within a certain range in order to avoid emittance exchange between non-equipartitioned planes of the beam.

6.2. *Space charge and beam halo*

The beam halo can be understood as a cloud of single particles, traveling with the beam but having much larger distances from the beam center than the majority of beam particles. The most important mechanism for halo development is based on parametric 2:1 resonances [56, 57], which develop between the oscillations of a mismatched beam core and the movements of single particles. The core oscillations are generally excited by a beam mismatch, and the resonant interaction with single particles can be understood with a simple 1D particle core model [58], which can be extrapolated to 3D bunched beams. Depending on the halo density, one observes rms emittance growth and beam loss on the beam pipe aperture. Other sources of halo and rms emittance growth are listed in the following:

- Resonances between beam envelopes and the focusing lattice (see above) can create rapid rms emittance growth and a beam halo;
- Intrabeam scattering can create a halo, but is usually of minor importance unless it causes charge stripping (see next section);
- Space charge resonances can yield emittance growth but generally no halo (explained in the following);
- Random lattice errors can create a halo via parametric resonances.

6.2.1. *Parametric resonances*

The core radius r_c of an unbunched, azimuthally symmetric beam (1D case) follows the equation of motion for the transverse plane,

$$\frac{d^2 r_c}{dz^2} + k_0^2 r_c - \frac{\varepsilon^2}{r_c^3} - \frac{K}{r_c} = 0, \tag{19}$$

with the averaged transverse focusing forces k_0, the total unnormalized beam emittance ε, and the space charge constant K, which contains charge, mass, axial velocity, and the number of particles. While a matched beam will maintain a constant core radius ($r_c = r_0$), any mismatch will cause core oscillations with the eigenfrequency of the 1D eigenmode. When considering a realistic 3D beam, one will obtain oscillations consisting of a superposition of the three 3D eigenmodes. If we now look at a single particle with amplitude x under the influence of the core's space charge field F_{SC}, we get

$$\frac{d^2 x}{dz^2} + k_0^2 - F_{\mathrm{SC}} = 0, \tag{20}$$

with

$$F_{\mathrm{SC}} = \begin{cases} \dfrac{Kx}{r_c^2}, & |x| < r_c, \\ \dfrac{K}{x}, & |x| \geq r_c. \end{cases} \tag{21}$$

Using Eqs. (20) and (21), the trajectory of a single particle traversing a mismatched beam core can be plotted as shown in Fig. 10 [59]. Looking closely, one can see that the core is oscillating twice as fast as the single particle, demonstrating the mechanism of a parametric 2:1 resonance: particles gain in energy until they get out of phase with the core oscillations. At that point their amplitude diminishes until once again the two oscillations are in phase. The same principle of parametric resonances can be observed in realistic 3D beams, when exciting one of the three beam eigenmodes [60]. Since the particle–core model assumes a constant beam emittance, we do not learn anything about halo-related emittance increase, but we can draw some important conclusions, which can be verified by 3D beam simulations [61]:

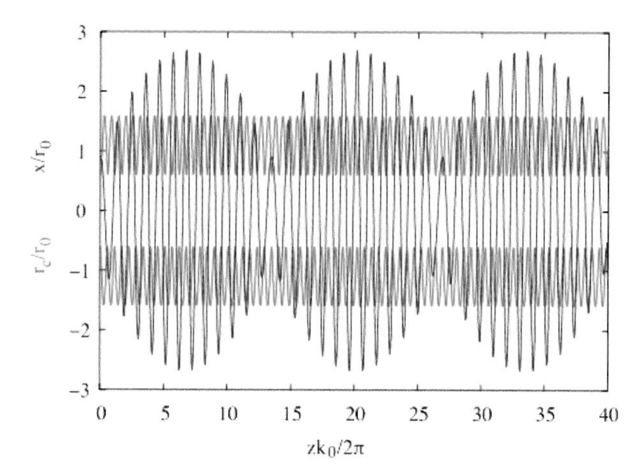

Fig. 10. Core and single-particle oscillations of an initially mismatched beam.

- An rms mismatch results in a beam halo via parametric resonances.
- Single particles oscillate around certain fixed points in phase space. The fixed-point–core distance is insensitive to the mismatch strength and tune depression [58].
- Tune depression defines the time constant for halo development [58]. A large tune spread increases the likelihood of single particles getting trapped in 2:1 resonances, thus creating a halo — meaning that high tune depression yields higher halo density.
- An increased mismatch increases the number of particles affected by the mismatch and thereby increases the halo density [59].
- The maximum halo radius is limited. 1D particle–core models predict a limit of ~7 rms radii, while 3D models predict more than 10 rms radii [61].
- Statistical errors (e.g. lattice errors) contribute to a beam halo also via parametric resonances [63, 64] and can lead to even larger single-particle radii.

Furthermore, it should be mentioned that halo scraping does remove halo particles but does not inhibit further halo development as long as the core is still oscillating due to a mismatch.

6.2.2. *Core–core resonances and equipartitioning*

Core–core resonances are coherent space charge resonances, which were identified in 1979 for 2D beams [65] and then found to be responsible for emittance transfer between the different beam planes [66]. Most likely for this reason linac designers in the past often applied equipartitioning in their lattices for high-intensity machines, which demands that the product of phase advance per period (tune) times the rms emittance has to be equal in x, y, and z:

$$\sigma_x \varepsilon_x = \sigma_y \varepsilon_y = \sigma_z \varepsilon_z. \quad (22)$$

In 2001 non-equipartitioned linac designs were systematically studied with 3D simulations [67] and compared to the theoretical predictions [66] for 2D beams. It was found that emittance exchange between the longitudinal and the two transverse planes can be predicted by Hofmann's instability charts, which have since been used for the design of high-intensity linacs (see Fig. 11).

An experimental proof of emittance exchange was conducted at GSI in 2009 [68].

The key to avoiding emittance exchange via core–core resonances is to maintain the ratio between the transverse and the longitudinal phase advance within certain "stable" bands, which are given by Hofmann's charts. Nevertheless, fast transitions of instable bands are often possible without having much effect on the emittances.

6.3. *Statistical lattice errors*

When considering lattice errors in beam simulations, one distinguishes between dynamic (unpredictable) errors and static (known) errors. In the latter case we have transverse alignment errors (some of which can be corrected by steerers), residual magnetic gradient errors (given by the precision of magnetic

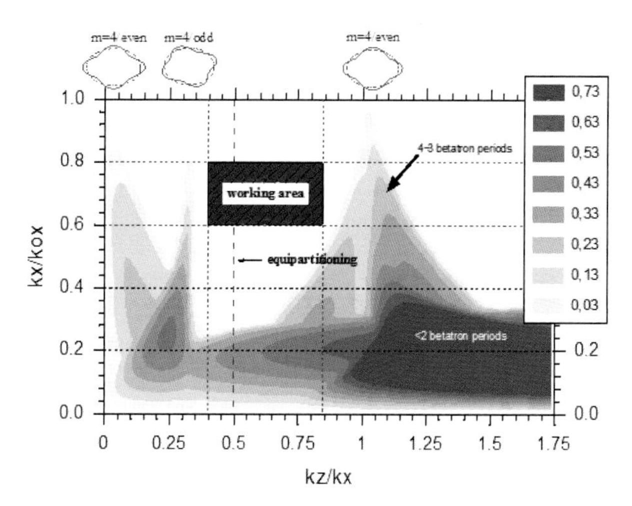

Fig. 11. An example of a stable working area in a Hofmann chart for an emittance ratio of 2 between longitudinal and transverse.

measurements and manufacturing tolerances), and residual electric field errors (errors from manufacturing tolerances which result in nonflat fields). Static errors also include slow "seasonal" changes of alignment and tuning. Dynamic errors change from pulse to pulse or even within a pulse and are caused, for instance, by the limited stability of RF power systems or magnet power supplies. For both types of errors statistical error simulations need to be performed so as to define the acceptable error limits. In most cases these limits translate directly into machine cost (e.g. machining tolerances) and should therefore be considered very carefully. Halo development due to statistical errors can be explained via parametric resonances [63] and is a function of lattice length, tune depression, and the error amplitudes.

6.4. H^- stripping

In high-power linacs, one uses H^- instead of protons if the beam needs to be injected into a circular machine. However, since the second electron of the H^- is only loosely bound (0.75 eV), it can easily be stripped and cause unacceptable losses. So far four stripping mechanisms have been considered in high-power linac designs, which are described in the following.

6.4.1. *Magnetic stripping*

Magnetic stripping has to be considered when one is defining the aperture of focusing magnets or in case the beam is bent before being injected into a subsequent circular machine. The principle of magnetic stripping is based on Ref. 69, and it was later on improved with experimental data to become the following formula, which is commonly used to estimate H^- stripping in magnetic fields:

$$\frac{\text{loss rate}}{\text{meter}} = 1 - e^{-\frac{1m}{\beta\gamma c\tau}}, \tag{23}$$

with τ being the lifetime of H^- in its rest frame.

$$\tau = \frac{a}{E} e^{\frac{b}{E}}, \tag{24}$$

where a and b are constants chosen for example using Ref. 70 with $a = 2.47 \cdot 10^{-6}$ Vs/m, $b = 4.49 \cdot 10^{9}$ V/m, and E is the electric field in the rest frame of the ion passing through a magnetic field with flux density B:

$$E = \beta\gamma c B. \tag{25}$$

Magnetic stripping puts a constraint on the minimum quadrupole apertures, which can be used in the linac, and severely limits bending radii for arcs through which the H^- beam may have to be transported.

6.4.2. *Stripping by blackbody radiation*

At higher energies (several GeV) the blackbody radiation of beam pipes at room temperature can substantially increase the photo detachment rate of electrons. Thermal photons are Doppler-shifted to energies capable of detaching the second electrons of the H^- ions. The effect is described in Ref. 71 and was studied for Project X at FNAL [72], where the dependency of blackbody photo detachment is plotted as a function of particle energy and beam pipe temperature. In the case of the FNAL 8 GeV beam, the stripping rate is estimated at $0.8 \cdot 10^{-6}$ m^{-1} and it is suggested to use a cold (150 K) beam screen as mitigation measure. Due to the lack of multi-GeV H^- beams, however, H^- stripping by blackbody radiation has not yet been experimentally validated in an accelerator environment.

6.4.3. *Rest gas stripping*

Rest gas stripping is usually the least dominant effect but will be mentioned for the sake of completeness. For rising H^- energies the cross-section for losing electrons via collisions with residual particles in the beam pipe vacuum decreases [73], which means that rest gas stripping is more pronounced at lower energies, where losses are less harmful. Using the energy dependence given in Ref. 73, one can extrapolate loss rates for typical energies of H^- linacs.

6.4.4. *Intrabeam stripping*

Here the second electron is stripped by Coulomb interaction H^- ions with each other. The process was first studied at CERN [74] over 20 years ago, and then proposed [75] as a loss mechanism in the SNS linac, where it was recently experimentally verified [76]. The intensity loss per unit length is given by [75]

$$\frac{1}{N}\frac{dN}{ds} = \frac{N\sigma_{\max}\sqrt{\gamma^2\theta_x^2+\gamma^2\theta_y^2+\theta_s^2}}{8\pi^2 r_x r_y r_s \gamma^2}F(\gamma), \quad (26)$$

with the rms bunch dimensions $r_{x,y,s}$, the rms angular spreads $\theta_{x,y}$, the relative rms momentum spread θ_s and the maximum stripping cross-section $\theta_{\max}$. The form factor

$$F_r(\gamma) = 1 - \left(1 + \frac{1}{5}\left(\frac{(\gamma-1)mc^2}{191.4\,\mathrm{MeV}}\right)^2\right)^{-0.69} \quad (27)$$

was introduced to account for the reduction of residual radiation for small particle energies [75]. The description above suggests an increased beam size as mitigation measure, which was confirmed at SNS by using reduced transverse focusing along the linac [76] and by comparing H^- and proton beam losses along the same lattice.

Recently, all mechanisms of H^- stripping have been implemented in the multiparticle beam dynamics simulation code TRACK [77].

6.5. *Higher-order modes*

How a high-intensity linac beam is affected by the higher-order modes (HOMs), excited in the SC cavities, strongly depends on the beam and cavity characteristics of the linac. Monopole modes have field components on the cavity axis and are therefore always excited by the beam. In the following we will explain the formalism to understand their excitation, and we refer to the literature for the excitation of dipole modes by off-axis particles [78]. The voltage per HOM (monopole), which is excited by a point charge q, is given by the fundamental theorem of beam loading as

$$\Delta V_{q,n} = -q\frac{\omega_n}{2}\left(\frac{R}{Q}\right)_n(\beta), \quad (28)$$

with ω_n being the angular mode frequency and $(R/Q)(\beta)$ being the velocity-dependent R/Q of a particular mode n. It is defined as

$$\left(\frac{R}{Q}\right)_n(\beta) = \frac{|\int_{-\infty}^{\infty} E_{z,n}(r=0,z)e^{i\omega_n\frac{z}{\beta c}}\,dz|^2}{\omega_n W_n}. \quad (29)$$

Here W_n is the stored energy of mode n. Once a HOM voltage is excited by the beam, it will decay according to

$$V_n(t) = \Delta V_{q,n} e^{-\frac{t\omega_n}{2Q_{l,n}}} e^{-i\omega_n t}, \quad (30)$$

where $Q_{l,n}$ is the loaded Q of mode n. By adding up the excitation caused by the bunches of a complete bunch train, and by taking into account the voltage decay over time, one can find the maximum possible HOM voltage per mode and cavity [78]. If the HOM frequency of a mode with high R/Q coincides with or is close to a machine line (such as a multiple of the bunch frequency or a multiple of any frequency, which is introduced through low-energy beam chopping or beam pulsing), it can yield a significant blowup of the effective beam emittance and even lead to complete beam loss. The emittance blowup is increased by the presence of bunch-to-bunch charge scatter but is decreased by the cavity-to-cavity frequency spread of the HOMs, which is introduced by construction tolerances. A good criterion to judge whether monopole HOMs should be damped is to compare their effect on the longitudinal emittance with the effect of amplitude and phase errors caused by the RF system.

The second criterion is to look at the additional surface losses in the cavity,

$$P_{d,n} = \frac{(V_n T)^2}{(\frac{R}{Q})_n(\beta)Q_n}, \quad (31)$$

and to compare them with the surface losses of the fundamental mode. If the additional load is larger than a certain fraction of the load of the fundamental mode (e.g. 10%), one should consider measures to reduce the HOM voltage.

In order to extract HOM power from the cavities, one can employ HOM couplers or, alternatively, HOMs can be damped using lossy material in the beam pipes between cavities. However, since HOM dampers also introduce an additional design complication, which can lead to multipacting, field emission, leaks, etc., one usually tries to avoid their use. A method that has shown success in the past and which can be used in case HOMs are close to machine lines is to detune and retune the cavities. In doing that, the frequency of the HOM concerned often shifts to a value with sufficient distance to existing machine lines. However, if the exact distribution of machine lines is not clear at the time of construction (for example because it is already clear that the linac will be used with a large variety of beam chopping patters), then the use of suitable HOM dampers is highly recommended.

Finally, it should be mentioned that the choice of cavities has a strong influence on the induced HOM voltages. Doubling the frequency and doubling the number of cells per cavity will each double the induced HOM voltage. Furthermore, a higher number of cells will yield a higher risk for trapped modes [79]. From Eq. (28) it is also clear that higher currents yield higher induced HOM voltages.

7. Overview of Operational and Future SC Ion Linacs

The first CW SC ion accelerators have been built to boost the energy after DC accelerators such as pelletrons or tandems. A review of such SC linacs is given in Ref. 1. In the past decade the scientific community has increasingly demanded accelerated CW stable and radioactive ion beams which can be efficiently provided by SC ion linacs. The existing facilities are being refurbished and upgraded to higher energies and beam intensities. Several new projects are under development or construction worldwide. Table 3 summarizes the main parameters of existing and planned SC ion linacs.

All existing SC ion linacs operate at 4 K. Similarly, low-energy linacs at GANIL and SNRC will operate at 4 K. The Facility for Rare Isotope Beams (FRIB) at MSU and RAON in South Korea are the most ambitious projects to accelerate uranium to 200 MeV/u and deliver 400 kW beams to targets. These large facilities are planned to operate at 2 K. This is consistent with recent studies finding that operation of TEM-class cavities is more economical at 2 K rather than at 4 K. Both FRIB and RAON projects are based on conservative parameters of peak surface electric and magnetic fields in all cavities. The 200 MeV/u uranium linac requires approximately 800 MV accelerating voltage. This leads to a large number of accelerating cavities.

8. Overview of Operational and Projected SC Proton and H$^-$ Linacs

The first SC proton linac for high-power beams that went into operation is the Spallation Neutron Source (SNS) H^- linac at ORNL [6]. SNS is a user facility built to provide 1.4 MW of beam power at an energy of 1 GeV. In the SNS case H^- is used to be able to inject into an accumulator ring via H^- stripping, which means that all the problems of unwanted H^- stripping during acceleration (see Subsec. 6.4) had to be considered. The phenomenon of intrabeam stripping was proposed **and** experimentally proven as the reason for an unexpectedly high rate of beam loss in the SC part of the linac [76]. Future H^- linac projects will have to consider this effect at

Table 3. Existing and planned SC ion accelerators.

Laboratory	Cavity types	Frequency (MHz)	Lowest q/A	Cavity β	Number of cavities (including bunchers)	Ref.
Argonne, after intensity upgrade	QWR, split rings	48.5, 72.75, 97, 109.25	1/7	0.025, 0.038, 0.077, 0.105, 0.15	53	1
INFN LNL	RFQ, QWR	80, 160	1/7	0.047, 0.11, 0.13	80	80
TRIUMF	QWR	106, 141	1/5	0.057, 0.071, 0.11	40	81
IUAC, New Delhi	QWR	97	1/7	0.051, 0.08	27	82
ReA3 (MSU)	QWR	80.5	1/4	0.041, 0.085	16	83
*SARAF, Phase II	HWR	176	1/2	0.08, 0.16	28	5
*GANIL	QWR	88	1/3	0.07, 0.12	26	2
*FRIB (MSU)	QWR, HWR	80.5, 322	1/7	0.041, 0.085, 0.29, 0.53	340	4
*HIE ISOLDE	QWR	101.28	1/4.5	0.063, 0.103	32	84
*IFMIF	HWR	175	1/2		42	3
*RAON	QWR, HWR, SSR	81.25, 162.5, 325	1/7	0.047, 0.12, 0.3, 0.53	386	85

*These linacs are being designed or constructed.

the design stage. Common to all planned projects is the use of multi-cell elliptical cavities above a specific beam energy. Designers usually try to minimize the number of different cavity types in order to reduce the engineering effort and production cost. Differences between the projects can be observed for the lower-energy end. Many projects favor spoke cavities as transition between the very-low-energy NC front ends and the SC elliptical cavities, while others exclude spoke cavities to save on engineering effort or simply because spoke cavities have so far not been employed in operational linacs. The transition energy between normal and superconducting cavities should be chosen according to a careful cost and performance comparison, and it should be customized for each particular application. Another feature of most designs is the upper frequency restriction to 700–800 MHz for the cavities of the high-beta section. A doubling of this frequency would imply a doubling of the number of cells per cavity if one wants to achieve the same "packing factor" (active cavity length per real-estate length of the linac) for both frequency regimes. High cell numbers combined with high frequencies, however, yield a higher risk of trapped modes and increase the sensitivity to HOMs (see Subsec. 6.5). Furthermore, the power density in RF equipment like circulators, loads, waveguides, or klystrons is much more difficult to handle when one is considering higher frequencies for high-duty-cycle linacs. These practical considerations usually outweigh the slightly higher efficiency of higher-frequency structures (>1 GHz). Table 4 summarizes existing and planned proton or H^- linacs.

9. R&D for Future Hadron Linacs

9.1. *Low-beta linacs*

As we discussed in Secs. 4 and 5, ion accelerators require four or five types of SC cavities to cover the velocity range up to $0.6c$. In the majority of proposed designs of future accelerators, this velocity range is covered by two-gap SC cavities. However, each cavity provides rather low accelerating voltage. Therefore, the path for future cost-effective linacs is through the improvement of the performance of TEM-class SC cavities, in terms of both accelerating gradient and residual resistance. In fact, there are examples where QWRs can operate at peak electric and magnetic fields higher than 117 MV/m and 165 mT [87]. The surface resistance below 3 nΩ has been demonstrated for operational peak magnetic fields up to 80 mT.

More radical reduction of the linac cost can be achieved if the operational cryogen temperature can be increased to save on the cost of the cryoplant. High-temperature SC film coating is being explored to increase cryogenic liquid temperatures. This direction of research is popular with TM-class cavities but there is no reported experience with TEM-class

Table 4. Overview of existing and planned superconducting proton linacs.

Machine	SC cavity types	f_{RF} of SC cavities (MHz)	Cavity geometric β	Energy (GeV)	P_{beam} (MW)	f_{rep} (Hz)	I_{pulse} (mA)	Application
SNS	Elliptical	805	0.61/0.81	1	1.4	60	26†	Neutron production
*Project X	HWR, spoke, ell.	162.5/325/650/1300	0.094/0.19/0.43/0.61/0.9/1.0	1/3/8	1/3/0.4	CW	1–5	Neutrino driver
*ESS	Spoke, ell.	352/707	0.5/0.65/0.86	2	5	14	61	Neutron production
*EURISOL	HWR, spoke, ell.	176/352/704	0.09/0.15/0.3/0.47/0.65/0.78	1–2	5	CW	6	RIB
*Myrrha	Spoke, ell.	352/704	0.35/0.47/0.65	0.6	2.4	CW	4	ADS
*HP-SPL	Ell.	704	0.65/1.0	5	4	50	40†	Neutrino driver
*LP-SPL	Ell.	704	0.65/1.0	4	0.14	2	20†	LHC injector
*India ADS	Spoke/ell.	325/650	t.b.c./0.61/t.b.c.	1	30	CW	30	ADS
*China ADS	HWR/ spoke/ell.	162.5/325/650	0.12/0.21/0.4/0.63/0.82	1.5	15	CW	10	ADS

*Under construction/planning; †pulse current after chopping.

cavities. Although worth mentioning, there has been limited success in this research field for the past several decades.

For many decades now, there have been efforts in several laboratories worldwide to use sputtered Nb on Cu as the SC surface. There are many TEM cavities built using this technique and installed in existing ion linacs. It seems unlikely that sputtered Nb will be superior to bulk metal, but it may reduce costs.

In the low-energy region, following the RFQ, high accelerating fields cannot be applied, primarily due to the long focusing periods provided by focusing elements external to SC cavities. This situation can be substantially improved with the development of RF focusing SC cavities. There were studies in this field a decade ago [30] but a technically viable solution does not exist yet.

The effect of transitions between cryomodules on beam quality can be significantly mitigated by using local matching, which requires an increased number of SC cavities and focusing elements. Further studies in this area are necessary in order to find more cost-effective solutions. The use of long cryomodules containing multiple accelerating and focusing units with a minimal distance between the cryomodules has been successful in low-intensity ion linacs. R&D work is being pursued to implement similar designs for future high-power CW proton linacs.

A multibeam pulsed SC linac as an injector to a synchrotron is required for future electron–ion colliders. The acceleration of any ion species up to 100 MeV/u for the heaviest ions and up to 200 MeV for protons in a pulsed mode can be realized on the basis of individually phased SC resonators. A cost-effective design of such SC injector linacs should be demonstrated.

9.2. *High-beta linacs*

At present most of the construction and surface treatment technologies for high-gradient SC cavities are focused on the developments for the International Linear Collider (ILC) [88]. This technology was developed for acceleration of low currents (5.8 mA) of electrons at duty cycles of up to 10 Hz with a beam pulse length of 1.4 ms.These cavities operate at 1.3 GHz, with an average gradient of 31.5 MV/m. For acceleration of high-power proton beams, the use of lower frequencies (600–800 MHz) is preferable (see Sec. 8) and, apart from SNS, there is little experience with high-gradient SC cavities in this frequency range. The big challenge for future high-power linacs is to transform the low-duty-cycle, high-frequency technology of ILC into high-duty-cycle, lower-frequency cavities suitable for high-intensity proton acceleration. While the ILC cavities and cryomodules have been optimized during the production of large numbers of cavities all over the world, this kind of statistics is missing for lower frequencies, and since proton linacs have much fewer cavities than ILC, it is unlikely that we will ever get to the same number of prototype cavities and cryomodules which were made for ILC.

Apart from the limitations of the cavities themselves, the cavity gradients are today mostly limited by the maximum power, which the high-power couplers can handle. Further research is needed to push these power limits, while still ensuring reliable operation.

For ADS applications, projects aim at tens of MW of beam power, and for these machines loss control is a major issue. So far high-power linacs do not consider any intermediate-beam collimation to limit particle loss in the accelerator itself. However, approaches for dedicated collimation areas in linacs or for "distributed collimation" (for example after each cryomodule within a transverse focusing section with NC quadrupoles) should be studied seriously for ADS requirements. Another concern for these systems is the overall accelerator reliability, which needs to ensure a minimum number of beam trips and basically requires redundant systems and methods to compensate for failing cavities or front ends.

Another issue is to raise the efficiency of the RF power systems. Klystrons, which are needed to cover high-peak-power needs of pulsed high-current machines, usually operate at efficiencies of 50–60%. Higher efficiencies would be possible when operating at saturation but then the LLRF system would no longer be able to control the voltage level in the cavities. The development of solid state amplifiers is certainly promising but present devices are still a factor of 10 below the MW-class performance, which is often needed for high-power machines. A promising alternative to klystrons would be the use of MW-class multibeam IOTs, and some first efforts have started to explore the feasibility of such systems [89].

10. Summary

The majority of new large hadron linear accelerator projects, both pulsed and CW, are based on SRF technology. Substantial research-and-development work toward future SC hadron linacs is being pursued in many laboratories worldwide. The R&D work is essential for the cost reduction of SC linacs and it is primarily focused on development of different types of accelerating structures and cryomodules required for hadron linacs. In contrast to electron linacs, the lack of radio frequency standards worldwide weakens the efficiency of the R&D work.

Acknowledgments

This work was partially supported by the US Department of Energy, Office of Nuclear Physics, under Contract No. DE-AC02-06CH11357.

References

[1] L. M. Bollinger, Low-β SC linacs: Past, present and future, in *Proc. Linac '98* (Chicago, USA, 1998), pp. 3–7.
[2] R. Ferdinand *et al.*, The SPIRAL 2 superconducting linac, in *Proc. Linac '08* (Vancouver, Canada, 2008), pp. 196–198.
[3] A. Mosnier, The IFMIF 5 MW linacs, in *Proc. Linac '08* (Vancouver, Canada, 2008), pp. 1114–1118.
[4] J. Wei *et al.*, FRIB accelerator status and challenges, in *Proc. Linac '12* (Tel-Aviv, Israel, 2012), pp. 417–421.
[5] D. Berkovits *et al.*, Operational experience and future goals of the SARAF linac at SOREQ, in *Proc. Linac '12* (Tel-Aviv, Israel, 2012), pp. 100–104.
[6] S. Henderson, Commissioning and initial operating experience with the SNS 1-GeV linac, in *Proc. Linac '06* (Knoxville, USA, 2006), pp. 1–5.
[7] P. Lebrun *et al.*, The CLIC Programme: Towards a staged $e^+ e^-$ linear collider exploring the terascale: CLIC conceptual design report. CERN-2012-005.
[8] K. F. Johnson *et al.*, Commissioning of the low-energy demonstration accelerator (LEDA) radio-frequency quadrupole (RFQ), in *Proc. PAC '99* (New York, USA), pp. 3528–3530.
[9] A. Grudiev, S. Calatroni and W. Wuensch, New local field quantity describing the high gradient limit of accelerating structures, *Phys. Rev. ST Accel. Beams* **12**, 102001 (2009).
[10] W. D. Kilpatrick, Criterion for vacuum sparking designed to include both rf and dc, *Rev. Sci. Instrum.* **28**, 824 (1957).
[11] T. J. Boyd, Jr., Kilpatrick's criterion. Los Alamos Group AT-1 report AT-1:82-28 (1982).
[12] P. N. Ostroumov *et al.*, in *Proc. PAC '09* (Vancouver, Canada), pp. 4869–4871.
[13] M. P. Kelly, *Reviews of Accelerator Science and Technology*, Vol. 5 (2012), pp. 185–203.
[14] S. Belomestnykh, *Reviews of Accelerator Science and Technology*, Vol. 5 (2012), pp. 147–184.
[15] G. W. Foster and J. A. MacLachlan, Mission 8 GeV injector linac as a Fermilab booster replacement, in *Proc. Linac '02* (Gyeongju, S. Korea), pp. 826–830.
[16] T. P. Wangler *et al.*, Basis for low beam loss in the high-current APT linac, in *Proc. Linac '98* (Chicago, USA, 1998), pp. 657–659.
[17] T. Wangler, *RF Linear Accelerators*, Wiley Series in Beam Physics and Accelerator Technology (2004), p. 285.
[18] J. R. Delayen, in *Proc. 4th Workshop on RF Superconductivity* (Tsukuba, Japan, Aug. 1989), KEK Report 89-21, p. 249.
[19] K. W. Shepard, C. H. Scheibelhut, R. Benaroya and L. M. Bollinger, Split-ring resonator for the Argonne superconducting heavy-ion booster, in *Proc. PAC '77* (Chicago, USA, 1977), pp. 1147–1149.
[20] K. W. Shepard, P. N. Ostroumov and J. R. Delayen, *Phys. Rev. ST Accel. Beams* **6**, 080101 (2003).
[21] P. N. Ostroumov, *New J. Phys.* **8**, 281 (2006). See also J.-P. Carneiro, B. Mustapha and P. N. Ostroumov, *NIM Phys. Res. A* **606**, 271 (2009).
[22] A. Facco and V. Zviagintsev, Study on beam steering in intermediate superconducting quarter wave resonators, in *Proc. PAC '01* (Chicago, USA, 2001), pp. 1095–1097.
[23] P. N. Ostroumov and K. W. Shepard, *Phys. Rev. ST Accel. Beams* **4**, 110101 (2001).
[24] R. E. Laxdal, ISAC-I and ISAC-II at TRIUMF: Achieved performance and new construction, in *Proc. Linac '02* (Gyeongju, S. Korea), pp. 294–298.
[25] P. N. Ostroumov and K. W. Shepard, Minimizing transverse-field effects in superconducting quarter-wave cavities, in *Proc. Linac '02* (Gyeongju, S. Korea), pp. 473–475.
[26] B. Mustapha, Z. A. Conway and P. N. Ostroumov, A ring-shaped center conductor geometry for a half-wave resonator, in *Proc. IPAC '12* (New Orleans, USA), pp. 2289–2291.
[27] P. Berrutti *et al.*, Multipole effects study for Project X front end cavities, in *Proc. IPAC '12* (New Orleans, USA), pp. 2309–2311.
[28] R. W. Garnett *et al.*, Conceptual design of a low-SC proton linac, in *Proc. PAC '01* (Chicago, USA, 2001), pp. 3293–3295.
[29] B. I. Bondarev, V. V. Kushin, B. P. Murin, L. Yu and A. P. Fedotov, *Atomnaya Energiya* **34**, 131 (1973).
[30] R. W. Garnett, F. L. Krawczyk, R. L. Wood and D. L. Schrage, RF-focused spoke resonator, in *Proc. Linac '02* (Gyeongju, S. Korea), pp. 84–86.

[31] P. N. Ostroumov, Design features of high-intensity medium-energy superconducting heavy-ion linac, in *Proc. Linac '02* (Gyeongju, S. Korea), pp. 64–66.
[32] I. M. Kapchinsky, *Theory of Resonance Linear Accelerators* (Harwood, 1985).
[33] F. Gerigk, Formulae to calculate the power consumption of the SPL SC cavities. AB-Note-2006-011 RF.
[34] V. Parma, Cryostat design, Lecture at CERN Accelerator School, Erice, Italy (2013).
[35] P. N. Ostroumov *et al.*, ATLAS upgrade, in *Proc. PAC '11* (New York, USA), pp. 2110–2112.
[36] S. Nagaitsev, Project X — New multi megawatt proton source at Fermilab, in *Proc. PAC '11* (New York, USA), pp. 2566–2569.
[37] F. Gerigk *et al.*, Layout and machine optimization for the SPL at CERN, in *Proc. Linac '10* (Tsukuba, Japan, 2010), pp. 761–763.
[38] http://europeanspallationsource.se
[39] W. J. Schneider *et al.*, Design of the SNS cryomodule, in *Proc. PAC '01* (Chicago, USA, 2001), pp. 1160–1162.
[40] C. Pagani *et al.*, TESLA Report 2001-36.
[41] I. M. Kapchinski and V. A. Teplyakov, in *Proc. Int. Conf. on High-Energy Accelerators* (Erevan, 1968). See also I. M. Kapchinskii and V. A. Teplyakov, *Prib. Tekh. Eksp.* **119**, 19 (1970).
[42] B. M. Gorshkov *et al.*, *Sov. Phys. Tech. Phys.* **22**(11) (1977).
[43] A. Pisent *et al.*, Results on the beam commissioning of the superconducting-RFQ of the new LNL injector, in *Proc. Linac '06* (Knoxville, USA, 2006), pp. 227–229.
[44] R. Poirier *et al.*, CW performance of the TRIUMF 8-meter-long RFQ for exotic ions, in *Proc. Linac '00* (Monterey, USA, 2000), pp. 1023–1027.
[45] P. N. Ostroumov *et al.*, *Phys. Rev. ST Accel. Beams* **15**, 110101 (2012).
[46] S. Abeyratne *et al.*, Science requirements and conceptual design for a polarized medium energy electron–ion collider at Jefferson Lab; http://arxiv.org/abs/1209.0757
[47] J. A. Nolen, The U.S. Rare Isotope Accelerator Project, in *Proc. Linac '02* (Gyeongju, S. Korea), pp. 29–33.
[48] J. Y. Benitez *et al.*, *Rev. Sci. Instrum.* **83**, 02A311 (2012).
[49] P. N. Ostroumov, V. N. Aseev and B. Mustapha, *Phys. Rev. ST Accel. Beams* **7**, 090101 (2006).
[50] P. N. Ostroumov, B. Mustapha and J. A. Nolen, A driver LINAC for the Advanced Exotic Beam Laboratory: Physics design and beam dynamics simulations, in *Proc. PAC '07* (Albuquerque, USA), pp. 1661–1663.
[51] P. N. Ostroumov, V. N. Aseev and B. Mustapha, http://www.phy.anl.gov/atlas/TRACK
[52] L. W. Smith and R. L. Gluckstern, Focusing in linear ion accelerators, *Rev. Sci. Instrum.* **26**, 220 (1955).
[53] I. Hofmann, L. J. Laslett, L. Smith and I. Haber, Stability of the Kapchinskij–Vladimirskij (K–V) distribution in long periodic transport systems, *Part. Accel.* **13**, 145 (1983).
[54] D. Jeon, L. Groening and G. Franchetti, Fourth order resonance of a high intensity linear accelerator, *Phys. Rev. ST Accel. Beams* **12**, 054204 (2009).
[55] L. Groening *et al.*, Experimental evidence of the 90° stop band in the GSI UNILAC, *Phys. Rev. Lett.* **102**, 234801 (2009).
[56] J. O'Connel, T. P. Wangler, R. Mills and K. R. Crandall, Beam halo formation from space-charge dominated beams in uniform focusing channels. IEEE Report CH3279-7 (1993).
[57] R. Gluckstern, Analytical model for halo formation in high current ion linacs, *Phys. Rev. Lett.* **73**, 1247 (1994).
[58] T. P. Wangler, K. R. Crandall, R. D. Ryne and T. S. Wang, Particle–core model for transverse dynamics of beam halo, *Phys. Rev. ST Accel. Beams* **1**, 084201 (1998).
[59] F. Gerigk, Beam halo in high intensity proton/H^- linear accelerators, *ICFA Beam Dynam. Newslett.* No. 36 (2005).
[60] F. Gerigk, Beam halo formation in linacs: Theory and experiment, in *Proc. HPSL* (Naperville, USA, 2005).
[61] F. Gerigk, Beam halo in high-intensity hadron linacs. PhD thesis, TU Berlin (2006).
[62] J. Qiang and R. D. Ryne, Beam halo studies using a three-dimensional particle–core model, *Phys. Rev. ST Accel. Beams* **3**, 064201 (2000).
[63] F. Gerigk, Beam halo in high-intensity hadron accelerators caused by statistical gradient errors, *Phys. Rev. ST Accel. Beams* **7**, 064202 (2004).
[64] I. V. Sideris and C. L. Bohn, Production of enhanced beam halos via collective modes and coloured noise, *Phys. Rev. ST Accel. Beams* **7**, 104202 (2004).
[65] I. Hofmann, Coherent space charge instability of a two-dimensional beam, in *Proc. HIF Workshop* (Berkeley, USA, 1979), p. 388.
[66] I. Hofmann and O. Boine-Frankenheim, Resonant emittance transfer driven by space charge, *Phys. Rev. Lett.* **87/3**, 034802 (2001).
[67] F. Gerigk and I. Hofmann, Beam dynamics of non-equipartitioned beams in the case of the SPL project at CERN, in *Proc. PAC '01* (Chicago, USA), pp. 2872–2874.
[68] L. Groening *et al.*, Experimental evidence of space charge driven emittance coupling in high intensity linear accelerators, *Phys. Rev. Lett.* **103**, 224801 (2009).

[69] L. R. Scherk, An improved value for the electron affinity of the negative hydrogen ion, *Can. J. Phys.* **57**, 558 (1979).

[70] A. J. Jason, D. W. Hudgings and O. B. van Dyck, Neutralisation of H^- beams by magnetic stripping, *IEEE Trans. Nucl. Sci.* **28**, 2704 (1981).

[71] H. C. Bryant and G. H. Herling, Atomic physics with a relativistic H^- beam, *J. Mod. Opt.* **53**, 45 (2006).

[72] W. Chou *et al.*, 8 GeV H^- ions: Transport and injection, in *Proc. PAC '05* (Knoxville, USA), pp. 1222–1224.

[73] G. H. Gillespie, Double closure calculation of the electron-loss cross section for H^- in high-energy collisions with H and He, *Phys. Rev. A* **15**, 563 (1977).

[74] M. Chanel *et al.*, Measurements of H^- intra-beam stripping cross section by observing a stored beam in LEAR, *Phys. Lett. B* **192**(3–4), 475 (1987).

[75] V. Lebedev *et al.*, Intrabeam stripping in H^- linacs, in *Proc. Linac '10* (Tsukuba, Japan), pp. 929–931.

[76] A. Shishlo *et al.*, First observation of intrabeam stripping of negative hydrogen in a superconducting linear accelerator, *Phys. Rev. Lett.* **108**, 114801 (2012).

[77] J.-P. Carneiro, B. Mustapha and P. N. Ostroumov, Implementation of H^- intrabeam stripping into TRACK, in *Proc. PAC '11* (New York, USA), pp. 1642–1644.

[78] M. Schuh *et al.*, Influence of higher order modes on the beam stability in the high-power Superconducting Proton Linac, *Phys. Rev. ST Accel. Beams* **14**, 051001 (2011).

[79] O. Brunner *et al.*, Assessment of the basic parameters of the CERN Superconducting Proton Linac, *Phys. Rev. ST Accel. Beams* **12**, 070402 (2009).

[80] E. Fagotti *et al.*, Operational experience in PIAVE-ALPI complex, in *Proc. HIAT '09* (Venice, Italy, 2009), pp. 208–212.

[81] R. E. Laxdal, Operating experience of the 20 MV upgrade linac, in *Proc. Linac '10* (Tsukuba, Japan, 2010), pp. 21–25.

[82] A. Roy, Superconducting linac and associated developments at IUAC Delhi, in *Proc. Linac '12* (Tel-Aviv, Israel, 2012), pp. 763–767.

[83] O. K. Kester *et al.*, ReA3 — The rare isotope reaccelerator at MSU, in *Proc. Linac '10* (Tsukuba, Japan, 2010), pp. 26–30.

[84] M. Pasini *et al.*, A SC upgrade for the REX-ISOLDE accelerator at CERN, in *Proc. Linac '08* (Vancouver, Canada, 2008), pp. 124–126.

[85] D. Jeon *et al.*, Overview of the superconducting linacs of the Rare Isotope Science Project, in *Proc. Linac '12* (Tel-Aviv, Israel, 2012), pp. 540–542.

[86] http://neutrons.ornl.gov/facilities/SNS

[87] Z. A. Conway *et al.*, Reduced-beta cavities for high-intensity compact accelerators, in *Proc. Linac '12* (Tel-Aviv, Israel, 2012), pp. 458–460.

[88] ILC Technical Design Report, June 2013: http://www.linearcollider.org

[89] D. McGinnis *et al.*, Applications of high power induction output tubes in high intensity superconducting proton linacs, in *Proc. IVEC* (Paris, France, 2013).

Peter N. Ostroumov is a Senior Accelerator Physicist and Chief of the Linac Development Group in the Physics Division of Argonne National Laboratory (ANL). He started his scientific career at the Institute for Nuclear Research (INR) in Moscow, Russia, after graduating from Moscow Engineering Physical Institute. At INR, he has led installation and commissioning of the 600 MeV high intensity proton linac. In 1999 he joined ANL's Physics Division to develop the design for Rare Isotope Accelerator (RIA) facility based on superconducting (SC) RF technology. Developed and experimentally demonstrated acceleration of multiple charge state beams in linacs. Currently, Dr. Ostroumov is the Principal Investigator (PI) for several R&D projects related to the development of CW hadron linacs.

Frank Gerigk joined CERN as a fellow in 1999 after his graduation from the Technical University of Berlin. He worked on RF cavity design and beam dynamics for high-intensity proton linacs until he joined the Rutherford Lab in the UK in 2002. There he performed in-depth studies on halo development in high-intensity hadron beams, which also became the subject of his PhD thesis, which he defended in 2006 at the Technical University of Berlin. Since 2005 he is based at CERN, where he is presently heading the Linac RF section carrying the responsibility for the operation and maintenance of the CERN hadron linacs as well as the development and construction of the accelerating cavities for the Linac4 project. Apart from his regular duties, Frank is a well-known lecturer on RF technology, chair of the Linac14 Scientific Program Committee, member of various technical advisory committees and also a regular reviewer for journals on accelerator physics.

Reviews of Accelerator Science and Technology
Vol. 6 (2013) 197–219

DOI: 10.1142/S1793626813300090

Ion Injectors for High-Intensity Accelerators

Martin P. Stockli
Spallation Neutron Source,
Oak Ridge National Laboratory,
Oak Ridge, TN 37831, USA
stockli@ornl.gov

Takahide Nakagawa
Nishina Center for Accelerator-Based Science,
RIKEN, Hirosawa 2-1,
Wako, Saitama 351-0198, Japan
nakagawa@riken.jp

There are a growing number of applications for ion accelerators, with increasingly complex beam requirements and progressively higher beam intensities. The performance of the ion injector is critical to the success of these projects. First, there is the ion source that has to produce the desired ion species, with a large variety of desired species requiring vastly different ion sources. In addition, the ion source has to produce those ions with the desired rate and without debilitating impurities, as well as with the desired duty factor. Several examples will show that very successful ion sources can fail when the duty factor is increased because their lifetime becomes too short or their failure rate too high. Equally important is the extraction of those ions and their transport to the next stage of acceleration, because the slow ion velocities pose a serious challenge to increasing the intensity. As the beam intensity is increased, its emittance, stability and controllability become more important. This article cannot cover this subject in depth. It tries to provide a flavor of the complexities and serve as an introduction to further reading and studies.

Keywords: Ion source; intense low-energy ion beam; ion injectors; ion accelerator.

1. Introduction

The demand for beams with higher intensity and the demand for increasing the variety of ion species continue to grow for operational ion accelerator facilities as well as for new ones being developed. Some facilities require intense beams of light ions, while other facilities require intense beams of multicharged heavy ions, or intense beams of highly charged ions (see e.g. Refs. 1 and 2). On the other hand, some facilities require intense beams of negative ions [3]. There is no universal ion source that can efficiently produce all these vastly different ions. Over the last many decades, different ion sources have been developed to efficiently produce certain classes of ions [4–6]. This article focuses on the classes of ion sources and low-energy ion injection systems developed for high-intensity accelerators.

Ion accelerator facilities are normally rated according to their production rate of the desired final product, and accordingly need to be subjected to a cost–benefit analysis. For example, electron cyclotron resonance ion sources (ECRISs) can efficiently produce the highly charged heavy ion beams needed for radioisotope beam factories (RIBFs) [1, 2]. Heavy ions with a higher charge-to-mass ratio are easier to accelerate and therefore enable lower-cost, more compact accelerators, or alternatively enable higher ion energies, which is an increase in benefits. For example, an increase in beam energy from 200 to 400 MeV/u requires very expensive accelerator equipment and provides only an increase of a factor of 5 in the production of ^{138}Sn with an in-flight uranium fission reaction [7]. On the other hand, the construction of a new ECRIS can increase the beam intensity by a factor of 5 at drastically lower costs. Over the past decade the RIKEN injector has increased the suitable $U^{33+\sim35+}$ ion beam currents by one order of magnitude.

For such reasons, ion injectors are in a prime position to significantly contribute to the desired increases in high intensity ion beams. These gains of course have to be accomplished without compromising, and preferably improving, the quality of the extracted ion beam, such as its brightness and stability, or the reliability of the ion injector.

Because this accelerator review volume features only this one article on ion injectors, it will start with a basic introduction to the generation of high ion densities, and continue with the production and transport of intense ion beams. Then it discusses specific ion sources, most of which have proven their reliability as production sources at existing high-intensity accelerators. This series starts with microwave ion sources, which easily exceed 100 mA of protons in a single small-diameter beam. It then continues with ion sources for negative light ions, which are closely related to proton sources. From there it moves to ion sources for heavy positive ions, such as the metal vapor vacuum arc ion sources (MEVVA). The article ends with ion sources for highly charged heavy ions, the Electron Beam Ion Source (EBIS), the Laser Ion Source (LIS) and the ECRIS. Throughout it, close attention is paid to the beam transport and the duty factor, which can present serious challenges when the ion source is enhanced to increase the production rate of the accelerator.

2. Generating High Densities of Positive Ions

Positive ions are atoms or molecules which lack one or more electrons. The outer electrons are bound to the neutral particles with the ionization energy, which is in the range between 3.9 eV for Cs and 24.6 eV for He [5]. The removal of that electron requires an electric field on the order of 10^{10} V/m. Such high field strengths are most effectively achieved in close collisions of charged particles. Collisions between equal-mass particles yield the highest energy and momentum transfer, and accordingly electrons are most efficient in removing electrons from atoms, molecules and ions. The electron impact ionization cross section has a threshold at the ionization energy, and peaks around 2–3 times the ionization energy [8], as shown in Fig. 1. For higher energies the cross section falls off inversely proportional to the electron velocity [8]. This means that electrons with ~50 eV very effectively ionize all

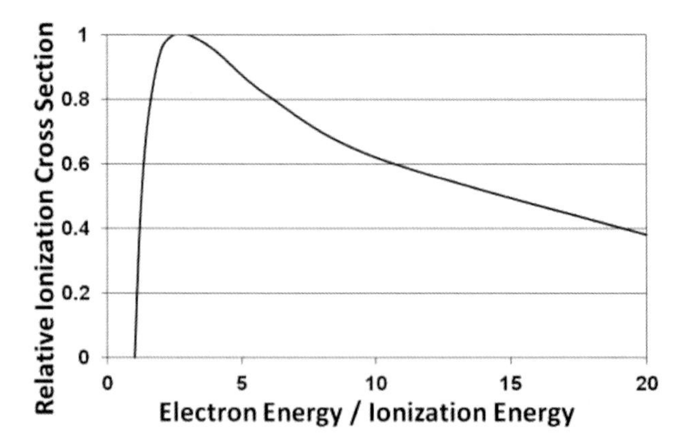

Fig. 1. Typical cross section for electron impact ionization.

neutral atoms and molecules found in a vapor or a low-pressure gas.

This fact is the basis of electron bombardment ion sources which are commonly used in ion analyzers, such as ion gauges, residual and other gas mass analyzers, etc. Being operated at low pressure and with modest electron currents supplied from a heated filament and extracted with ~70 eV, they provide ion signals that are proportional to the sample particle densities, but yield rather small beam currents.

A drastic increase in the ion density requires foremost a drastic increase in the particles to be ionized — the neutral density. This is as simple as increasing the opening of the valve which leaks the most suitable gas or vapor into the ion source, or increasing the vaporization rate by raising the oven temperature.

This needs to be accompanied by a drastic increase in the electron current, which can be achieved either with massive filaments and large discharge currents, or with high-current, high-voltage discharges that multiply the initial electron density, or with high electric fields induced by RF antennas, or with a cavity being energized with microwaves.

When about 10% of the particles are ionized, the gas becomes a plasma, where the highly interactive and mobile charged particles govern the system's behavior. Most importantly, the plasma imposes a condition of neutrality by balancing the positive ion density with the electron density plus the density of negative ions. This condition excludes the presence of extended, static electric fields inside the plasma. Electric fields are displaced to thin layers near potential carrying electrodes where imbalances of charge densities enable the electric fields.

Plasmas depend on steady feeds of energy to compensate for the loss of charged particles through recombination and through losses at the walls of the container. The wall losses can be reduced, but not eliminated, with magnetic fields generated by large coils or permanent magnets.

Electrons are 1836 times lighter than protons, and are normally hotter than ions, and therefore travel more than 43 times faster than ions. Near conductive walls some of the electrons escape and are absorbed by the wall. The charge imbalance causes the positive ions to follow, although at a smaller speed. The resulting net space charge causes an electric field in the boundary layer, which drives the plasma normally to a positive potential, called the plasma potential, which depends on many factors.

The boundary layer is called the plasma sheath and is several Debye lengths thick. The Debye length λ_D is

$$\lambda_D = \left(\frac{\varepsilon_0 \cdot k \cdot T_e}{e^2 \cdot n_e}\right)^{1/2}, \qquad (1)$$

where ε_0 is the vacuum permittivity, k the Boltzmann constant, T_e the absolute electron temperature in K, e the elementary charge and n_e the electron density [9]. In most ion sources the plasma sheath is on the order of 10^{-5} m thick.

3. Generating High-Intensity Ion Beams

To form an ion beam, the ions need to be extracted from the low-pressure plasma environment inside the source through an outlet opening into a vacuum space where coescaping neutral particles are rapidly pumped away to minimize collisional losses.

The escape of the ions is supported by the extraction voltage, which generates a strong electric field between the ion source and the extractor, as shown in Fig. 2. This field accelerates the charged particles with the desired polarity while it pushes the particles with opposite polarity back into the plasma. The location where the plasma ends and the acceleration of the desired polarity starts is called the meniscus, which is pictured as a line in Fig. 2. Depending on the plasma density in the source and the strength of the extraction field, the meniscus can be convex [Fig. 2(a)], planar [Fig. 2(b)] or concave [Fig. 2(c)]. Planar or slightly concave menisci are normally preferred, because expanding beams from convex menisci often lead to undesirable beam losses.

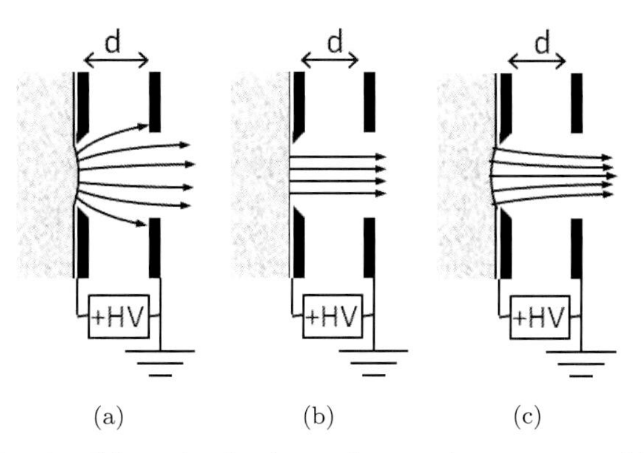

Fig. 2. Schematic drawings of extraction systems with (a) a convex, (b) a planar and (c) a concave meniscus.

If the plasma instantaneously replaces the extracted ions, the extracted current density j from diode style extraction systems (shown in Fig. 2) follows the Child–Langmuir equation [10, 11]:

$$j = \frac{(4 \cdot \varepsilon_0/9) \cdot (2 \cdot q/m)^{1/2} \cdot V^{3/2}}{d^2}, \qquad (2)$$

where ε_0 is the vacuum permittivity, q/m the charge-to-mass ratio of the extracted ions, and V the extraction voltage applied to the extraction gap d.

Equation (2) is derived from the Poisson equation in the gap of the diode, where the charge density is the space charge of the ions that enter from the meniscus and are accelerated through the gap. This value is obtained when the field generated by the ions slowly entering the acceleration gap cancels the external field V/d at the meniscus, but the resulting field increases rapidly with the cube root of the distance from the meniscus.

The high voltage applied to the ion source is normally a fixed design parameter of the system determined by the ion beam acceptance energy of the next stage of acceleration. However, applying a voltage of opposite polarity to the extractor allows for increasing the extraction voltage, and possibly for extracting more ion current. After passing the extractor, the ion beam is decelerated back to ground. Applying an opposite polarity to the extractor also reflects charged particles of opposite polarity generated downstream of the extractor and keeps them from being drawn into the ion source, where they may cause damage.

Equation (2) shows the critical role played by the extraction gap d, which needs to be as short as possible without arcing excessively. A survey of existing extraction systems led to the empirical formula [12]

$$d = 0.014 \cdot V^{3/2}[mm/kV^{3/2}]. \tag{3}$$

Substituting Eq. (3) into Eq. (2) predicts that the extracted current density decreases with increasing extraction voltage proportional to $V^{-3/2}$.

However, increasing the gap d allows for increasing the outlet diameter d_O without changing the level of aberrations. Selecting an outlet diameter equal to the gap distance $d = d_O$ yields a total extracted ion current J,

$$J = (\pi \cdot \varepsilon/9) \cdot (2 \cdot q/m)^{1/2} \cdot V^{3/2}, \tag{4}$$

if the extracted current density is uniform and the losses on the extractor are negligible. Equation (4) yields the well-known $V^{3/2}$ dependence, because gap increases are compensated for by increases in the outlet area.

Equations (2) and (4) also show the dependence of the current density on the charge-to-mass ratio of the extracted ions due to their acceleration in the extraction field. This explains the lower currents obtained for heavier isotopes when using the same ion source.

However, the plasma density can limit the extractable ion current, which is described by the saturation current j_s,

$$j_s = n_i q \cdot (kT_e/m)^{1/2}, \tag{5}$$

where n_i is the extractable ion density, q the ion charge and m the ion mass [13, 14]. The resulting extractable ion currents are shown in Fig. 3. The solid line shows the extraction-limited ion current from a very dense plasma, increasing with $V^{3/2}$. The dashed line shows a gradual transition to the saturation current for a less dense plasma, and finally the dotted line shows the transition for an even thinner plasma.

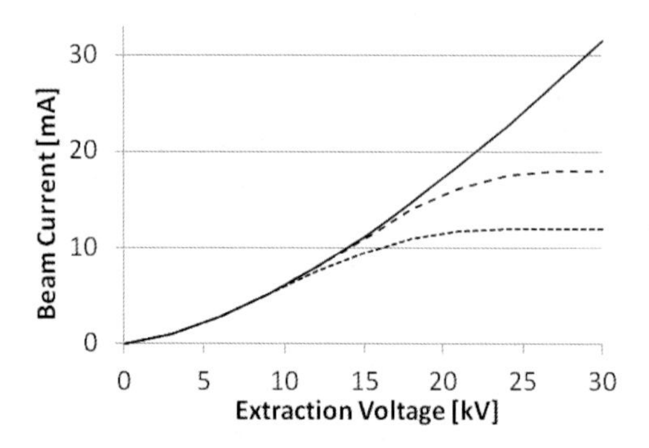

Fig. 3. Typical extraction-limited (solid line), plasma-density-limited (dashed line) and lower-plasma-density-limited (dotted line) ion beam currents versus the extraction voltage.

4. Transverse Emittance of Ion Beams

If all the extracted ions would emerge normal to the spherical menisci, as suggested in Fig. 2, and there were no aberrations and space charge, the ion beams could be transported without losses and focused into tiny spots.

However, in reality the ions emerge from the meniscus with transverse velocity components due to their finite temperature. Second, the menisci are only approximately spherical. Third, the openings in the ion source, in the extractor and in other electrodes cause aberrations. Finally, there is a contribution from the repulsive self-fields of the ion beam, i.e. the space charge effect.

Figure 2(c) shows the most likely extraction case, where the locally averaged ions are extracted normal to the meniscus with a trajectory slightly pointing toward the center of the beam, establishing a correlation between the distance from the axis, x, and the trajectory angle x'. The temperature distribution of the ions causes this correlation to be a distribution around the average ion rather than an exact correlation. When all those distributions are integrated over the entire meniscus, one obtains an ellipse tilted to the left in an x–x' diagram representing the converging beam shown on the left side of Fig. 4.

At the waist of the ion beam, the ellipse is upright before it tilts to the right to represent an expanding beam. Figure 4 also shows how a lens can change the trajectory angles and allow for a smaller waist.

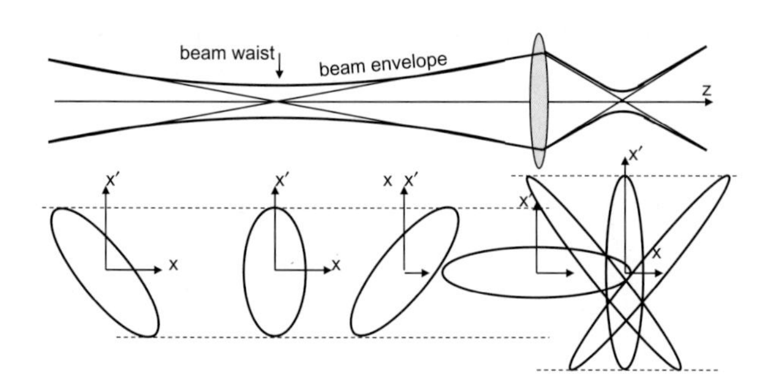

Fig. 4. The emittance ellipse and its rotation as a converging ion beam drifts through a waist and then is focused with a lens.

The area of the ellipse is called the x–x' area emittance A^x, which is conserved along the axis of propagation, according to Liouville's theorem [15].

The actual emittance E^x is defined as the product of the semiaxes of the ellipse, and can be determined from the diameter of the beam waist d_W and the corresponding maximum trajectory angle x'_{max}, as shown in Eq. (6) and Fig. 5:

$$E^x = d_W x'_{\mathrm{max}}/2 = d_W \frac{(d^2 - d_W^2)^{1/2}}{4L}, \qquad (6)$$

where x'_{max} can be determined from the beam diameter d measured a drift distance L from the waist [16].

However, it is more common to give the normalized emittance, E^x_{norm}:

$$E^x_{\mathrm{norm}} = \beta\gamma E^x, \quad \text{with } \beta = v/c \text{ and } \gamma = (1-\beta^2)^{-1/2}, \qquad (7)$$

where v is the ion velocity, c the speed of light and γ the Lorentz factor. This normalization factors out the shrinkage of the trajectory angles when the ions are accelerated and accordingly allows comparing the emittance of ion beams independent of the ion energy.

Because the trajectory angles are expressed as dimensionless tangents, emittances have the dimension of distances. Common units are mm · mrad and μm, which are dimensionally equivalent. In accelerator physics it is common to write this unit as π · mm · mrad, where π is not a factor but a symbol to indicate that the value would have to be multiplied by π to obtain the area emittance [16, 17].

The normalized emittance E^x_{norm} of an ion beam extracted from a stable plasma with an aberration-free extraction system can be estimated as

$$E^x_{\mathrm{norm}} \approx d_O \left(\frac{kT_i}{mc^2}\right)^{1/2}, \qquad (8)$$

where d_O is the outlet diameter, T_i the temperature and m the mass of the ions [5].

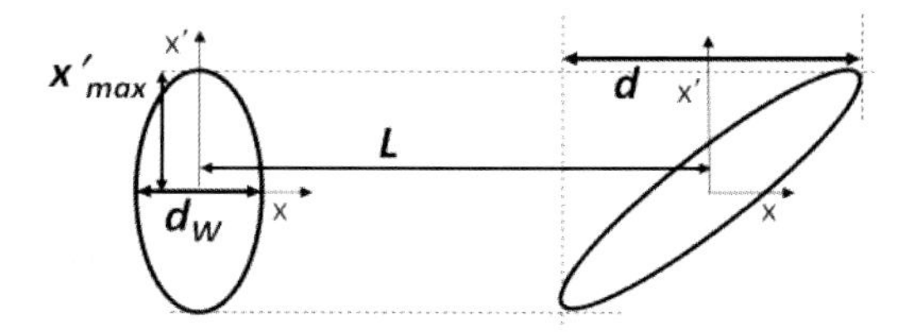

Fig. 5. The emittance ellipse in a beam waist and a drift distance L downstream of the waist.

This introduction to emittances is drastically simplified to foster the understanding of the basic requirement for ion sources to have a small emittance, which does not exceed the acceptance of the next accelerator component. This becomes a growing challenge as ion beam intensities are increased. More elaborate discussions on the various aspects of emittances can be found in Refs. 16–18.

The bottom line is that extraction systems need to be designed to minimize their aberrations [19], which often involve intermediate electrodes to more accurately control the acceleration field [12]. Experimental measurements are necessary because the models rarely include the interaction of the ions and neutrals with the surfaces of the outlet, electrodes and extractor [20].

The tuning of the source and the low-energy beam transport is equally important, because increasing the plasma power or increasing the extraction field is likely to increase the total beam current, but it is also likely to increase the emittance. Tuning on the beam current emerging from the next, acceptance-limited accelerator component is very effective in finding the optimum between high beam current and acceptable emittance [21].

5. Transporting High-Intensity, Low-Energy Ion Beams

5.1. *The physics issues*

Low-energy ions normally emerge from the extractor as diverging beams and therefore require refocusing to successfully inject into the next stage of acceleration, such as a radio frequency quadrupole accelerator (RFQ). RFQs prefer a certain combination of beam size and convergence, which can be obtained with the common two-lens transport system shown in Fig. 6. As indicated by the solid and dashed beam envelopes, relaxing the first lens while strengthening the second lens allows increasing the

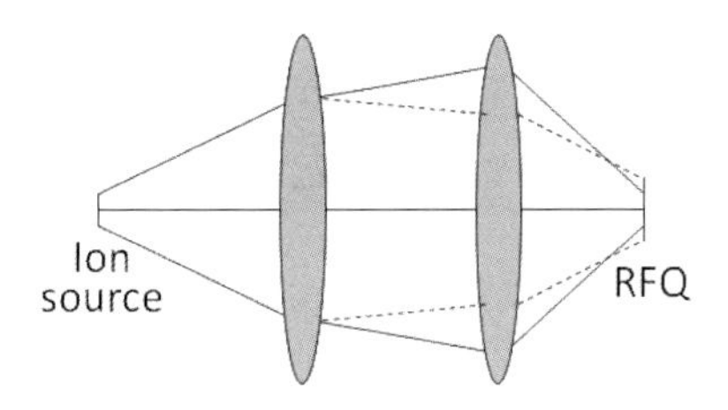

Fig. 6. A simple two-lens low-energy beam transport system.

beam convergence and reducing the diameter of the waist inside the RFQ.

The placement of the lenses is important. The second lens needs to be close to the RFQ in order to supply it with the desired strongly convergent beam. The first lens needs to be close to the source in order to capture the entire diverging beam before the lens lowers the beam divergence to reach the second lens with the desired beam diameter. However, optimal control suggests the beam to fill about 50% of the lens aperture, which means that some drift distance is needed between the extractor and the first lens.

The space charge ρ_i of an ion beam is

$$\rho_i = \frac{j}{v_i}, \tag{9}$$

the ratio of the current density j and the ion velocity v_i. The slow ion velocities of low-energy ion beams, roughly 1% of the speed of light for protons and about 10 times slower for heavy ions, combined with high ion current densities increases the space charge to levels where the repulsion between the equal charges causes noticeable changes in the ion trajectories.

Figure 7 shows Trace-3D beam envelope calculations for 65 keV proton beams emerging from an ~6 mm diameter extractor with maximum trajectory angles of ~5 mrad for the case of ~0 mA and 100 mA of beam current. During the 0.2 m to the first lens, the 100 mA beam diameter grows to more than twice the diameter of the ~0 mA beam. The calculation also shows that increasing mainly the first lens allows achieving the same beam envelope in front of the RFQ as one would with a small-current ion beam. The weak waist of the 100 mA envelope between the lenses is not the usual waist accompanying a crossover like in Fig. 4, but shows the necessity of forming a weakly converging beam to counteract the continued self-repulsion as the beam drifts to the second lens.

For circular ion beams with uniform density, the Gauss law predicts the ions to be repelled with a radial force that is proportional to the distance from the axis, as shown in Fig. 8(a). This causes all ions to drift apart uniformly and thus maintain a uniform density. It just appears that the ions are emerging from a shorter distance than they really are.

However, ion beams typically feature Gaussian-like distributions, like the one assumed for Fig. 8(b). The resulting radial force is approximately linear near the center and accordingly will uniformly dilute the central (peak) density. However, with increasing

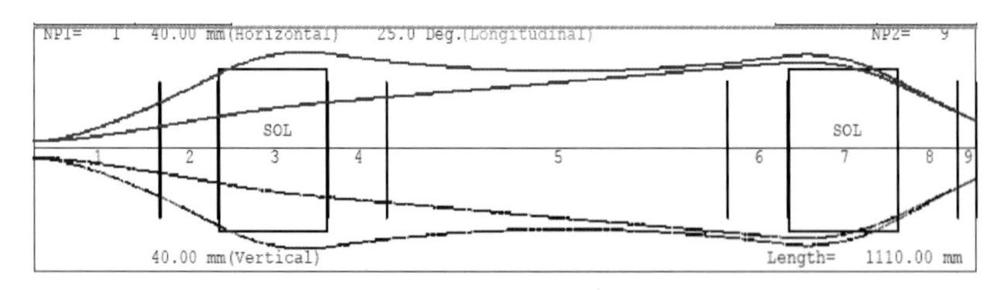

Fig. 7. The horizontal (top traces) and vertical (bottom traces) beam envelopes of 0 mA (smaller radii) and 100 mA (larger radii) uniform density proton beams calculated with the beam envelop code Trace-3D.

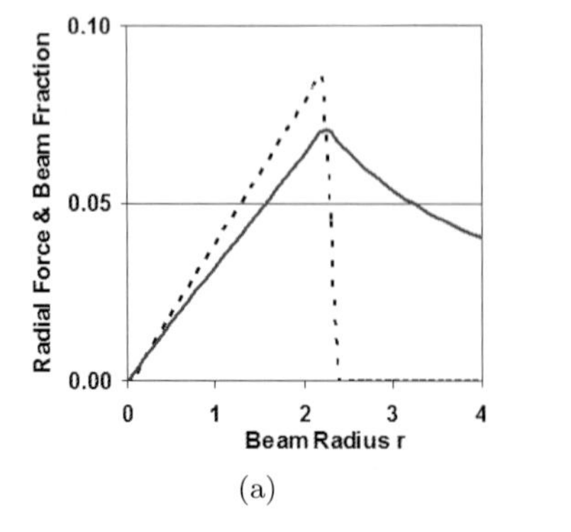

(a)

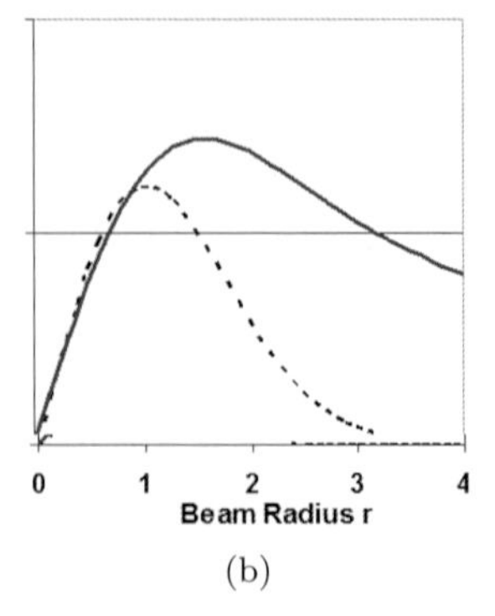

(b)

Fig. 8. The fraction of the beam inside a hollow cylinder with radii r and $r + dr$ (dashed line), and the resulting radial force (solid line) (a) for ion beams of uniform density and (b) Gaussian-distributed ion beams.

radius the additional contributing charges start to saturate and then drop, and so does the radial force. This means that the charge density grows for some larger radii, flattening out the Gaussian distribution.

While many details remain to be fully understood, it is generally accepted that the space charge leads to emittance growth and downstream losses, such as reduced transmission through the RFQ [22].

5.2. *Electrostatic low-energy beam transport*

In electrostatic beam transport systems, all ions follow exactly the same trajectories independent of their mass and charge if those were acquired in the ion source at the voltage V. Different ions just travel with different velocities. A simple aperture lens energized with a voltage V_a with grounded apertures on both sides separated by gaps g has a focal length f:

$$f \approx \frac{2gV}{V_a}. \tag{10}$$

Equation (10) shows that short focal lengths require lens voltages that are comparable to the source voltage. Opposite-polarity accel–decel lenses feature low aberrations but can easily require excessively high voltages. Same-polarity decel–accel lenses are more powerful but suffer from aberration. Filling only about half of the aperture with beam limits the aberrations, but increasing the aperture to compensate weakens the lens.

Electrostatic beam transport systems are very efficient in terms of initial fabrication and operational costs, but they are clearly limited to low-energy ions.

An attractive feature is the compactness with which electrostatic transport systems can be designed. The compactness can have significant advantages for high-intensity ion beams, because the brevity of space charge interaction limits the emittance growth of the beam.

For example, Fig. 9 shows the small-radius beam ($\leq$5 mm) emerge from the source and an intermediate electrode before passing the extractor and entering the 0.1-m-long, two-lens electrostatic low-energy beam transport (LEBT) of the Spallation Neutron Source. After 70 ns the ions enter the RFQ, where their acceleration rapidly reduces the space charge. The second lens is split into four quadrants, which allow steering and fast chopping of the beam.

The SNS LEBT routinely transports up to 65 mA of 65 keV H^- beams, with up to a 5.4% duty factor. 90% RFQ transmission has been measured for ~45 mA. After some initial learning curves [23, 24], the SNS LEBT reached and now remains at 100% availability.

Some emissions from the ion source and corona discharges from the negatively charged lenses, especially from the first lens which gets partially covered with Cs, heat the center ground electrode [25]. Much higher duty factors may require water cooling, which is feasible for the grounded electrode.

5.3. *Magnetic low-energy beam transport*

Magnetic solenoids focus ion beams as a second order effect with a focusing power $1/f$:

$$\begin{aligned} 1/f &= q^2(2\,mv)^{-2}\int B^2(z)dz, \\ \theta &= q(mv)^{-1}\int B(z)dz, \end{aligned} \tag{11}$$

where q, m and v are the ion charge, mass and velocity, and B is the magnetic field along the z axis, the direction of the ion propagation. In addition, the

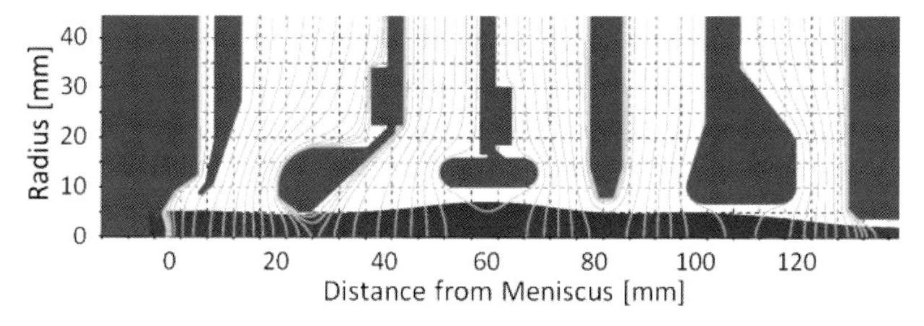

Fig. 9. The ion beam (small radii) emerging from the source outlet passing the e-dump, the extractor, the first lens, the center ground and the second lens before entering the RFQ.

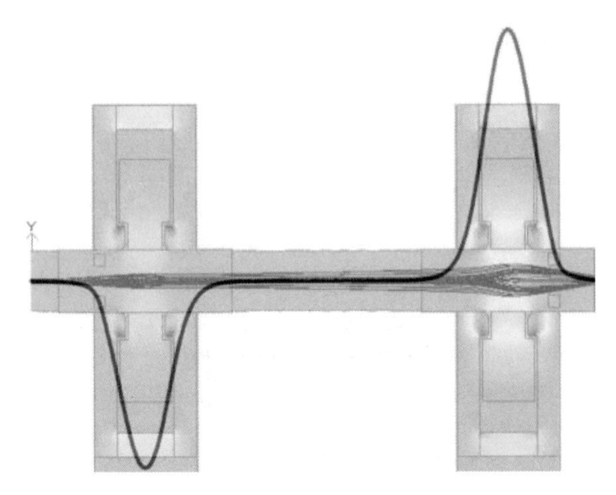

Fig. 10. The axial magnetic field (thick line) and many ion trajectories near the axis of a two-solenoid LEBT.

magnetic field rotates the beam by an angle θ, which can be a concern for noncircular beams. The v^{-2} proportionality limits solenoidal focusing to low-energy ion beams. Magnetic fields tend to trap low-energy charged particles, which can be minimized by using short, high-field solenoids featuring return yokes. An example of such a magnetic two-lens LEBT is shown in Fig. 10 [26].

Simple two-solenoid LEBTs typically feature apertures of 5–10 cm and are roughly 1 m long. This increase in dimension allows the LEBT to accommodate diagnostics and other features, such as a Y-magnet to switch between different ion sources, as shown in Fig. 11. The bends in the dipole Y-magnet yield horizontal focusing, which is compensated for with defocusing quadrupoles [27].

Dipole magnets are invaluable for separating heavy ions, which often come with many different isotope masses m and a range of charges q. The desired ions are selected by dialing a field B where the desired m/q ratio matches the design radius r_0:

$$B = \frac{(2V\,m/q)^{1/2}}{r_0}, \tag{12}$$

where V is the ion source voltage and qV the energy of the ions. Lenses are often placed near the extractor to control the divergence of the beam. However, the m/q dependence of solenoid lenses can significantly complicate the ion optics.

5.4. *Beam neutralization*

The charges of a uniform-density ion beam with current J and radius r form a potential well with potential ϕ_0 on axis:

$$\phi_0 = J(4\pi\varepsilon_0 v)^{-1}\left(1 + 2\ln\left(\frac{R}{r}\right)\right), \tag{13}$$

where v is the ion velocity and the beam is centered in a beam pipe with radius R. A rapidly launched high-intensity ion beam initially blows up, as discussed under Subsec. 5.1. However, some of the ions will ionize the (residual) gas in the LEBT, producing low-energy electrons and low-energy ions inside the potential well. The electric fields in electrostatic LEBTs continuously drain most of these charged particles in the axial direction. However, in purely magnetic LEBTs the beam potential well will trap oppositely charged particles while repelling particles of the same charge. This neutralization mechanism gradually reduces the potential well. Neutralizations of 95–99% have been measured for protons [28]. The characteristic time for this process is called the neutralization time τ:

$$\tau = (\sigma_{\text{ioni}} n_n v)^{-1}, \tag{14}$$

where σ_{ioni} is the ionization cross section of the neutral gas with density n_n in the LEBT and v is the

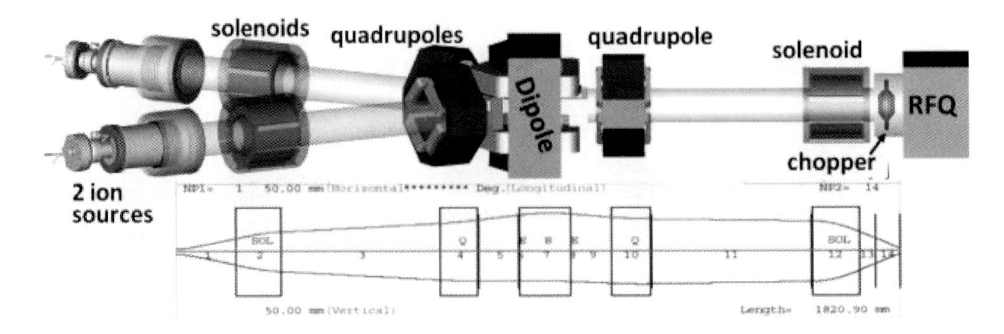

Fig. 11. A model (top) and a beam envelope calculation (bottom) of a magnetic LEBT with a Y-magnet.

ion velocity. During this time the ion trajectories adjust to the decreasing space charge and stabilize after about 2–3τ.

However, with the (at least partial) loss of the potential well, some charged particles can get lost, especially in the axial direction. These particles are either drawn back into the ion source or they drift into the RFQ, where the converging beam generates an attractive electric field. Repelling apertures are needed in many cases.

Simulations suggest that space charge compensation works well in large beams but not as well in small beams in the vicinity of beam waists [22]. For this reason, low-energy, high-intensity beams are normally transported as large beams, as shown in Figs. 6, 7, 10 and 11, and only focused as required by the next accelerator element.

Systems requiring an m/q analysis can be designed with quadrupoles, which can place the x and y focus in different locations.

5.5. *Chopping low-energy beams*

Low-energy choppers are essential for high-intensity accelerators which frequently switch the beam. Beam chopping protects switchyards from excessive activation. High-intensity beams rapidly acquire high power, and therefore most of the beam needs to be chopped and dumped in the low-energy beam transport section. Such choppers can also protect the accelerator from unstable beams that are generated during building up or turning off the plasma and the beam neutralization. Various chopping schemes have been developed, with transverse deflections being most common. Applying a voltage difference ΔV to two deflector plates of length L separated by a gap d deflects ions with energy qU by an angle α with

$$\tan \alpha = \frac{(\Delta V/U)(L/d)}{2}. \tag{15}$$

The length and gap of the deflector can be leveraged to reduce the required deflection voltage. The beam switching time cannot exceed the sum of the electronic switching time and the ion transit time. Systems designed to minimize the required deflection can achieve significantly faster beam switching times [29, 30].

However, a neutralizing cloud of low-energy charged particles does not act with the same speed as the fast-moving beam ions. Accordingly, fast

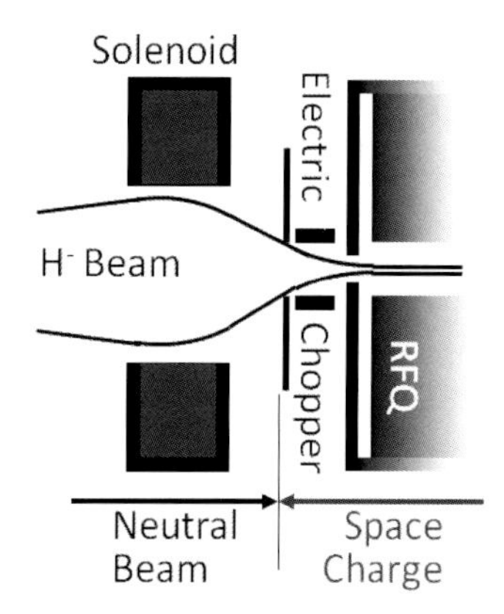

Fig. 12. A chopper at the entrance to the RFQ, with the neutralization controlled by a repelling aperture.

chopping can displace the neutralizing cloud against the switching beam and compromise the neutralization. When a fast time-of-flight chopper was placed between two LEBT solenoids, the emittance became distorted to a degree, where the transmission through the RFQ dropped by a factor of 2 [31].

Accordingly, the chopper should be placed next to the entrance of the RFQ, as shown in Fig. 12.

An aperture charged with the same polarity as the neutralizing cloud keeps the neutralizing cloud from drifting into the chopper. The potential well of the converging beam drains the locally produced neutralizing particles into the RFQ, thus preventing significant neutralization [22].

6. Ion Sources for Positive Light Ions

Studies with light ions go back to 1886, after Eugen Goldstein discovered the canal rays [32]. In 1956 Manfred von Ardenne described the duoplasmatron [33], an amazingly efficient ion source. Over 50 years later, the Large Hadron Collider (LHC) is being stacked with protons from a duoplasmatron source, which outputs over 200 mA of protons for almost a year with 99.8% availability. This amazingly long lifetime is achieved because the accelerator requires protons for less than 150 μs every 1.2 s, and in between the discharge is switched off [34].

However, duoplasmatron sources are not so successful on accelerators requiring continuous high-intensity beams, because sputtering of the cathode reduces the lifetime below use levels [35]. Significant lifetime improvements for high-duty-factor sources have been achieved by replacing direct current discharges with high-frequency discharges. One such breakthrough was achieved with microwave ion sources, which are now increasingly called 2.45 GHz ECR ion sources (ECRISs).

The first 2.45 GHz sources were developed for industrial applications about 30 years ago [36]. In the early 1990s, Chalk River National Laboratory developed a simple and robust 2.45 GHz ECRIS. Two solenoid coils placed at the two ends of the plasma chamber can produce a flexible magnetic field distribution that establishes an ECR zone, where electrons are resonantly heated. To maximize the beam intensity, various experiments explored the key parameters of the ion source. More than 95 mA of hydrogen ions were extracted through a single 5 mm diameter aperture with an RF power of only 500 W [37]. This prototype source continues to be duplicated with various modifications to be tested and used at many laboratories worldwide [38].

The source of light ions with high intensities (SILHI) was developed to reach a continuous H^+ beam of more than 135 mA with an energy of 95 keV and high reliability and reproducibility [38]. To meet the requirements, several modifications were successfully implemented; these are shown in Fig. 13.

To protect the vacuum window from the highly energetic electrons from the LEBT, the window was placed behind a 90° bend waveguide. The electromagnetic field at the entrance of the plasma chamber was increased with a ridged waveguide. The plasma density was increased by placing ceramic disks at both ends of the plasma chamber. The magnetic field was designed to place an ECR zone at both the entrance and the exit of the source, which improves the absorption efficiency of the microwaves. This enhances the plasma density and accordingly extracted beam intensity.

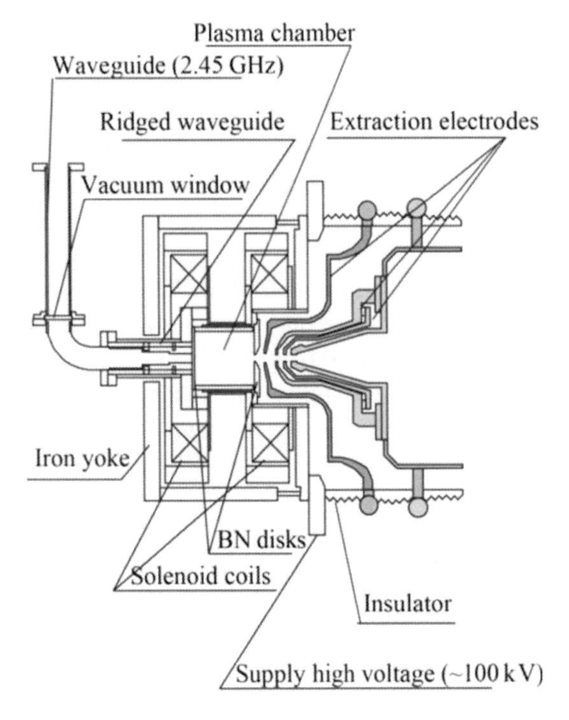

Fig. 13. Schematic drawing of a microwave ion source.

To reduce the space charge effects, various gases were leaked into the LEBT [39]. As seen in Fig. 14, the emittance decreases with increasing gas pressure and with increasing Z of the added gas.

When ^{84}Kr or Ar gas was injected into the LEBT, the beam emittance decreased by a factor of 3, and only 5% of the beam current was lost in the LEBT. The rms normalized emittance in the r–r' plane was less than $0.15\pi \cdot \text{mm} \cdot \text{mrad}$ for a 75 mA total beam current with a proton fraction of 88%.

This successful source continues to be employed and developed for other projects. The international fusion material facility project requires 140 mA of a continuous deuteron beam [38]. To meet this requirement, a magnetic configuration was adopted that allows online tuning during conditioning and operation. The extraction system was optimized with four electrodes. A pulsed, 125 mA deuteron beam was recently demonstrated at 100 keV with a 1% duty factor. The FAIR injector requires 100 mA with a pulse width of less than 10 μs and an energy of 9 keV. A slow chopper at the RFQ entrance is being developed to obtain short rise times [38].

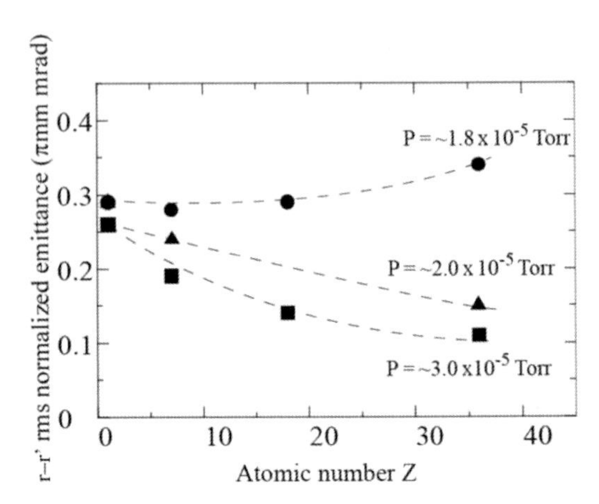

Fig. 14. r–r' emittance as a function of atomic number for several gas pressures [39].

The TRASCO project is a research and development program with the goal of designing an accelerator-driven system for nuclear waste transmutation [40]. The project requires 35 mA protons at 80 keV to be injected into the RFQ with a normalized rms emittance of less than $0.2\pi \cdot \mathrm{mm} \cdot \mathrm{mrad}$. A proton fraction above 70% is required to avoid large beam losses in the beam transport line. To meet these requirements, a new ion source with a 2.45 GHz microwave (TRIPS) [36, 40] was designed. To allow for a very flexible magnetic configuration, the two solenoids are independently powered and can be moved independently. Microwave coupling with a matching transformer and an automatic tuning unit enables operation with less than 5% reflected power and high electric fields on the axis. This increases the proton fraction and the current density up to $200\,\mathrm{mA/cm^2}$, which is near the Child–Langmuir limit. The source performance depends strongly on the solenoid coil parameters near the extraction, and the best performance is clearly obtained when the two ECR zones are located exactly on the boron nitride (BN) disks at either end of the plasma chamber. The BN disks inside the plasma chamber add cold electrons to the plasma, which increases the extracted beam currents. The effect of the chamber surface on the beam intensity was tested using TRIPS. A 5-mm-thick Al_2O_3 tube was inserted in the plasma chamber. The use of a thick alumina tube increased the current and proton fraction; it also yielded a lower beam ripple and improved the stability.

To generate intense light ion beams for a neutron imaging facility, a permanent magnet ECRIS with 2.45 GHz microwaves was constructed at Peking University. They produced 120 mA of H^+ and 83 mA of D^+ ions. The normalized rms emittance was less than $0.2\pi \cdot \mathrm{mm} \cdot \mathrm{mrad}$ [41].

7. Ion Sources for Negative Light Ions

7.1. *Introduction*

Negative ions have the unique feature that their charge can be dramatically changed. Forming high-intensity beams of negative ions and then stripping the loosely bound, extra electron produces powerful neutral beams. These beams can be injected through the magnetic confinement walls of power producing fusion plasma, yielding the phenomenon known as neutral beam heating. Forming high-intensity beams of negative ions and then stripping two or more electrons changes the polarity of the ion beam. This allows the positive high-voltage terminal in tandem accelerators to first attract the negative ions, then strip them, and repel out the resulting positive ions, multiplying their energy. A change in polarity also allows changing the direction of the curvature of the ion trajectories in magnetic dipole fields, enabling the stacking of ion beams in accumulator rings, or simplifying the extraction from cyclotrons.

The extra electron is bound to the atom by the electron affinity, which ranges from 0.08 eV for Ti^- to 3.6 eV for Cl^-, and is 0.75 eV for H^-. Negative ions are very fragile, because those energies are much smaller than the roughly 15 eV ionization energies for atoms and molecules.

Electrons with energies of roughly 40 eV are needed to efficiently generate hydrogen plasma. For such electrons and hydrogen atoms, the radiative electron attachment cross section is only $3 \cdot 10^{-19}\,\mathrm{cm^2}$, compared with a $3 \cdot 10^{-15}\,\mathrm{cm^2}$ electron detachment cross section for negative hydrogen ions [42]. Not many negative ions can be expected to survive in plasma that is so skewed toward their destruction.

7.2. *Volume-produced H^- ion beams*

In 1977, large signals from negative ions were unexpectedly discovered in hydrogen plasma [43]. Many years of research have revealed that the majority of those ions are produced by highly rovibrationally excited molecules ($4 \leq \nu \leq 9$) which disintegrate when colliding with slow electrons (~1 eV) [44].

Hot electrons (>20 eV) excite H_2 molecules easily ($5 \cdot 10^{-18}\,\mathrm{cm^2}$), but such fast electrons (>5 eV) destroy resulting H^- ions very rapidly ($4 \cdot 10^{-15}\,\mathrm{cm^2}$) [5]. While the H^- lifetime is extremely short in a hot plasma, they survive much longer in cold plasma due to their 0.75 eV ionization threshold.

The problem of short H^- lifetimes was overcome with tandem sources [45]. As shown in Fig. 15, they contain a magnetic filter between the hot plasma, where the molecules get excited, and the much colder plasma near the outlet, where the H^- ions survive long enough to be extracted.

Several "volume sources" have been developed for the production of intense H^- and D^- beams. They all feature plasma confinement using

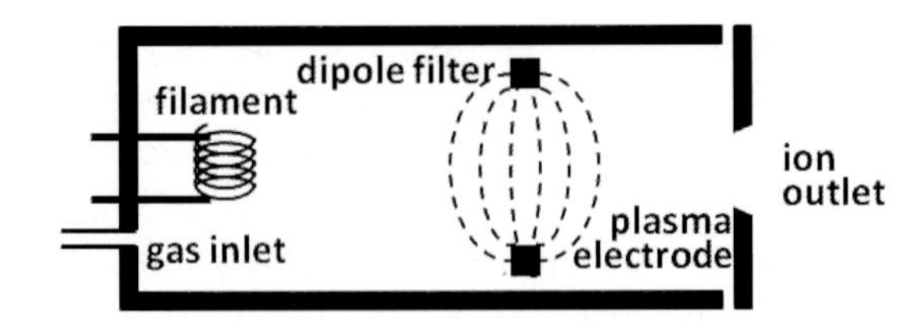

Fig. 15. A schematic tandem source with a magnetic filter.

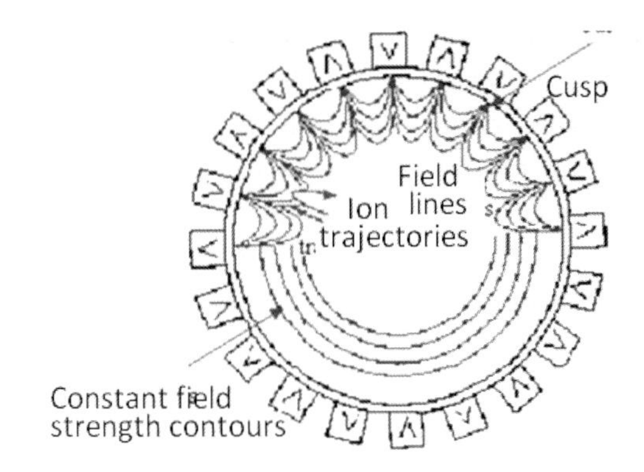

Fig. 16. Twenty alternating bar magnets forming a multicusp magnetic wall.

permanent magnets arranged in alternating direction on the outside of the plasma chamber, as shown in Fig. 16 [46]. This reduces the plasma losses to the loss lines around the center of the magnets. The strength of the cusp field decreases with the distance from the wall. This yields a minimum field in the center of the chamber, forming a magnetic bucket. Low magnetic fields yield a quiescent plasma and accordingly more stable ion beams.

The first generation of "volume" H^- sources were all driven by filament discharges and delivered up to 4.2 mA with an ~2 mm diameter outlet [47]. The most successful filament-driven volume H^- source was developed by TRIUMF [48]. It delivers up to 15 mA in a continuous fashion for at least two weeks, after which the filament needs to be replaced. It is commercially available [49].

In 1990, to overcome the limitation of the filament lifetime, a filament was replaced with a three-turn inductive coil. When powered with 2 MHz RF, the H^- beam current increased [47], but not necessarily the lifetime [50]. However, antenna research, development [51], and quality assurance efforts [52] have largely resolved the problem. The SNS operates with improved antennas for up to six weeks at high RF power and with a 5.3% duty factor. There are no old-age failures, only the occasional infant mortality [53].

As an alternative, DESY developed an external antenna source using an alumina plasma chamber. This very successful source delivered up to 40 mA H^- with a 0.1% duty factor and no apparent lifetime limitations [54]. Efforts to develop an external antenna source for a 6% duty factor and high RF power have so far failed, due to the limited thermal conductivity of alumina and the limited vacuum compatibility of the higher-conductivity aluminum nitrate [55].

There were multiple efforts to use microwave sources for the production of H^- beams, none of which ever produced more than a few mA of H^- beam. However, recently, Peking University reported 15 mA of H^- beam with 180 W 2.45 GHz and 40 kV extraction [56]. A tungsten grid was installed in the plasma chamber to prevent the microwave from penetrating into the negative ion production area.

7.3. *Surface-produced H^- ion beams*

Interest in Cs ions for space propulsion led to the 1962 discovery that Cs increases the rate of sputtered negative ions by an order of magnitude [57].

Years of research have shown that Cs, the largest stable atom, can significantly reduce the work function when being desorbed on certain metal surfaces [58]. As shown in Fig. 17, for example the work function of a Mo substrate is 4.6 eV, which drops to ~1.6 eV when it is being covered with a little more than half a monolayer of Cs. The lower work function greatly increases the probability that atoms capture an extra electron when they leave the surface with sufficient speed [59].

In the early 1970s, long before this mechanism was understood, a Novosibirsk team added Cs to

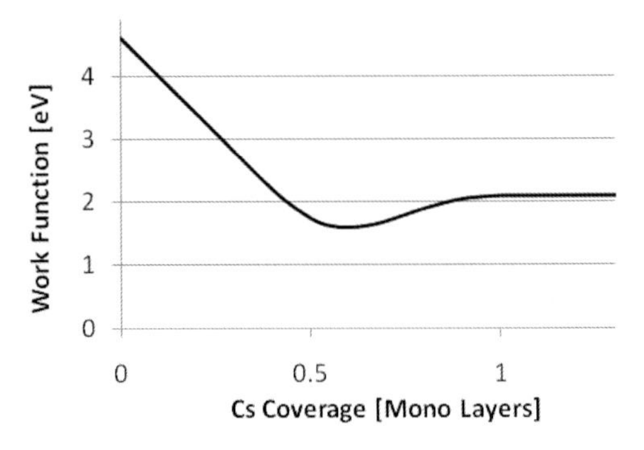

Fig. 17. The Mo work function versus a fractional Cs layer.

their magnetron source and obtained 150 mA of H^- beam [60]. Magnetrons have been adopted by DESY, FNAL and BNL to produce up to ~80 mA with lifetimes exceeding six months. However, magnetrons cannot support duty factors in excess of 0.5% without losing the optimal Cs coverage of the cathode.

The Novosibirsk team also developed a Penning source which delivered up to 150 mA continuously [61]. A similar result was obtained by LANL [62]. RAL ISIS uses a Penning source to produce 35 mA at a 1% duty factor to yield a 180 kW beam for neutron production with a lifetime of ~4 weeks. ISIS is developing a 5%-duty-factor, 60 mA Penning source [63].

These surface plasma sources (SPSs) are very compact and efficient. Sputtering of the cathode gradually changes the beam and eventually limits their lifetime.

LANL uses a large Mo converter illuminated by a filament-driven plasma to produce ~18 mA H^- beam with a 5% duty factor, consuming ~20 g of Cs during its ~4-week service cycle [64].

7.4. *Surface-enhanced volume H^- sources*

Volume-produced H^- ion beams can be augmented with surface-produced H^- ions by surrounding a roughly 1-cm-diameter outlet with a conical converter, as shown in Fig. 18.

The SNS H^- source, an RF-driven, Cs-enhanced volume-type multicusp H^- source, injects up to 65 mA, with 50–55 mA being more typical, into the RFQ. After proper conditioning, Cs_2CrO_4 cartridges are heated to release a few mg of Cs. The beam normally grows a few mA over the first few days, after which it becomes persistent without adding any Cs.

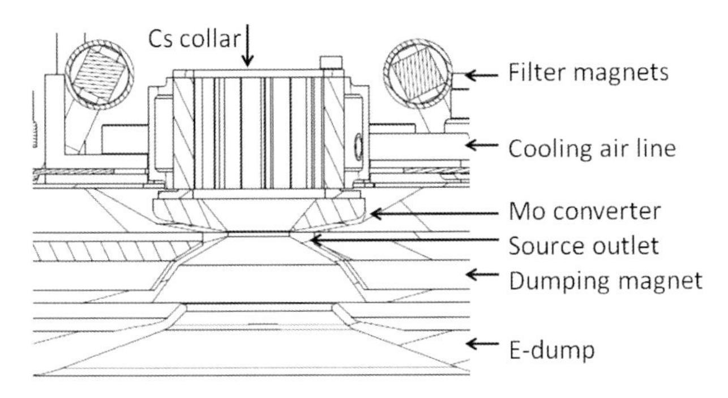

Fig. 18. The SNS source outlet is surrounded by a thermally isolated Mo converter.

Experience suggests that the persistence is due to a lack of any heavy impurities in the hydrogen plasma that could sputter the Cs. Depending on the performance, the source is left in operation for up to six weeks. The beam does not decay unless the e-dump voltage is lowered to improve its stability [53].

A similar conical converter is used in the LaB_6 filament-driven, volume-type, multicusp H^- source at J-PARC. It produces up to 38 mA with 50 kW RF. However, it is normally operated with only 17 mA and a 1.25% duty factor, where it easily meets the 50-day lifetime requirement [65].

7.5. *Which H^- ion source?*

Currently, there is no prototype H^- source with the capability of producing very intense H^- beams of arbitrary duty factor. However, H^- sources have been developed that can very successfully deliver very intense beams for low duty factors, intense beams for moderate duty factors, and moderate currents in continuous operation, all with satisfactory lifetimes.

8. Ion Sources for Heavy Ions

In the past, when cyclotrons were operated with internal Penning ion sources, those sources were developed to produce continuous beams of single- and multiple-charged ions of heavy atoms, especially metals. However, the decreasing number of cyclotrons with internal ion sources and the increasing number of pulsed accelerators have directed recent attention toward vacuum arc ion sources. These are now commonly used as external ion sources for particle accelerators and also for various other applications, such as ion implantation and surface modification. More than 50 ion species have been produced from solid metals with beam currents of up to several amperes [6]. The pulse width of the beam is up to some milliseconds, and the duty cycle is as high as 10%. Multiply charged ions can be produced.

The mechanisms for ion production in a vacuum arc plasma are reported in detail in several review papers [6, 66, 67]. The ion source consists of two components: one for plasma production with a vacuum arc and the other for ion extraction. The cathode is a simple cylindrical metal rod. A high-voltage pulse is applied to a trigger electrode that encloses the cathode, and a thin alumina insulator is installed between the cathode and the trigger electrode. The vacuum arc is produced between the trigger electrode and the cathode. The plasma created at the cathode goes through the anode hole and then through a drift space to the extractor grids. Three grids are typically used for expanding the dense plasma. A negative high voltage is applied to the middle grid to suppress the backflow of low-energy electrons. The longitudinal magnetic field in the arc region can be used to transport the plasma from the cathode to the extractor. The current density is on the order of 10^6–10^8 A/cm^2. The current for one spot is typically on the order of a few amperes to approximately a few tens of amperes. Therefore, several tens of spots may be created when the discharge current is several hundreds of amperes. The plasma parameters of the arc discharge are strongly affected by the residual gas, especially the charge distribution of the produced ions. The arc current and voltage of the MEVVA ion source are typically several hundreds of amperes and 20–30 V, respectively. The plasma ion current is proportional to the arc current and is ~10% of the arc current over a wide range of conditions. Depending on the operational arc parameters, the structure of the extraction system and the conditions of the ion source, the typical beam current might be several hundreds of mA.

MEVVA ion sources can produce multiply charged heavy ions. The mean charge state of the produced ions is 1–3 and depends on the metallic species. Figure 19 shows the mean charge state of heavy ions and the charge state distributions for aluminum and tungsten ion beams produced from a MEVVA ion source [4]. The mean charge state increases slightly with increasing atomic number and depends strongly on the material (lower-boiling-point metals yield a lower mean charge state). An empirical expression that provides a reasonable prediction for the mean charge state Q_p was reported as

$$Q_p = 0.38\left(\frac{T_{\mathrm{BP}}}{1000}\right) + 1, \tag{16}$$

where T_{BP} is the boiling point of the metal [68].

To increase the charge state, it is essential to increase the electron temperature in the plasma. GSI investigated the effect of the magnetic field on the charge state distribution and observed enhanced beam intensity of multicharged heavy ions. Figure 20

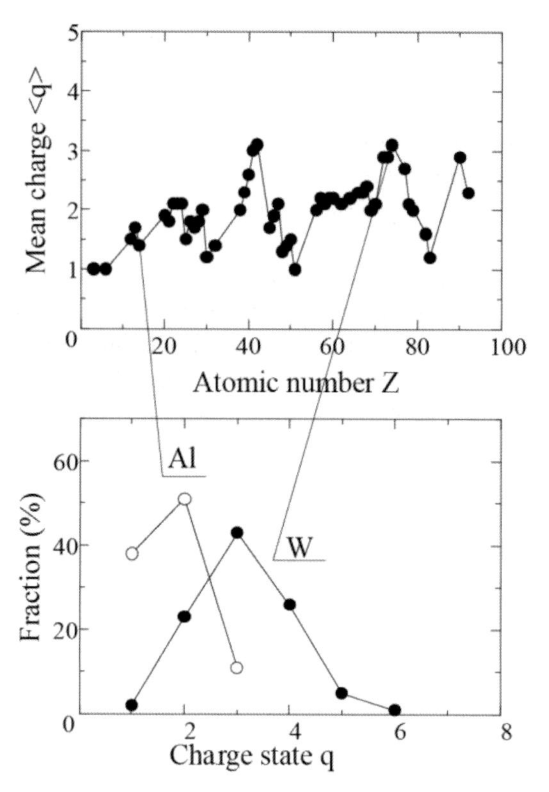

Fig. 19. The mean charge state of heavy ions and charge distribution for Al and W ion beam.

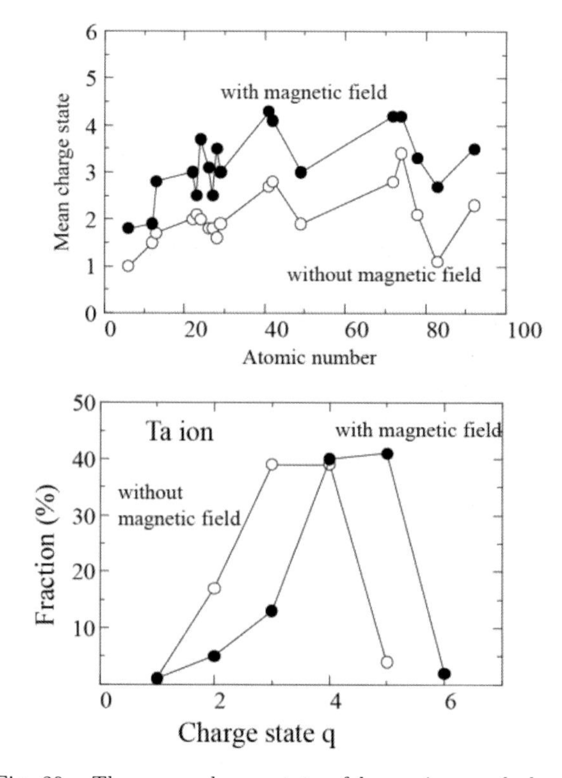

Fig. 20. The mean charge state of heavy ions and charge distribution of Ta ions with and without a magnetic field.

shows the mean charge state of heavy ions with and without a magnetic field [69]. The enhancement factors of the mean charge state were between 1.2 and 2.45, depending on the magnetic field strength and cathode materials.

A numerical simulation shows that a small increase in the electron temperature of the cathode spot plasma (1–2 eV) significantly increases the multiply charged ion fraction in vacuum arc ion sources [70]. The electron temperature in the arc plasma is governed by the discharge power. Therefore, it is crucial to increase the discharge voltage. Following this assumption, several methods have been attempted [71–73]. An alternate method is to use an additional ionization. An intense electron beam was injected into the vacuum arc plasma, and it was observed that the mean charge state significantly increased for lead ion beams [74]. A combination of ECR heating and a vacuum arc plasma was also applied to boost the mean charge state at a high electron temperature [75].

The beam stability and pulse-to-pulse reproducibility are very important for particle accelerator applications. However, the ion beam from a MEVVA ion source is relatively noisy, and the reproducibility is not very high. Electrostatic grids can be used to separate the ion flow from the electrons in the plasma. In this case, the ion flow in the gap was controlled by space charge effects, as it would be with a thermionic ion source. A constant extracted current is observed even with large variations in the source flux. Following this success, solenoid coils were added to the grid to reduce the beam noise. Reliable, long-lifetime arc triggering is also a key issue for accelerator applications, and good progress has been made in this direction. To increase the lifetime of arc triggering, several methods have been proposed. One is the "triggerless" arc initiation, which increases the triggering lifetime in excess of 10^6 pulses [76]. The MEVVA performance was improved at GSI in the past few decades. The methods described above (mesh grids, additional magnetic field, optimization of the beam extraction conditions) were used to produce acceptable conditions [77, 78]. The typical normalized rms emittance of the heavy ions was $\sim 0.2\pi$ mm · mrad. Consequently, the vacuum arc ion source is now routinely used for high-current metal ion injection into the heavy ion accelerators at GSI.

9. Ion Sources for Highly Charged Ions

9.1. *Physics of producing highly charged ions*

While neutral atoms have ionization energies in the order of 10 eV, removing additional electrons from the resulting ions requires increasingly high ionization energies, especially when electrons have to be removed from inner shells such as the L- or K-shell. As Fig. 21 shows, removing many electrons from the outer shells of heavy ions can be accomplished with electrons having hundreds of eV, while stripping the inner shells will take electrons with many keV, and over 100 keV for the K-electrons of the heaviest elements. Such energetic electrons cannot be produced with ordinary plasmas, but require special heating.

The Lotz approximation [8] estimates the direct electron impact ionization cross section $\sigma_{(k\to k+1)}$ for ionizing an ion X^{k+} with atomic number Z,

$$\sigma_{(k\to k+1)} = 4.5\cdot 10^{-14}\sum_{[i=k,Z]}(\ln(E_e/I_i))/(E_e\cdot I_i)[\mathrm{cm}^2], \tag{17}$$

which is the sum of the cross sections for all remaining electrons, where E_e is the electron energy and I_i is the ionization energy of the ith electron.

As seen in Fig. 21, the ionization energies vary over many orders of magnitude, which causes the ionization cross sections to vary over even more orders of magnitude. Figure 22 shows the specific case of Ar [79]. Multiple-charged Ar^{5+} can be produced with $\sim$300 eV electrons with a cross section of 10^{-17} cm^2. Ar^{16+} will take 3 keV electrons for $\sim 10^{-20}$ cm^2, while bare Ar^{18+} requires $\sim$10 keV electrons for a cross section of 10^{-21} cm^2.

The cross section is linked to the time it takes to remove the kth electron:

$$\tau_{(k\to k+1)} = \frac{e}{j\cdot\sigma_{(k\to k+1)}}, \tag{18}$$

where e is the electron charge and j is the electron current density. Multiple ionization of highly charged ions is normally negligible, and therefore the time summation has to be included to reflect the stepwise ionization to produce ions with charge state k, X^{k+}.

$$\tau_{(0\to k)} = \sum_{[i=0,k-1]}\left(\frac{e}{j\cdot\sigma_{(i\to i+1)}}\right). \tag{19}$$

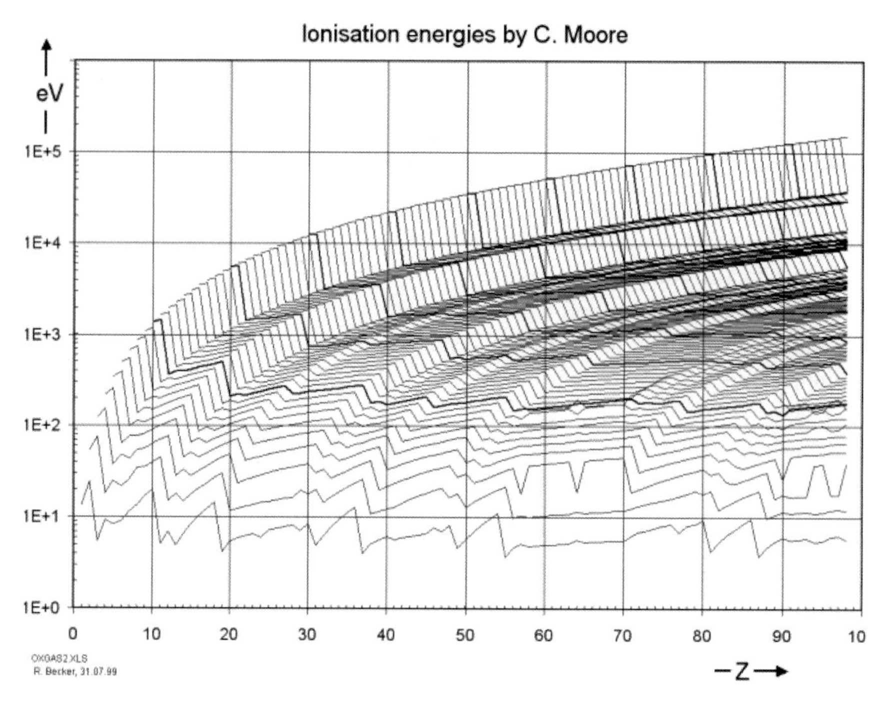

Fig. 21. Ionization energies for the successive removal of electrons versus the atomic number Z. (Courtesy of R. Becker, University of Frankfurt.)

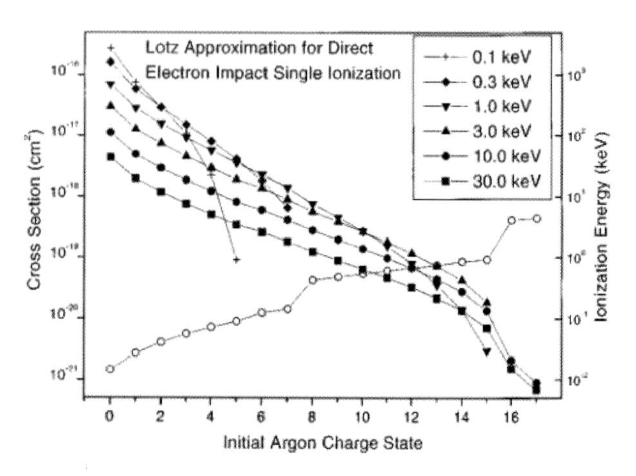

Fig. 22. Ionization energies (open symbols) and ionization cross section for all Ar charge states.

As the cross section for ionization decreases rapidly with increasing charge state, losses and recombination processes start to play a more and more significant role. This increases the duration of time necessary for reaching high charge states, and eventually leads to equilibrium charge state distributions.

As the electron current density j can vary, it is common to quote the product $j \cdot \tau_k$ required to produce ions X^{k+} with charge state k:

$$j \cdot \tau_k = \sum_{[i=0,k-1]} \left(\frac{e}{\sigma_{(i \to i+1)}} \right). \tag{20}$$

For hydrogen-like ions $X^{(Z-1)+}$, and optimum electron energies, this can be approximated with

$$j \cdot \tau_{Z-1} \approx \left(\frac{Z}{5} \right)^4, \tag{21}$$

which shows the true challenge of producing few-electron, heavy ions [4].

9.2. *Electron beam ion sources*

The electron beam ion source (EBIS), shown in Fig. 23, was invented by Donets in the 1960s, and used as an external ion source for accelerators in the 1970s [80]. In the past three decades, more than 30 EBISs have been constructed for various applications owing to their excellent performance, as described in past review papers [80–84].

The ion capacity of the trap of the ion source is determined as follows:

$$Q = 3.36 \cdot 10^{11} JLE_e^{-1/2}, \tag{22}$$

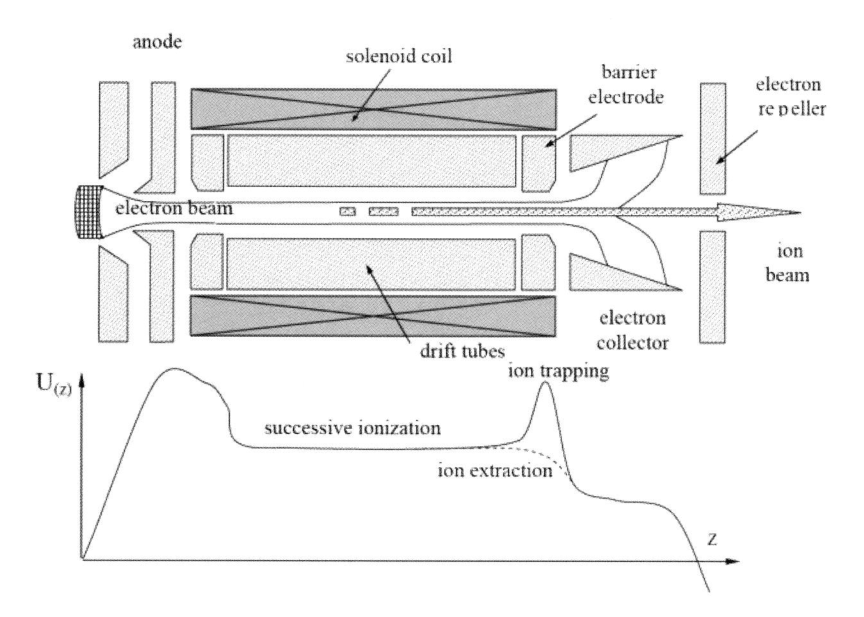

Fig. 23. Schematic drawing of an EBIS.

where J and E_e are the electron beam current and energy, and L is the trap length. After a certain confinement time, when the desired charge state becomes intense, the barrier electrode shown in Fig. 23 is lowered so that the highly charged ions can exit the trap. The unique feature is that the total extracted charge per pulse is almost independent of the ion species or charge state. Furthermore, the beam pulse width can be controlled by manipulating the extraction barrier voltage; therefore, short pulses ($\sim$10 μs) of high current (several mA) can be obtained. Thus, the EBIS is an ion source suitable for single-turn synchrotron injection.

To improve the efficiency, Donets replaced in the electron collector with an electron reflector and succeeded with operating the modified KRION-II as an electron string ion source (ESIS), delivering N^{7+}, Ar^{16+} and Fe^{24+} to the JINR Nuclotron for experiments [85].

Donets continues to develop more efficient and higher-yield EBISs. Using a tubular beam geometry (TESIS) allows a much larger trap capacity and accordingly higher ion currents [86].

The most advanced EBISs use refrigerator-cooled SC magnets, which produce high magnetic fields (typically several teslas) and several-ampere electron beams. These achieve extremely high electron current densities at high electron beam energies and produce the desired charge states of heavy ions. The advanced RHIC-EBIS [87] was designed to produce mA currents of any ion species with a pulse width of $\sim$10 μs, allowing single-turn injection for both the RHIC and the National Aeronautics and Space Administration Space Radiation Laboratory (NSRL). Because it may be necessary to supply heavy ion beams to multiple users simultaneously, it has to be able to switch the ion species within 1 s. The species from the EBIS can be changed on a pulse-to-pulse basis by changing the 1^+ ion injected into the EBIS trap from external ion sources. For example, it can produce more than 1.7 mA of Au^{32+} at a 10 μs pulse width and a 5 Hz repetition rate for the RHIC [88]. The NSRL requires He^{2+}, Si^{13+}, Fe^{20+} and others at 2–3 mA with a pulse width of 10 μs. An electron current of 10 A and a trap length of 1.5 m are sufficient to produce the required total extraction ion charge of 5×10^{11}. An extraction energy of 17 keV/u was chosen to minimize the space charge effect in the LEBT. The pulse-to-pulse fluctuation of the beam current is lower than 1%. The EBIS has been operated with alternating Au^{32+} and Fe^{20+} beam pulses at a 0.5 Hz repetition rate.

As an electron beam device, the EBIS can also generate RF from the energy of the electron beam. This occurs very likely with high beam power and when the distance between the beam and the surrounding structure is small. The latter can be a design issue of using too-small drift tubes or the result of radial or magnetic misalignments. In the past that happened in the Berkeley EBIS [89] as well as in other installations. This interpretation is consistent with the RHIC-EBIS using drift tubes with a large inner diameter and operating with 20 kW electron beams without observing such instabilities.

9.3. *Laser ion sources*

Laser ion sources (LISs) were proposed in 1969 for the production of highly charged ions. The first LIS was used in 1977 as an injector for the JINR synchrotron in Dubna [90]. Since 2000 a LIS has been in routine operation for the TWAC facility at the Institute for Theoretical and Experimental Physics in Moscow [91].

The main energy transfer mechanism from the laser to the plasma is inverse bremsstrahlung for laser power densities of up to 10^{13} W/cm^2 [92]. The transfer efficiency depends strongly on the electron–ion collision frequency, which is governed by the critical density of the plasma, the electron temperature and the atomic number of the material. A detailed analysis of the absorption process indicated that the efficiency decreases with increasing laser power and wavelength. Accordingly, lasers with power densities of 10^{10}–10^{13} W/cm^2, wavelengths of >1000 nm and pulse widths of 1–100 ns are normally used for LIS designs. In this case, the experimental results and theoretical calculations show that the absorption efficiency ranges from 70 to 90% [92]. Also in the laser plasma, successive ionization by electron impact is the dominant process. For this reason, the electron temperature, density and exposure time of ions in the plasma are the key parameters for producing highly charged, heavy ions. The electron temperature increases with increasing laser power and decreases with increasing critical plasma density (wavelength of the laser).

In the early stage of free expansion of the laser plasma, recombination processes are the dominant mechanism for reducing the mean charge state of heavy ions in the plasma. This is written as

$$R_{3B} \cdot \propto \frac{Z_{\text{ion}}^3 N_e}{T_e^{9/2}}, \tag{23}$$

where Z_{ion}, N_e and T_e are the atomic number of the ions and the density and temperature of the electrons, respectively. To maximize the intensity of the highly charged ions, they should be made into a high-temperature, low-density plasma [93].

For injecting intense beams into the synchrotron, a 100 J CO_2 laser (1 Hz repetition) has been used in the most advanced LIS [94]. With this laser, intense pulsed Pb^{27+} ion beams with a few-mA peak current have been successfully produced.

The laser ablation plasma has a very high density and an initial expanding velocity. Therefore, the intense ion beam can be transported in (neutralized) plasma condition to the first stage of the accelerator. Taking into account the advantages of the laser plasma, the direct plasma injection scheme (DPIS) was proposed [95]. The obtained peak current on the extraction side of the RFQ was more than 9 mA from a carbon graphite target using a 4 J CO_2 laser already in the early stage of test experiments. After the test experiments, a new RFQ linac was fabricated to accelerate high-intensity heavy-ion beams (~100 mA). A 400 mJ Nd-YAG laser was tested to produce a fully stripped carbon beam, and the accelerated peak current reached 17 mA [96]. In 2005, intense C beams (>60 mA) were accelerated using a 4 J CO_2 laser [97]. They also obtained 70 mA of Al ions with a 2.3 J commercial Nd-YAG laser. The stability of the beam intensity and pulse width was demonstrated in 2006 [98]. The reported fluctuations of the RFQ output current and pulse duration were ±6% and ±11%, respectively. Although the peak current is sufficiently high, the pulse width of the beam is sometimes too short for synchrotron injection. The pulse width (τ) and beam intensity (J) are described by $\tau \propto L$ and $J \propto L^{-3}$ [98], where L is the distance from the target to the extraction system. It is easy to increase the pulse width by increasing the drift distance between the ion source and the RFQ. However, the injected current to the RFQ becomes very small because the intensity is proportional to L^{-3}, as shown in the formulas. Recently, to minimize the reduction of the current, a solenoid magnet was successfully used to focus the beam [99]. As an example of the applications of this method, a research and development program was initiated using the DPIS as an injector for compact carbon ion cancer therapy and for intense heavy-ion beam injection for the Cooler-Storage Ring of the Heavy Ion Research Facility in Lanzhou [100].

9.4. *Electron cyclotron resonance ion sources*

In the past three decades, the beam intensities of highly charged heavy ions from high-performance electron cyclotron resonance ion sources (ECRISs) have increased dramatically. For example, the beam intensity of Ar^{8+} increased from several $10\,\mu A$ to $2\,mA$ [101]. The improvement is due mainly to a better understanding of the ECR plasma and to the use of modern technology, such as improved superconducting and permanent magnet technologies.

The ECRIS uses very complex, high-field magnetic confinement and high-power, short-wavelength microwaves to heat the electrons to the energies required for the production of highly charged ions, as discussed in Subsec. 9.1. Accordingly, there is no longer a current density j of monoenergetic electrons, but instead a cloud of electrons with a distribution of energies and directions. Nonetheless, the ECR plasma is still dominated by stepwise electron impact ionization, but the rate R_k of producing ions with charge state k, X^{k+}, is

$$R_k = n_e \langle \sigma_{(k-1 \to k)} v_e \rangle n_{k-1}, \qquad (24)$$

where n_e is the electron density and n_{k-1} the density of ions with charge state $k-1$. The rate coefficient is the product of the ionization cross section and the electron velocity v_e averaged over the electron energy distribution. Again, charge exchange and losses lead to an equilibrium charge state distribution, which can be tuned with the magnetic confinement.

To reach high charge states, the magnetic confinement, characterized by the ion confinement time τ_k, needs to be very good. Instead of the jt values used for EBISs, ECR sources are characterized by the product $n_e\tau_k$. For example, to produce Xe^{20+} requires an $n_e\tau_k$ of $\sim 10^9\,cm^{-3}s$, while Xe^{27+} requires an $n_e\tau_k$ of $\sim 10^{10}\,cm^{-3}s$, which is 10 times higher [102].

The beam current I_k for ions with charge state k from an ECR in CW mode can be written as

$$I_k = \left(\frac{n_k k V}{\tau_k}\right) f_{\text{ext}}, \qquad (25)$$

where V is the plasma volume and f_{ext} the efficiency of the plasma flow from the main plasma to the extraction hole. To maximize the beam intensity, τ_k needs to be minimized, while n_k and f_{ext} are maximized. Clearly, creating these conditions relies on an understanding of how the main components of the ion source affect these parameters. The improved performance of ion sources can often be attributed to a better understanding of their physics.

The criterion for manipulating the confinement time is governed by magnetic mirror confinement techniques. The magnetic field strength on both the beam extraction side (B_{ext}) and the injection side (B_{inj}), as well as the radial magnetic field strength on the inner surface of the plasma chamber (B_r), strongly affect the charge distribution and beam intensity. The beam intensity increases with an increase in $B_{\text{inj}}, B_{\text{ext}}$ and B_r, and then becomes saturated above certain values ($B_{\text{inj}} > 4B_{\text{ECR}}$, $B_r > 2B_{\text{ECR}}$ and $B_r > B_{\text{ext}} > 2B_{\text{ECR}}$, for high B-mode operation) [103]. Additionally, it is obvious that the minimum strength of the mirror magnetic field (B_{min}) affects the beam intensity because the magnetic field configuration determines the plasma confinement and the efficiency of electron heating in the resonance zone [104, 105].

Simulation codes based on the Fokker–Planck equation have developed rapidly in the past decade. Since 2000, the code proposed by Girard has provided information about the effects of components of the ion source (the RF power, mirror ratio and gas pressure) on both the key parameters (n_e, T_e and τ_c) of the ECR plasma and the beam intensity [106]. The simulation code also shows a frequency effect, implying that higher frequencies enable a longer electron confinement time in the plasma. Such results were seen with SERSE (14 and 18 GHz) [107] and SECRAL (18 and 24 GHz) [108]. It is obvious that there are limitations on the beam intensity at high RF powers. The effect of the RF power on the plasma parameters and high-RF-power instability were demonstrated using FAR-TECH's generalized ECRIS model [109]. The origin of the instability was pitch angle scattering of electrons. They showed that the instability threshold of the RF power increased with an increase in gas pressure. Despite this important information, few experimental results for beam intensity saturation are currently available. The beam intensity of a high-performance SC-ECRIS that has a larger plasma chamber volume (larger than several liters) increases linearly and does not saturate at high power [108]. Further investigations under various conditions are needed for clarification.

All the results presented so far have been described as either dimensionless or one-dimensional in real space. In fact, it is well known that the geometry of the ion source strongly affects the ECR plasma and beam intensity. Several studies have shown the effect of the size of the ECR zone on the beam intensity and the effectiveness of ECR heating in the past decade [110–112]. Two new ECRISs, the RIKEN SC-ECRIS at RIKEN [113] and SuSI at MSU [114], can produce flexible magnetic field distributions. In these ion sources, the magnetic field gradient and ECR zone size can be changed independently. So far it has been shown that a gentler field gradient and a larger zone yield higher beam intensities [111, 112, 115].

Changes in the ECRIS performance have recently been obtained by slightly varying the microwave feed frequency (frequency tuning) [116], which produces strong fluctuations in the beam intensities even for frequency variations in the MHz range. Experiments suggest that frequency tuning affects the heating process as well as the efficiency of beam extraction (f_{ext}).

The fully SC-ECRIS has been one of the main developments in producing intense, highly charged ion beams in the past few years. The design, construction and development of high-magnetic-field SC-ECRISs such as SERSE, VENUS [117] and SECRAL have addressed many of the technological challenges and opened new avenues for further development of new ECRISs.

VENUS was the first high-magnetic-field SC-ECRIS developed for operation at 28 GHz and has broken numerous records for the beam currents of highly charged ions [118]. Recently they reported a record-breaking 440 eμA of U^{33+} with 8 kW (18 + 28 GHz) using a high-temperature oven [119, 120].

VENUS pioneered a number of new solutions, such as a new clamping technique for the hexapole magnet to increase the radial magnetic field. Such new solutions are being incorporated into the design of the new SC-ECRISs.

SECRAL is a compact SC-ECRIS designed to operate at microwave frequencies of 18–28 GHz. The unique feature of the SECRAL source is its unconventional magnetic structure, in which SC solenoid coils are placed inside the SC sextupole. An advantage of this structure is that the magnet assembly can be compact in size compared to similar high-magnetic-field ECRISs with conventional magnetic structures and yield higher RF power densities. SECRAL produced 50 eμA of Bi^{40+} at an RF power of $<$4 kW (24 + 18 GHz) [121].

The RIKEN SC-ECRIS can be operated with flexible axial field distributions from the classical B_{min} to the flat B_{min} [111]. It produced $\sim$183 eμA of U^{35+} and $\sim$230 eμA of U^{33+} with the sputtering method at an injected RF power of $\sim$4 kW (28 GHz) in 2012. Figure 24 shows a schematic drawing of the RIKEN SC-ECRIS with its 28 GHz gyrotron.

There is a consensus [122, 123] that the most critical technology for developing an ion source is a high-field SC magnet system capable of confining the plasma at high frequencies ($>$18 GHz). The maximum field produced in SC magnets is generally limited by quenching. To avoid quenching, the magnet design must keep the current densities and

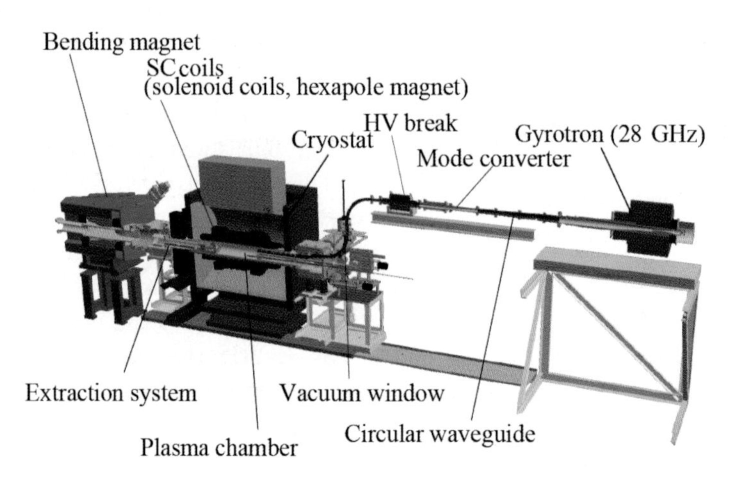

Fig. 24. Schematic drawing of the RIKEN SC-ECRIS with a 28 GHz gyrotron.

local magnetic fields at the coils below the short-sample critical current in the SC wire. All the current ECRISs use NbTi. However, the performance of NbTi is limited by its upper critical field of about 10 T at 4.2 K. The maximum magnetic field for an SC-ECRIS at the coils must be ~7 T to produce the designed magnetic field. In this case, the operational current (I_{op}) is ~80% of the critical current (I_c), which is almost the limiting value for long-term operation. Therefore, to employ a frequency higher than 28 GHz, a new magnet structure using advanced SC magnet technology may be needed. For example, at an operating frequency of ~50 GHz, reasonable estimates for the axial and radial fields are ~7 and ~4 T, respectively. When one is using higher microwave frequencies (>18 GHz), the large heat load of highly energetic X-rays emitted from the plasma becomes another challenge that needs to be met.

In the past few years, SC-ECRISs operating at high microwave frequencies (>18 GHz) have been considered for practical use in various fields, and the beam intensity of highly charged heavy ions continues to increase owing to optimization of the ion source performances. However, large beam current increases will require a new generation of ECR sources like the ones that have been proposed [122, 123].

Acknowledgments

A part of this manuscript was prepared at Oak Ridge National Laboratory, which is managed by UT-Battelle, LLC, under contract DE-AC05-00OR22725 for the US Department of Energy. The authors appreciate the very helpful suggestions from E. D. Donets, R. Becker and the referees, and the high-quality proofreading by S. Cousineau.

References

[1] J. Wei *et al.*, HIAT 2012, MOB01 (Chicago, June 2012); http://www.JACoW.org
[2] Y. Yano, *Nucl. Instrum. Methods B* **261**, 1009 (2007).
[3] D. Olsen, APAC2004 (Gyeongyi, Mar. 2004), p. 248; http://www.JACoW.org
[4] *Handbook of Ion Sources*, ed. B. Wolf (CRC, New York, 1995).
[5] H.-S. Zhang, *Ion Sources* (Springer, 1999).
[6] *The Physics and Technology of Ion Sources*, ed. I. G. Brown (Wiley-VCH, Weinheim, 2004).
[7] C. Jiang *et al.*, *Nucl. Instrum. Methods A* **492**, 57 (2002).
[8] W. Lotz, *Z. Phys.* **216**, 241 (1968).
[9] P. Debye and W. Huckel, *Phys. Z.* **24**, 183, 305 (1923).
[10] C. D. Child, *Phys. Rev.* (*Ser. 1*) **32**, 492 (1911).
[11] I. Langmuir and K. T. Compton, *Rev. Mod. Phys.* **3**, 251 (1931).
[12] R. Keller, *Nucl. Instrum. Methods A* **298**, 247 (1990).
[13] J. R. Coupland, R. S. Green, D. P. Hammond and A. C. Riviere, *Rev. Sci. Instrum.* **44**, 9 (1973).
[14] E. Thompson, *Physica* **1040**, 199 (1981).
[15] J. Liouville, *J. de Math.* **3**, 349 (1838).
[16] M. P. Stockli, Measuring and analyzing the transverse emittance of charged particle beams, *AIP Conf. Proc.* **868**, 25 (2006).
[17] C. Lejeune and J. Aubert, in *Charged Particle Optics Part A*, ed. A. Septier (Academic, New York, 1980), pp. 159–259.
[18] O. R. Sander, Transverse emittance: its definition, applications, and measurement, *AIP Conf. Proc.* **212**, 127 (1991).
[19] J. Reijonen, R. Thomae and R. Keller, Evolution of the LEBT layout for SNS, in *Proc. 2000 Linac Conf.* (2000), pp. 253–255.
[20] E.g. K. Saadatmand, J. E. Hebert and N. C. Okay, *Rev. Sci. Instrum.* **65**, 1173 (1994).
[21] M. P. Stockli, B. Han, S. N. Murray, T. R. Pennisi, M. Santana and R. F. Welton, *Rev. Sci. Instrum.* **81**, 02A729 (2010).
[22] N. Chauvin, O. Delferriere, R. Duperrier, R. Gobin, P. A. P. Nghiem and D. Uriot, *Rev. Sci. Instrum.* **83**, 02B320 (2012).
[23] M. P. Stockli, B. X. Han, S. N. Murray, D. Newland, T. R. Pennisi, M. Santana and R. F. Welton, Ramping up the SNS beam power with the LBNL baseline H^- source, *AIP Conf. Proc.* **1097**, 223 (2009).
[24] B. X. Han, T. Hardek, Y. Kang, S. N. Murray Jr., T. R. Pennisi, C. Piller, M. Santana, R. F. Welton and M. P. Stockli, Performance of the H^- ion source supporting 1-MW beam operations at SNS, *AIP Conf. Proc.* **1390**, 216 (2011).
[25] M. P. Stockli, K. D. Ewald, B. X. Han, S. N. Murray Jr., T. R. Pennisi, C. Piller, M. Santana, J. Tang and R. F. Welton, *Rev. Sci. Instrum.* **85** (2014).
[26] B. Han, D. J. Newland, W. T. Hunter and M. P. Stockli, Physics design of a prototype 2-solenoid LEBT for the SNS injector, in *Proc. 2011 Part. Accel. Conf.* (2011), pp. 1564–1566.
[27] B. Han and M. P. Stockli, The new LEBT for the spallation neutron source power upgrade project, in *Proc. 1997 Part. Accel. Conf.* (1997), pp. 2723–2725.
[28] R. Ferdinand, J. Sherman, R. Stevens and T. Zaugg, Space-charge neutralization measurement

of a 75 keV, 130 mA hydrogen ion beam, in *Proc. 2007 Part. Accel. Conf.* (2007), pp. 1823–1825.
[29] J. W. Staples, J. J. Ayers, D. W. Cheng, G. B. Greer, M. D. Hoff and A. Ratti, The SNS four-phase LEBT chopper, in *Proc. 1999 Part. Accel. Conf.* (1999), pp. 1961–1963.
[30] B. X. Han and M. P. Stockli, Model of a SNS electrostatic LEBT with a near-ground beam chopper, *AIP Conf. Proc.* **1097**, 395 (2009).
[31] J. G. Alessi, J. M. Brennan and A. Kponou, *Rev. Sci. Instrum.* **61**, 625 (1990).
[32] E. Goldstein, *Berlin Akd. Monatsber.* **II**, 691 (1886).
[33] M. V. Ardenne, *Technik* **11**, 65 (1956).
[34] R. Scrivens, M. Kronberger, D. Kuchler, J. Lettry, C. Mastrostephano, O. Midttun, M. O'Neil, H. Pereira and C. Schmitzer, Overview of the status, and developments on primary ion sources at CERN, in *Proc. 2011 Int. Part. Accel. Conf.* (2011), pp. 3472–3474.
[35] D. I. Kim, H. J. Kwon, H. S. Kim, K. T. Seol, I. S. Hong and Y. S. Cho, *J. Kor. Phys. Soc.* **59**, 615 (2011).
[36] S. Gammino *et al.*, *Rev. Sci. Instrum.* **81**, 02B313 (2010).
[37] T. Taylor *et al.*, *Nucl. Instrum. Methods A* **336**, 1 (1993).
[38] R. Gobin *et al.*, LINAC2012 (Tel-Aviv, Sep. FR1A02); http://www.JACoW.org
[39] R. Gobin *et al.*, *Rev. Sci. Instrum.* **70**, 2652 (1999).
[40] L. Celona *et al.*, *Rev. Sci. Instrum.* **75**, 1423 (2004).
[41] H. Ren *et al.*, *Rev. Sci. Instrum.* **83**, 02B905 (2012).
[42] R. K. Janev, D. Reiter and U. Samm, *Julich Rep.* **4105** (2003).
[43] M. Bacal *et al.*, *Phys. Rev. Lett.* **42**, 1538; *J. Phys. (Paris)* **38**, 1399 (1977).
[44] M. Bacal, A. Hatayama and J. Peters, *IEEE Trans. Plasma Sci.* **33**, 1845 (2005).
[45] K. N. Leung *et al.*, *Rev. Sci. Instrum.* **54**, 56 (1983).
[46] K. W. Ehlers and K. N. Leung, *Rev. Sci. Instrum.* **51**, 721 (1980).
[47] K. N. Leung *et al.*, *Rev. Sci. Instrum.* **62**, 100 (1991).
[48] T. Kuo *et al.*, *Rev. Sci. Instrum.* **67**, 1314 (1996).
[49] D-Pace, at www.d-pace.com/products_hion.html
[50] J. Peters, *Rev. Sci. Instrum.* **71**, 1069 (2000).
[51] R. Welton *et al.*, *Rev. Sci. Instrum.* **73**, 1008 (2002).
[52] R. F. Welton *et al.*, *AIP Conf. Proc.* **1515**, 341 (2013).
[53] M. P. Stockli *et al.*, *AIP Conf. Proc.* **1515**, 292 (2013).
[54] J. Peters, *Rev. Sci. Instrum.* **79**, 02A515 (2008).
[55] R. F. Welton *et al.*, *Rev. Sci. Instrum.* **83**, 02A725-1 (2012).
[56] H. Ren *et al.*, IPAC2013 (Shanghai, May 2013), MOPFI03; http://www.JACoW.org
[57] V. E. Krohn Jr., *J. Appl. Phys.* **33**, 3523 (1962).
[58] W. G. Graham, in *Proc. 2nd Int. Symp. on the Production and Neutralization of Negative Ions and Beams* (Brookhaven National Laboratory, New York, 1980), pp. 126–132.
[59] J. N. M. Wunnik, B. Rasser and J. Los, *Phys. Lett. A* **87**, 288 (1982).
[60] Yu. I. Belchenko, G. I. Dimov and V. G. Dudnikov, *Nucl. Fusion* **14**, 113 (1974).
[61] V. G. Dudnikov, *Nauka, Moscow*, Vol. 1 (1975), pp. 323–325.
[62] H. Vernon Smith *et al.*, *Rev. Sci. Instrum.* **65**, 123 (1994).
[63] D. C. Faircloth *et al.*, *AIP Conf. Proc.* **1390**, 205 (2011).
[64] R. Keller *et al.*, *AIP Conf. Proc.* **1097**, 161 (2009).
[65] H. Oguri *et al.*, *AIP Conf. Proc.* **1515**, 379 (2013).
[66] I. Brown and E. Oks, *IEEE Trans. Plasma Sci.* **25**, 1222 (1997).
[67] I. Brown, *Rev. Sci. Instrum.* **65**, 3061 (1994).
[68] I. Brown *et al.*, PAC1991 (San Francisco, May 1943); http://www.JACoW.org
[69] E. M. Oks *et al.*, *Appl. Phys. Lett.* **67**, 200 (1995).
[70] A. Anders, *Phys. Rev. E* **55**, 969 (1997).
[71] A. Anders *et al.*, *Rev. Sci. Instrum.* **69**, 1332 (2000).
[72] A. S. Bugaev *et al.*, *Rev. Sci. Instrum.* **71**, 701 (2000).
[73] T. V. Kulevoy *et al.*, *Nucl. Instrum. Methods A* **552**, 171 (2004).
[74] V. Batalin *et al.*, *J. Appl. Phys.* **92**, 2884 (2002).
[75] A. Vodopyanov *et al.*, *Rev. Sci. Instrum.* **75**, 1888 (2004).
[76] A. Anders *et al.*, *Rev. Sci. Instrum.* **71**, 827 (2000).
[77] E. Oks *et al.*, *Rev. Sci. Instrum.* **65**, 3109 (1994).
[78] E. Oks *et al.*, *Rev. Sci. Instrum.* **73**, 735 (2002).
[79] M. P. Stockli *et al.*, *Rev. Sci. Instrum.* **71**, 1052 (2000).
[80] E. D. Donets, *Nucl. Instrum. Methods B* **9**, 522 (1985).
[81] J. Faure, *AIP Conf. Proc.* **3**, 188 (1989).
[82] E. Donets, *Rev. Sci. Instrum.* **69**, 614 (1998).
[83] B. Becker, *Handbook of Ion Sources* (CRC, New York, 1995), p. 157.
[84] R. Becker *et al.*, *Rev. Sci. Instrum.* **81**, 02A513 (2010).
[85] E. D. Donets *et al.*, *Rev. Sci. Instrum.* **75**, 1543 (2004).
[86] D. E. Donets *et al.*, IPAC '10 (Kyoto, May 2010), THPEC067, 4208 (2010); http://www.JACoW.org
[87] A. Pikin *et al.*, PAC '11 (New York, Mar. 2011), WEP261, 1966 (2011); http://www.JACoW.org
[88] J. Alessi *et al.*, LINAC2010 (Tsukuba, Sep. 2010), FR103, 1034 (2010); http://www.JACoW.org
[89] M. Levine *et al.*, *Nucl. Instrum. Methods A* **237**, 429 (1985).

[90] O. P. B. Anan'in *et al.*, *Sov. Phys. JETP Lett.* **17**(69), 460 (1973).
[91] N. N. Alexeev *et al.*, in *Proc. Int. Part. Accel. Conf.* (2011), pp. 2193–2195.
[92] B. Sharkov, in *The Physics and Technology of Ion Sources* (John Wiley & Sons, New York, 1989), p. 233.
[93] B. Sharkov and R. Scrivens, *IEEE Trans. Plasma Sci.* **33**, 1778 (2005).
[94] S. Kondrashev *et al.*, EPAC04 (Lucerne, July 2004), 1402 (2004); http://www.JACoW.org
[95] M. Okamura *et al.*, EPAC '2000 (Vienna, June 2000), THP5A05, 848 (2000); http://www.JACoW.org
[96] M. Okamura and S. Kondrashv, PAC '07 (Albuquerque, June 2007), FRXAB02, p. 2206; http://www.JACoW.org
[97] M. Okamura *et al.*, PAC '05 (Knoxville, May 2005), p. 2206; http://www.JACoW.org
[98] S. Kondrashev *et al.*, HB2006 (Tsukuba, June 2006), THBY01, p. 341; http://www.JACoW.org
[99] K. Kondo *et al.*, *Rev. Sci. Instrum.* **83**, 02B319 (2012).
[100] Z. Zhang *et al.*, IPAC '11 (San Sebastian), MOPC028, p. 130; http://www.JACoW.org
[101] T. Nakagawa *et al.*, *Nucl. Instrum. Methods B* **226**, 392 (2004).
[102] R. Geller, *Electron Cyclotron Resonance Ion Sources and ECR Plasma* (Institute of Physics, Bristol, 1996), p. 89.
[103] T. Antaya and S. Gammino, *Rev. Sci. Instrum.* **65**, 1723 (1994).
[104] D. Hitz *et al.*, *Rev. Sci. Instrum.* **73**, 509 (2002).
[105] H. Arai *et al.*, *Nucl. Instrum. Methods A* **491**, 9 (2002).
[106] A. Girard *et al.*, *Phys. Rev. E* **62**, 1182 (2000).
[107] S. Gammino *et al.*, *Rev. Sci. Instrum.* **70**, 3577 (1999).
[108] H. Zhao *et al.*, *Rev. Sci. Instrum.* **81**, 02A202 (2010).
[109] B. Cluggish *et al.*, *Nucl. Instrum. Methods A* **631**, 111 (2011).
[110] X. Q. Xie and C. Lyneis, *Rev. Sci. Instrum.* **66**, 4218 (1995).
[111] G. D. Alton and D. N. Smithe, *Rev. Sci. Instrum.* **65**, 775 (1994).
[112] Y. Kawai *et al.*, *Phys. Lett. A* **371**, 307 (2007).
[113] T. Nakagawa *et al.*, *Rev. Sci. Instrum.* **81**, 02A320 (2010).
[114] P. Zavodszky *et al.*, *Rev. Sci. Instrum.* **79**, 02A302 (2008).
[115] T. Nakagawa *et al.*, ECRIS2010 (Grenoble, Aug. 2010), MOCOAK03; http://www.JACoW.org
[116] D. Mascali *et al.*, ECRIS2010 (Grenoble, Aug. 2010), p. 165; http://www.JACoW.org
[117] D. Leitner *et al.*, ECRIS2010 (Grenoble, Aug. 2010), p. 11; http://www.JACoW.org
[118] D. Leitner *et al.*, *Cyclotron 2004* (2005), p. 272.
[119] J. Benitez *et al.*, ECRIS2012, THX002.
[120] http://www.frib.msu.edu/content/collaboration-lawrence-berkeley-national-laboratory-sets-new-record
[121] H. W. Zhao *et al.*, *Rev. Sci. Instrum.* **83**, 02A320 (2012).
[122] Z. Q. Xie, *Rev. Sci. Instrum.* **83**, 02A302 (2012).
[123] C. Lyneis *et al.*, *Rev. Sci. Instrum.* **83**, 2A301 (2012).

Martin Stockli was born in Switzerland, where he completed his education with a PhD in Atomic Physics from the Eidgenossischen Technischen Hochschule. After a year at Western Michigan University, he did research and developed an electron beam ion source at Kansas State University. In 2000, he joined Oak Ridge National Laboratory to lead the Ion Source Group for the Spallation Neutron Source (SNS). The Cs-enhanced, multicusp, RF-driven H^- source at the SNS is the first source to routinely deliver ~50 mA H^- at a 5% duty factor for up to six weeks. He has published over 200 papers and lectures on ion injectors and volume-type H^- sources at particle accelerator schools.

Takahide Nakagawa was born in Japan, where he completed his education with a PhD in Nuclear Physics from the University of Tsukuba. In 1987, he performed research on heavy-ion nuclear reactions at the Hahn Meitner Institute. In 1989, he joined the Cyclotron Laboratory, RIKEN, to conduct research on heavy-ion nuclear reactions and to develop ECR ion sources. Since 2007, he has been the RIKEN Ion Source Team leader. He has developed and constructed several types of high-performance ECR ion sources. He has a strong publishing record, with over 100 papers.

Reviews of Accelerator Science and Technology
Vol. 6 (2013) 221–236

DOI: 10.1142/S1793626813300107

Charge Strippers of Heavy Ions for High Intensity Accelerators

Jerry A. Nolen
Physics Division,
Argonne National Laboratory,
Argonne, IL 60439, USA
Facility for Rare Isotope Beams,
Michigan State University, 640 S. Shaw Lane,
East Lansing, MI 48824, USA
nolen@anl.gov

Felix Marti
Facility for Rare Isotope Beams,
Michigan State University, 640 S. Shaw Lane,
East Lansing, MI 48824, USA
marti@frib.msu.edu

Charge strippers play a critical role in many high intensity heavy ion accelerators. Here we present some history of recent stripper technology development and indicate the capabilities and limitations of the various approaches. The properties of solid, gaseous, and liquid strippers are covered. In particular, the limitations of solid strippers for high intensity, high atomic number heavy ions and the unique features of helium gas and liquid lithium for high intensity applications are covered. The need for high quality simulation of stripper performance as important input for system optimization is explained and examples of the current simulation codes are given.

Keywords: Strippers; ion beams; liquid lithium; helium gas; charge state distributions; high intensity; radiation damage; atomic physics simulations; uranium beams; heavy ion accelerators.

1. Introduction

The use of strippers in high energy hadron accelerators is becoming almost universal. Most heavy ion accelerators start from a low charge state from the ion source and strip once or twice during acceleration to increase the energy gain (e.g. RIKEN in Japan, FAIR in Germany, RHIC and FRIB in the USA). Proton accelerators utilize H^- to accelerate initially and inject into a booster or storage ring by stripping the two electrons and continue with H^+ (e.g. Spallation Neutron Source, SNS, at Oak Ridge National Laboratory). In high power radioisotope-producing cyclotrons, stripping foils are used to extract the beam without the complications of an electrostatic deflector (a recent example in Ref. 1). However, the specific technology of H^- ion stripping will not be addressed in this article.

In the majority of cases some form of carbon has been the preferred material for preparing thin foils. It works well at high temperatures (the sublimation temperature is 3642°C) and it is relatively easy to prepare the required thin foils needed for strippers (~0.1 μm and thicker).

Heavy ions present a more difficult problem than protons, because of the much higher energy deposition per unit length (see Fig. 1). For the same velocity the heavy ions stop in a much shorter distance than the protons. This very high power deposition presents two problems: how to deal with the heat deposited on the stripper and how to deal with the radiation damage of the material. The heat problem appears in both protons and heavy ions, because what counts is the product of dE/dx and the number of ions per second. The radiation damage is

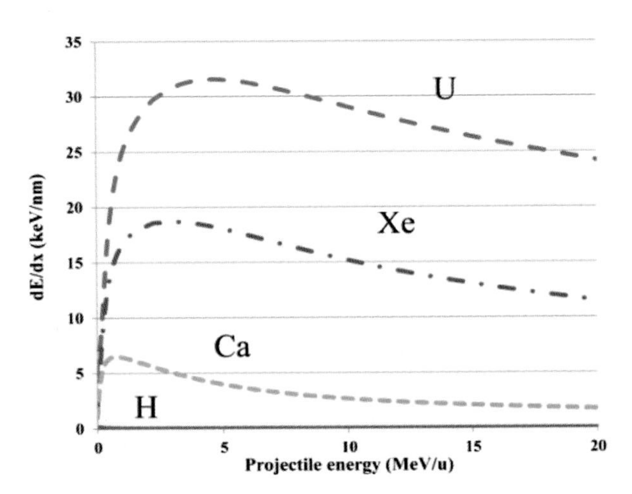

Fig. 1. Energy loss dE/dx of U, Xe, Ca, and H when traversing a carbon stripper as a function of the ion velocity (energy per nucleon). The proton energy loss is so small that it almost overlaps the abscissa axis in this graph. Calculations performed with SRIM [2].

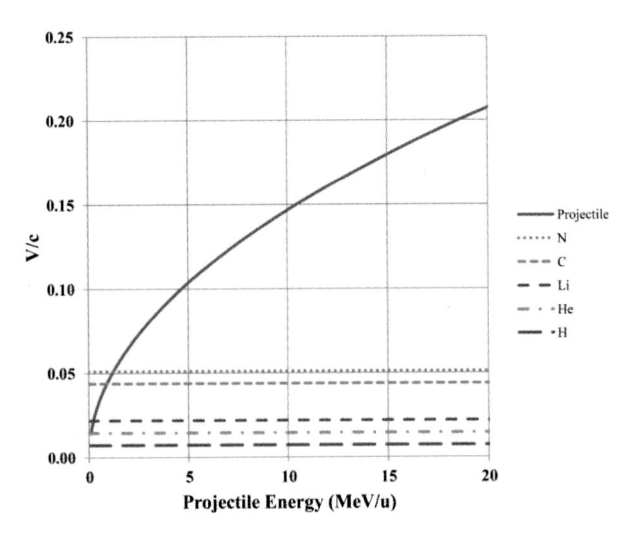

Fig. 2. Comparison of the projectile velocity and the 1s electron velocities in different stripping media. The electron velocities are calculated from the simple formula $\beta = Z/137$, where Z is the atomic number of the stripping media. The velocity mismatch that depresses the electron capture cross section is more important for the lighter media H and He.

significantly different, since a single heavy ion can modify the solid lattice.

2. Charge State Distributions

2.1. *Ion–matter interaction*

Excellent early treatments of the charge exchange process of ions in matter were given by Bohr [3] and Bohr and Lindhard [4]. A significant difference exists between the behavior of protons and of heavy ions in traversing matter. In the case of the proton the electron capture cross section by the nucleus is very small compared to the loss cross section for atomic H. Consequently, most of the projectile path will not have an electron attached to the nucleus. For heavy ions the situation is different and the number of electrons attached will fluctuate and is determined by the competition between the loss and capture cross sections, which depend on the projectile velocity, its atomic number, and the medium being traversed. The electron loss can occur as a result of the interaction of just two particles, while the electron capture requires the interaction of at least three particles due to the need to conserve both energy and momentum.

For ions passing through a stripping medium, it is very improbable to remove electrons with orbital velocities much larger than the projectile velocity. On the other hand, electrons are easily captured into orbits with velocities similar to the projectile velocity but improbable if the projectile velocity is much larger than the orbital velocity of the electrons in the stripper medium. To illustrate this difference we show in Fig. 2 a comparison of the relativistic velocity β of the projectile with the β of the 1s electrons in the different stripping media as a function of the projectile energy. We see that the largest mismatch occurs for H and He gases, supporting the idea of the depression of the electron capture cross section. For the stripping media we have used the simplified equation $\beta = Z/137$. This approximate equation can be obtained by using Bohr's model of the atom, equating the electric force on the electron to the centripetal force and recalling the quantization of the angular momentum. $1/137$ is an approximation to the fine structure constant.

Initial treatments of the ionization process considered only ions in the ground state, but the differences found between solids and gases as stripping media [5] showed the need to include the intermediate excitations of the projectiles. Higher charge states in solids relative to gases can be the result of the importance of ionization arising from sequential excitations due to the time between collisions being less than the lifetimes of the intermediate excitations, whereas in gases the ionization must be in single step collisions.

An alternative explanation was proposed by Betz and Grodzins [6, 7], where the projectile inside the solid has several electrons excited simultaneously and multiple Auger emissions occur when leaving the solid. The magnitude of this so-called "exit effect" is not definitively known at this time.

A systematic measurement of the stopping power of noble gases is summarized in Ref. 8. The authors found that the enhancement of the charge state (solids versus gases) is greater for lighter stripping media, but that in both media the average charge state was higher for the lighter materials (see Figs. 7 and 8 of Ref. 8). In heavy ions the energy differences between excited states are significantly smaller than the full ionization energy. Also, it is probable that several electrons can be captured, lost, or excited at the same time.

2.2. *Charge state distribution models*

Generally, when a heavy ion beam is accelerated in some initial charge state q_i and then stripped by some medium, the average charge state increases to a value $\langle q_f \rangle$ but the beam then contains a distribution of charge states. For heavy ion beams of high atomic number, Z, and intermediate energies such that final ions are much less than fully stripped, the charge state distributions are approximately Gaussian, with the central most probable charge states having ~15–20% of the total distribution. At higher energies the distributions evolve toward fully stripped ions and the distributions become narrower; examples are shown in the data and simulations below. The optimal choice of strippers for various applications is a complex process. And since there is relatively little systematic data covering the many parameters, it is desirable to have comprehensive simulation methods. The need for data and simulations increased dramatically in the 1970s and 1980s as large heavy ion accelerators were built for nuclear physics, including the Hilac and Bevalac at LBNL, GANIL in Caen, GSI in Darmstadt, NSCL at MSU, and ATLAS at Argonne. Many tandem Van de Graaff accelerators worldwide were also accelerating heavy ions and employing strippers (both carbon foil and gas strippers were used). The RHIC complex at Brookhaven, commissioned in the year 2000, initially used four ion strippers in its accelerator chain. The new Nishina heavy ion facility at RIKEN uses two strippers and has special optimization issues associated with its intense uranium beams. And at this time simulations are needed to guide the detailed design of the FRIB high power driver linac.

For each heavy ion accelerator there are optimization issues in both the design and operation phases. The experimental programs at heavy ion facilities also require detailed simulations of charge state distributions for a wide variety of beams, targets, and reaction products.

The parameters to be studied and optimized in the simulations are diverse. It is essential to be able to predict the evolution of charge state distributions following stripper foils along with the emittance growth caused by multiple scattering in the transverse dimensions and energy straggling in the longitudinal dimension. The cross sections for electron pickup and loss and the underlying atomic processes vary with the beam energy and atomic number, as well as with the stripper medium properties, such as its atomic number and density.

Strippers are used with ion beams from energies of a few keV/u to 10 GeV/u and higher, for ions as light as H^- and as heavy as uranium, and with intensities from a few particles per second up to ~10 particle microamps of uranium or many milliamps of light ions. Stripper media can be gases (such as H_2, He, N_2, or Ar), solids (such as Be, C, Ti, Nb, or W), or liquids (oils or liquid metals such as lithium). Simulations benchmarked against data where available can guide the optimization process, including the choice of the optimal stripper medium for a given application. However, for high intensity ion beams, further constraints on the media exist, leading to the need for simulations for new materials for which there is little or no existing data.

Sometimes data in new regimes of parameters point to the need for improved simulations. For example, game-changing data came soon after the availability of relativistic uranium beams from the Bevalac at LBNL. Figure 3, from Ref. 9, shows the charge state distributions following stripping of 437 MeV/u and 962 MeV/u uranium beams by foils of mylar, copper, and tantalum. The data show that at 437 MeV/u the optimal yields of fully stripped uranium beams (~50%) came from the copper foils, with a smaller yield from both the lower atomic number mylar and the higher atomic number tantalum foils. This feature required improved understanding of the underlying relativistic atomic physics and the energy dependence of the electron loss and capture

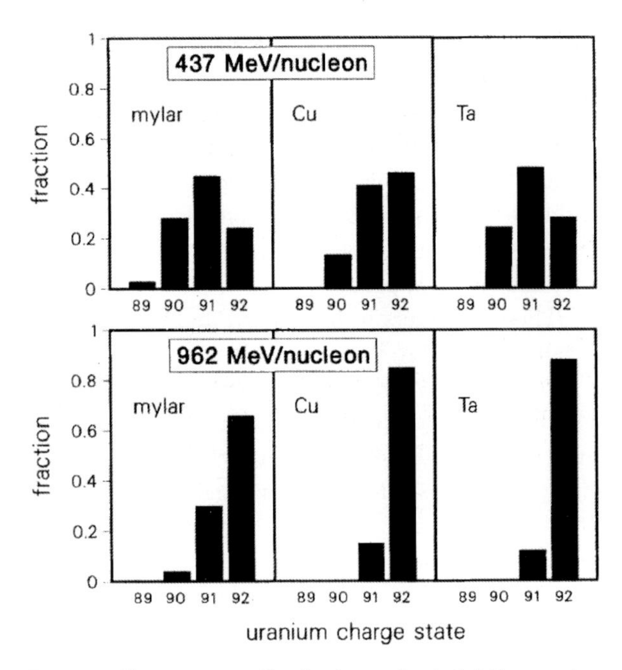

Fig. 3. Charge state distributions of relativistic uranium beams following stripper foils of low, medium, and high atomic numbers, illustrating for the first time the variation of the yields of fully stripped uranium as a function of the atomic number of the stripper [9].

cross sections in stripper media of various atomic numbers.

In addition to the dependence on the atomic number of the stripper medium, variables to be accounted for in simulations include the density effect (mostly solid/liquid versus gas), the energies of the incoming and outgoing ions, the initial charge state of the incoming ions, the stripper thickness, and the nonuniformity of the stripper material. Sometimes the optimal stripper thickness for a given application is less than the so-called equilibrium thickness when tradeoffs such as emittance growth, energy loss, charge state distribution width, and peak fraction in a charge state of interest must be made. In the following, various simulation methods from semiempirical parametrizations to models based on relativistic atomic physics are discussed and some examples are given.

2.2.1. *Empirical parametrizations*

Most of the early data on charge state distributions were measured with heavy ion beams produced by tandem Van de Graaff accelerators, i.e. total ion energies in the ~1–200 MeV range, or with lower energies directly from ion sources. One of the earliest parametrizations of charge state distributions in terms of the equilibrium charge states and distribution widths was done at the Institute for Nuclear Physics, Moscow State University [10]. This parametrization was also later shown to be consistent with data by Datz *et al.* [11] for beams of Br and I ions covering energies from 15 MeV to ~150 MeV. Several different solid media and gaseous media were used. For ions in this energy and atomic number range the equilibrium charge states for solid media are systematically ~30% or more higher than those for gaseous media, while the atomic number dependence within solid or gaseous media is much smaller — in the few-percent range. A thorough compilation of all known modeling and systematics of charge pickup and loss cross sections and the parameter dependence of charge state distributions for all known stripper systems was given in the review by H.-D. Betz in 1972 [12].

In the following 20 years a great deal of additional data on charge state distributions became available, including the evolution of the distributions with stripper thickness and beams covering a much broader energy, ~1–30 MeV/u, and atomic number range, up to uranium at 16 MeV/u. New data came from the accelerators at Oak Ridge, GSI, and GANIL, and most of these were compiled by E. Baron *et al.* in a series of papers [13]. These papers gave a simple equation to parametrize the equilibrium charge states and distribution widths for beams with atomic number 54 and above and energies of 1.3 MeV/u up to ~40 MeV/u. The fits are restricted to foil strippers, mostly carbon, although a small correction factor showing the dependence on the atomic number of the stripper foil is included. A small adjustment to the fits of Baron *et al.* was made by Leon *et al.* [14], including an adjustment to the foil atomic number dependence by including a wider variety of foils. Note that neither Ref. 13 nor Ref. 14 includes shell effects of the atomic structure of the projectiles. A summary and comparison of the parametrizations by Baron and Leon is included in Subsec. 3.3 of the Ph.D. thesis of M. Portillo [15]. He also discusses simulations with the codes GLOBAL and ETACHA, which are introduced below.

2.2.2. *Atomic-physics-based computer simulations*

Physical models for the calculation of cross sections for electron loss and pickup have been reported in

Refs. 3 and 4 and many subsequent papers. However, no simulations of the evolution of charge state distributions of ion beams with a large number of active electrons including important effects such as atomic shell structure and applicable from low energies, <1 MeV/u, and up to relativistic energies over 1 GeV/u, were attempted prior to the late 1990s. The need for such models increased greatly with the advent of large scale heavy ion accelerator facilities such as the Bevatron at LBNL, the Unilac and SIS18 at GSI, the Nuclotron in Dubna, and the construction of RHIC at BNL. When relativistic ion beams of high atomic number became available and produced data such as those presented in Ref. 9, where fully stripped, hydrogen-like and helium-like uranium ions were produced in charge stripper foils, a need surfaced for simulations that include large numbers of active electrons and atomic shell effects. The ultimate simulation would be based on microscopic relativistic atomic wave functions and dynamics, and would be applicable over the full range of ion energies, all atomic numbers for ion beams and stripper media, and for solids, liquids, and gases. The codes described in the following subsections represent significant progress toward this goal.

2.2.2.1. CHARGE and GLOBAL

Scheidenberger *et al.* developed an atomic-physics-based model for the simulation of the detailed evolution of charge states of relativistic heavy ions through stripper media [16]. To perform computer simulations based on this model, two different codes were developed. The first, CHARGE, is based on the three-charge-state model, which applies to almost fully stripped heavy ions, i.e. it is applicable at velocities where the projectiles are bare, H-like, or He-like (zero, one, or two electrons). The second code, GLOBAL, is applicable for ions with up to 28 electrons active. It can, for example, simulate the evolution of uranium ions starting with charge state 64+. The applicable energy range is considered to be ~80 MeV/u to over 1 GeV/u. Many examples of simulations of the detailed evolution of high atomic number beams through a variety of solid charge stripper media are shown in Ref. 16. At the high energies treated in Ref. 16 gaseous media are not practical as strippers, but the charge-changing cross sections are still important as a source of beam loss effects in synchrotrons and storage rings. In that reference it is also argued that at these energies and for high Z ions the solid–gas effective cross section differences are minimized due to very short atomic excitation lifetimes and minimal Auger electron effects.

An important use of GLOBAL was in the simulation of the charge stripper at RHIC for the 100 MeV/u gold 31+ beams extracted from the booster synchrotron and stripped to 77+ (helium-like) for injection into the AGS, the "BTA" stripper. The development of an optimized two-step stripper, an aluminum foil followed by a "glassy" carbon foil, is described in Ref. 17. Since the cross section for electron loss is higher for strippers with a larger atomic number, the aluminum foil causes a rapid increase in the charge state of the Au beam. However, the peak fraction in the helium-like charge state is lower for aluminum than for carbon, so it is advantageous to follow the aluminum layer with a carbon layer. Figure 4, taken from Ref. 17, illustrates the more rapid rise of the 77+ fraction for aluminum versus carbon and beryllium. The result is that a high yield of 77+ gold is achieved with a thinner overall stripper and a significantly lower energy loss.

2.2.2.2. ETACHA

ETACHA is a code somewhat like GLOBAL except that it is aimed at application to nonrelativistic ion beams of up to ~30 MeV/u, which are appropriate for many ion beams used at GANIL. The initial model, developed by Rozet *et al.* in 1989, was limited to 10 electrons and had some deficiencies, which

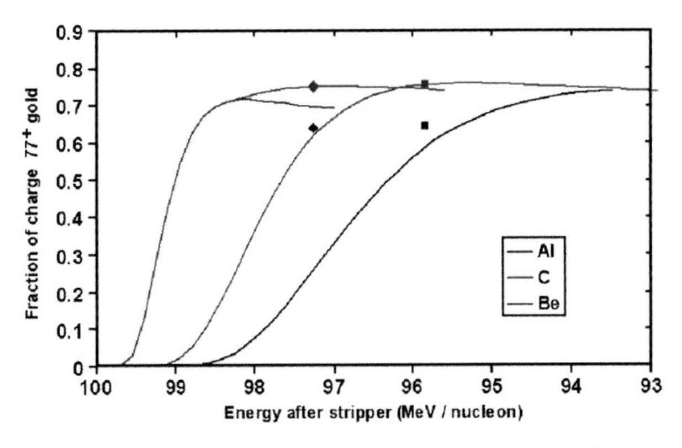

Fig. 4. Simulations done with GLOBAL of options for the RHIC BTA stripper. The upper pair of points represent the simulated yields of Au77+ using a pure carbon stripper (*right*) and a two-stage Al/C stripper (*left*). The lower pair of dots indicate the actual achieved fractions. (The figure is from Ref. 17.)

were corrected in a new version that permits up to 28 electrons in the $n = 1, 2, 3$ subshells [18].

The goal is to simulate the evolution of charge state distributions through both solid and gaseous stripper media by appropriately including successive multiple excitations and ionization based on the density effect. Shell effects in the projectiles and the dependence on the atomic number of the stripper are also modeled. However, this version of the code overestimates the charge states of high atomic number beams, so there is now an effort for additional improvements to include the $n = 4$ subshell, increasing the number of active electrons to 60, ETACHA4 [19].

Preliminary results with ETACHA4 are shown in Figs. 5 and 6 [20]. Figure 6 illustrates the evolution of the charge state distributions of 10.5 MeV/u uranium in carbon foils of various thicknesses [21] approaching equilibrium. In both figures some adjustment of the simulation parameters was required. A goal of the development is to understand the need for or justify these adjustments.

3. Solid Strippers

3.1. *Materials explored as strippers*

An important issue to consider is the thickness uniformity of the stripper. Nonuniformities in solid stripper foils can introduce a variable energy loss and increase the energy spread of the beam. This has been observed at RIKEN (see comments below when discussing the gas stripper) and at BNL [17]. The nonuniformities in the carbon strippers by far dominated the beam energy spread at the entrance to the AGS prior to the introduction of glassy carbon, as discussed in Ref. 17. Other stripper materials tested (mica, titanium, and fused silica) at BNL are also discussed in that reference. Finally, the aluminum–glassy-carbon combination as illustrated in Fig. 4 was adopted for the RHIC BTA stripper. A tungsten stripper was chosen to replace the alumina stripper for the stripper at the higher energy following the AGS (ATR) since it could be much thinner and had a correspondingly smaller energy loss for the gold beam.

A large effort has been invested in the fabrication of carbon stripper foils to extend their lifetime or minimize the nonuniformity. Sugai at KEK [22], P. Maier-Komor in Munich [23], Hasebe at RIKEN [24], Shaw at ORNL [25], and others have developed several variations of the traditional carbon foils, which have increased their endurance when used as H^- or light ion strippers. But these developments have had only small effects when used with very heavy ions.

In recent years the availability of carbon nanotubes has made possible experimentation with this medium as a stripper [26]. The properties that make it attractive are its superior thermal conductivity and strength. Beam tests were performed at ORNL HRIBF, with disappointing results. The nanotube structure survived the light and medium mass ions, but heavy ions appear to disintegrate the bundles of nanotubes. In addition, there was not enough

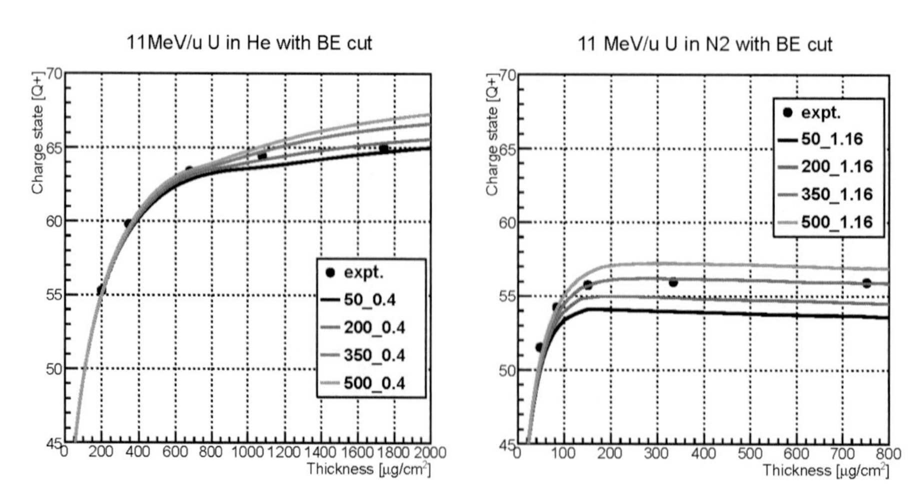

Fig. 5. Preliminary results of a simulation with ETACHA4 comparing predictions for charge state evolution of a uranium beam in a helium gas stripper (*left*) to that for an N_2 gas stripper (*right*). To obtain these results some parameter adjustments were done [20]. The data points are from the RIKEN work described in Sec. 4.

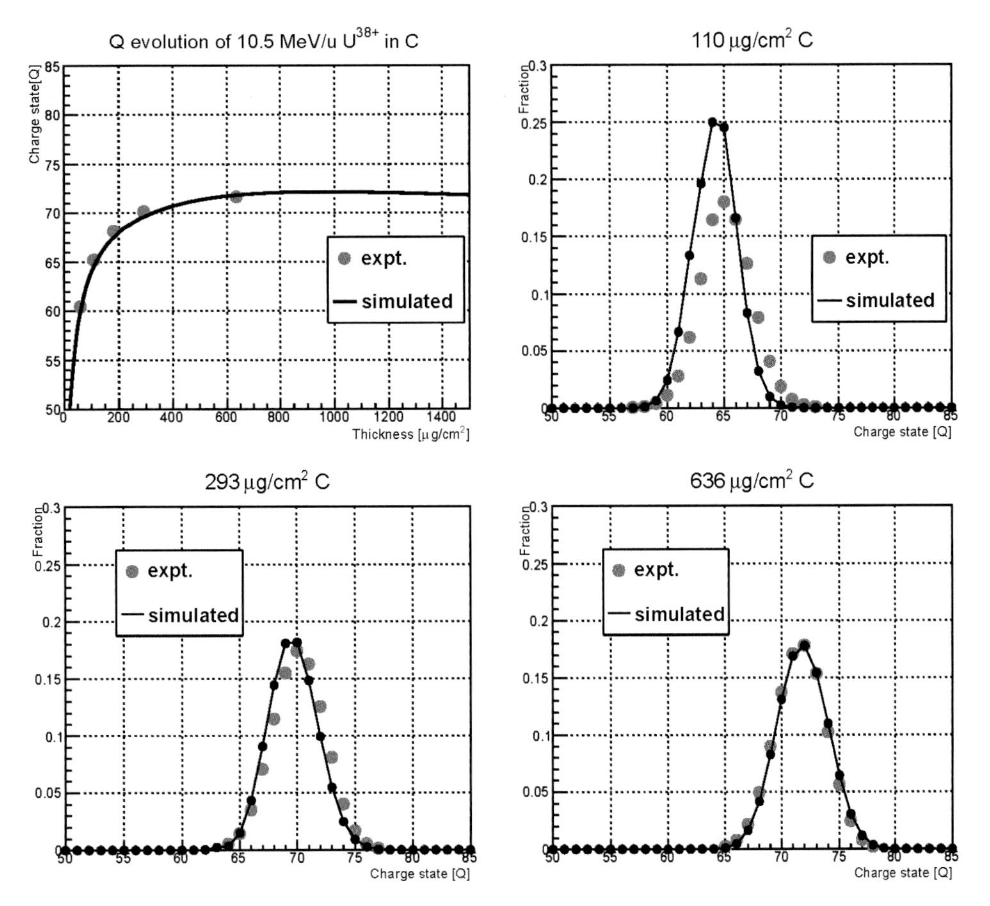

Fig. 6. ETACHA4 simulations of the charge state evolution of a uranium beam through a carbon stripper showing the average charge state versus thickness and the simulated distribution at three different foil thicknesses in the other panels [20]. The larger circular dots indicate data from Ref. 21. Residual disagreements such as for the thin foil (*upper right*) may be partially explained by foil thickness uncertainty which was not quantified well in the measurements.

area coverage of nanotube bundles, reducing the area that contributes to the stripping, and reducing the yield.

Stripper foils made of graphene are becoming available in sizes large enough to be used in rotating disks for high intensity accelerators [27]. Preliminary tests at NSCL/MSU in stationary strippers inside the cyclotron showed mean lifetimes two or three times the average lifetime of amorphous carbon foils. This was a very small sample, but the promising results justify more experimentation.

3.2. *Carbon foil damage*

Carbon foils have the advantage of being mechanically stable in the thicknesses needed for strippers but their lifetimes are short when they are exposed to intense beams of heavy ions. The two main mechanisms that damage the foils are sublimation and radiation damage.

An early analysis of carbon foil damage can be found in Ref. 28, which describes foil behavior in tandem terminals and attempts to extend the foil's lifetime by heating it with the purpose of annealing the radiation damage. The foil thickening at the beginning of the irradiation was interpreted as being due to deposits from hydrocarbon cracking in the residual vacuum.

Baron [29] has proposed the following formula to estimate the lifetime of carbon foils:

$$T = C\frac{E/A}{jNZ_p^2},$$

where $C = 3.6 \times 10^4\,\mathrm{h} \cdot \mathrm{p}\mu\mathrm{A} \cdot \mathrm{cm}^{-2}\,(\mathrm{MeV/nucleon})^{-1}$, j is the current density ($\mathrm{p}\mu\mathrm{A} \cdot \mathrm{cm}^{-2}$), and N is the average number of atoms displaced per collision ($N \sim 5$–8). This value for N was obtained by fitting to experimental lifetimes. Z_p is the projectile atomic number.

A more complicated formula that includes the effect of the foil temperature was derived by Nickel [30] by fitting to experimental data. The temperature dependence appears through the term $e^{(T-300)/230}$, with T in K, or using a different fitting curve as $e^{-\beta/T}$, where β is an empirical constant.

The damage to carbon foils in the Dubna cyclotrons has been studied in Ref. 31, reaching the conclusion that for thin targets (<1 μm) and current densities of ions <20 pμA/cm^2 the principal cause of the carbon target lifetime limit is radiation damage, but for higher current densities sublimation limits the lifetime.

Lebedev and Lebedev [32] have described in detail the derivation of the following formula for the foil lifetime:

$$t = 50K_d^{-5/4}\exp\left(\frac{-870}{T}\right),$$

where t is the lifetime in seconds, K_d is the rate of atom displacements per atom per second. They concluded that for temperatures below 2500 K radiation damage is the dominant mechanism of destruction, while for higher temperatures foil sublimation defines the lifetime. The experience with beams at the NSCL cyclotron stripper indicates lifetimes much shorter than predicted by the formula above.

The energy loss of the ions in the solid is due mostly to interactions with the electrons in the solid. The typical energies of the ions at the stripper are high enough that the contribution of the nuclear stopping is small. The nuclear stopping peaks at around 100 keV/u, near the end of the range, with the electronic stopping becoming dominant above that energy. The lower importance of the electronic stopping at lower energies is expected because the ion will immediately pick up electrons from the stripping media and reduce the effective charge that interacts with the media electrons.

An ion with energy of 20 MeV/u will traverse the stripper media in a very short time (1.6×10^{-14} s/μm). The effect of an ion on the solid it traverses can be considered as happening in three stages with different timescales and characteristic distances. In the first stage the ion transfers energy to the electrons in the solid. This happens obviously at a time scale shorter than the traversal time: $\sim\leq 10^{-14}$ s. In the second stage these electrons produce cascades of secondary delta electrons ($d \leq 1\,\mu$m), and in the third phase the atoms left with an excess charge suffer a Coulomb repulsion or "explosion" ($d \leq 0.01\,\mu$m, $t \geq 10^{-12}$ s) [33]. The damage done by a single ion appears in a high resolution electron microscope (HREM) image of a thin layer of the mica as a well-defined region of amorphous material within the undamaged crystalline mica (see Fig. 7).

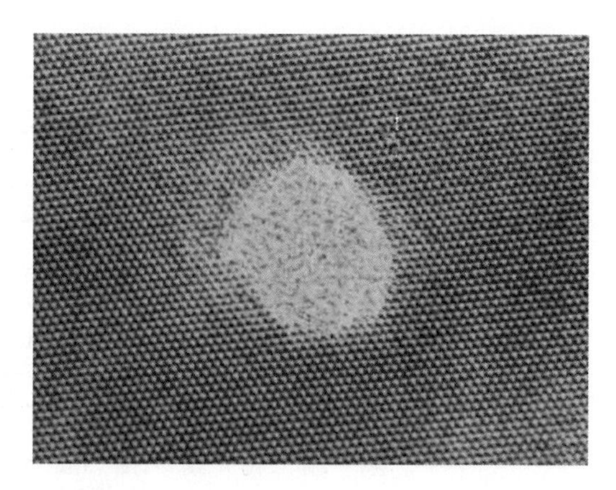

Fig. 7. HREM micrographs of damage in mica created by a single lead ion. The damaged region is ~8 nm in diameter and the peak electronic energy loss rate is ~27 keV/nm per ion at the Bragg peak at ~6 MeV/u kinetic enegy.

Another interpretation of the formation of tracks in solids is the *thermal spike model*, proposed by Seitz and Koehler [34] and extended into the inelastic spike model [35]. In this model the electrons transfer the energy to the atoms via electron–phonon coupling after being thermalized among the electrons.

It is interesting to note that even though the dE/dx curves in Fig. 1 are double-valued (the same dE/dx for two different ion energies) the emitted electron energy spectra are different for the low and high energy values. The electron energies could be several times higher on the high energy side, as shown in Fig. 21 of Ref. 36.

There appears to be a critical threshold for the formation of tracks in graphite of the order of 7 keV/nm, with a track produced for every ion with energy above 18 keV/nm [37]. From Fig. 1 we can estimate the likelihood of radiation damage to a graphite stripper by the heavy ions. The light ions, Ca and lighter do not appear to be a problem, but heavier ions will produce substantial damage.

An example of radiation damaged foils is shown in Fig. 8, where the foils have been exposed to an 8.1 MeV/u Pb beam. The foils are diamond-like carbon (DLC) and have a thickness of ~0.3 mg/cm^2, the accumulated dose was $4–6 \times 10^{14}$ ions on a beam spot of approximately 2 mm diameter. We see on

Fig. 8. Damage produced by a $^{208}Pb^{27+}$ beam, 8.1 MeV/u, on DLC foils in the NSCL K1200 Cyclotron. The leftmost photo shows an unused foil, and the middle and rightmost photos show two different foils exposed to the beam.

the left an unused foil where the edge of the foil is aligned with the edge of the C frame where it is mounted. After exposure to the beam the foil has thinned where the beam went through and the material has been displaced beyond the edge of the C frame. Thickness measurements with an alpha source confirmed the significant reduction in the foil thickness where the projectiles hit the foil (from approximately 300 down to 80 μg/cm^2). This effect can be attributed to the "ion hammering effect" described in Ref. 38. This anisotropic plastic deformation was observed in amorphous materials. It consists of an increase of the foils in the dimensions perpendicular to the beam trajectory and a decrease in the dimension parallel to the beam. The stripper thickness is normally chosen below the thickness that produces the equilibrium charge state to minimize the energy spread introduced by the stripper. This implies that a thickness reduction of the stripper by the beam will shift the mean charge state toward lower values.

4. Gas Strippers

When stripping at low energies as in the case of the high voltage terminal of a tandem, a thin carbon foil is typically used, of ~10 μg/cm^2 and thinner for heavy ions [39]. The high dE/dx of the heavy ions puts the foils under high stress and they break in short times. Gas strippers have the advantage of autorenovation compared to the solids but provide a lower mean charge state and load the vacuum of the accelerator. A vacuum pump is located on the tandem terminal. Dual systems with combined foil and gas capabilities have been implemented [40].

For intense heavy ion beams, gas strippers can be the preferred stripping media even when providing lower mean charge states [41, 42]; see Fig. 9. The

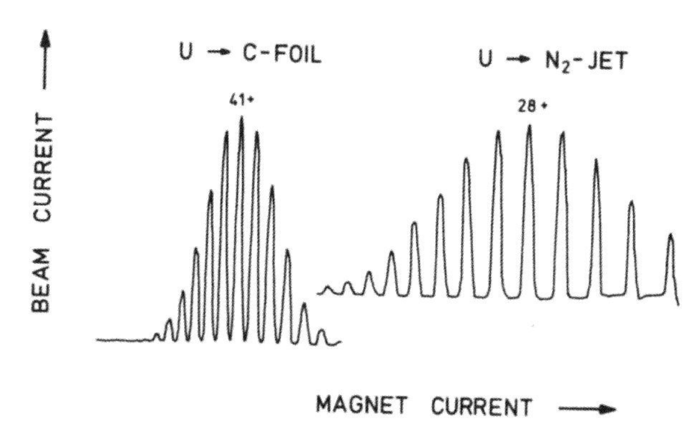

Fig. 9. Equilibrium charge state distribution for 1.4 MeV/u uranium ions after stripping in N_2 gas (*right*) and in a carbon foil (*left*). (Graphics from Ref. 41.)

N_2 gas stripper with the charge state distribution shown in this figure has been used at the GSI UNILAC successfully for over 30 years.

4.1. *RIKEN recirculating He stripper*

The work at the RIKEN Nishina Center in Japan deserves special mention in our review. The chain of accelerators at RIKEN consists of a linear accelerator (RILAC) followed by four cyclotrons (RRC, fRC, IRC, and SRC). Two strippers are used for the heavy ions, located between the RRC and the fRC, and between the fRC and the IRC [43]. The first stripper operates at an energy of approximately 11 MeV/u, stripping the incoming U beam to a charge state that has to be accepted by the next cyclotron. The required charge state must be equal to or higher than 71+. Traditional carbon foils had been used but the short lifetime prevented the facility from operating at the design intensity. A lifetime of 10 h was typical and mean beam intensities were limited to about 0.16% of the design goal.

An accelerated program was established to develop a gas stripper that would allow higher intensities to be reached. Initial studies [43] utilized N_2, Ar, and CO_2 as stripping media. The U ions were stripped to charge states around 56+. The thickness required to achieve equilibrium was determined to be 125, 79, and 126 μg/cm^2 for N_2, Ar, and CO_2. The RIKEN group in the paper mentioned above proposed the following formula for the equilibrium charge state:

$$\frac{q_{eq}}{Z} = 1 - 1.01 \exp\left[-\left(\frac{v}{v_o}\right)^{0.99} Z^{-0.676}\right],$$

where v_o is the Bohr velocity (2.188×10^8 cm/s). As the charge state was below what the next cyclotron could accept, the next step was to plan for a lower Z gas with the potential of obtaining a higher average charge state. The feature of light gases had been demonstrated before for low energy beams in Refs. 8, 44 and 45. With the purpose of estimating the average charge state of a He gas stripper, the RIKEN team started by measuring the electron loss and capture cross section for uranium at 11, 14, and 15 MeV/u. Although a stripper cell capable of accumulating the equilibrium thickness of He was not available, a thin cell (13.27 μg/cm^2 [46]) was used

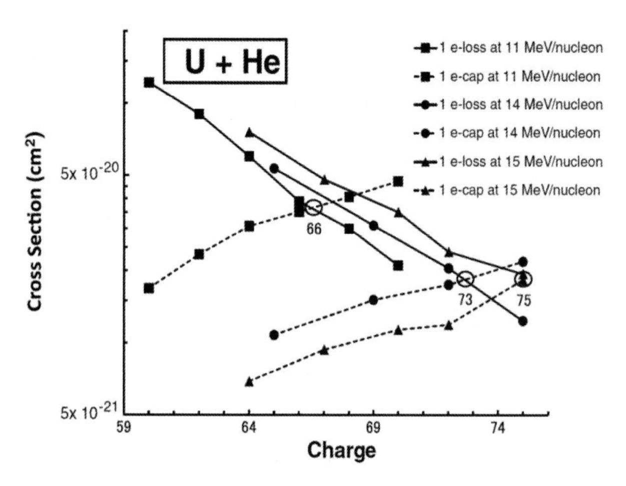

Fig. 10. Measured cross sections of electron loss and capture as a function of the charge state of uranium ions at 11, 14, and 15 MeV/u in He gas. The intersections between the two lines for the cross sections of loss and capture are indicated by the circles, with the corresponding numbers of ion charge below. (Reproduced from Ref. 46.)

in the experiment. The results are reproduced in Fig. 10, where the intersections of the loss and capture cross sections determine the average charge state expected in the equilibrium thickness (if achieved before the energy loss is significant and displaces the intersection to a lower charge state).

Two different approaches have been used at RIKEN to contain the helium gas; a differential pumping system and a plasma window (PW) contained helium cell. The second system was proposed by Peter Thieberger (BNL) at the 2009 FRIB/MSU workshop on high power targets and strippers. We will describe it in the next section, when presenting the FRIB approach.

A very successful differential pumping system has been built and operated for the RIKEN stripper (see Fig. 11) [47, 48]. Apertures of 10 mm allow the beam to enter the windowless helium main chamber. The system is implemented "... with a powerful multistage mechanical booster pump (MBP) array consisting of four foreline MBPs and three back MBPs with a total nominal pumping speed of 11,900 m^3/h...." The pressure in the central cell is approximately 7 kPa, and in five stages of differential pumping a final pressure of 5×10^{-6} Pa is achieved. The differential pumping system includes six turbomolecular pumps backed by six rotary pumps. As a significant amount of helium is escaping from the main cell toward the differential pumping stages, the mechanical booster pump exhaust is sent through a heat exchanger and a foreline trap before reinserting it into the main stripper cell. The exhaust from the turbomolecular pumps is sent to the helium recovery

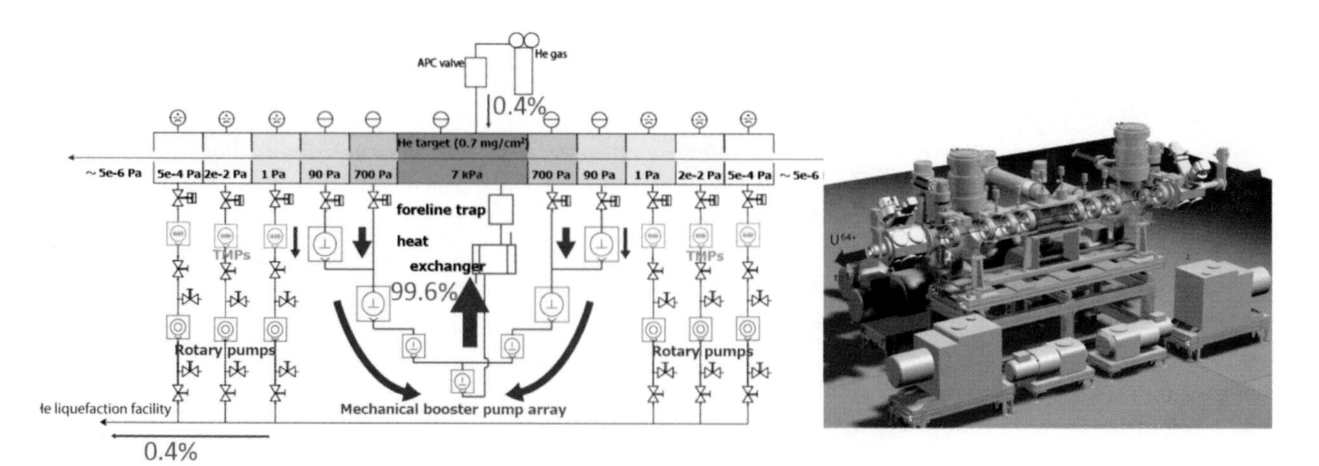

Fig. 11. Schematic of the He recirculating system at RIKEN [48]. The diagram on the left shows the multiple mechanical booster pumps needed to recirculate the He gas that leaks toward the differential pumping stages on both sides of the stripper cell.

system. The system is designed to recirculate the helium with 99.5% efficiency.

An interesting point is that in spite of the larger integrated areal density of the helium stripper with respect to the carbon foil stripper, the energy spread of the beam is smaller in the helium system. The reason appears to be the nonuniformity in thickness of the carbon foils.

The fRC was modified for use with the new helium gas stripper, increasing the magnetic rigidity for operating with 64+ uranium ions from the initial limit of 69+ or higher. The average beam intensity on target was increased to ~1% of the design value from the previous 0.16%.

We expect that when the intensity is increased the heat deposited in the gas by the beam will produce a density rarefaction that will increase the energy spread of the beam. At this moment no attempt at circulating large volumes of helium to minimize the temperature variations has been made. See the next section for the FRIB plans to diminish this effect.

4.2. *FRIB recirculating He stripper*

Encouraged by the experiments performed at RIKEN and following the suggestion by Thieberger at the December 2009 FRIB workshop on high power strippers and targets, the FRIB project started an R&D program on a plasma window [49] contained helium stripper. This program has been going on mostly at Brookhaven National Laboratory, under the guidance of A. Hershcovitch and Thieberger. As described above, helium gas would provide a higher mean charge state than nitrogen, reducing the number of accelerating structures.

As the stripping energy at FRIB is approximately 16.5 MeV/u, compared to RIKEN's 11 MeV/u, a thicker stripper is needed to reach equilibrium charge state distributions. The difficulties associated with pumping helium, as shown by the very complex system at RIKEN, would be exacerbated at FRIB. The plasma windows have the potential to provide a significant conductance reduction and make the problem more manageable. They can be interpreted as a gas heater that increases the temperature and reduces the density of the gas flowing through a small tube until a choked flow condition is reached, limiting the mass flow of helium

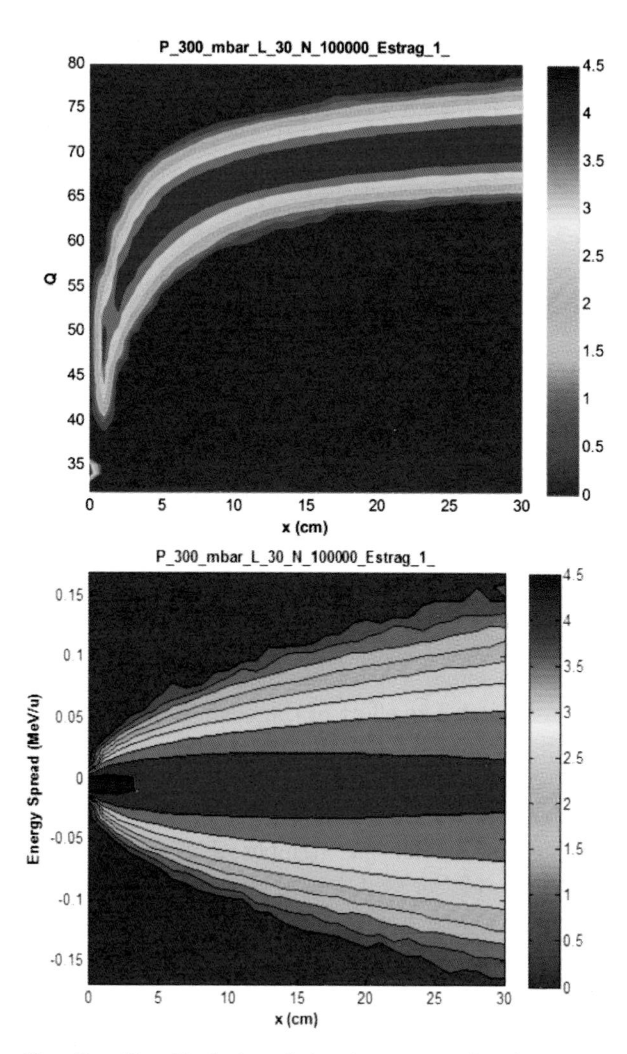

Fig. 12. *Top*: Evolution of the charge state distribution in a 300 mbar helium gas cell as a function of the cell length. *Bottom*: Evolution of the energy spread of the same beam. The contours show the intensity on a logarithmic scale.

between the main stripping chamber and the surrounding chambers.

We show in Fig. 12 the simulation of the charge state evolution in a gas cell appropriate for the FRIB uranium conditions. A 30-cm-long cell with a pressure of 300 mbar can produce an average charge state of 71+, starting from 33+. The simulation was performed with the code described in Ref. 50 and utilizing the cross sections measured at RIKEN as calibration. A 30-cm-long cell has not achieved equilibrium distribution yet but has reached most of the gains. Increasing the length would increase the charge state very little but increase significantly the energy spread of the beam (see bottom graph). This simulation did not include the density change

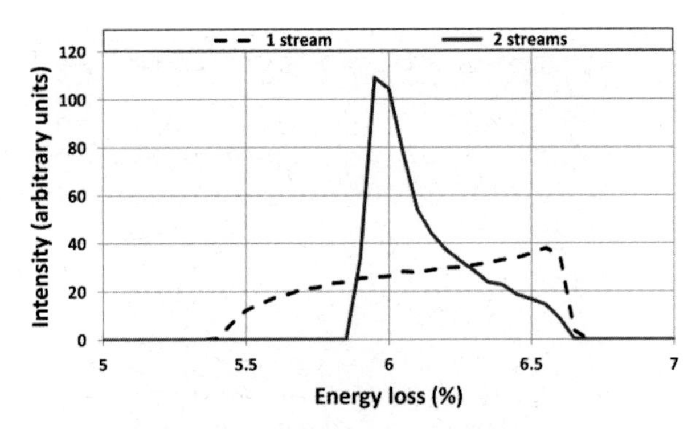

Fig. 13. Energy loss of a 16.5 MeV/u U beam (rms diameter 3 mm) in a 2 mg/cm^2 stripper cell produced by the differential heating of the gas. The helium is flowing in one direction across the beam (one stream) or in counterflows to partially compensate for the effect. (Courtesy of P. Thieberger, BNL.) The FRIB accelerator accepts a variation in energy loss at the stripper of ±20%.

produced by the power deposited by the beam in the gas; a uniform density along and across the cell was assumed. The effect of the beam heating can be seen in the simple simulation by Thieberger (Fig. 13). The simulation corresponds to a uranium 3 mm rms diameter beam with a power of 42 kW traversing a 2 mg/cm^2 helium gas stripper. The helium flows transverse to the beam either in one direction (one stream) or each half in an opposite direction (two streams). The gas velocity is 100 m/s. We estimate that the helium flow rate is 450 l/s. We can see that the counterflow solution improves significantly the energy spread. A simplified sketch of the proposed stripper cell is shown in Fig. 14. Only one half of the system is represented. A similar pumping arrangement should be implemented on the left hand side.

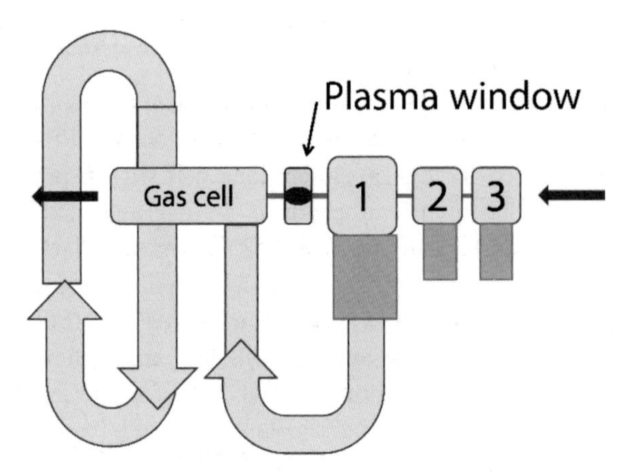

Fig. 14. Sketch of the stripper cell proposed as an alternative stripper for FRIB. Only one half of the system is shown. The green rectangles represent vacuum pumps.

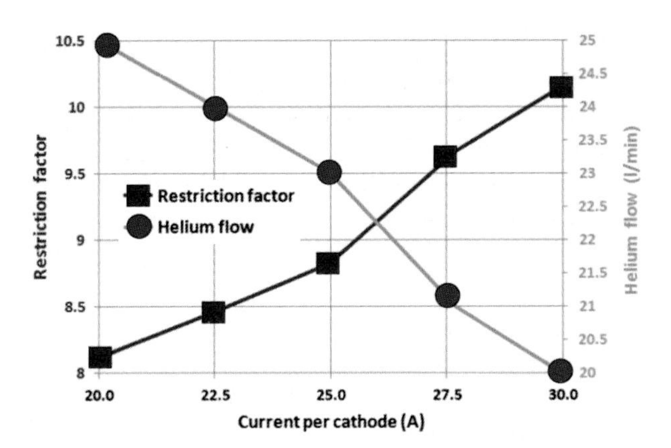

Fig. 15. Conductance reduction factor (squares) and helium flow (circles) for the 4.5 mm diameter plasma window and 0.5 atmospheres in the gas cell [51]. Three cathodes are used with the current specified in the figure.

The experiments at BNL [51] have shown that with a plasma window with an aperture of 6 mm a conductance reduction factor of approximately 10 can be achieved. Figure 15 shows the conductance reduction factor and the helium flow (l/min) for a 4.5 mm aperture plasma window. Three cathodes were used simultaneously, with the current specified in the abscissa in each of them. Improved reduction can be expected from higher currents if a new design of the plasma window with improved cooling is achieved.

5. Liquid Strippers

Liquid films for both targets and charge strippers for intense heavy ion beams have the advantage of being immune to radiation damage of the type discussed in Sec. 3. Liquid lithium has many desirable physical properties and is the target material chosen for the 10 MW scale target for the International Fusion Materials Irradiation Facility. A large liquid lithium loop was constructed and operated for over a year at Hanford to test system reliability and impurity control. There was no ion beam used during these tests, however. Engineering design studies have progressed [52, 53] and support the case for adapting this technology for high intensity heavy ion accelerator targets and charge strippers.

5.1. *Oil film strippers*

In the 1980s there were several studies and prototypes made for potential use as ion beam strippers at low energies at ion sources such as at the Hilac at LBNL [54–56]. A prototype oil film stripper was also built and tested at RIKEN [57]. However, none of these films was used as a practical stripper. One problem was that the film velocity was too small to be useful for the necessary ion beam currents, and there were also charging effects due to the electrically nonconducting nature of the oil films. Furthermore, these films were envisioned to be used at low energies to boost the charge states of ion sources; this concept became obsolete, as in this same time frame ECR ion sources advanced to the point of providing intense currents of high charge states.

5.2. *Liquid lithium strippers*

During the R&D phase for the Rare Isotope Accelerator, liquid lithium technology was proposed and developed at Argonne to be used as the coolant for high power spallation targets and as windowless thick targets for heavy ion fragmentation [58]. A prototype windowless target 1 cm thick was built and tested at high power density [59]. In this test a 20 kW, 1 MeV electron beam was stopped in the target. With the 1 mm diameter electron beam the deposited power density was in ~10 MW per cm^3.

With the successful test of the thick liquid lithium target concept, it was decided to initiate R&D on a high speed thin liquid lithium film for use as the uranium beam charge stripper in the FRIB driver linac. The need was for a stripper film to increase the charge state of the 34+ uranium beam from the ECR ion source by stripping at ~16 MeV/u. The required beam current at this stripper is 10 particle microamperes, in order to achieve the goal of 400 kW of uranium beam out of the driver. The beam dynamics of the linac can tolerate a beam spot diameter of up to 3 mm at the stripper. A thermal analysis was done and a prototype stripper loop constructed at Argonne with financial support from the RIA/FRIB R&D program. The ability to create an apparently stable, high speed, ~50 m/s film was demonstrated [60].

Continuing with funding through the MSU FRIB R&D program, the film uniformity was quantitatively demonstrated via an electron beam thickness monitor [62].

To demonstrate the power density capability of the liquid lithium film, the ion source of the Low Energy Demonstration Accelerator (LEDA) [63] was borrowed from LANL, recommissioned at MSU, and then installed at the liquid lithium loop at Argonne. A beam power of 300 W was deposited into a beam spot with a sigma of 0.7 mm, as shown in the pictures in Fig. 16. The proton range in the ~10 μm lithium film was only 2 μm. The results are presented in an FRIB Technical Report [64], and are to be published.

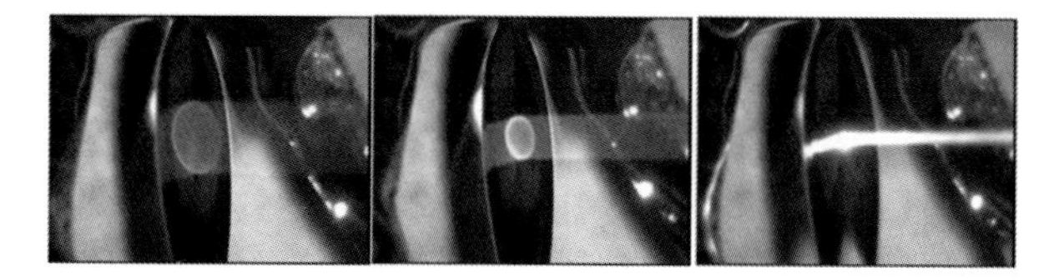

Fig. 16. Photos of the high speed lithium film being impacted by a 70 keV, 300 W proton beam. The beam spot diameter is adjusted via an Einzel lens, with the spot on the right side having a sigma of 0.7 mm. On the left of each beam spot is a reflection of the incoming beam in the mirror-like lithium surface. Neon gas at ~1E-6 Torr enabled viewing the proton beam with a CCD camera.

6. Plasma Stripping

The use of plasma strippers has not been implemented routinely yet but the Frankfurt group working at GSI is making important steps in that direction. The idea behind a plasma stripper is that direct capture of an electron by a moving ion violates the simultaneous conservation of energy and momentum [65–67]. The excess binding energy must be taken away by radiative recombination, three-body recombination, or dielectronic recombination, all low probability processes. This restriction lowers the electron capture cross section compared to the bound electron case, increasing the charge of the ion compared with a cold target. This effect is more important at low energies. At higher energies the high relative velocity is already depressing the capture process.

An experiment is scheduled at GSI [68] to test a plasma stripper with a 4 MeV/u U beam. The ions will interact with a θ-pinch plasma produced by an induction coil. The process is geared toward low energy beams and in pulsed mode, not continuous beams.

Acknowledgments

We would like to thank the numerous colleagues who contributed information for this article, including

W. Barth, H. Imao, E. Kanter, H. Kuboki, H. Okuno, M. Portillo, J.-P. Rozet, J.-S. Song, P. Thieberger, and D. Verhnet.

The authors were supported by the US Department of Energy Office of Science under Cooperative Agreement DE-SC0000661.

References

[1] V. Sabaiduc *et al.*, New high intensity compact negative hydrogen ion cyclotrons, in *Proc. Cyclotrons 2010*, p. 81.

[2] *The Stopping and Range of Ions in Solids*, eds. J. F. Ziegler, J. P. Biersack and U. Littmark (Pergamon, New York, 1985). http://www.srim.org, and J. F. Ziegler, M. D. Ziegler and J. P. Biersack, SRIM — the stopping and range of ions in matter, *Nucl. Instrum. Methods Phys. Res. Sec. B* **268**, 1818 (2010).

[3] N. Bohr, The penetration of atomic particles through matter, *Dan. Mat. Fys. Medd.* **18**(8) (1948).

[4] N. Bohr and J. Lindhard, Electron capture and loss by heavy ions penetrating through matter, *Dan. Mat. Fys. Medd.* **28**(7) (1954).

[5] N. O. Lassen, Total charges of fission fragments in gaseous and solid media, *Phys. Rev.* **79**, 1016 (1950).

[6] H.-D. Betz and L. Grodzins, Charge states and excitation of fast heavy ions passing through solids: A new model for the density effect, *Phys. Rev. Lett.* **25**, 211 (1970).

[7] H.-D. Betz, Heavy ion charge states, Chap. 1 of *App. Atom. Coll. Phys.*, Vol. 4 (Academic, 1983), ed. S. Datz.

[8] R. Bimbot *et al.*, Stopping power of gases for heavy ions: Gas–solid effect II, *Nucl. Instrum. Methods B* **44**, 19 (1989).

[9] H. Crawford, H. Gould, D. Greiner, P. Lindstrom and J. Symons, LBL-16241 (June 1983); H. Gould, D. Greiner, P. Lindstrom, T. J. M. Symons and H. Crawford, *Phys. Rev. Lett.* **52**, 180 (1984).

[10] I. S. Dimitriev and V. S. Nikolaev, *Zh. Eksp. Teor. Fiz.* **47**, 615 (1964), *Sov. Phys. JETP*, **20**, 409 (1965); V. S. Nikolaev and I. S. Dimitriev, *Phys. Lett. A* **28**, 277 (1968). Note: The Moscow State University group maintains a database of experimental and theoretical charge-changing cross sections and charge state distributions at: http://cdfe.sinp.msu.ru/services/cccs/HTM/main.htm

[11] S. H. Datz, C. D. Moak, H. O. Lutz, L. C. Northcliffe and L. B. Bridwell, *Atom. Data* **2**, 273 (1971).

[12] H. D. Betz, Charge states and charge-changing cross sections of fast heavy ions penetrating through gaseous and solid media, *Rev. Mod. Phys.* **44**, 465 (1972).

[13] E. Baron, M. Bajard and Ch. Ricaud, *Nucl. Instrum. Methods Phys. Res. A* **328** 177 (1993); E. Baron and B. Delaunay, *Phys. Rev. A* **12**, 40 (1975); E. Baron, *IEEE Trans. Nucl. Sci.* **NS-26**(2) (1979); E. Baron, M. Bajard and Ch. Ricaud, in 6th Conference on Electrostatic Accelerators and Associated Boosters (Montegrotto Terme, Padova, Italy; 1–5 June 1992).

[14] A. Leon, S. Melki, D. Lisfi, J. P. Grandin, P. Jardin, M. G. Suraud and A. Cassimi, *Atom. Data Nucl. Data Tables* **69**, 217 (1998).

[15] M. Portillo, Beam Physics, "Beam physics developments for a rare isotope accelerator," Ph.D. thesis (Physics and Astronomy, Michigan State University, 2002). www.nscl.msu.edu/ourlab/publications/year/2002/Theses.

[16] C. Scheidenberger, Th. Stoehlker, W. E. Meyerhof, H. Geissel, P. H. Mokler and B. Blank, *Nucl. Instrum. Methods Phys. Res. B* **142**, 441 (1998). Note: The codes GLOBAL and CHARGE can be downloaded from the GSI website and are also functional within the general purpose heavy ion code LISE++ (http://lise.nscl.msu.edu/lise.html)

[17] P. Thieberger, L. Ahrens, J. Alessi, J. Benjamin, M. Blaskiewicz, J. M. Brennan, K. Brown, C. Carlson, C. Gardner, W. Fischer, D. Gassner, J. Glenn, W. MacKay, G. Marr, T. Roser, K. Smith, L. Snydstrup, D. Steski, D. Trbojevic, N. Tsoupas, V. Zajic and K. Zeno, Improved gold ion stripping at 0.1 and 10 GeV/nucleon for the Relativistic Heavy Ion Collider, *Phys. Rev. Accel. Beams* **11**, 011001 (2008).

[18] J. P. Rozet, C. Stephan and D. Vernhet, *Nucl. Instrum. Methods Phys. Res. B* **107**, 67 (1996); D. Vernhet, J. P. Rozet, K. Wohrer, L. Adoui, C. Stéphan, A. Cassimi and J. M. Ramillon, *Nucl. Instrum. Methods Phys. Res. B* **107**, 71 (1996).

[19] J. P. Rozet, Towards improvement of ETACHA, Super Separator Spectrometer Workshop (GANIL, France, 2009); http://s3ws.ganil.fr/Presentations/wednesday/target/rozet_EtachaS32009.pdf

[20] J. P. Rozet, D. Vernhet and J.-S. Song, Preliminary results. Private communication (2013).

[21] J. Livesay, U. Greife, E. Kanter, J. Nolen, R. Watson and D. Youngblood, Evaluation of charge state distribution and energy straggling data for the first stripper section of the RIA driver, American Physical Society, Division of Nuclear Physics Fall Meeting (Tucson, Arizona, USA; 30 Oct.–1 Nov., 2003).

[22] I. Sugai *et al.*, Suppression of carbon buildup and lifetime improvement by heating carbon stripper foils, *Nucl. Instrum. Methods B* **269**, 223 (2011).

[23] P. Maier-Komor, G. Dollinger and H. J. Korner, Reproducibility and simplification of the preparation procedure for carbon stripper foils by laser plasma ablation deposition, *Nucl. Instrum. Methods A* **438**, 73 (1999).

[24] H. Hasebe *et al.*, Development of long-life carbon stripper foils for uranium ion beams, *Nucl. Instrum. Methods A* **613**, 44 (2007).

[25] R. W. Shaw *et al.*, Diamond stripper foil experience at SNS and PSR, in *Proc. EPAC08* (2008), p. 3563.

[26] K. von Reden *et al.*, Carbon nanotube foils for electron stripping in tandem accelerators, *Nucl. Instrum. Methods B* **261**, 44 (2007).

[27] I. Pavlovsky and R. L. Fink, Graphene stripper foils, *J. Vac. Sci. Technol. B* **30**, 03D106-1 (2012).

[28] J. L. Yntema and F. Nickel, Targets for heavy ion beams, in *Experimental Methods in Heavy Ion Physics*, ed. K. Bethge, Lecture Notes in Physics, Vol. 83 (Springer-Verlag, 1978).

[29] E. Baron, The beam-stripper interaction studies for GANIL, *IEEE Trans. Nucl. Sci.* **NS-26**, 2411 (1979).

[30] F. Nickel, The influence of temperature on stripper foil lifetimes, *Nucl. Instrum. Methods* **195**, 457 (1982).

[31] B. P. Gikal *et al.*, Lifetime calculation of charge-exchange carbon targets in intense heavy ion beams. JINR Communication No. P9-2005-110 (2005). (English translation provided to us by courtesy of O. Tarasov, NSCL.)

[32] S. G. Lebedev and A. S. Lebedev, Calculation of the lifetimes of thin stripper targets under bombardment of intense pulsed ions. PRST-AB **11**, 020401 (2008).

[33] J. Vetter, R. Scholz, D. Dobrev and L. Nistor, HREM investigation of latent tracks in GeS and mica induced by high energy ions, *Nucl. Instrum. Methods Phys. Res. B* **141**, 747 (1998). Figure 7 from http://upload.wikimedia.org/wikipedia/commons/6/6e/Latent_Track_Vetter_Scholz.jpg

[34] F. Seitz and J. S. Koehler, Displacement of atoms during irradiation, in *Solid State Physics*, Vol. 2, eds. F. Seitz and D. Turnbull (Academic, 1956).

[35] W. Assmann, M. Toulemonde and C. Trautmann, Electronic sputtering with swift heavy ions; R. Behrisch and W. Eckstein (eds.), *Sputtering by Particle Bombardment, Topics in Applied Physics* **110**, 401 (2007) (Springer-Verlag, Berlin, Heidelberg).

[36] P. Sigmund, Stopping of swift ions: solved and unsolved problems, *Dan. Mat. Fys. Medd.* **52**, 557 (2006).

[37] J. Liu *et al.*, Tracks of swift heavy ions in graphite studied by scanning tunneling microscopy, *Phys. Rev. B* **64**, 184115 (2001).

[38] A. Benyagoub and S. Klaumunzer, *Radiat. Eff. Defects Solid.* **126**, 105 (1993).

[39] K. Shima *et al.*, Appropriate carbon stripper foils of tandem accelerator in thickness, lifetime and transmission, in *Proc. 1999 Symp. North East. Accel. Pers.* (World Scientific, 2001), p. 266.

[40] P. Thieberger and H. E. Wegner, Test of heavy-ion gas-foil stripping for improved foil lifetime in tandem Van de Graaff accelerators, *Nucl. Instrum. Methods* **126**, 231 (1975).

[41] N. Angert *et al.*, The stripper-section of the UNILAC, in *Proc. 1976 Proton Linear Accel. Conf.* (Ontario, Canada), p. 286.

[42] W. Barth and P. Forck, The new gas stripper and charge state separator of the GSI high current injector, in *Proc. XX Int. Linac Conf.* (Monterrey, Mexico, 2000), p. 235.

[43] H. Kuboki *et al.*, Charge-state distribution measurements of 238U and 136Xe at 11 MeV/nucleon using gas charge stripper, *Phys. Rev. ST Accel. Beams* **13**, 093501 (2010).

[44] J. A. Nolen, Overview of LINAC applications at future radioactive beam facilities, in *Proc. 1996 LINAC Conference*, p. 32.

[45] P. Decrock, E. P. Kanter and J. A. Nolen, Low-energy stripping of Kr^+, Xe^+ and Pb^+ beams in helium and nitrogen, *Rev. Sci. Instrum.* **68**, 2322 (1997).

[46] H. Okuno *et al.*, Low-Z gas stripper as an alternative to carbon foils for the accelerations of high-power uranium beams, *Phys. Rev. ST Accel. Beams* **14**, 033503 (2011).

[47] H. Imao *et al.*, Charge stripping of 238U ion beam by helium gas stripper, *Phys. Rev. ST Accel. Beams* **15**, 123501 (2012).

[48] H. Imao *et al.*, Electron stripping of high-intensity 238U ion beam with recirculating He gas, in *Proc. 2013 Int. Part. Accel. Conf.* (Beijing, China, 2013) (THPWO038).

[49] A. Hershcovitch, High-pressure arcs as vacuum–atmosphere interface and plasma lens for nonvacuum electron welding machines, electron beam melting, and nonvacuum ion material modification, *J. Appl. Phys.* **79**, 5283 (1995).

[50] F. Marti *et al.*, Stopping of energetic radioactive ions using cyclotron principles, in *Proc. 2007 Conf. on Cyclotrons and Their Applications* (Giardini Naxos, Italy, 2007), p. 487.

[51] A. Hershcovitch and P. Thieberger, personal communication (2013).

[52] H. Nakamura, M. Ida, M. Sugimoto, H. Takeuchi and T. Yutani (International IFMIF Team), Status of lithium target system for international fusion materials irradiation facility (IFMIF), *Fusion Eng. Des.* **58–59**, 919 (2001).

[53] IFMIF Comprehensive Design Report (2003), http://www.iea.org/techno/technologies/fusion/IFMIF-CDR_partA.pdf

[54] J. G. Cramer, D. F. Burch and R. Rodenberg, Production of optically thin free-standing oil films from the edge of a rotating disc, *Nucl. Instrum. Methods* **185**, 29 (1981).

[55] B. T. Leemann *et al.*, A liquid film stripper for high intensity heavy ion beams, *IEEE Trans. Nucl. Sci.* **NS-28**, 2794 (1981).

[56] B. Gavin *et al.*, A continuous liquid sheet generator for ion stripping, LBL-17996 (1984) and *Nucl. Instrum. Methods B* **10/11**, 788 (1985).

[57] H. Ryuto *et al.*, Liquid film stripper for intense heavy-ion beams, *Jpn. J. Appl. Phys.* **43**, 7753 (2004).
[58] J. A. Nolen, C. B. Reed, A. Hassanein and I. C. Gomes, Liquid lithium cooling for 100 kW ISOL and fragmentation targets, *Nucl. Phys. A* **701**, 312 (2002).
[59] J. A. Nolen *et al.*, Behavior of liquid lithium jet irradiated by 1 MeV electron beams up to 20 kW, *Rev. Sci. Instrum.* **76**, 073501 (2005).
[60] Y. Momozaki, J. A. Nolen, C. B. Reed, V. Novick and J. R. Specht, Development of a liquid lithium thin film for use as a heavy ion beam stripper, *J. Instrum.* **4**, P04005 (2009).
[61] C. B. Reed, J. Nolen, Y. Momozaki, J. Specht, D. Chojnowski and R. Lanham, FRIB lithium stripper thickness and stability measurements. ANL Nuclear Engineering Division Technical Report ANL/NE-11/01 (2011).
[62] S. Kondrashev *et al.*, *Proc. Particle Accelerator Conference 2009*, Tu6RFP048, 1656–58, www.jacow.org.
[63] J. D. Sherman *et al.*, Status report on a dc 130 mA, 7S keV proton injector, *Rev. Sci. Instrum.* **69**, 1003 (1998).
[64] J. A. Nolen, J. Brandon, D. Chojnowski, P. Guetschow, S. Hitchcock, R. Lanham, F. Marti, R. McDaniel, Y. Momozaki, C. B. Reed, J. Ringold, J. Sherman, J.-S. Song and J. Specht, Report on the proton beam on liquid lithium film experiment. FRIB technical Note T30705-TD-000450-R001 (Mar. 2013).
[65] E. Nardi and Z. Zinamon, *Phys. Rev. Lett.* **49**, 1251 (1982).
[66] K. G. Dietrich *et al.*, Charge state of fast heavy ions in a hydrogen plasma, *Phys. Rev. Lett.* **69**, 3623 (1992).
[67] J. Jacoby *et al.*, *Phys. Rev. Lett.* **74**, 1550 (1995).
[68] C. Teske, G. Xu, J. Jacoby and T. Rienecker, Frankfurt, Beam Time Proposal: U275.

Jerry A. Nolen did his Ph.D. in Physics at Princeton. He was previously an Associate Director of the MSU National Superconducting Cyclotron Laboratory and is presently a Distinguished Fellow in the Physics Division of ANL and a Senior Advisor to the FRIB Project at MSU. His current research is in the field of radioactive beams, which involves high power, heavy ion superconducting linacs, high power density production targets and strippers, large scale electromagnetic separators, and other beam-optical instruments. He was a pioneer in the development of superconducting beam transport and magnetic spectrographs, including the NSCL S800 and the GANIL S^3.

Felix Marti obtained his Ph.D. in Physics from Michigan State University in 1977. He returned in 1979 to work on the design and construction of the superconducting cyclotrons K500 and K1200 at MSU. He was later involved in the design of the medical cyclotrons spearheaded by MSU. He has worked on beam dynamics, magnetic field design and measurements, and other areas, with strippers being the focus in recent years. From 1993 to 2009 he was head of the accelerator R&D department at NSCL. He is currently in charge of the charge stripping area for the Facility for Rare Isotope Beams being built at MSU.

Reviews of Accelerator Science and Technology
Vol. 6 (2013) 237–258

DOI: 10.1142/S1793626813300119

Targets and Secondary Beam Extraction

Etam Noah
DPNC – Department of Nuclear and Particle Physics,
University of Geneva,
24 Quai Ernest-Ansermet,
Geneva, CH-1211, Switzerland
Etam.Noah@unige.ch

Several applications make use of secondary beams of particles generated by the interaction of a primary beam of particles with a target. Spallation neutrons, bremsstrahlung photon-produced neutrons, radioactive ions and neutrinos are available to users at state-of-the-art facilities worldwide. Plans for even higher secondary beam intensities place severe constraints on the design of targets. This article reports on the main targetry challenges and highlights a variety of solutions for targetry and secondary beam extraction. Issues related to target station layout, instrumentation at the beam–target interface, safety and radioprotection are also discussed.

Keywords: High power target; spallation neutron; neutrino; radioactive ion beam; ISOL; in-flight.

1. Introduction

The comprehensive Canadian ING proposal was drafted in the 1960s for a 60 MW proton beam impinging on a liquid metal target, motivated by the need for intense fluxes of neutrons [1]. A wide range of applications was considered, including studies of radiation effects on structural and target materials, production of radioactive isotopes, studies of structure and dynamic processes in matter, and studies of the production of economical and safe nuclear power with accelerator-driven systems (ADSs). Fifty years later, the ING proton beam power still exceeds by over one order of magnitude the most powerful operational facilities. Limitations on available beam power on target have mostly come from accelerators, although increasingly the ability of targets and target stations to accommodate high power primary beams has been tested to extremes. We may not have reached 60 MW of beam power on target, but arguably more challenging conditions have been met, with highly pulsed beams depositing instantaneous power densities in the GW/cm^3 regime. High power targets serve several facilities that provide spallation neutrons, radioactive ion beams and neutrinos. They have also been proposed for other applications, such as transmutation of radioactive waste, accelerator production of tritium (APT) and the linear collider positron source.

This article reviews non-commercial targets exposed to beams of electrons, protons and ions in the energy range of 40 MeV/u to 400 GeV/u which are considered representative of the high power targetry field and the associated secondary particle extraction schemes. Emphasis is placed on operational and planned facilities with the assumption that the historical thread can be followed through the listed references.

2. Overview of Facilities

Accelerated beams of charged particles, the primary beams, are used to generate interactions in the target material, resulting in a flux of secondary particles. A fraction of these will be of interest to the user and may undergo further transformation/manipulation. The remaining secondaries usually contribute unwanted effects such as contamination (background) of the particles of interest, excess heat deposition, activation of the target station and radiation damage to the target structures. Optimization for a given application entails selecting the combination of primary particle beam and target material that maximizes the flux of useful secondaries whilst keeping unwanted products of beam interactions at a minimum.

When quoting primary beam power, the notation adopted here is with the subscript "b" appended to the unit (e.g. MW_b), to distinguish the primary beam power from the power deposited in the target, as done for example in Ref. 2. The subscript "th" is used to indicate target thermal power.

2.1. *Radioactive ion beams*

Radioactive ion beams (RIBs) are of interest for nuclear structure, nuclear astrophysics and fundamental interactions, as well as solid state physics and medical applications. Over 3000 nuclei of 118 elements are known to exist, with a further 3000–4000 predicted to exist. A fraction of these are thought to be discoverable. Several separation methods exist to produce radioactive ion beams, of which two are discussed here:

- In-flight;
- Isotope separation on-line (ISOL).

The in-flight method consists in accelerating light to heavy ion beams with energies in the range of ~100 MeV/u to 5 GeV/u toward a primary target in which they undergo fragmentation or fission. The resulting highly ionized fragments are selected in-flight by a fragment separator according to their masses and charges. Facilities include FAIR-GSI, RIBF-RIKEN and FRIB-MSU [3]. The in-flight method is well suited for the production of very-short-lived radioactive isotopes and, among all techniques, it has in recent years contributed the most to the discovery of new exotic nuclides.

Beyond an isotope half-life of ~1 ms, the ISOL method produces higher yields, and is suitable for re-acceleration of RIBs [4]. In this method, first used in the 1950s at the Niels Bohr Institute [5], a thick primary target is bombarded by a beam of primary particles (typically protons) or with neutrons from a reactor or spallation neutron source. The cascade model best describes the interaction of the energetic primaries with individual nucleons in the nucleus of the target material. The initial intranuclear cascade (or hadron cascade, since it involves charged pions, protons and neutrons) is characterized by kinetic energy transfer from primaries to target nucleons via a series of collisions. The resulting particles can leave the nucleus of origin and set off further spallation reactions in other nuclei, a process referred to as the internuclear cascade. The target nuclei are then left in a highly excited state, following which nuclear evaporation occurs, and with it the emission of neutrons (used in spallation neutron sources; see next section). The nuclear evaporation process can be treated with statistical models applied to fission. Final state fragmentation, fission or spallation reactions, illustrated in Fig. 1, lead to the production of a wide range of stable and radioactive isotopes of

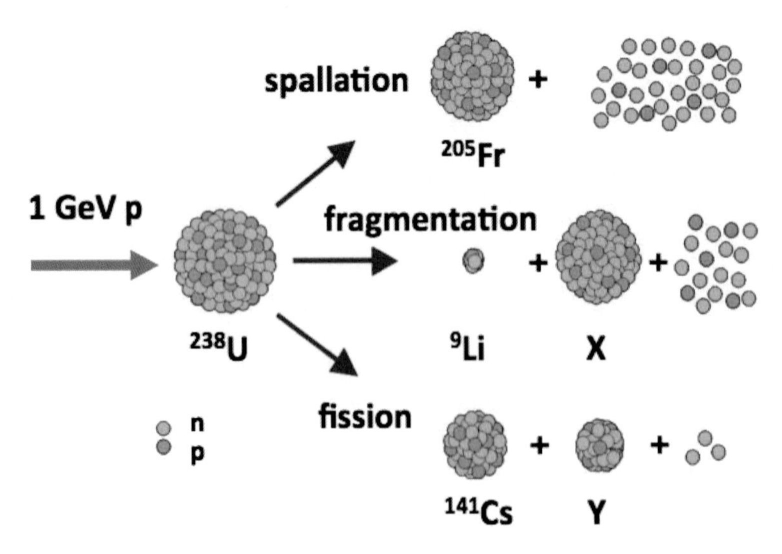

Fig. 1. The main production mechanisms for incident ~GeV protons on ^{238}U.

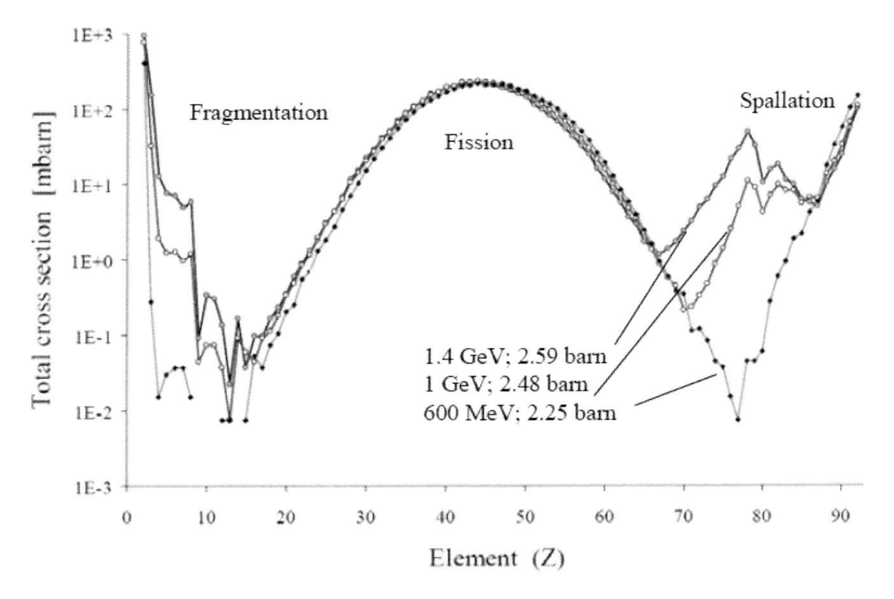

Fig. 2. The range of isotopes produced through interaction of ~GeV protons on ^{238}U.

elements from hydrogen to the target element and above $(Z + 2)$; see Fig. 2. A fraction diffuse out of the target matrix and effuse out of the target container through the transfer line to the ion source. They are ionized, extracted toward the secondary beam line, mass-separated, and in some cases re-accelerated toward experimental areas. More than 1000 different RIBs of 70 elements have been produced in this way with driver beams of protons or light ions from 40 MeV/A to 1.4 GeV/A at a dozen or so facilities worldwide including ISOLDE-CERN, ISAC-TRIUMF and IRIS-Gatchina.

The figure of merit is the RIB intensity at the experimental setup, given by

$$I_{\mathrm{RIB}} = (I_{\mathrm{prim}}\sigma_{\mathrm{prod}}N_{\mathrm{targ}})\epsilon_{\mathrm{rel}}\epsilon_{\mathrm{ion}}\epsilon_{\mathrm{trans}}, \quad (1)$$

where I_{prim} is the primary beam intensity, σ_{prod} the isotope production cross-section in the target material, N_{targ} the target thickness, ϵ_{rel} the isotope release efficiency, ϵ_{ion} the isotope ionization efficiency and $\epsilon_{\mathrm{trans}}$ the transport efficiency (which includes all stages from mass separation to post-acceleration if applicable).

A large variety of targets are employed for the production of RIBs with the ISOL method. They include:

- Refractory metal foils (e.g. Ta, Ti, Nb);
- Carbides (e.g. SiC, UC, ThC);
- Oxides (e.g. CaO, MgO, ZrO, CeO);
- Molten metals (e.g. La, Pb, Sn).

One crucial parameter concerning RIBs is contamination by unwanted isotopes, which are often produced in quantities orders of magnitude higher than the requested RIB. Mass separation alone is not sufficient, since isobars with essentially identical masses are extracted from the target along with the isotopes of interest. For the ISOL method, the requirements for beam purity (defined simply as the ratio of wanted RIB to total RIB sent to the experimental setup) are partially satisfied with combinations of molecular side bands, active transfer lines and selective ion sources [6, 7]. In practice, this has led to the adoption of over 100 different combinations of highly specialized secondary beam extraction systems optimized for a given range of isotopes, the widest variety of extraction systems by far of any facility discussed here. An example target unit from ISOLDE–CERN is shown in Fig. 3.

2.2. *Spallation neutron sources*

Spallation neutron sources are a viable alternative to research reactors for the provision of high fluxes of thermal, cold and ultracold neutrons. In the early days of proton-driven neutron sources, with access to technologies enabling the construction of proton accelerators with the required intensity, pulse length and repetition frequency, it was recognized that their developments would provide increases in peak thermal neutron fluxes more rapidly than further developments of research reactors, which were more costly,

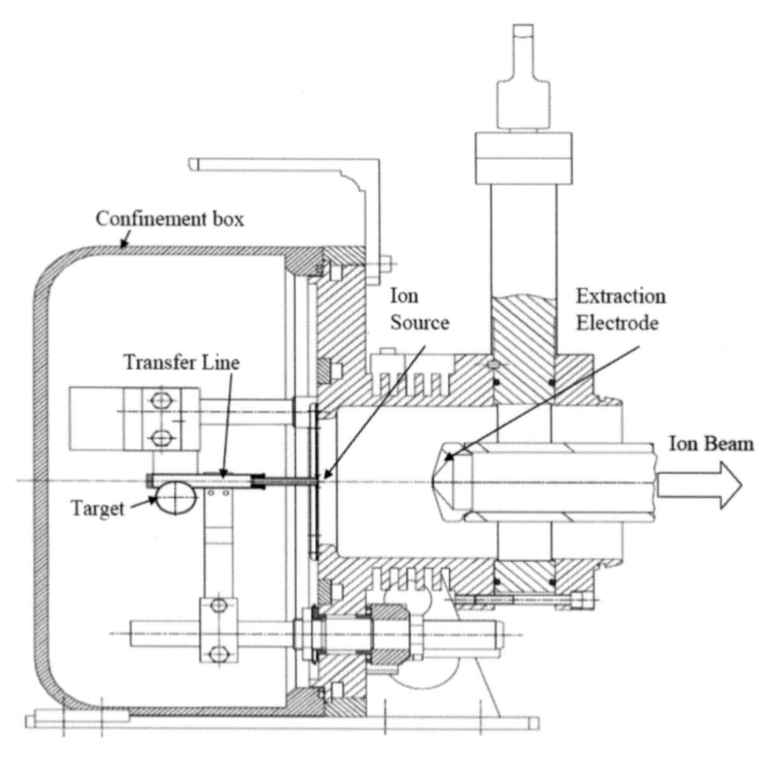

Fig. 3. A typical ISOLDE target unit system. Isotopes are produced in the target through interactions with a 1.4 GeV proton beam. The intensity and purity of the ion beam are determined by production and extraction processes.

and limited by the attainable power density in the core [8]. Operational facilities include the quasi-continuous SINQ-PSI, and the pulsed ISIS-RAL, SNS-ORNL and JSNS-J-PARC. Proposed future facilities include ESS-Lund, and CSNS in China. Thorough accounts of the physics and technology of such sources can be found elsewhere [8–10]. R&D for spallation facilities is relevant to ADS and APT. They also produce fluxes that are of interest to the fusion and fission communities for radiation damage studies on materials.

Typically, 0.5–5 GeV protons are used as primary particles to drive the spallation reaction in high Z target materials for the production of neutrons via the same process as that used for the production of RIBs. Ninety percent of produced neutrons are emitted isotropically around 1–2 MeV, with a long tail to the distribution, which extends up to the energy of the primaries.

A large fraction (50% or above) of the beam power will be deposited in the immediate surroundings of the cascade sites. Thus, neutron production and power deposition are intrinsically related. In the plane perpendicular to the beam axis, the power distribution is related to the incident beam profile. This distribution widens as the beam travels into the target, mostly due to the hadronic cascade from secondary protons. The effect is more pronounced at lower beam energies, as secondaries are emitted at wider angles with respect to the beam axis.

The ∼MeV neutrons produced in the initial interaction must be slowed to ∼meV for use by the experiments through elastic collisions in suitable moderators. Reflectors surround the target and moderators, and steer unmoderated neutrons back into the moderator. A host of combinations of moderators and reflectors are employed to maximize fluxes at three main classes of facilities distinguished by their primary proton beams:

- Quasi-continuous primary beam (e.g. SINQ-PSI);
- Short pulse beam (e.g. ∼μs ISIS-RAL, SNS-ORNL, JSNS-J-PARC);
- Long pulse beam (e.g. ∼ms ESS-Lund).

For continuous spallation sources, the time-averaged flux is to be optimized. Materials with low absorption cross-sections are used to ensure a long lifetime of moderated neutrons. Moderators are large for complete thermalization; for example, the SINQ D_2O tank surrounds the target and fulfills both moderator and reflector functions. Cold moderators are of the rethermalization type embedded in the D_2O tank, similar to research reactors, e.g. liquid D_2 [9].

For short pulse facilities, the thermal neutron flux is produced in a short burst so as not to contribute significantly to a spread in the arrival times of moderated neutrons. Neutron absorbing decouplers are placed around the moderators to prevent the return of slow neutrons from the reflector. Occasionally a poison (neutron-absorbing material) is placed inside the moderator to limit the time that slow neutrons spend diffusing out of the moderator, which also reduces the overall intensity by an order of magnitude.

Long pulse facilities take their beam directly from a pulsed linac without the need for a ring to compress the pulse duration. The proton pulse is significantly longer than characteristic times for the slowing down of neutrons in the moderator. Therefore the pulse width of moderated neutrons is similar to the proton pulse, typically ~1 ms. Choppers can be inserted in the neutron beam line to produce a neutron beam adequate for time-of-flight measurements such as those done by short pulse facilities [11].

The figure of merit used in optimization studies for the ESS is based on maximizing both thermal and cold neutron spectra with two separate Maxwellian distributions [12]. The combination of these distributions is characterized by the peak position of each, by the value of the neutron flux at their peaks, by the crossing point between the two distributions, and by the value of the neutron flux at the crossing point; see Fig. 4.

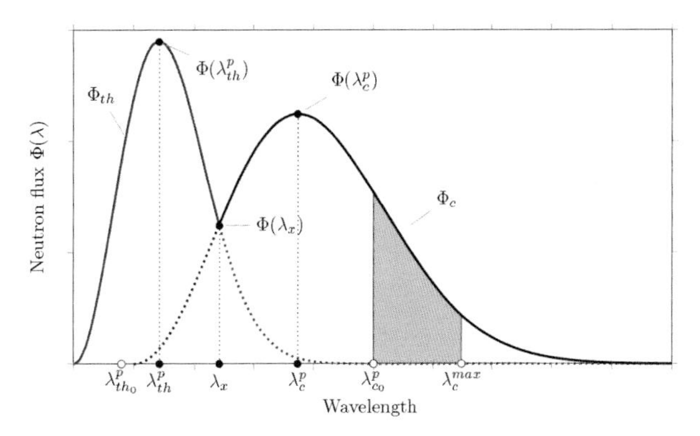

Fig. 4. Graphical description of the bispectral spectrum used in ESS optimization studies, from Ref. 12.

2.3. *Neutrino sources*

Detectors optimized for the reconstruction of neutrino interactions have measured neutrinos in so-called beam stop experiments such as Karmen-RAL or LSND-LANL, where neutrinos studied emerged directly from a target conceived for another main goal (such as production of spallation neutrons). Although there is some potential for beam stop neutrino studies at existing or planned spallation neutron sources (SNS, ESS), the discussion in this section centers on targets specifically built to provide neutrinos and optimize fluxes at a certain distance from the target. Such targets include:

- Conventional targets;
- Targets for muon storage rings;
- Beta beam targets based on RIBs.

2.3.1. *Conventional neutrino beams*

In a first stage, accelerated proton beams impinge on a target producing muons and mesons (pions and kaons), which are focused along the beam axis by magnetic horns (two in the case of CNGS-CERN and NuMI-FNAL, three in the case of T2K-J-PARC). As they travel down a long decay tunnel, the mesons decay "in-flight" into muons and neutrinos. The muons further decay into electrons and more neutrinos, e.g. $\pi^+ \to \mu^+ + \nu_\mu$ and $\mu^+ \to e^+ + \nu_e + \overline{\nu}_\mu$ or the corresponding (anti)leptons from a π^- beam. At the end of the decay tunnel, a hadron absorber is used to remove the remaining protons and mesons from the beam. The large muon flux is suppressed in a muon absorber and the earth shield that follows. A significant muon background can still be expected from the high energy tail of the muon spectrum at the near detector location (when there is one) ~100–1000 m. A background-free pure neutrino beam (of potentially two flavors) is then directed to the far detector location (a few kilometers for short baseline beams to a few hundred kilometers or even a few thousand kilometers for long baseline beams).

The adoption of horn-focusing systems is motivated by the need to have a directional beam with high flux, especially important given the distances to detectors. The following approximations for the kinematics of pion decay are valid here:

$$E_\nu = \frac{0.427E_\pi}{1+\gamma^2\theta^2}, \quad (2)$$

$$Flux = \frac{4\gamma^2}{(1+\gamma^2\theta^2)^2}\frac{A}{4\pi r^2}, \quad (3)$$

where γ is the Lorentz boost of the pion, θ the angle between the pion and neutrino flight paths, A the cross-sectional area of the detector, and r the distance to it. Neutrino energy and flux peak at $\theta = 0$. The pion production angle is typically $\sim 2/\gamma$, which is the same as θ above for zero-degree targeting. There is thus up to a factor of 25 ($\theta = 0$ case) in flux to be gained in adopting a horn-focusing system. The neutrino energy spectrum and flux are optimized by selecting appropriate target and horn combinations and by choosing θ. As an example, the T2K beam line is optimized for an off-axis beam (CNGS and NuMI are on-axis wideband beams) with $\theta = 2.5°$ and a three-horn system for a narrowband neutrino beam peaked at 0.6 GeV [13].

2.3.2. *Neutrinos from stored muons*

The Neutrino Factory (NF) relies on a muon storage ring as a source of neutrinos. A potential precursor to the NF is the nuSTORM facility, see Fig. 5, a muon storage ring–based facility designed to search for sterile neutrinos, provide precision measurements of neutrino cross-sections and act as a technology test bed for proposed neutrino factory and muon collider experiments [14].

A crucial figure of merit for these facilities is the number of muons, either produced directly in the target or decayed from pions and kaons, that end up within the muon accelerator or storage ring acceptance (for the NF, 30 mm transversely, 150 mm longitudinally, longitudinal momenta between 100 and 300 MeV/c). By tracking muons and their parent particles from the point of production through the decay channel to the accelerator or storage ring, it was confirmed that the dependence of muon yield on primary beam energy was relatively flat (within 10%) between 4 and 11 GeV for the NF configuration with a mercury jet target [15].

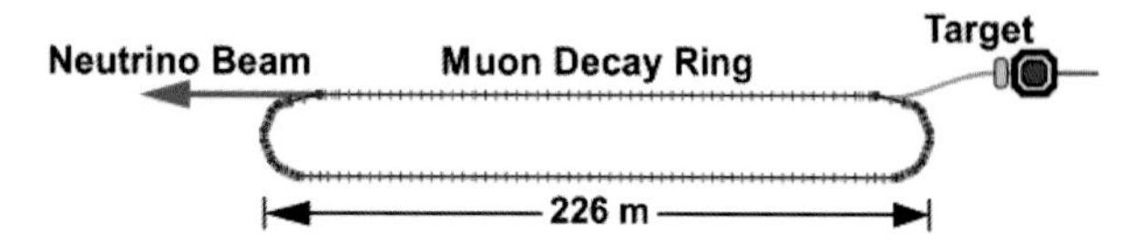

Fig. 5. Schematic of the nuSTORM neutrino source with its muon storage ring.

2.3.3. *Beta beam neutrino sources*

Beta beams explore the possibility of harnessing neutrinos from the beta decay of radioactive isotopes with an $\sim$1 s half-life [16]. These are produced in targets, ionized, accelerated to relativistic speeds and fed into a storage ring, where they undergo decay. The neutrino energy range of 0.1–5 GeV is given by the end point of the β spectrum and by the Lorentz boost, $E_{\text{max}} = \gamma Q_\beta$. Four isotopes are considered: ^{18}Ne, ^{8}B for neutrinos and ^{6}He, ^{8}Li for antineutrinos. Options include high $\gamma = 350$, where ^{8}B and ^{8}Li would be preferred since they have a higher Q_β and therefore lead to higher neutrino energies for the same γ (but less flux for the same neutrino energy since it is proportional to γ^2), and low $\gamma = 100$, which has been the focus of studies and where fluxes of 1.1×10^{18} ions per year for ^{18}Ne and 2.9×10^{18} ions per year for ^{6}He are needed.

2.4. *Electron linac-driven sources*

Several photon/neutron sources are driven by electron linacs, as opposed to hadron accelerators. They are economically viable and their good beam quality is suitable for cross-section measurements using the time of flight (TOF) method. For example, in the range of 50 keV–10 MeV, of interest for fission and fusion reactors and for the transmutation of minor actinides in nuclear waste, ORELA at ORNL has contributed 80% of US Evaluated Nuclear Data File (ENDF/B) evaluations. At electron-driven facilities, a chain of reactions leads to the production of the particles of interest. First, electrons are stopped in the target material producing photons by bremsstrahlung. These photons then strip neutrons from target nuclei by photoneutron or photofission processes from the same target or from a secondary target. Many facilities provide high neutron fluxes (e.g. ORELA-ORNL, ELBE-HZDR, GELINA-JRC). Some facilities plan to use the produced neutrons to initiate further reactions, such as photofission to produce radioactive ion beams (e.g. ARIEL-TRIUMF).

3. Technical Challenges for High Power Targetry

Challenges in the design and operation of targets exposed to high-power-accelerated beams are the following:

- Extraction of the heat deposited by the beam;
- For pulsed beams and to a lesser extent cw beams that cycle ON and OFF — transient effects, pressure waves, fatigue;
- Radiation damage to structures, which affects lifetime prediction, and hence exchange strategies and availability of beam;
- Knowledge of material properties for operation conditions of interest.

These challenges have to be met against a set of requirements for physics performance and for safety, both under normal conditions and accidental scenarios, minimizing at all times exposure to people, the environment and the machine.

3.1. *Technical challenges: Materials*

Unsurprisingly the overwhelming majority of studies concerning R&D on high power targets address materials issues. Knowledge of target and containment material behavior, their compatibility and the evolution of their properties under irradiation is crucial for design choices, target performance, lifetime estimations, accidental scenario definition and safety.

Attention is directed at both the target material itself (solid or liquid) and the target vessel and protective shroud material. The main target materials are graphite (low Z, excellent mechanical stress properties, high melting point), tungsten (high Z, high density, high melting point, much lower afterheat than the nearest equivalent tantalum) amongst the solids, and mercury, lead, lead bismuth eutectic (LBE) and lithium amongst the liquids. The main vessel materials are austenitic stainless steels (e.g. SS316L), martensitic stainless steels (e.g. T91), tantalum (for low power, high temperature applications), aluminum alloys, titanium alloys and beryllium (as a beam entrance window in some cases).

The thermomechanical properties of solid materials exposed to energetic primary beams are altered by irradiation-induced changes. These take two forms:

- Alteration of the chemical composition of the material;
- Kinetic energy transfer to atoms of the material.

The change in chemical composition is quantified by beam-induced impurities that have a significant impact on properties and are therefore material-dependent, such as hydrogen and helium in steels or silicon in aluminum.

Kinetic energy transfer will cause an atom to be displaced from its lattice site. The scale for this effect is quantified by the notion of displacements per atom (dpa) and is the most widely used unit of material damage due to irradiation. Dedicated irradiation programs such as STIP at PSI [17] address issues of radiation damage in a spallation environment, allowing some comparisons with the more established fission damage environment.

One example of the difficulty in setting a lifetime for target vessels is provided by the MEGAPIE project, an extensive program for the design, operation and post-irradiation examination of an LBE loop target at the MW_b level at SINQ-PSI [18]. Irradiation induces significant hardening and embrittlement in the T91 martensitic steel, which is the beam entrance window material in contact with flowing LBE. Moreover, flowing LBE leads to corrosion. With strong recommendations for operation of the T91 steel in a ductile rather than brittle condition, it took extensive studies to set the lifetime to 6 dpa, corresponding to 2.4 Ah of beam for MEGAPIE [19]. The corresponding peak beam-induced helium concentration is 600 appm. The He/dpa ratio of ~100 is typical of components directly exposed to the proton beam in a spallation environment and much higher than would be experienced in fission ~0.2 or fusion ~15 reactor environments [20].

Another example of lifetime estimation is that of SNS-ORNL. It is well known that stainless-steel-type 316L loses uniform elongation with increasing levels of dpa due to radiation effects in the target material. At around 10 dpa, the remaining ductility is low enough for there to be a risk of fracture to the target vessel. The target vessel itself is usually surrounded by a shroud. In the beam entrance window zone, either vessel or shroud, or both, are double-walled, so up to four layers (SNS) of containment can be present, depending on the design, which provides some redundancy. However, the simultaneous rupture of all layers would entail expensive cleanup of the spilled radioactive mercury. Failure probabilities were estimated by JSNS to be 10^{-6} for the target shroud, and 99.9% for the target vessel for 2500 h at 1 MW [21]. The initial allowable exposure for the beam entrance window was set to 5 dpa for JSNS [22]. At SNS, the operational limit, defined as a target vessel failure (the safety shroud does not fail), is estimated to be somewhere between 5 and 10 dpa [23]. The target exchange strategy is to replace the target either when it reaches its operational limit or during planned shutdowns of the facility. In practice, to avoid unplanned interruptions to neutron users, exchanges during planned shutdowns are preferred. As will be described in another section, damage due to cavitation-induced erosion of the target vessel competes with radiation damage to set allowable vessel lifetimes.

For low Z targets using graphite, the choice of grade can be an issue, with significantly different degradation of thermal conductivity as a function of radiation damage.

Tungsten is one of the more widely used high Z target materials, and yet very little data exists on its mechanical properties under irradiation. It is known to be corroded by water, so for water-cooled systems cladding or canning tungsten with another material such as tantalum is widely adopted [24].

For the large variety of solid targets used to produce RIBs, parameters such as grain size and porosity play a crucial role in the release of isotopes from the target matrix. Operating temperatures are typically 1400–2200°C. Under such conditions, release parameters are greatly affected by high temperature sintering and radiation damage, and yields of RIBs drop below useable values, at which point the target reaches its end of life. Typical lifetimes are of the order of 1.5×10^{19} protons at ISOLDE with a pulsed beam (a few weeks), and 3.2×10^{20} protons at TRIUMF with a cw beam. For comparison, 100 kW_b beam at a next-generation facility such as EURISOL corresponds to 6.25×10^{14} 1 GeV protons/s or 5.4×10^{19} protons/day [2].

Typical heavy liquid metals used in high power targetry are mercury, lead and lead bismuth eutectic (LBE). Not only do they not suffer structural damage from radiation, they also act as the primary coolant, transporting the deposited heat from the proton beam interaction zone to a heat exchanger located elsewhere. Their specific heats at typical operating temperatures are all within 10% of ~0.13 J/gK. The more notable differences in their properties are listed in Table 1. Hg has a higher density (13.4 g/cm^3) than Pb (10.47 g/cm^3) or LBE (10.4 g/cm^3) at operating temperatures of 90°C, 450°C and 250°C, respectively. This higher density for Hg does not, however, lead to a proportionally higher neutron yield, partly because of the much higher thermal neutron absorption coefficients for Hg. For 2.5 GeV protons on target, neutron yields were calculated to be ~6% higher in Hg compared to Pb. Hg is liquid at room temperature, which is a crucial advantage for operation, but it has a low boiling point, so the range of operation is limited. Pb and LBE have much larger operating ranges but are solid at room temperature, and so require auxiliary heating systems. That they are solid at room temperature is an advantage, however, for final disposal of ~m^3 radioactive material compared with Hg, for which a long term disposal solution must be further investigated. LBE expands slowly upon solidification, which is an issue for design of containment, along with its corrosive properties (which can to some extent be mitigated by controlling the oxygen content). Another issue for LBE is the production of α-active polonium isotopes. Data exist on the evaporation rates of polonium, which are low, under 600°C. Additional polonium release paths are less well understood and could include more volatile polonium compounds, or the decay of highly volatile astatine isotopes into polonium once the astatine is released from the LBE [25]. Other alloys of lead have been suggested as alternatives with orders of magnitude less Po production, such as lead gold eutectic (LGE) [26], but none come close to lead and LBE in terms of operational experience and available data.

Table 1. Some properties of typical liquid metals used in targetry. LBE: lead bismuth eutectic.

Property	Unit	Hg	Pb	LBE
Composition	%	Elem.	Elem.	45(Pb)/55(Bi)
Melting point	(°C)	−39	327	124
Boiling point	(°C)	357	1750	1670
Th. neutron abs.	(b)	389	0.17	0.11
Po production		Low	Med.	High
Hg production		Med.	High	High

Experience with mercury exposed to high beam powers is being gained at SNS and JSNS. Experimental studies of the radiochemistry of mercury have shown that the radionuclides produced by the proton beam tend to accumulate on vessel walls and free surfaces rather than be distributed homogeneously in the bulk liquid [27]. During maintenance of Hg loop components, the Hg is drained to a storage tank. Residual activation of empty pipes can be underestimated with the assumption of homogeneous distribution of isotopes.

3.2. *Technical challenges: Energy deposition*

From a targetry perspective, the definition of a high intensity beam must take into account the pulse structure of the beam, its spot size, and the fraction of beam power deposited in the target. Such an approach, combined with other operating conditions, highlights the range of challenges met in targetry beyond a discussion on time-averaged power levels on target. MW_b-class targets are all subject to stringent safety requirements; however, the technical challenges of some kW_b targets are worth mentioning here. MW_b-class proton beams may deposit <5% of their power in the target, for example with the T2K pion production target, where 23.4 kW of beam is deposited in the graphite target by the 750 kW_b proton beam. That is not to play down the complexity of such targets, but rather to highlight that of other targets exposed to kW_b beams, where most of the beam is deposited in the target. Targets exposed to kW_b electron beams in the 10–200 MeV range receive a large fraction of the beam power in $\sim$cm^3 volumes, leading to deposited power densities of GW/cm^3 within pulse durations as low as 2 ps; see Table 2. Complexity in design for electron beam facilities is demonstrated by a variety of operating targets, such as the GELINA-JRC target, a rotating U-10%Mo target cooled with a liquid Hg loop; the ELBE-HZDR target, a liquid lead loop; and the ORELA-ORNL target, a water-cooled tantalum plate target.

Requirements for high secondary beam brilliance from small target volumes lead almost inevitably to high deposited power densities in the target. Once the power is deposited in the target, heat transfer and removal can occur in several ways. Where high temperatures are required or permissible (>1000°C), radiation cooling to a surface located further away can be an option. Critical parameters for radiation cooling are the temperature itself (since it scales as T^4), available surface area and emissivities of the surfaces involved. For lower temperatures, natural convection has been proposed even for relatively high beam powers (e.g. ADS) but forced convection with a liquid or gaseous coolant in a loop is most frequently adopted. For some applications, liquid metals are favored if they can act as both production target and coolant. Many targets function with a combination of solid target material and suitable coolant (water, helium gas). Radiation cooling, natural convection and heat conduction to surrounding components are the default choices in accidental scenarios with loss of coolant, where afterheat from activated target material must be managed.

Figure 6 illustrates the main loads on a rotating target, pressure loads, static and dynamic loads, thermal loads and fatigue loads due to a pulsed proton beam, and beam trips during the operation phase. These loads lead to a complex stress profile for which data do not always exist. For example, fatigue data in a radiation environment are lacking to predict the behavior of targets under intense cyclic stress loads. One feature to note is the importance of beam trips: switching the beam off can introduce larger swings in stress levels than loads due to a pulsed beam. The number and duration of beam trip cycles must therefore be estimated and taken into account during the design phase transient load analyses.

Stresses from the pulsed nature of some beams in particular affect materials in different ways. The energy deposited in a target material during a pulse will introduce a temperature jump in the material. Depending on the target geometry and cooling scheme, the temperature profile will be nonuniform, introducing further stresses within the target material (solids). If the pulse is short, of the order of microseconds or less, heat conduction across the target material within the pulse is negligible, leading to a further source of stress due to reaction to thermal expansion by mass inertia of the target material [28]. This type of stress will depend on the state of the material, whether it remains elastic or whether it is heated enough to become plastic.

Cavitation-induced erosion of containment vessels is a significant and at times dominant damage mechanism for liquid metal targets at pulsed beam

Table 2. Targets at representative particle accelerator facilities. Parameters are nominal design parameters; some existing facilities do not operate at nominal values. Target coolant is the main heat transfer medium. Transfer by radiation is indicated (Rad.). Abbreviations for target types are as follows: liquid metal loop (LML), rotating solid target (ROT), stationary solid target (STAT). The time-averaged beam power is listed along with the pulse power density. The pulse power density represents the peak power deposition in the target material, taking into account the pulse length and repetition frequency, i.e. it is equivalent to the peak pulse energy density divided by the pulse length. Most of the facilities listed are operational. SINQ-MEGAPIE operated for several months in 2006 and was dismantled. Four of the facilities are proposals: the Neutrino Factory, ESS-Lund, EURISOL and FRIB.

Facility	Material	Target coolant	Type	Primary beam parameters				Primary beam energy		
				Energy (GeV/A)	Pulse length (μs)	Rep. rate (Hz)	Beam spot (mm)	Avg. power (MW)	Pulse energy (J/pulse)	Pulse pow. dens. (MW/cm^3) pulse
Neutrino targets										
NuMI	Graphite	Water	STAT	120	10	0.535	1	0.411	769,000	70
CNGS	Graphite	Rad.	STAT	400	10.5 × 2	0.167	0.53	0.750	2,250,000	
J-PARC-T2K	Graphite	Helium	STAT	30	4.2	0.47	$\sigma 4$	0.750	1,584,000	
Neutrino Factory	Hg	Hg	Jet	8	0.002	50	$\sigma 0.5$	4	80,000	
Spallation neutron sources										
SNS	Hg	Hg	LML	1	0.695	60	70 × 200	1.4	23,333	19
JSNS	Hg	Hg	LML	3	1	25	70 × 180	1	40,000	
LANSCE	W	Water	STAT	0.8	0.25	20		0.1	5000	70
SINQ-solid	Pb	Water	Solid	0.575	cw	cw	100 × 100	0.7	cw	N/A
SINQ-MEGAPIE	LBE	LBE	LML	0.575	cw	cw	100 × 100	0.7	cw	N/A
ESS-Lund	W	Helium	ROT	2.5	2860	14	60 × 160	5	357,143	0.08
Radioactive isotope targets ISOL and in-flight										
TRIUMF-ISAC	Many	Rad.	STAT	0.5	cw	cw	$\sigma 3$	0.05		N/A
ISOLDE	Many	Rad.	STAT	1.4	2.4	0.8	$\sigma 4$	0.003	7400	40 (Pb)
EURISOL MW	Hg	Hg	LML	1.0	cw	cw		4	cw	N/A
FAIR	Graphite		ROT	2.7				0.036		0.6 (stat.)
FRIB	Graphite	Rad.	ROT	0.266				0.4		60 (stat.)
RIBF	Graph./Be/W	Water	ROT	0.350	cw	cw	1	0.100		5.7 (stat.)
Electron linac-driven sources										
GELINA	U, 10% Mo	Hg	ROT + LML	0.1	0.00067	800	1	0.007	9	
ELBE	Pb	Pb	LML	0.04	0.000002	13,000,000	0.3	0.040	0.003	1600
ORELA	Ta	Water	Plates	0.18	0.024	1000	1	0.060	60	1000

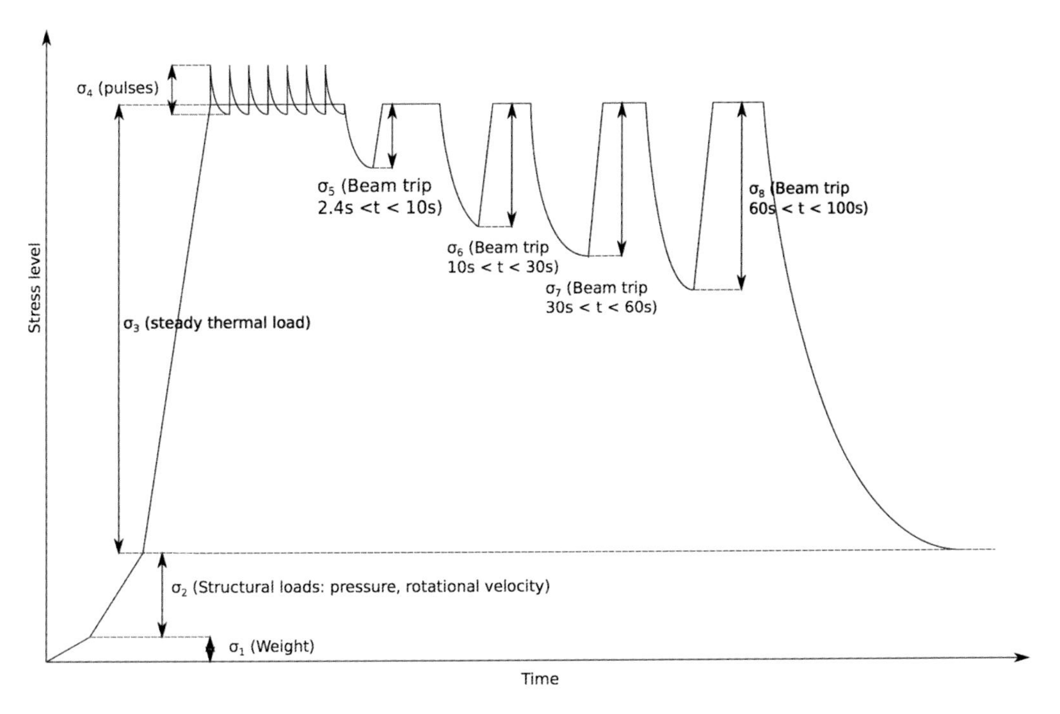

Fig. 6. Stresses due to pulses and beam trips in the ESS rotating target from Ref. 12.

facilities. When the energy deposition rate within a proton pulse is high enough to induce a thermal shock, the resulting compressive pressure wave is reflected by the container wall, creating a rarefaction zone in the liquid metal, where a cavitation bubble forms. In a free surface jet, the collapse of a cavitation bubble would generate a liquid jet that goes through the bubble itself (referred to as the microjet), and one that travels in the opposite direction, away from the bubble (referred to as the counterjet). In the confined liquid metal vessels of spallation neutron sources, the microjets traveling through the bubble at high velocity impinge on the vessel wall, creating pits that erode it and can lead to cracks and holes in the vessel. Stresses resulting from the cavitation bubble collapse lead to additional fatigue of the vessel wall.

One particularity for ISOL RIB targets is that the zone where the production of isotopes occurs must be kept at high temperatures to promote diffusion and effusion and suppress adsorption onto the target container walls. This is achieved by the beam itself for high power beams on direct targets, but external heating must be provided for lower beam power operation or for secondary targets not directly exposed to the primary beam, when isotope production occurs via neutron- or photon-induced fission.

4. Target Station Layout

The target station has the function of housing the target and its ancillary systems. It also has to interface with primary beam systems and secondary beam lines. The design of a target station is heavily constrained by requirements to handle both prompt radiation from the interaction of the primary beam with the target, and decay radiation from radioactive inventories. Shielding and handling concepts ensure that exposure to personnel and the environment meets safety rules during the operational and maintenance phases.

The more recent operational and planned MW spallation neutron sources adopt a layout with a central monolith containing the target, moderators and reflectors and from which the neutron beam lines are distributed along the horizontal plane and cover 110°–120° on either side of the horizontal proton beam axis; see Fig. 7. Because of the number of neutron beam lines tens to hundreds of meters feeding experiments in adjacent halls, the monolith is at ground level. Zoning concepts usually have the proton beam interface area and active cells used to process radioactive components at the same level as the monolith. A floor above the monolith, the high bay, is dedicated to transport into and out of each of the

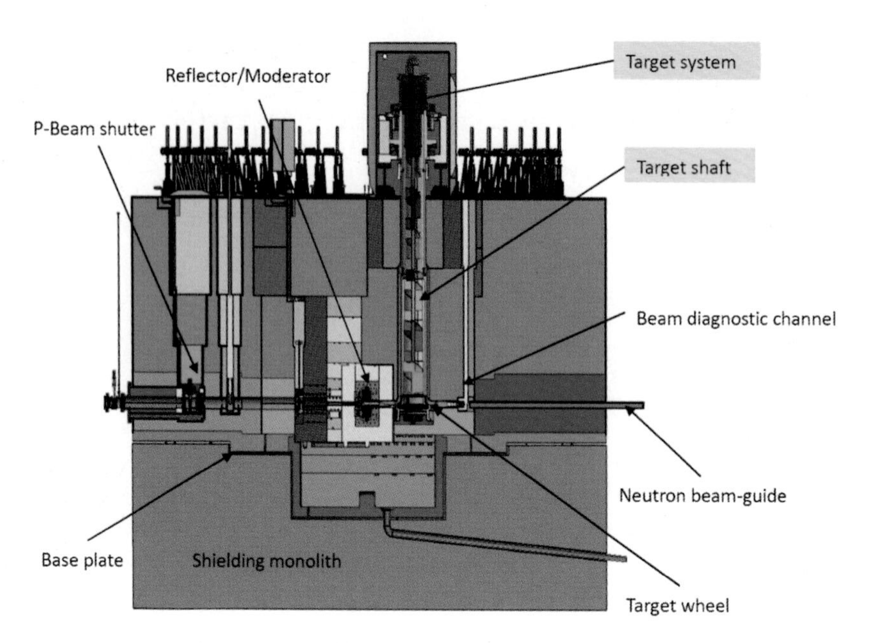

Fig. 7. The ESS target station monolith from Ref. 12.

zones. One example of a vertical proton beam line in the target zone is at the SINQ-PSI facility, where the proton beam hits the target vertically from below, a design choice motivated by full 360° for neutron beam lines and the now-obsolete plan to adopt the natural convection-cooled liquid metal target [9].

5. Safety and Radioprotection

As primary and secondary particle beam intensities increase, so does the importance of safety aspects. At MW facilities, these become the dominant criteria in the design of the target and target station. The definition of accident scenarios is particularly important. Some incidents and accidents are reported in the literature, providing a valuable source of information for designing future systems.

Facility safety can be defined by the set of safety systems, protocols, and the safety culture of the personnel. Each facility is therefore unique, although some general safety principles are universal. The main safety objective is the protection of individuals and the environment from harm arising from the construction, operation, maintenance and decommissioning of a facility. It is met through principles such as the requirement that exposure to hazards be "as low as reasonably achievable" (ALARA), including activation of buildings, systems, fluids, soil and ground water. The defence-in-depth concept, developed through years of experience at nuclear facilities, is applicable to high power targets. It relies on several levels of redundant protection through the implementation of independent sequential confinement barriers.

Passive safety systems relying on natural laws and material properties can provide protection independently of power or human intervention. Unlike traditional nuclear reactors, there is no risk of targets going "critical," so that when the proton beam is switched off the heat source is limited to residual afterheat, typically a few kW at MW facilities.

Safety, radioprotection and environmental aspects must be integrated at an early phase of the design. When possible, safety authorities are drafted into the design process for guidance (it goes both ways; safety authorities are not always ready for proposals with exotic beams, in contrast to more traditional facilities such as reactors). As an example, the ESS preliminary design work centered on Hg as a target material. Due to local regulations, however, it had to be demonstrated that there were no viable alternatives to Hg. In addition to Hg being chemically hazardous, the motivation for banning it in Sweden, disposal of activated Hg is nontrivial. It requires several processes for solidification before long term disposal, which are not demonstrated.

For next-generation ISOL facilities, direct irradiation of actinide targets leading to alpha-active products such as ^{210}Po and activation of post-accelerated beamlines in experimental areas are significant safety issues.

6. Beam–Target Interface Instrumentation

One of the most likely causes of incidents at a high power accelerator-driven facility is an off-normal proton beam on target related to the following beam parameters:

- Spot size,
- Position,
- Pulse duration,
- Intensity.

Overfocusing of the quasi-cw proton beam by up to a factor of almost 2 has been reported at SINQ-PSI [29, 30], with some damage to components within the target shroud, but no release of radioactivity.

On 23 May 2013 an accident at J-PARC led to accidental release of radioactivity, exposing some workers in the hadron facility experimental hall to higher-than-normal levels of radiation (although still within allowable limits). This was due to a malfunction of the beam extraction system, which sent 2×10^{13} 50 GeV protons on a gold target in 5 ms instead of the nominal 3×10^{13} protons over 2 s, resulting in overheating of the gold target with release of radioactivity [31]. This led to closure of the facility for a number of months.

Specific instrumentation is required to steer and monitor beam and target. Taking the T2K primary beamline as an example, its final focusing section consists of four steering, two dipole and four quadrupole normal-conducting magnets which position and focus the beam onto the target [32]. Five current transformers (CTs), 21 electrostatic monitors (ESMs), 19 segmented secondary emission monitors (SSEMs) and 50 beam loss monitors (BLMs) are used to monitor the intensity, position, profile and loss of the proton beam. The CT measures the absolute beam intensity with a 2% uncertainty and beam timing with a precision better than 10 ns. The beam position is measured with better than 450 μm precision. The systematic uncertainty on beam width measurement is 200 μm.

At the operational spallation neutron sources with proton pulses in the microsecond regime, cavitation-induced erosion scales as high as P^4, as discussed previously. At SNS, the last beam profile monitor is a multiwire harp located 9.5 m upstream of the target. Unfortunately, this device has an estimated uncertainty of 25%. At JSNS, the last beam profile monitor is significantly closer to the target, affording better accuracy. SNS developed an optical system, viewing the profile directly on the target shroud [33]. The system can be used at full power and is based on a flame-sprayed Al_2O_3:Cr luminescent coating of the beam entrance window portion of the shroud. An imaging system records the change in light yield of the coating as a function of beam power. JSNS developed and deployed a system to monitor target vibrations based on a laser Doppler vibrometer [34].

SINQ-PSI uses a thermal incandescence-based technique, which, though not proportional to beam intensity, nevertheless has good sensitivity to off-normal conditions [35].

7. Worldwide R&D Targetry Efforts

R&D efforts on targetry have been ongoing through the need to optimize the performance and lifetime of existing targets (e.g. the liquid metal targetry program at ISOLDE from 1992, SNS) and through programs to investigate the feasibility of advanced facilities (SNQ, ESS Germany, ESS-Lund, EURISOL, Neutrino Factory, ADS, APT). A review of selected examples from different types of facilities is given here.

7.1. *ISOL RIB R&D*

The EURISOL-DS (European Isotope Separation Online Design Study — 2005–2009) addressed some of the main challenges for ISOL RIB targets subjected to high intensity primary beams with the aim of producing orders of magnitude higher intensities of RIBs than are currently available, as well as new rare isotopes. The EURISOL-DS basic parameters were a 4 MW 1 GeV cw proton beam incident on up to four target stations in parallel, one housing a two-step 4 MW_b mercury converter target for the production of neutrons, which then induce fission in a secondary actinide target, releasing RIBs. Three additional target stations each receive 100 kW_b for the production of RIBs, mainly via spallation and fragmentation reactions, in a large variety of direct targets combined with specific transfer lines and ion sources.

7.1.1. *Direct targets*

Amongst the different RIB target classes, oxide targets are challenging due to their low thermal conductivities. High operating temperatures are required to reduce diffusion and effusion time constants. For short-lived species, compact target bodies lead to fast extraction times since they minimize effusion times. This, though, is incompatible with the requirement for larger surface areas to dissipate the high heat loads through radiative heat transfer, which is preferable at these high temperatures. With a 20-cm-long, 2 cm radius target container, 10–50 kW can be radiated away at high temperatures between 1400°C and 2200°C for a container emissivity of 0.9. A segmented approach, with multiple targets each taking a fraction of the beam (e.g. 25 kW each for four targets), was therefore adopted to handle the 100 kW_b of beam power. This has the advantage of spreading the heat load whilst keeping target dimensions compact for fast effusion of isotopes from production site to ion source. Developments of oxide target pills brazed (or diffusion-bonded) onto niobium plates and then using the principle of differential dilatation (whereby the niobium plates heat up, and expand more rapidly than the tantalum container), ensuring a good thermal contact conductance, and therefore heat transfer from the zone where the beam deposits its energy to the tantalum container surface. These oxide target pills were loaded into a special target container capable of dissipating >10 kW of power by radiation through a set of fins (that bring the effective emissivity for the tantalum container surface to ~0.9) using a design operational at ISAC-TRIUMF [4], an evolution of the RIST target [36].

The merging of non-ionized isotopic fluxes from multiple target bodies into a single ion source was benchmarked at ISOLDE with the Bivalve experiment. The test involved two target containers (each with a transfer line), which were equipped with remotely controlled tight valves and connected to one ion source [37]. Release properties for the system were measured, taking into account effusion delays due to the probability that a fraction of isotopes revisit the other transfer line and target and can decay in the process.

The current limit for liquid metals directly exposed to a 1 GeV pulsed proton beam in a confined static container for RIB production is ~5 kW. For the EURISOL 100 kW_b, a target assembly separating the functions of isotope production and diffusion out of the bulk target material was proposed: the beam is incident on an irradiation cell, and then the liquid metal flows through a sieve structure, which creates a shower or droplets of liquid metal into the diffusion chamber, promoting faster release of isotopes. The liquid metal is in a loop, which recirculates the liquid collected in the diffusion chamber back into the irradiation chamber, and dissipates the deposited beam energy by forced convection cooling through a heat exchanger. Flow velocity must be matched to the cooling requirements, but also to the requirement that the mass flow in the diffusion chamber be slow enough to allow isotopes enough time to diffuse out of the metal. The diffusion chamber concept is applicable to any target material that can be fluidized, and it could also serve to purify liquid metal loops at spallation neutron facilities.

Similar examples of the optimization of the lifetime of targets producing isotopes can be found elsewhere, such as for the production of medical radioisotopes at the Isotope Production Facility (IPF) at the Los Alamos Neutron Science Center (LANSCE), where an extensive program was conducted to increase the lifetime of the Nb capsules containing Ga targets from which ^{68}Ge is produced [38]. The Nb capsules are subjected to corrosive attack of the Ga and to the pulsed 90 MeV 230 μA beam, 12.7 mm FWHM spot diameter, swept across the face of the target in circular fashion with a sweep radius of 12.7 mm at 5 kHz.

7.1.2. *Two-step targets*

The main target station at EURISOL operates with a two-step target concept. It houses a mercury target based on the neutron converter ISOL concept, which was first employed at low power in the 1950s. For high power applications, the idea that the power could be dissipated in the converter target, leaving the function of RIB production to a secondary target, was espoused by Nolen [39], and explored by others [5, 40]. The neutron converter is a spallation neutron target, in this case mercury, with 2.3 MW of the 4 MW 1 GeV proton beam deposited in the target. Up to six actinide secondary target modules are placed close to the mercury converter target following the MAFF (Munich Accelerator for Fission Fragments) and PIAFE (Projet d'Ionisation et d'Acceleration de Faisceau Exotiques) configurations, for the production of fission fragments, which are then extracted using the schemes outlined previously for RIBs. Possibilities for the actinide target include highly enriched uranium dispersed in a graphite matrix with a ^{235}U density of 12 g/cm^3, for which a thermal neutron flux is required (each target unit would handle up to 10^{15} fissions/s and 30 kW of heat), or ^{238}U, for which a harder neutron flux is required ($\sim 10^{14}$ fissions/s expected). The feasibility of high density pellets of uranium carbide is the subject of dedicated studies [41].

The EURISOL-DS mercury converter adopts the coaxial guided stream (CGS) design, where the mercury flows within a double-walled tube with a proton beam window at one end [42]. The mercury flows in the outer part of the tube toward the proton beam, executes a u-turn at the window and flows back along the inner part of the tube in the same direction as the proton beam. Peak energy deposition is 1.9 kW/cm^3/MW$_b$ and 0.9 kW/cm^3/MW$_b$ in the Hg and beam window, respectively [2]. An optimization process was conducted over several iterations aimed at minimizing pressure losses and preventing vaporization and cavitation in the fluid whilst lowering temperature and thermal stresses in the beam window below allowable stress limits for irradiated materials. This resulted in the insertion of special annular flow guides to accelerate flow, increase local cooling and reduce the pressure drop at the u-turn close to the proton/target beam window; a full-size mockup was built and tested [43]. With a bulk pressure of 7.5 bar, the mercury peak temperature is 180°C and its maximum velocity is 6 m/s at the u-turn. The maximum beam window temperature and von Mises stress are 200°C and 135 MPa, respectively. An alternative converter was proposed, with a windowless design that could accommodate much higher power densities of 25 kW/cm^3. The resulting harder neutron spectrum is also better suited for ^{238}U.

7.1.3. *R&D facilities*

Prototyping work on targets and ion sources at oversubscribed operational facilities is not straightforward. The potential advantages of improvements in RIBs for the prototypes are offset by the risk of not delivering beams to users. More convenience is afforded by dedicated test facilities such as the Holifield Radioactive Ion Beam Facility (HRIBF) at ORNL [44]. Here novel ISOL concepts can be easily tested in an environment that allows hands-on access to the target and ion source assembly for fast modification and further testing if required. The primary 50 MeV proton beam is limited to a current of 50 nA (2.5 W) for safety considerations. Nevertheless, this is sufficient to measure yields from ISOL target and ion source combinations by gamma ray spectroscopy.

7.2. *In-flight RIB R&D*

For a review of the challenges associated with in-flight high power density targets, see for example Ref. 3. The maximum ion energies are 2700, 350 and 266 MeV/A for FAIR-GSI, RIBF-RIKEN and FRIB-MSU, respectively. The maximum primary beam powers are 36, 100 and 400 kW$_b$ for FAIR-GSI, RIBF-RIKEN and FRIB-MSU, respectively. Heavy ion beams have a short range in the target, leading to high deposited power densities, the highest for uranium beams ranging from 0.6 (FAIR, although it is pulsed so much higher peak power densities) to 60 MW/cm^3 (FRIB) for static targets with no rotation.

Targets are typically low Z graphite operating at high temperature (750°C for FAIR to more than 2000°C for FRIB). All facilities have adopted a rotating wheel design to spread the deposited power over a larger area. RIBF also used tungsten, notably in prototyping activities to validate the target concept with ^{48}Ca beams.

A windowless liquid lithium target had been proposed for the RIA facility: the 20 m/s flow ensured that the 4 MW/cm^3 power deposition for a 400 kW U beam on static lithium was reduced to an acceptable 200 W/cm^3 with a hot spot temperature of 300°C [45].

One important step in the selection of target material was the testing of graphite at high temperature showing that for polycrystalline graphite irradiated at high temperature, significant annealing of damage occurs above 1200°C [46].

7.3. *Spallation neutron targetry R&D*

Much work has been devoted to this field in the last two decades or so with the design, commissioning and operation of SNS and JSNS, and more recently with the resurgence of the ESS project, originally proposed in Germany and now set to be constructed in Lund, Sweden.

The R&D efforts for liquid mercury targets operational at the highest power pulsed beam facilities SNS and JSNS center on the mitigation of cavitation-induced erosion damage to the target vessels [47]. Known mitigation strategies include surface hardening [48] and the introduction of gas bubbles into the liquid metal stream [49] or close to the vessel surface [50, 51].

Recently, some post-irradiation examinations have been carried out on targets at SNS and JSNS. The main goal is to provide information to the target R&D program, in order to develop targets that are less sensitive to cavitation-induced erosion damage. The targets were exposed to 3055 MW.h (Target 1) and 3215 MW.h (Target 2) at SNS [52] and 475 MW.h at JSNS [53]. These targets were operated at different power levels: SNS Target 1 <0.7 MW mostly, SNS Target 2 >0.7 MW mostly and JSNS target up to 0.2 MW (from May 2008 to 11 March 2011). The JSNS Target 1 was exchanged well before its end of life, due to the Great East Japan Earthquake of March 2011, which caused damage to ancillary equipment, having received calculated damage on the beam window part of 0.67 dpa.

Previous R&D work had identified possible characteristics of cavitation-induced erosion in liquid metal targets due to exposure to short pulse beams:

- An incubation period is observed, during which surfaces display pitting damage, but no weight loss;
- For a given beam profile the erosion rate post-incubation is heavily dependent on beam power, as low as P^2 or as high as P^4;
- Crucially, erosion rate is higher for stagnant fluid, compared with flowing fluid;
- There is a power threshold below which no damage occurs (depending on various factors, such as material condition, mercury flow on beam intensity).

A Kolsterizing surface-hardening treatment producing a 33–47 μm surface layer of supersaturated carbon is applied to SNS stainless steel (SS)-type 316 L mercury vessels to increase the power threshold and incubation period. The JSNS 3-mm-thick SS-type 316 L mercury vessel also receives a special surface-hardening treatment in the form of plasma carburizing and plasma nitriding (PCN) [54].

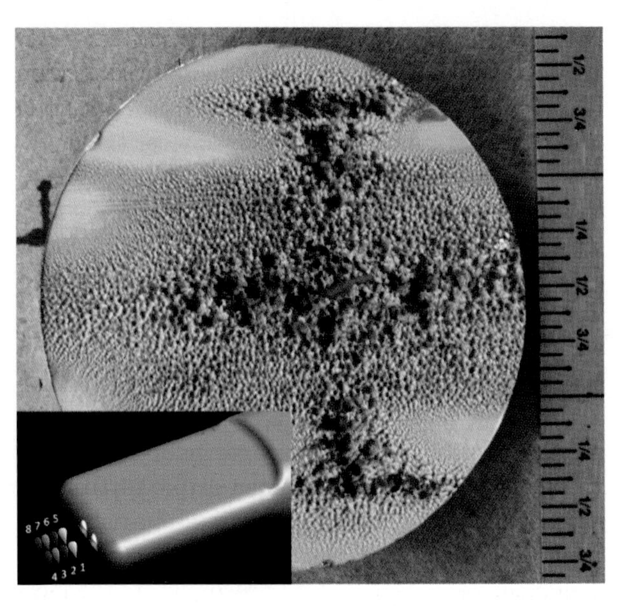

Fig. 8. Cavitation-induced erosion damage to the SNS target vessel, disk 5 of the inset image is shown here, from Ref. 52.

The cavitation-induced erosion damage observed on the beam entrance area of both SNS target vessels was very significant, with several holes going through the 3-mm-thick stainless steel, see Fig. 8. The higher operational power level of Target 2 is believed to be responsible for the greater degradation compared with Target 1. It is postulated that the damage pattern is partly the result of the quasi-stagnation zone in the mercury flow. The damage-mitigating advantage of local fluid flow is further supported by the lack of mass loss on the surface of the vessel exposed to the fast-flowing channel ($\sim$2.5 m/s) that cools the beam entrance window, the so-called window flow, compared with the extensive damage on the surface of the same beam entrance window sample exposed to the bulk mercury flowing with irregular lower velocity ($<$1 m/s). JSNS adopted a cross-flow configuration for its bulk mercury flow, which results in few stagnation points close to the beam entrance window than the return-flow configuration of SNS. The damage to the JSNS target vessel was far less significant, with shallow pits of $\sim$250 μm measured. However, in view of the different operational power levels (a factor 3$\times$ more for SNS), direct comparisons are difficult to make. Damage patterns are arguably similar, suggesting a dependence not only on the flow pattern, but also on the negative pressure distribution in the mercury [53].

This class of target yet again provides an example of the difficulties posed in choosing concepts. Cavitation-induced erosion due to pulsed beams was known to be a very damaging effect a couple of decades ago, when it was first identified as a problem with a kW proton beam at ISOLDE [55]. R&D efforts to study and mitigate the effects of pressure waves helped further understanding of the main damage characteristics [56, 57]. But, for these first-in-class facility prototypes, it is only when operating the first targets at close to nominal power that the full consequences can be finally measured. For SNS, currently the highest power pulsed spallation neutron source, the challenge is in finding the right operational balance, with the main parameters being maximum beam power and target lifetime.

For ESS, various targets have been proposed based on stationary and rotating solid targets water-, LBE- or helium-cooled and liquid metal loops of mercury, lead, and LBE with/without beam windows. The baseline choice is a RoTating Helium-cooled Tungsten tArget — RoTHeTa. A backup being studied is a water-cooled option [30]. A third option which also received significant attention was a lead (or LBE) loop with specific design features to mitigate the effects of beam-induced pressure waves. With very little difference in terms of neutronic performance, the main criteria for the choice of the baseline were safety and lifetime of the target vessel.

The 2.5 m diameter target wheel at ESS consists of 33 sectors designed for mechanical integrity and neutronic performance [12]. The proton beam pulses will be synchronized with the target rotation speed. The number of sectors has been chosen in order to have an arc length for each sector that is large enough to accommodate the beam footprint. The odd number of sectors prevents direct line of sight through the target wheel via the helium cooling channels in the event of a de-synchronized proton pulse hitting the target between two sectors. Helium flow channels have been designed to provide optimum cooling near the outer rim of the target, where the heat deposition is highest, taking into account the helium inlet, which is from the shaft through the back of each tungsten sector.

The steady state thermo-mechanical analysis based on a parabolic proton beam profile with a conservative peak current density of $64\,\mu A/cm^2$ shows maximum temperatures of 175°C and 462°C in the beam entrance window and tungsten, respectively. Transient analysis shows a change in peak temperature of 20°C and 110°C for the beam entrance window and tungsten, respectively. For the tungsten, the von Mises stress is 225 MPa in the steady state, with a stress range of 120 MPa during operation in pulsed mode.

The estimated 47 kW of afterheat can be dissipated by radiation heat transfer alone from the target to the monolith (which acts as the heat sink), leading to maximum tungsten temperatures of 550°C. If heat conduction through helium that fills the gap between target and shielding, and free convection are included as heat transfer paths, the maximum tungsten temperature is reduced to 480°C.

A long vertical shaft supports the rotating target. It reaches out through the top of the target monolith, where the driving and bearing systems are located; see Fig. 7. This concept was considered by previous studies and has the advantage that all system-relevant components are located at a significant distance (~5 m) from the high radiation area of the spallation zone [58]. A double labyrinth seal and a pressurized seal gas guarantee that leakage always flows from non-contaminated to contaminated volumes, and interfaces the rotating helium circuits with the stationary helium loop.

High cycle thermal fatigue was identified as the main materials issue for the SNQ rotating target in the 1980s [59], and is an important element in studies for the ESS target material and its structural vessel. Large volumes of helium are used extensively in various target facilities, but there is little experience in using relatively high velocity helium flow in a spallation environment. Materials issues at ESS will therefore also address tungsten and its interaction with flowing helium. Although tungsten slabs will not have a structural role, they must nevertheless retain structural integrity over the lifetime of the target to avoid blocking cooling channels. Another requirement for the tungsten is to retain as much as possible its radioactive inventory to minimize contamination of the helium cooling loop.

One potential problem is activated tungsten dust production, which would be entrained in the helium flow. Studies are required to understand the mechanisms and potential consequences of such an issue.

Pure tungsten grains that might detach from the tungsten bulk are above the micron range in size and can thus be handled by cyclonic filters with high efficiency. Less well known is the size of tungsten oxide particles, which could form if significant levels of oxygen impurities are present in the helium loop. If this is confirmed as a potential hazard, mitigation methods include containment of the individual tungsten slabs (cladding, canning, surface treatment) and helium purity control.

The workers involved with operational liquid mercury targets at SNS and JSNS also explored in detail solid target alternatives, notably water-cooled tantalum-clad tungsten, motivated by experience of this material combination at ISIS, LANSCE [24] and IPNS. Designed for a 3-MW-long pulse beam (~1 ms) at 1.3 GeV and 20 Hz rep. rate, the SNS STS (Second Target Station) target is a rotating 1.2 m diameter target made of tantalum-clad tungsten 70 mm thick, cooled by water through top, bottom and middle planes [58]. This middle plane cooling channel introduces only a 3% loss in neutronic performance. The cold neutron flux below 5 meV, 10 m from a $100 \times 120\,\text{mm}^2$ area of the moderator, is $5.7 \times 10^{-8}\,\text{n/cm}^2/\text{proton}$, equivalent to a mercury system. The lifetime is estimated to be 5–6 years for 10 dpa on the steel target container. The peak temperature was calculated to be 700°C following a loss of coolant accident with radiation cooling to surrounding reflector assemblies, remaining below the steam–tungsten interaction threshold of ~800°C.

Detailed studies showed that for a 1 MW stationary water-cooled tantalum-clad tungsten plates target (ISIS design) irradiated for one year, the decay heat was 7.6 kW, above the safe limit of 1.6 kW set by the ISIS target during target exchange after one week of cooling. Managing the decay heat would require a separate cooling loop based on helium, to cool the target for a minimum of 100 days after it has been removed from the proton beam line [60]. Substituting tantalum for steel mitigates this issue.

The Chinese Spallation Neutron Source is another example of a facility considering a rotating water-cooled solid tungsten target [61].

7.4. *Neutrino factory targetry R&D*

Second generation conventional neutrino facilities (NUMI, CNGS, J-PARC, etc.) all use relatively high energy proton beams (30–400 GeV) impinging on low Z target materials such as graphite. Next generation facilities such as the Neutrino Factory or the Muon Collider plan to optimize production and capture of pions by sending a more focused, higher power, lower energy beam onto a much higher Z target. In terms of engineering challenges, these beam and material considerations all translate into orders of magnitude higher deposited power densities, and hence higher static and dynamic stresses on the target material. The envelope ($E = 8\,\text{GeV}$, beam spot rms $= 1.2\,\text{mm}$, pulse duration $= 2\,\text{ns}$, rep. rate $= 50\,\text{Hz}$, beam power $= 4\,\text{MW}_b$) is being pushed much further than the extremes currently experienced at spallation neutron sources (e.g. SNS $E = 1\,\text{GeV}$, spot size $= 70 \times 200\,\text{mm}$, pulse duration $= 695\,\text{ns}$, rep. rate $= 20\,\text{Hz}$, beam power $= 1.4\,\text{MW}_b$). Whilst several workers are actively engaged in finding solutions with high Z materials, it is not improbable that a lower Z alternative will be found to be better suited.

Targets include a variety of proposals at the conceptual design stage, such as:

- A liquid metal jet target;
- A stationary target made of tungsten spheres;
- A fluidized powder jet;
- A radiation-cooled rotating ring of solid tungsten rods;
- A conducting target.

The fifth option has a pulsed axial current to increase the brightness of the pion beam [62]. The fourth option requires the solid tungsten rods to be levitated and driven by a magnetic lift and drive system due to high operating temperatures [63]. The second option is gas-cooled, for example with helium [64].

The International Design Study for the Neutrino Factory (IDS-NF) has a free-flowing liquid mercury jet target as the baseline (see Fig. 9), with a fluidized powder jet (tungsten) and solid tungsten rods as backups.

The MERIT experiment showed that a 50 Hz, $4\,\text{MW}_b$ pulsed beam would induce pressure waves that would disperse the Hg jet, resulting in droplets with a peak velocity of 30 m/s in a 20 T field [65]. Calculations of energy deposition show that roughly 10% of the 4 MW beam power is deposited in the mercury jet, and a further 12.5% is deposited in the mercury pool downstream of the jet [15]; see Fig. 10.

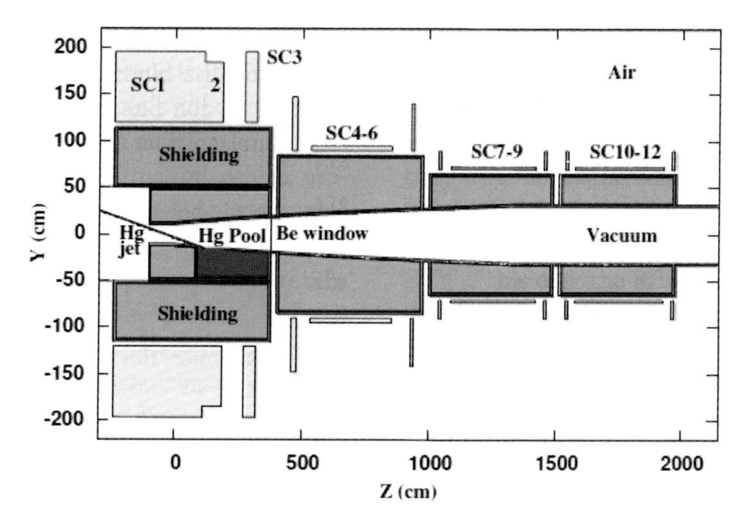

Fig. 9. Schematic of the Neutrino Factory mercury jet target, shielding and magnet assembly from Ref. 15. SC indicates superconducting magnets.

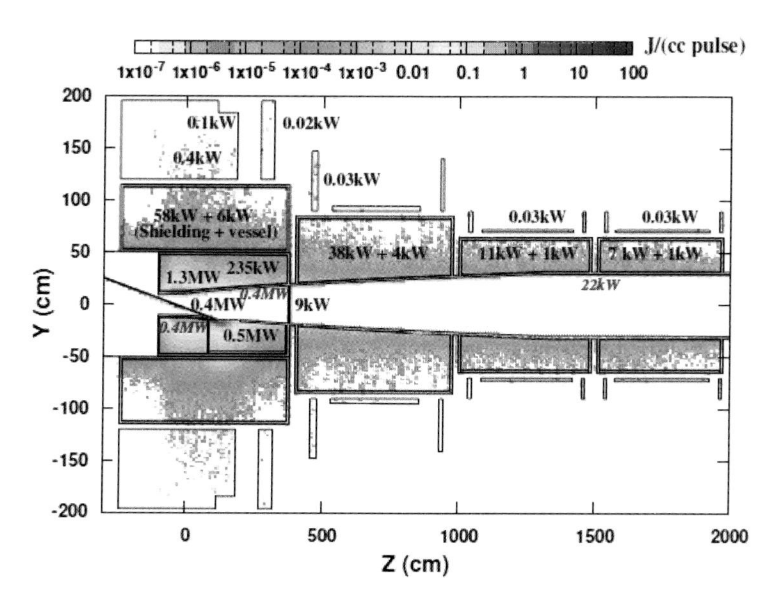

Fig. 10. Simulations of the impact of the proton beam on the Neutrino Factory mercury jet target from Ref. 15: deposited energy density (J/cc) per beam pulse for a 50 Hz pulse repetition frequency. Also shown is the time-averaged power deposited in different elements of the target, shielding and magnet assembly.

The energy densities remain high in the pool, raising issues of potential cavitation-induced erosion and disturbances at the pool surface. The engineering of the shielding and beam pipe sections will be challenging, with a combined time-averaged power deposition in those elements of 2.4 MW.

For solid targets, studies have indicated that on the basis of static stress analysis alone, the beam sigma would have to be increased to 6.5 mm in order for a tungsten target rod of 40 mm diameter to operate within its yield strength [66]. Dynamic stress investigations have also been carried out with 3% strain on tungsten and up to a $1000\,s^{-1}$ strain rate [67].

7.5. *ADS*

The multipurpose hybrid research reactor for high tech applications (MYRRHA) is an accelerator-driven system planned for operation in 2023, to replace the ageing BR2 reactor at SCK-CEN. It is a flexible fast spectrum research reactor (50–100 MW_{th}) able to operate in subcritical and critical modes. The main systems include a 600 MeV proton accelerator, a spallation target and a neutron multiplying core with MOX fuel, cooled by lead bismuth eutectic [68].

7.6. *IFMIF*

The International Fusion Materials Irradiation Facility (IFMIF) is designed to provide high energy neutrons at sufficient fluxes and volumes to test samples of candidate materials to an equivalent full lifetime of operation at a fusion reactor [69]. It is based on an accelerated deuteron beam impinging on a liquid lithium target. Two 125 mA quasicontinuous deuteron beams will deliver a total of 250 mA at between 32 and 40 MeV to the target(s). The flux provided over a 0.5 L volume is $2\,\text{MW/m}^2$ ($0.9 \times 10^{18}\,\text{n/m}^2/\text{s}$). A full mockup of the lithium loop has been built and tested [70].

8. Outlook

With advances in accelerator technologies leading to ever-increasing beam intensities for a variety of applications, challenges in targetry and secondary beam extraction are numerous. The target in particular is increasingly a major cost driver for high power facilities in terms of the R&D and in terms of the target station construction, where safety and radioprotection call for extensive shielding, remote handling and maintenance infrastructure.

As target materials continue to be pushed to their limits, knowledge of their properties under irradiation and representative operational conditions requires specific test beam facilities, which are currently limited. Efforts to consolidate materials databases for targets exposed to charged particle beams are ongoing. Specific standards, such as nuclear industry standards (ASME, RCC), for design of targetry components are lacking. These would give a framework for design that would benefit not only the designer of accelerator facilities but also the dialog with regulatory authorities and third-party contractors involved in the construction of systems.

References

[1] G. A. Bartholomew and P. R. Tunnicliffe, The AECL study for an intense neutron-generator. Technical report, AECL-2600 (1966).
[2] Y. Kadi *et al.*, EURISOL high power targets, *Nucl. Phys. News* **18**(3), 19–25 (2008).
[3] F. Pellemoine, High power density targets, *Nucl. Instrum. Methods Phys. Res. B*; http://dx.doi.org/10.1016/j.nimb.2013.06.038 (2013).
[4] M. Dombsky and P. Bricault, High intensity targets for ISOL: Historical and practical perspectives, *Nucl. Instrum. Methods Phys. Res. B* **266**, 4240–4246 (2008).
[5] H. L. Ravn, Advanced target concepts for production of radioactive ions and neutrino beams, *Nucl. Instrum. Methods Phys. Res. B* **204**, 197–204 (2003).
[6] R. Kirchner, Review of ISOL target-ion-source systems, *Nucl. Instrum. Methods Phys. Res. B* **204**, 179–190 (2003).
[7] U. Koster *et al.*, Progress in ISOL target-ion source systems, *Nucl. Instrum. Methods Phys. Res. B* **266**, 4229–4239 (2008).
[8] J. M. Carpenter, Pulsed spallation neutron sources for slow neutron scattering, *Nucl. Instrum Methods* **145**, 91–113 (1977).
[9] G. S. Bauer, The physics and technology of spallation neutron sources, *Nucl. Instrum. Methods Phys. Res. A* **463**, 505–543 (2001).
[10] N. Watanabe, Neutronics of pulsed spallation neutron sources, *Rep. Prog. Phys.* **66**, 339–381 (2003).
[11] F. Mezei, Accelerator requirements for next generation neutron sources, *Nucl. Instrum. Methods Phys. Res. B* **562**, 553–556 (2006).
[12] S. Peggs *et al.*, ESS Technical Design Report, ISBN 978-91-980173-2-8 (European Spallation Source, Apr. 2013).
[13] A. K. Ichikawa, Design concept of the magnetic horn system for the T2K neutrino beam, *Nucl. Instrum. Methods Phys. Res. A* **690**, 27–33 (2012).
[14] D. Adey *et al.*, Neutrinos from stored muons (nuSTORM): Expression of interest. arXiv:1305.1419v1 (2013).
[15] J. J. Back, C. Densham, R. Edgecock and G. Prior, Particle production and energy deposition studies for the neutrino factory target station. arXiv:1210.5131v2 (2013).
[16] M. Benedikt *et al.*, Conceptual design report for a beta-beam facility, *Eur. Phys. J. A* 47 (2011).
[17] Y. Dai *et al.*, Materials researches at the Paul Scherrer Institute for developing high power spallation targets, *J. Nucl. Mater.* **389**, 288–296 (2009).
[18] W. Wagner *et al.*, MEGAPIE at SINQ — The first liquid metal target driven by a megawatt class proton beam, *J. Nucl. Mater.* **377**, 12–16 (2008).
[19] Y. Dai *et al.*, Assessment of the lifetime of the beam window of the MEGAPIE target liquid metal container, *J. Nucl. Mater.* **356**, 308–320 (2006).
[20] G. A. Cottrell and L. J. Baker, Structural materials for fusion and spallation sources, *J. Nucl. Mater.* **318**, 260–266 (2003).
[21] T. Wakui *et al.*, Failure probability estimation of multi-walled vessels for mercury target, *J. Nucl. Sci. Technol.* **44**(4), 530–536 (2007).

[22] M. Harada *et al.*, DPA calculation for Japanese spallation neutron source, *J. Nucl. Mater.* **343**, 197–204 (2005).
[23] D. A. McClintock *et al.*, Post-irradiation examination of the Spallation Neutron Source target module, *J. Nucl. Mater.* **398**, 73–80 (2010).
[24] A. T. Nelson *et al.*, Fabrication of a tantalum-clad tungsten target for LANSCE, *J. Nucl. Mater.* **431**, 172–184 (2012).
[25] Y. Tall *et al.*, Volatile elements production rates in a proton-irradiated molten lead–bismuth target, in *Proc. Int. Conf. on Nuclear Data for Science and Technology* (EDP Sciences, 2008).
[26] M. Medarde *et al.*, Lead–gold eutectic: An alternative liquid target material candidate for high power spallation neutron sources, *J. Nucl. Mater.* **411**, 72–82 (2011).
[27] J. Neuhausen *et al.*, Radiochemical aspects of liquid mercury spallation targets, *J. Nucl. Mater.* **431**, 224–234 (2012).
[28] P. Sievers, Elastic stress waves in matter due to rapid heating by an intense high-energy particle beam. LAB.II/BT/74-2 (CERN: European Organization for Nuclear Research, June 1974).
[29] Y. Dai *et al.*, The second SINQ target irradiation program, STIP-II, *J. Nucl. Mater.* **343**, 33–44 (2005).
[30] K. Thomsen *et al.*, A case for a SINQ-type cannelloni target at the ESS power level, *Nucl. Instrum. Methods Phys. Res. A* **625**, 5–10 (2011).
[31] JAEA, Accelerator Facility Accident Report. Technical report (Japan Atomic Energy Agency, 2013).
[32] K. Abe *et al.*, The T2K experiment, *Nucl. Instrum. Methods Phys. Res. A* **659**, 106–135 (2011).
[33] T. J. Shea *et al.*, Status of beam imaging developments for the SNS target, in *Proc. DIPAC09* (2009).
[34] M. Teshigawara *et al.*, Development of JSNS target vessel diagnosis system using laser Doppler method, *J. Nucl. Mater.* **398**, 238–243 (2010).
[35] K. Thomsen, Advanced on-target beam monitoring for spallation sources, *Nucl. Instrum. Methods Phys. Res. A* **600**, 38–40 (2009).
[36] J. R. J. Bennett *et al.*, The design and development of the RIST target, *Nucl. Instrum. Methods Phys. Res. B* **126**, 117–120 (1997).
[37] E. J. A. Bouquerel, *Atomic beam merging and suppression of alkali contaminants in multi-body high power targets.* PhD thesis, Universite Paris XI Orsay, CERN-THESIS-2010-057, 2009.
[38] H. T. Bach *et al.*, Improving the survivability of Nb-encapsulated Ga targets for the production of ^{68}Ge, *Nucl. Instrum. Methods Phys. Res. B* **299**, 32–41 (2013).
[39] J. A. Nolen *et al.*, Liquid-lithium cooling for 100-kW ISOL and fragmentation targets, *Nucl. Phys. A* **701**, 312c–322c (2002).
[40] W. L. Talbert *et al.*, Developments in high-intensity two-step target design, *Nucl. Instrum. Methods Phys. Res. B* **204**, 314–318 (2003).
[41] V. N. Panteleev *et al.*, Studies of uranium carbide targets of a high density, *Nucl. Instrum. Methods Phys. Res. B* **266**, 4247–4251 (2008).
[42] K. Samec *et al.*, Design of a compact high-power neutron source — The EURISOL converter target, *Nucl. Instrum. Methods Phys. Res. A* **606**, 281–290 (2009).
[43] K. Samec *et al.*, Measurement and analysis of turbulent liquid metal flow in a high-power spallation neutron source — EURISOL, *Nucl. Instrum. Methods Phys. Res. A* **638**, 1–10 (2011).
[44] H. K. Carter and D. W. Stracener, Radioactive target and ion source test facilities at HRIBF, *Nucl. Instrum. Methods Phys. Res. B* **266**, 4702–4705 (2008).
[45] J. A. Nolen *et al.*, Development of windowless liquid lithium targets for fragmentation and fission of 400-kW uranium beams, *Nucl. Instrum. Methods Phys. Res. B* **204**, 293–297 (2003).
[46] S. Fernandes *et al.*, *In-situ* electric resistance measurements and annealing effects of graphite exposed to swift heavy ions, *Nucl. Instrum. Methods Phys. Res. B*; http://dx.doi.org/10.1016/j.nimb.2013.04.060 (2013).
[47] L. K. Mansur, Materials research and development for the spallation neutron source mercury target, *J. Nucl. Mater.* **318**, 14–25 (2003).
[48] T. Koppitz *et al.*, Improved cavitation resistance of structural materials in pulsed liquid metal targets by surface hardening, *J. Nucl. Mater.* **343**, 92–100 (2005).
[49] M. Futakawa *et al.*, Mitigation technologies for damage induced by pressure waves in high-power mercury spallation neutron sources (II) — Bubbling effect to reduce pressure wave, *J. Nucl. Sci. Technol.* doi: 10.1080/18811248.2008.9711890.
[50] J. R. Haines *et al.*, Summary of cavitation erosion investigations for the SNS mercury target, *J. Nucl. Mater.* **343**, 58–69 (2005).
[51] B. W. Riemer *et al.*, Status of R&D on mitigating the effects of pressure waves for the Spallation Neutron Source mercury target, *J. Nucl. Mater.* **431**, 160–171 (2012).
[52] D. A. McClintock *et al.*, Initial observations of cavitation-induced erosion of liquid metal spallation target vessels at the Spallation Neutron Source, *J. Nucl. Mater.* **431**, 147–159 (2012).
[53] T. Naoe *et al.*, http://dx.doi.org/10.1016/j.jnucmat.2013.04.049; *J. Nucl. Mater.* (2013).
[54] M. Futakawa *et al.*, Development of the Hg target in the J-PARC neutron source, *Nucl. Instrum. Methods Phys. Res. A* **600**, 18–21 (2009).
[55] J. Lettry *et al.*, Release from ISOLDE molten metal targets under pulsed proton beam conditions,

Nucl. Instrum. Methods Phys. Res. B **126**, 170–175 (1997).
[56] B. W. Riemer *et al.*, SNS target tests at the LANSCE-WNR in 2001 — Part I, *J. Nucl. Mater.* **318**, 92–101 (2003).
[57] B. W. Riemer *et al.*, SNS target tests at the LANSCE-WNR in 2001 — Part II, *J. Nucl. Mater.* **318**, 102–108 (2003).
[58] T. McManamy *et al.*, 3 MW solid rotating target design, *J. Nucl. Mater.* **398**, 35–42 (2010).
[59] W. Lohmann and K.-H. Graf, Materials investigations for the target of the German Spallation Neutron Source and their relations to fusion reactor materials research, *J. Nucl. Mater.* **122–123**, 1033–1035 (1984).
[60] N. Takenaka *et al.*, Thermal hydraulic design and decay heat removal of a solid target for a spallation neutron source, *J. Nucl. Mater.* **343**, 169–177 (2005).
[61] X. J. Jia *et al.*, Mock-up stands for a rotating target for CSNS project, *J. Nucl. Mater.* **398**, 28–34 (2010).
[62] B. Autin *et al.*, Conducting target for pion production, *Nucl. Instrum. Methods Phys. Res. A* **503**, 348–353 (2003).
[63] P. Drumm *et al.*, Neutrino factory target based on levitating tungsten rods, *Nucl. Instrum. Methods Phys. Res. A* **472**, 627–631 (2001).
[64] P. Sievers, A stationary target for the CERN neutrino factory, *Nucl. Instrum. Methods Phys. Res. A* **503**, 344–347 (2003).
[65] K. T. McDonald *et al.*, The MERIT high-power target experiment at the CERN PS, in *Proc. IPAC10* (2010), p. 3527.
[66] O. Caretta *et al.*, Engineering considerations on targets for a neutrino factory and muon collider, *J. Nucl. Mater.* **433**, 538–542 (2013).
[67] G. P. Skoro *et al.*, Yield strength of molybdenum, tantalum and tungsten at high strain rates and very high temperatures, *J. Nucl. Mater.* **426**, 45–51 (2012).
[68] H. A. Abderrahim *et al.*, MYRRHA: A multipurpose accelerator driven system for research & development, *Nucl. Instrum. Methods Phys. Res. A* **463**, 487–494 (2001).
[69] T. Kondo, IFMIF, its facility concept and technology, *J. Nucl. Mater.* **258–263**, 47–55 (1998).
[70] N. Loginov *et al.*, Experimental investigation of the IFMIF target mock-up, *J. Nucl. Mater.* **386–388**, 958–962 (2009).

Etam Noah has a keen interest in the interaction of particle beams with matter and systems. He received his PhD from Imperial College London in 2003 for experimental studies of the radiation effects on CMS Tracker electronics. He then joined CERN, where he worked on radiation hardened optical links, before spending several years as a target physicist at ISOLDE, contributing to its operation and to future ISOL concepts through the EURISOL-DS project. In 2009 he joined the European Spallation Source – Lund, where, as head of the Materials Group, he played a leading role in the target selection process and in setting up the targetry collaboration. He is currrently at the University of Geneva, responsible for neutrino detector prototypes under the AIDA project.

Reviews of Accelerator Science and Technology
Vol. 6 (2013) 259–274

DOI: 10.1142/S1793626813300120

High Intensity Neutron Beamlines

Phillip M. Bentley,* Carsten P. Cooper-Jensen† and Ken H. Andersen‡

European Spallation Source ESS AB,
Box 176, 22410 Lund, Sweden
**phillip.bentley@esss.se*
†carsten.cooper-jensen@esss.se
‡ken.andersen@esss.se

We describe the current technology available for producing and delivering high performance neutron beams at modern facilities for neutron scattering experiments. This overview includes optimal neutron production, optical techniques and devices, the use of the beams on the instruments, and also methods of simulation and modeling.

Keywords: Neutron beams; neutron optics; neutron scattering.

1. Introduction

Shortly after the discovery of the neutron [1], it became apparent that these particles would provide an extremely powerful tool for studying the structure and dynamics of condensed matter systems. Thermal neutron beams have wavelengths comparable to typical interatomic spacings, and kinetic energies comparable to excitation and relaxation processes in condensed matter systems.

Research reactors were constructed around the world to harness these new techniques, especially after the 1950s. Neutron beams were initially apertures in the side of reactors, but it quickly became apparent that significant performance gains were available through the use of evacuated guide tubes with metal coatings [2]. In addition to preventing people from inadvertently interacting with the beam, and reducing air scattering, these drastically increased the total number of neutrons that were transmitted for beams that were more than several meters in length.

Neutron scattering has become an invaluable, multidisciplinary tool of the study of condensed matter. Furthermore, the efficient manipulation of neutron beams forms one of the three main technologies — the other being production and detection — that govern the quantity and quality of data that are obtained from instrumentation. Ultimately, these are the drivers of feasibility of experiments at the fringes of scientific knowledge. In this article, we will briefly review modern neutron sources, before describing the current technology for coupling of neutron sources to the neutron beam guides and handling the neutron beams. Finally, we will also provide an overview of the most important types of experimental stations at the end of the neutron beamlines: the neutron instruments.

2. Neutron Production

There are several methods of generating neutrons, but the most productive sources are based on nuclear fission and nuclear spallation. Nuclear fission is the well-known process by which uranium or plutonium releases energy in power stations or weapons [3]. Neutrons are liberated in the fragmentation of nuclei, typically with MeV energies.

The term "nuclear spallation" was famously coined by Seaborg in his PhD thesis [4]. Modern spallation sources [5–8] are based on ~1 GeV proton beams interacting with high Z target materials, such as W or Hg. This produces copious quantities of

*Corresponding author.

neutrons in a small volume. Spallation sources currently offer the brightest source of neutrons that is practically achievable, especially for time resolution studies [9].

3. Sources and Moderators

Fast neutrons created in the fission or spallation process have energies which are much too high to be useful for most condensed matter studies. They are slowed down by repeated collisions, usually in hydrogenous materials, and it is these neutron moderators which are viewed by beam tubes for extraction of intense neutron beams. In fission reactors, most beam tubes view the reflector around the fuel element(s), consisting in most cases of light or heavy water or solid beryllium. These elements are kept at approximately room temperature, resulting in neutron spectra with a quasi-Maxwellian distribution centered at around 300 K, corresponding to a neutron kinetic energy of about 25 meV or a wavelength of about 1.8 Å. In order to enhance the beam brightness of shorter or longer wavelengths, moderators of different temperatures can be placed near or inside the reflector. Of particular importance is the use of cold sources, consisting of refrigerated hydrogenous or deuterated materials at temperatures around 20–25 K, dramatically enhancing the flux of neutrons for wavelengths longer than about 2 Å.

At present-day pulsed spallation sources, the proton pulse incident on the neutron-producing target is less than 1 μs in length, resulting in an extremely intense burst of fast neutrons over a 1 μs timescale. They are known as short pulse sources. The target is surrounded by a small number of neutron moderators embedded in a reflector, usually composed of beryllium. As for steady state sources, the temperature of each neutron moderator is carefully controlled to enhance the neutron production in particular energy regions. Unlike the case of steady state sources, the moderators in a short pulse source are designed to tailor the duration of the thermalized neutron pulses, usually by the deliberate incorporation of materials which capture thermal neutrons, while remaining transparent to fast neutrons. A thin sheet of cadmium placed between the moderator and the reflector, known as a decoupler, allows fast neutrons to enter the moderator, but prevents partially thermalized neutrons from traveling between moderator and reflector, ensuring that the time structure of neutrons emitted from the moderator reflects the travel time of thermal neutrons inside the moderator, rather than the much larger structure of the reflector. The thermal neutron pulse width can be further reduced by incorporating absorbing sheets inside the moderator itself — known as poisoning.

Neutron sources are characterized by their spectral brightness (or brilliance):

$$B = \frac{\partial^4 N}{\partial A \partial t \partial \Omega \partial \lambda}, \quad (1)$$

i.e. the number of neutrons per unit area, time, solid angle and wavelength. This is a property which does not change with distance or collimation, unlike the most commonly measured source-related quantities, such as particle flux or capture flux. Pulsed sources distinguish between peak and time average brightnesses. The former is the maximum instantaneous brightness during a pulse, while the latter is the instantaneous brightness averaged over a full repetition period. They are measured in the same units.

A comparison of some of the leading neutron research facilities is made in Table 1, and Fig. 1.

The currently world-leading neutron source, by almost any measure, is the Institut Laue-Langevin (ILL) in Grenoble, France. It is based on a 57 MW fission reactor, offering a suite of 38 public access instruments [10], and has been in operation since the early 1970s. A hot neutron source provides a high brightness of short wavelength neutrons, while the heavy water reflector around the fuel element acts as the thermal source and is viewed by a large number of beam tubes. Two liquid deuterium cold sources placed near the peak flux region of the reflector provide the world's highest intensity of cold neutrons. The spectral brightness curves of the ILL moderators are shown in Fig. 2.

The ISIS facility in the UK is the most scientifically productive pulsed-neutron source today, currently operating 25 instruments distributed over two target stations [11]. The ISIS accelerator provides 200 μA of protons at an energy of 800 MeV and a frequency of 50 Hz, and is in the process of ramping up to 300 μA. Four consecutive pulses are directed toward ISIS Target Station 1 (TS1), and the fifth pulse is used by TS2, which is optimized for cold neutrons. The moderators on TS1 are primarily optimized for high resolution with extensive use of decouplers and poisoning. The

Table 1. A selection of the leading facilities for research with thermal neutron beams, summarizing their main performance parameters, and the year of first operation. The number of neutron instruments listed includes those which are operational and under construction or commissioning at each facility. The facility power is shown at both the current level of operation and projected to when the full design specification has been reached. Peak (PkB) and time average brightnesses (AvB) are given in units of n/cm^2/s/sr/Å. The thermal and cold brightness numbers are given for wavelengths of 1.5 Å and 3 Å, respectively. In each case, the moderator with the highest brightness at that wavelength has been represented, corresponding to the current facility power. In the case of the ESS, the moderator brightnesses correspond to the design value of the facility power [8].

Facility	Power	Rep. rate	Start	*Instr.*	*Thml.* AvB	*Thml.* PkB	Cold AvB	Cold PkB
ILL	57/57 MW	—	1971	38	2.6×10^{13}	2.6×10^{13}	7×10^{12}	7×10^{12}
ISIS-TS1	128/192 kW	50 Hz	1984	18	4×10^{10}	5×10^{13}	1.5×10^{10}	7×10^{12}
ISIS-TS2	32/48 kW	10 Hz	2009	11	1.1×10^{10}	4×10^{13}	2.7×10^{10}	1.8×10^{13}
SNS	0.9/1.4 MW	60 Hz	2006	20	2.7×10^{11}	1.5×10^{14}	5×10^{11}	5×10^{13}
JPARC	0.3/1.0 MW	25 Hz	2009	21	1.4×10^{11}	2.0×10^{14}	5×10^{11}	1.3×10^{14}
ESS	–/5 MW	14 Hz	2019	22	1.1×10^{13}	2.8×10^{14}	8.6×10^{12}	2.2×10^{14}

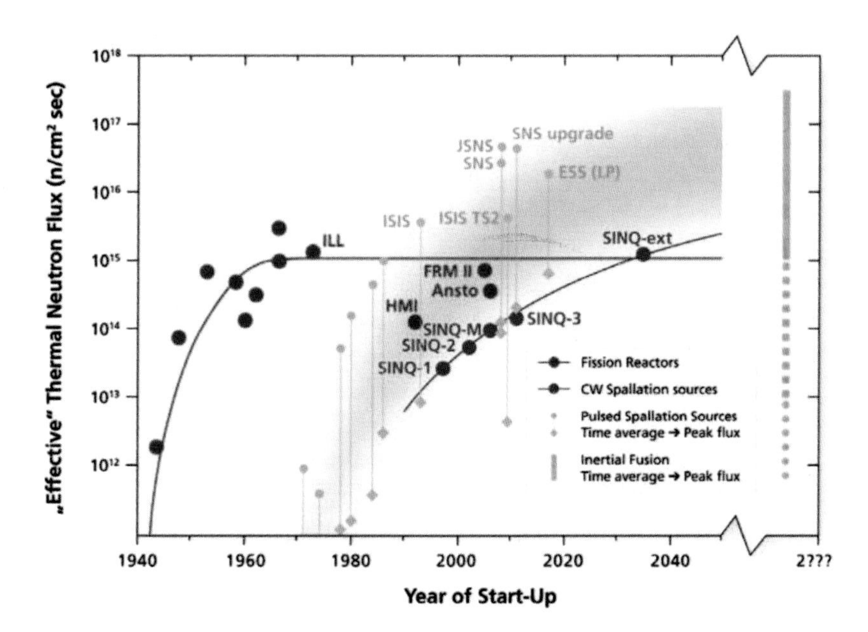

Fig. 1. The brightness of neutron sources as a function of time, after Klausen [9]. The pulsed sources shown in gray can significantly outperform the continuous sources when the instruments are properly optimized to the source time structure.

TS2 reflector–moderator design is optimized more strongly for coupled moderators, providing the very highest peak and time average brightnesses, particularly at long wavelengths. The time average and peak spectral brightness curves of the ISIS moderators are shown in Fig. 3.

In the last few years, two MW-class short pulse spallation sources have come on-line: SNS in Oak Ridge, USA, and the MLF facility at J-PARC in Tokai, Japan (labeled "JSNS" in Fig. 1). Both facilities are in the process of ramping up source power, reliability and instrument numbers. Their performance characteristics are summarized in Table 1, and these offer significant brightness increases over ISIS and ILL.

Most continuous neutron sources are fission reactors, and most pulsed sources are short pulse spallation sources. There are a few exceptions to this rule: the SINQ source at the Paul Scherrer Institute in Switzerland is a continuous spallation source, while the IBR-2 reactor in Dubna, Russia, is a pulsed reactor.

The European Spallation Source (ESS), currently in its construction phase and expected to come on-line in Lund, Sweden, in 2019, will be a long pulse spallation source. The duration of the

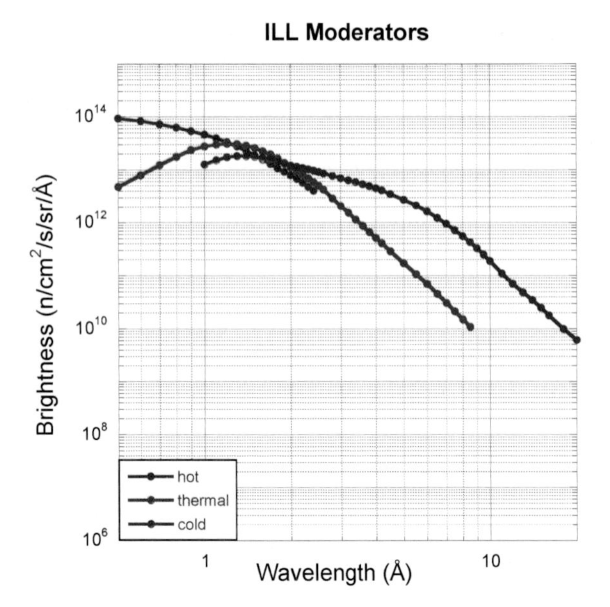

Fig. 2. Spectral brightness curves of the ILL hot, thermal and cold moderators.

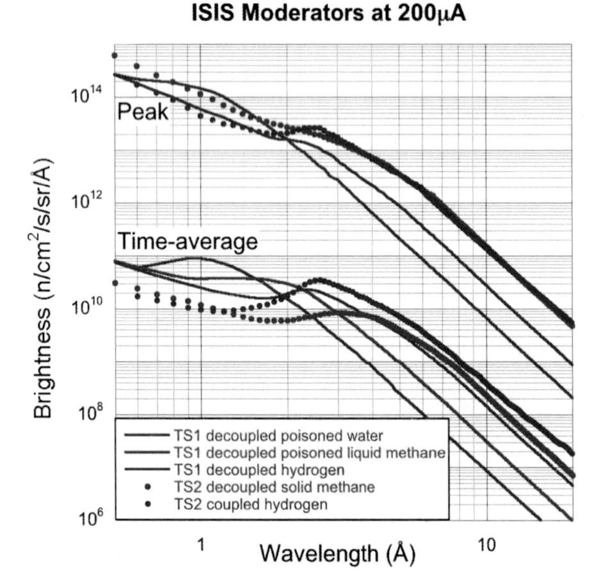

Fig. 3. Spectral brightness curves of the ISIS moderators.

proton pulse incident on the target is 2.86 ms in length, resulting in a time structure of the thermalized neutrons which is mainly determined by the length of the proton pulse, rather than the design of the moderator–reflector assembly. There is no use of decouplers or poisoners in the design of the moderators, resulting in very high peak and time average neutron brightnesses. When the ESS reaches its performance specifications, it will offer a time average neutron brightness matching that of the ILL, combined with a peak brightness which is higher than for any of the existing neutron sources.

4. Neutron Guide Optics

Neutron guide optics is based on principles very similar to those of photons. The refractive index for neutrons, n, is given by [12]

$$n = 1 - \lambda^2 \frac{Nb}{2\pi} + i\lambda \frac{N\sigma_a}{4\pi}, \tag{2}$$

where λ is the neutron wavelength, N the atomic number density of the material, b the average coherent neutron scattering length and σ_a the absorption cross-section. Nb is usually referred to as the neutron *scattering length density*.

4.1. *Monolayer reflection*

For most materials, the index of refraction, n, is slightly less than 1 for cold neutrons, which means that there is a critical angle, θ_c, below which total *external* reflection of neutrons takes place. The critical angle for reflection is given by [12]

$$\sin\theta_c = \lambda\sqrt{\frac{Nb}{\pi}}. \tag{3}$$

Total external reflection of slow neutrons was first demonstrated by Fermi and coworkers [13, 14] in the mid-1940s. This allows the construction of *neutron guides*, in which neutron beams are transported by total external reflection. From Eq. (3), we see trivially that the properties of a reflective neutron optical device for a given wavelength are governed mainly by the scattering length density.

The neutron scattering length density varies from element to element and from isotope to isotope. The neutron scattering length density for a "natural" element refers to the average over all isotopes with the natural abundances of that element. Ni has both the isotope with the lowest (^{62}Ni) and the second-highest (^{58}Ni) scattering length density of all the stable isotopes. Natural Ni has one of the highest scattering length densities for an element in the periodic table, which is 9.4×10^{-6} Å^{-2}. This gives the convenient rule of thumb that the critical angle (in degrees) for Ni is $\theta_c = 0.1 \times \lambda$, where the wavelength of the neutrons is in Å. Indeed, nickel is the standard reflectivity benchmark for neutron mirrors, known in the industry as $m = 1$.

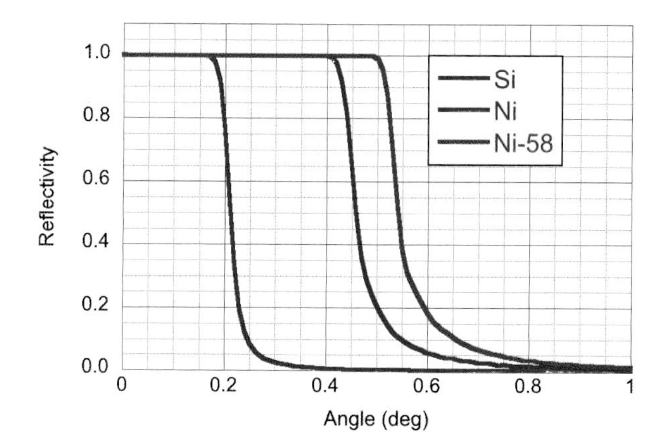

Fig. 4. Calculated neutron reflectivity curves for silicon, natural nickel and isotopically enriched ^{58}Ni, at a neutron wavelength of 4.4 Å.

Figure 4 shows the reflectivity curve of 4.4 Å neutrons in different materials. This illustrates the advantage of coating the neutron guide with a material with a high neutron scattering length density.

Neglecting surface roughness, a neutron hitting the surface of the guide below the critical angle will be reflected, and above the critical angle the neutron will be transmitted into the coating. The critical angle for short wavelength neutrons is small, and these neutrons pass through the mirror and are absorbed in the shielding, which allows the system to be tailored to the wavelength range that is desired.

Many neutron guides continue to be made by Ni coating on glass substrates, because of the high neutron scattering length density of Ni. Ni is relatively easy to coat using magnetron sputtering, and glass makes an excellent substrate in that it is easy to prepare glass surfaces with low roughness. In the past, ^{58}Ni has been used as a neutron guide coating, to yield a higher critical angle due to the greater neutron scattering length density over natural nickel (see Fig. 4), corresponding to $m = 1.2$. The drawback of ^{58}Ni is the high financial cost of isotopic enrichment, producing a sale price that is at present around US$1000 per gram. This cost is much higher than for the modern alternatives described in the next section.

4.2. *Supermirrors*

To increase the critical angle beyond that of Ni, Mezei proposed a specially designed multilayer, often called a neutron supermirror [15]. A supermirror coating consists of a repetition of bilayers that are thin layers of coatings of two altering materials, one with a high neutron scattering length density and the other with a low one. This creates an artificial structure that exhibits constructive interference for neutrons that satisfy the Bragg equation, namely

$$n\lambda = 2d\sin(\theta), \tag{4}$$

where n is a positive integer, d the thickness of the bilayer in the coating, and θ the angle between the incoming neutrons and the coating. Constant d spacing coatings can be used to reflect monochromatic neutron beams by selecting a particular reflection angle corresponding to the desired wavelength. Varying the bilayer thickness causes the associated neutron reflection angle to change, or — when keeping the angle fixed — changes the associated neutron wavelength that the mirror selects for a given direction. Mezei's invention is to increase the bilayer thickness from layer to layer throughout the multilayer structure, so that a range of neutron wavelengths and angles can find a bilayer structure at some depth in the mirror that satisfies Eq. (4). This results in an efficient mirror over a wide range of angles and wavelengths.

The reflectivity of each bilayer is proportional to the sharpness of the interface between the high contrast materials. However, roughness and interdiffusion between the layers conspire to produce an effective gradient of the scattering length density between the layers. Since the spatial extent of this gradient region should be considered as a fraction of the bilayer total thickness, these gradients become the driving quality factor for the thinnest coatings. As such, to achieve the highest quality coatings, one needs a substrate with a very small surface roughness, preferable with a RMS roughness of a few angstroms. The roughness of subsequent layers is normally greater than or equal to the roughness of the layer deposited immediately beforehand, for nearly all materials. Furthermore, it is worth noting that the absolute roughness tends to be higher in thicker coatings, even though it is the roughness relative to the bilayer thickness that is important for quality.

To achieve efficient and cost-effective neutron beam transport, there are other important factors beyond maximizing the neutron scattering length density between layers. For example, it should be easy to coat large areas during manufacture; the coatings should be stable over time in high radiation fields; the materials in the supermirrors should have a relatively low absorption cross section for neutrons; and the materials should be easy to deposit with low interface roughness between the layers, because surface roughness decreases the reflectivity of the coating. Ni–Ti supermirrors fulfill these requirements fairly well and are seen currently as a reasonable compromise, and hence have become the standard for guide optics.

Almost all supermirrors are capped with a thick, natural nickel layer on top of the graded bilayers, to ensure that there is total reflection below the critical angle for Ni. A neutron multilayer will therefore in general exhibit total external reflection below $m = 1$, and — above the critical angle for Ni — the effective reflectivity of the diffractive component gradually decreases up to a second feature, where the reflectivity decreases sharply. This sharp decline is related to the thinnest bilayer thickness of the supermirror, and the angle can be calculated from Eq. (4). This angle where the effective reflectivity suddenly decreases is referred to as the m value for the multilayer coating, and provides a convenient measure of the mirror performance relative to $m = 1$ of natural nickel. This angle, referred to as θ_c, is approximated by

$$\theta_c \approx 0.1\lambda m, \tag{5}$$

where θ_c is in degrees and λ is in angstroms. It is plotted in Fig. 5 for a number of supermirrors with different m values.

From Eq. (4) we see that if we decrease the bilayer thickness we can achieve a larger m value for the multilayer. Another thing that is very clear from Fig. 5 is that a coating with $m = 7$ yields a lower reflectivity around $m = 4$ than does an $m = 4$ coating. For an $m = 4$ supermirror the thinnest coating is on the substrate, whereas the bilayer with the same thickness in an $m = 7$ supermirror is coated on top of thousands of other layers. This creates imperfections in the higher m mirrors, and reduces the reflectivity.

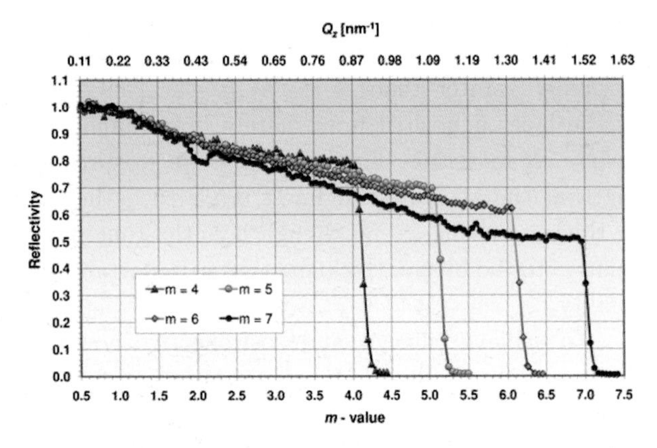

Fig. 5. State-of-the-art reflectivity curves for different Ni/Ti supermirror coatings [16].

The number of layers, N, in a supermirror increases rapidly with the m value, roughly following $N \approx 4m^4$ [17], and there are a number of algorithms to calculate the optimum of each bilayer pair in the supermirror as a function of the desired m value and bilayer number [18, 19]. The performance of a multilayer system can be computed from the same iterative mathematical approach as for X-rays [20].

Table 2 lists the number of layers and the thinnest bilayer thickness for some Ni–Ti supermirrors with different m values. Taken together with Fig. 5, it is clear that one should use only the minimum required m value for the coating, as exceeding the required m significantly degrades the performance of the coating at lower angles, and at the same time increases the number of layers and thus the cost.

Normally, neutron guides are coated using magnetron sputtering, since this technique can coat optics with the required quality and with a fairly good throughput. The standard material combination for neutron guides is Ni–Ti coatings, which have been done by many different groups. To increase the reflectivity of the Ni–Ti coatings, the roughness between the layers has to be reduced by either reducing the interdiffusion between Ni and Ti and/or by reducing the crystal size in the different layers. It has been shown that coating NiC–Ti multilayers results in smaller Ni crystals than in Ni–Ti coatings, and thereby reduces the diffuse scattering of

Table 2. Typical parameters for Ni–Ti supermirrors with various m values. "N layers" is the number of single layers in the supermirror, and $d_{\min}$ is the smallest d spacing of the layers in the supermirror.

Supermirror m value	N layers	$d_{\min}$ (Å)
1	1	2000
2	64	140
4	1024	71
6	5184	48
7	9604	41

the coating [21]. Also, varying the C contents in the Ni layer and/or adding a different amount of H to the Ti layers have been investigated [22]. To reduce the interdiffussion of materials, and to smoothen out the roughness between the Ni and Ti layers, interface engineering can be used. Interface engineering is the practice of coating very thin layers, of only a few angstroms, between all the Ni and Ti layers, and both Cr [23] and C [24] have been used to good effect.

Other material combinations than Ni–Ti have been suggested. Be is the only element, except for C in the form of diamond, that has a higher neutron scattering length density than Ni, and it has a very low neutron absorption cross section. However, only a little development work has been performed on Be–Ti coatings [25], mainly because of the health risk from Be dust associated with the deposition process. Another interesting alternative is Ni–Mn, which has been investigated for use in neutron guides [26]. Mn has the lowest neutron scattering length density of any element, and so provides good contrast with the adjacent Ni layer, but the main drawback of Mn is the absorption cross section, which is double that for Ti. In light of these considerations, supermirrors based on Ni–Ti remain the dominant technology at present.

4.3. *Beam transport modeling*

Modeling the behavior of neutron beams is physically straightforward but computationally intensive, due to the number of interactions and parameters required to determine the beam propagation accurately. Two principal methods are used.

4.3.1. *Monte Carlo methods*

By far the most popular method is to sample the neutron probability distributions by Monte Carlo. Several good packages exist with numerous pros and cons [27–31]. Among them, MCSTAS and VITESS are the most widely used packages.

The simulation follows the neutron from the surface of the moderator to the location of interest, sampling numerical distributions randomly. The level of detail available in modern packages is impressive, and has recently reached the stage where it is possible to use the simulation fully alongside data analysis and other simulation packages [32], thus leading to the idea of a virtual neutron experiment.

For the purposes of designing beamlines, it is clear that for highly lossy beam systems (e.g. strongly collimated beams) a clever strategy is required to avoid wasting CPU time on the majority of the neutrons that do not pass through the guide system. In most of these packages, various trajectory weighting strategies and parallelization methods are therefore employed to speed up the simulation time. For example, the beam-splitting action of a neutron mirror is normally modeled as shown in Fig. 6. Instead of simulating both components of the wave function, only the reflected component of the wave function is propagated with a statistical weight multiplied by the reflectivity R of the mirror, since the transmitted component is normally absorbed in the optical systems and shielding. Recent developments [33] hand the transmitted component $(1-R)\psi$ over to nuclear engineering codes, such as MCNP, so that beam transport, dosimetry and experimental backgrounds can be calculated accurately at the same time.

The main drawback with the Monte Carlo method for pure beam transport calculations is that, despite weight-biassing efforts, it can nonetheless be intrinsically slow when one needs to simulate a beam with a low transport efficiency, such as when one wishes to model a highly collimated beam.

4.3.2. *Phase space methods*

The second neutron beam modeling method involves *acceptance diagrams*, based on the work by Carpenter and Mildner [34, 35], and more recently by Copley [36], Cussen [37] and Bentley [38].

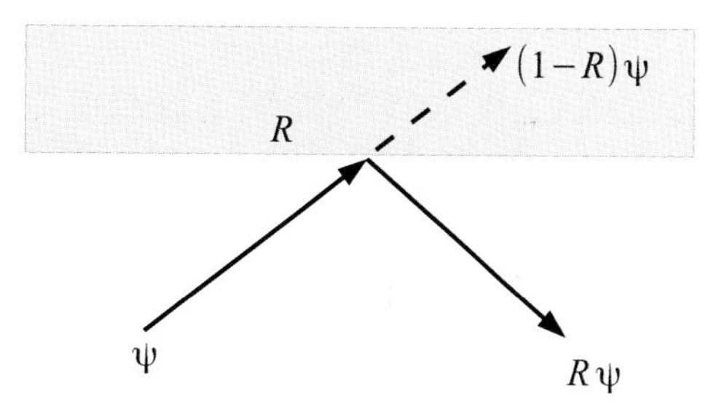

Fig. 6. The reflection of a neutron wave function ψ at a mirror interface with reflectivity R. The absorbed component $(1-R)\psi$ is normally discarded by most Monte Carlo packages.

The original idea behind acceptance diagrams is to calculate the location of boundaries in a phase space of position, divergence and wavelength for vertical and horizontal beam components. By defining these boundaries, the geometrical aspects of neutron beams can be completely understood. The computational extension of this method, "neutron acceptance diagram shading" (NADS) [38], simply adds a numerical weighting algorithm to the regions of phase space so that the beam intensity is accurately modeled in addition to the beam geometry. Here we will only summarize the acceptance diagram method, as it is covered in detail in the referenced articles.

The first stage of the process is to define the initial boundaries of the beam divergence and spatial dimensions, as defined by the entrance to the neutron beam system. This is shown in Fig. 7.

Temporarily neglecting gravitational distortion, by considering that an undisturbed neutron will continue in a straight line until it meets a mirror, beam propagation is handled trivially by applying a shear operation to the parallelogram in the vertical direction. Moreover, if we approximate the curve shapes in Fig. 5 with a piecewise function of three straight lines, and subdivide the parallelogram in Fig. 7 into triangles, then we can handle our phase space regions using well-established computational techniques. This makes NADS very fast for some guide shapes, because shear operations have a low computational burden.

To handle neutron mirrors, for each reflection in the physical neutron mirror, the phase space region is subject to a reflection in the spatial axis to reverse the direction, whilst at the same time subdividing the phase space regions into triangles in the appropriate place to follow our straight line approximation to the reflectivity curves. These phase space regions are then reflected in the plane corresponding to the position of the mirror in the spatial axis. This process is illustrated in Fig. 8 for a straight guide, although more complex mirror shapes are possible [38].

If necessary, gravity can be modeled by a parabolic curvature of the system in the vertical direction, but this correction is only infrequently required.

The drawback with the acceptance diagram method is that for divergent beams and curved optics with a relatively high transmission efficiency, the

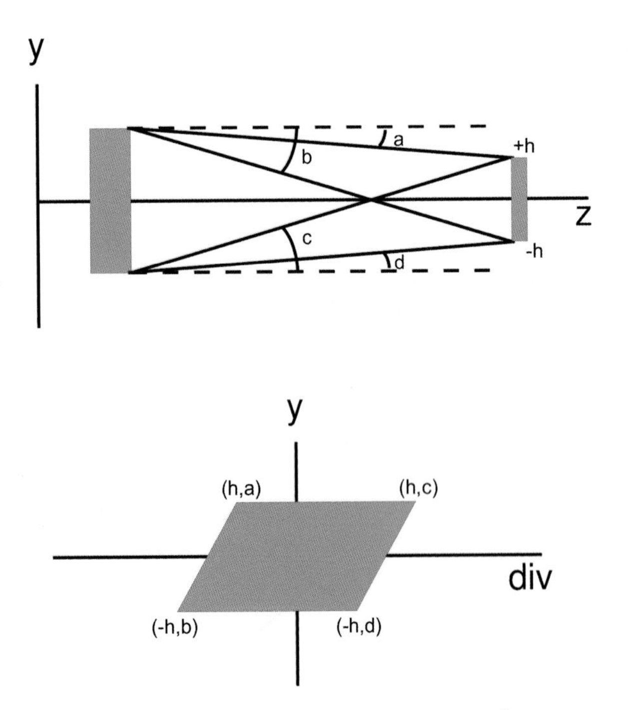

Fig. 7. Condensing a beam of neutrons onto a phase space defined by four points. The upper figure shows a side view of a large source illuminating a small guide entrance, and the lower figure shows the resulting phase space map, or acceptance diagram.

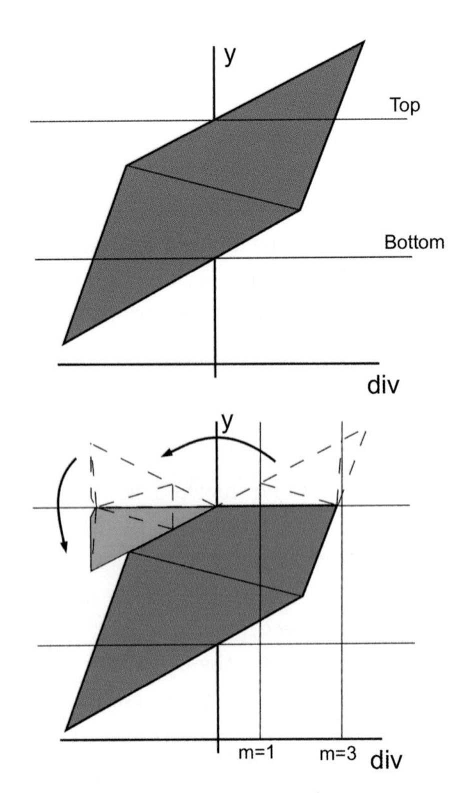

Fig. 8. Neutron beam propagation down a straight guide by subdividing and folding phase space.

number of triangles required in the model rapidly increases. There comes a point where Monte Carlo methods are faster. In practice, the two techniques should be used together for the optimization of a guide system.

5. Neutron Beam Guide Geometry

There are at present a number of neutron guide geometries in use around the world, and they are summarized in Fig. 9. The earliest neutron beam guides were simple tubes [2]. This guide is fine for short distances, or well-collimated beams, which results in relatively few bounces of the neutron particles on the mirrors, and this geometry is labeled A in Fig. 9.

Long guides and/or divergent beams result in many reflections, and — as we have seen in Fig. 5 — this would prove to be a very inefficient process for supermirror guides at large grazing angles.

In 1997, Mezei described a vision for a long pulse spallation source that required large distances between the source and the instrument stations for correct optimization, and introduced the concept of a "ballistic" guide shape [39] to deal with this issue. Ballistic guides have a wide section in the middle to reduce the number of reflections, as shown in Fig. 9 as guide type B with linear tapering, or as C with parabolic tapering.

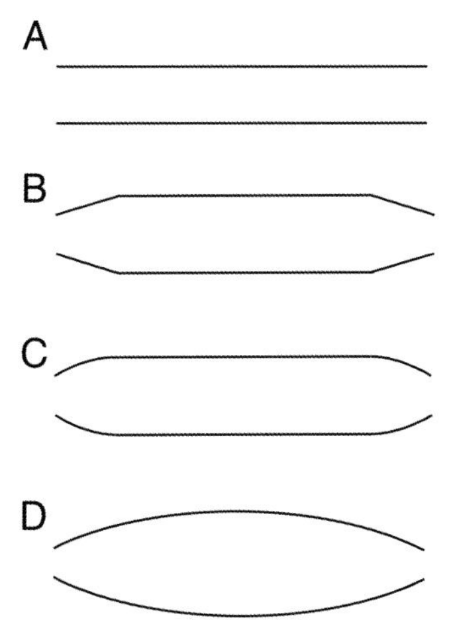

Fig. 9. Current neutron beam guide geometries that are widely deployed at sources around the world. A is a straight guide, or constant cross section guide; B is a ballistic guide with linear tapering; C is a ballistic guide with parabolic tapering; and D is a fully elliptic guide geometry.

More recently, developments by Schanzer *et al.* have pushed to a pure conic section geometry for the entire beam delivery system in the form of elliptic neutron guides [40], labeled D in Fig. 9. These are suggested to offer a smoother divergence distribution in the neutron beam at the focal point compared to the parabolic and linear tapered ballistic guides. Since this work was published, suggestions for a wide range of applications have been made using elliptic guides [41].

The principal drawback of elliptic guides from a user's point of view is the cost of the guide system compared to the ballistic guides. Examining Fig. 9 closely shows that for the ballistic geometries with a constant cross section in the center — namely B and C — we see that they can be fabricated from standardized, low m value straight sections in the middle, whereas each section in the fully elliptic geometry D requires a custom piece with only one twin on the opposite side of the central symmetry point of the optical system.

A recent study by Klenø [42] examined all of the geometries illustrated in Fig. 9 for different guide lengths in the range of 50–300 m. The conclusions of this study were twofold. Firstly, guide geometry C and D offer almost identical performance for all distances studied, *if* a fully optimized system is compared in each case. This also implies that large performance gains reported for D over C could be due to an inferior design of C. This is especially likely when one considers that the optimization process for D is more straightforward. The driving optimization is to establish the semiminor axes and two foci of the ellipse on the source and beam target for both the horizontal and vertical planes, whereas C requires additionally the definition of the lengths of each of the three sections in both planes. The second result from Klenø is that for geometry C the required m value for the straight section in the middle of the beam delivery is quite low. Certainly for cold neutrons, around $m = 2$ is required at most, and probably $m = 1$ would still be very attractive.

In the most recent years, two technologies have emerged that push neutron beam optic technology to the same level as for photon beam optics. The "selene" beam concept proposed by Stahn [43] uses the midpart of elliptical mirror pairs to provide an extremely high quality neutron beam phase

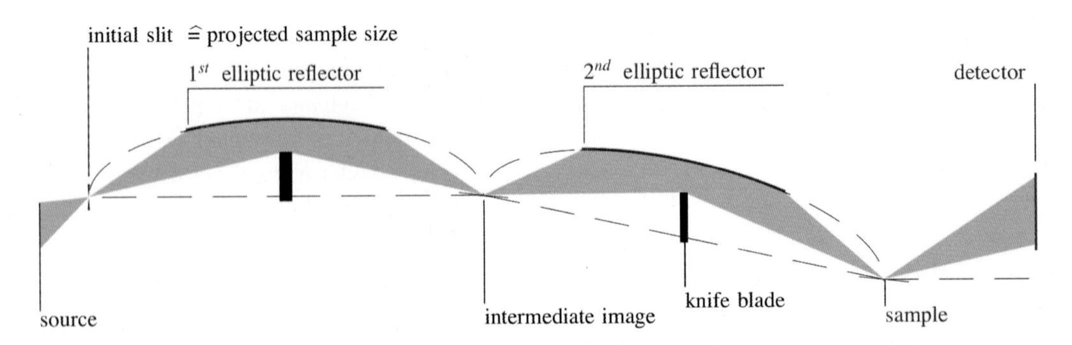

Fig. 10. The selene neutron beam concept, after Stahn *et al.* [43].

space with minimum optical aberration, as shown in Fig. 10.

This beam geometry involves only the minimum number of reflections, and is particularly designed to increase the performance of instruments measuring small samples.

Wolter optics [44] are established, demonstrated devices in the X-ray community [45] that have recently been demonstrated with neutrons also [46, 47]. Figure 11 shows the optical principles behind one of these focusing devices, which exploit a double reflection from axially symmetric curved mirrors to provide a beam that is excellent in optical quality.

Both the selene and Wolter optics represent the pinnacle of current reflective neutron beam concepts

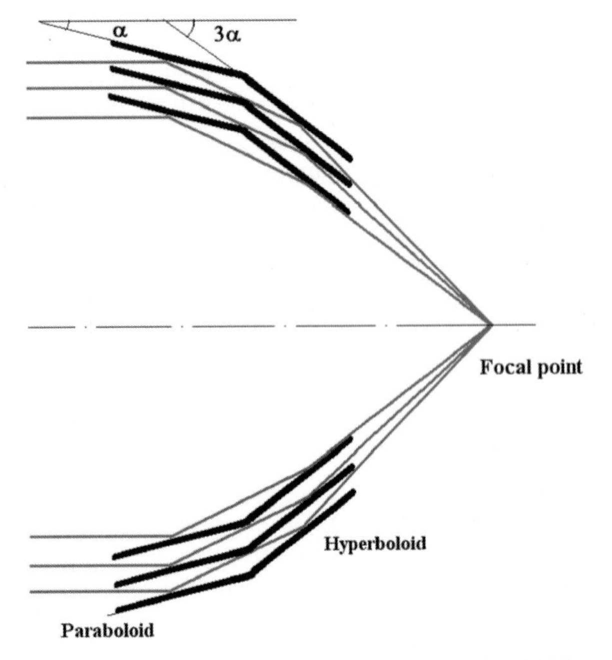

Fig. 11. The Wolter optics concept implemented by Mildner *et al.* [43].

both conceptually and theoretically, and the development of second generation prototypes from ongoing research in these areas is of high interest to the neutron optics community.

Further numerical optimization, beyond idealized shapes and basic trigonometry, has been explored by a number of teams, and offers potential gains in performance and robustness. After initial demonstrations of metaheuristic algorithms for optimizing high-dimensional parameter spaces for focusing optics [48], followed by simultaneously optimizing common beam extraction systems and instrument suites as a whole [49], these methods are readily available in a number of packages tailored to neutron scattering, such as iFit by Farhi *et al.* [50]. However, unguided metaheuristic strategies almost always fail in optimizing a completely free neutron guide geometry. Recent work by Bertelsen *et al.* shines, due to a double-pronged strategy of (a) restricting the search to a sequence of logical elements in the optics and (b) ensuring that the phase space requirements from any given module are fed backward to preceding modules [51]. This allows a user to optimize the system globally for a given set of requirements, and provides a very useful means of estimating the maximum beam transport efficiency that is possible before cost–benefit and robustness analyses begin.

6. Experimental Stations and Instrumentation

The scientific output of a neutron facility is the result of experiments performed on the instruments in which samples of the material under study are placed in the neutron beam and the scattered and/or transmitted neutron beam is measured. The neutron instrument comprises the beam manipulation optics both before and after the sample, leading up to the detectors, in which the scattered or transmitted neutrons are absorbed, as well as the beam stop for the transmitted beam and all the associated mechanical supports and shielding.

The interaction of a beam of thermal neutrons with a condensed matter sample can usually be well approximated as a neutron plane wave incident on a quasiperiodic assembly of weak, point-like scatterers. The resulting interference pattern allows the determination of the arrangement and movement of atoms in the sample. The diffraction theory is the same as for electron or X-ray scattering, but neutron scattering offers a number of distinct advantages:

- Thermal neutrons have wavelengths (a few Å) and energies (a few meV) which are well matched to probe both the interatomic distances and the fluctuations in condensed matter systems.
- They interact weakly with matter. They therefore penetrate deep into the bulk of the material and probe its internal structure, not just the surface.
- They interact via a simple point-like potential. This is because the nucleus is small compared to the neutron wavelength. As a result, the observed scattering intensities are straightforward to interpret in a quantitative manner.
- They have a magnetic moment, which means that they are scattered by the magnetic field of unpaired electrons in the material. They are thus sensitive to magnetic structures and fluctuations.
- They see a completely different contrast to other probes. Because they interact with the nucleus, they are sensitive to differences between isotopes. They can see light atoms in the presence of heavy atoms and can easily distinguish between elements which are close to each other in the periodic table.
- They are nondestructive.

Neutron instruments can broadly be categorized into two types: those that measure structure via *imaging*, *activation analysis* or *diffraction* methods and those that measure *dynamics*.

6.1. *Neutron imaging and activation analysis*

Neutron imaging allows the determination of the internal structure of an object by placing it in a neutron beam and recording the two-dimensional intensity of the transmitted beam. The simplest and most-used technique is neutron *radiography*, which produces a 2D attenuation map of the object, similarly to the 2D X-ray maps which we are familiar with from medical applications. In neutron *tomography*, the object is rotated and a series of attenuation maps are taken which are combined to produce a 3D reconstruction of the object. Neutron imaging requires good control of the divergence of the incoming beam and a detector with high spatial resolution placed a short distance behind the sample. It is one of the more accessible types of neutron instrumentation and can be found both at low power neutron sources and at high performance fission and spallation sources. The key areas of application are in the imaging of fuel cells, hydrogen storage devices, engineering, cultural heritage and biological systems.

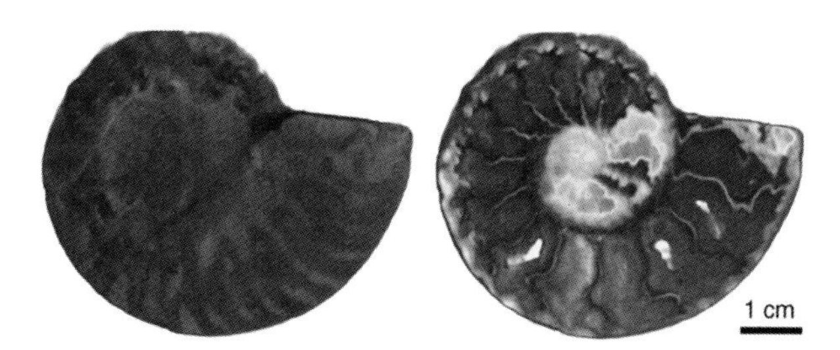

Fig. 12. Example of neutron imaging results, after Hilger *et al.* [52]. On the left is a radiograph of a fossil, and on the right is a tomographic slice through the fossil.

Neutron activation analysis provides a quantitative analysis of the elemental composition of the sample by exposing it to a thermal neutron beam and then measuring the decay of the radionucleides, usually by measuring the γ spectrum. It typically averages over the full sample volume, but can also produce a depth profile or a 2D map of the isotopic composition. It has applications in many areas, including archeometry and medicine, and as it often does not require very high neutron intensity, the technique is accessible for small and medium flux neutron sources.

6.2. *Neutron diffraction*

By far the majority of instruments for structural measurement extract the information by diffraction, in which the scattered intensity is recorded as a function of momentum transfer $\hbar\mathbf{Q}$, where $\mathbf{Q}$ is the wave vector transferred to the sample during the scattering process by $\mathbf{k}_i = \mathbf{k}_f + \mathbf{Q}$. $\mathbf{k}_i$ and $\mathbf{k}_f$ are the wave vectors of the incoming and scattered neutron beams, respectively, and the magnitude of the wave vector is given by $k = 2\pi/\lambda$. Since the scattering cross section for elastic ($k_i = k_f$) scattering usually dominates, these instruments tend to only measure k_i (or, in some cases, a weighted average of k_i and k_f), making the implicit assumption that the other is unchanged. Such instruments are known as *diffractometers*. The incident wave vector k_i can be determined by reflecting the incident beam off a large single crystal, which selects a single wavelength as described by Bragg's law [Eq. (4)] with $n = 1$, creating a *monochromatic* beam, provided that the higher harmonics ($n \geq 2$) can be appropriately removed. Graphite is a popular monochromator material, with a d-spacing corresponding to the [002] reflection of 3.355 Å. The diffraction pattern of the sample is then obtained by recording the scattered intensity as a function of the scattering angle.

The main alternative to crystal monochromators for the determination of the neutron wavelength is the *time of flight* (TOF) method. This is straightforward at pulsed sources, as the emission time of the neutrons from the source is well-known, and the neutron speed can be simply determined by noting the arrival time of each neutron at the detector. The neutron wavelength is then given by the de Broglie relation $\lambda = h/m_n v$, where h is Planck's constant, m_n the neutron mass and v the neutron speed. Using the TOF method, the diffraction pattern can be obtained at a single scattering angle, by recording the scattered intensity as a function of time.

Several types of neutron diffractometers have been developed, each specialized for the measurement of a particular type of system. *Small angle neutron scattering* (SANS) provides access to the largest length scales possible with neutron scattering. As the term implies, these instruments are optimized for measuring at small angles, typically only a few degrees away from the transmitted beam. They probe length scales from the nm range up to about a micron and are typically used for the measurement of macromolecules in aqueous solution, such as polymers, colloids or biological molecules. *Neutron reflectometry* is used to probe the structure of surfaces and interfaces over a similar range of length scales. Instruments for *powder diffraction* make up the largest group of neutron instruments. Many are workhorse instruments for structural characterization of polycrystalline samples by Rietveld refinement and play an important role in materials research. In addition to the bulk of the chemical crystallography instruments, some instruments for powder diffraction are specialized in measuring strain

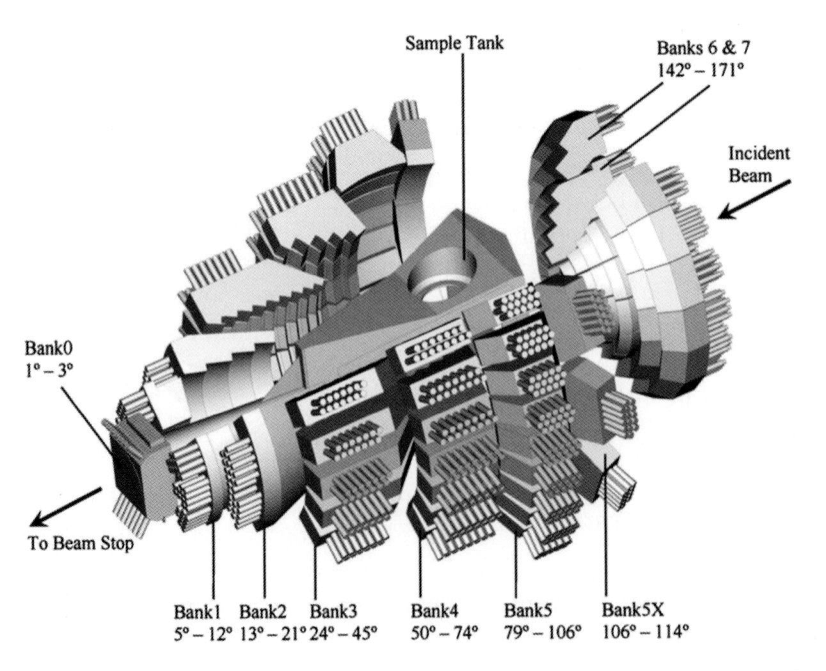

Fig. 13. An example of a neutron diffractometer, GEM at ISIS, after Hannon [53].

in engineering materials, while others are optimized for probing materials under extreme conditions of magnetic field or pressure. Materials which are available as large (usually mm-sized) single crystals, can be studied by *single crystal diffraction*, which provides a more direct and unambiguous structural determination than powder diffraction, particularly useful for complex magnetic structures and materials with very large unit cells. Unlike the other diffraction instruments, which use crystal monochromators or TOF to determine the neutron wavelength, instruments for single crystal diffraction often use the *Laue method*, which makes use of a wide wavelength spectrum.

6.3. *Neutron spectroscopy*

Neutron spectrometers measure the double differential scattering cross section $d^2\sigma/d\Omega d\omega = b^2 S \times (\mathbf{Q}, \omega) k_f/k_i$, where b is the scattering length and $\hbar\mathbf{Q}$ and $\hbar\omega$ are the momentum and energy transferred in the scattering process. The dynamic structure factor $S(\mathbf{Q}, \omega)$ is the Fourier transform of the Van Hove correlation function $G(\mathbf{r}, t)$, which is the pair correlation function quantifying the probability of finding an atom at position $\mathbf{r}$ and time t with respect to an atom at $\mathbf{r} = 0$ and $t = 0$. For magnetic scattering, it probes a similar spin–spin correlation function. Inelastic neutron scattering is thus a very direct measurement of the density and spin fluctuations of condensed matter systems.

Triple axis spectrometers can be tuned to access any position in $\mathbf{Q}$–ω space with adjustable resolution. They employ independent crystal monochromators for selecting both $\mathbf{k}_i$ and $\mathbf{k}_f$ and have been in use for more than 50 years. They are predominantly used for measuring coherent excitations in single crystals. *Chopper spectrometers* have established themselves as a workhorse technique for mapping out large areas of $\mathbf{Q}$–ω space in a single measurement, with applications in many areas of science. TOF or crystal monochromators are used for selecting $\mathbf{k}_i$, and $\mathbf{k}_f$ is obtained by TOF. *Backscattering spectrometers* employ monochromator crystals for analyzing $\mathbf{k}_f$, arranged with the Bragg angle θ as close to 90° as possible to achieve very high resolution, below 1 μeV in some cases. *Spin echo spectrometers* access the longest timescales possible with neutrons, by making use of Larmor precession of a polarized neutron beam to determine the neutron wavelength. They are typically used for polymer and protein dynamics.

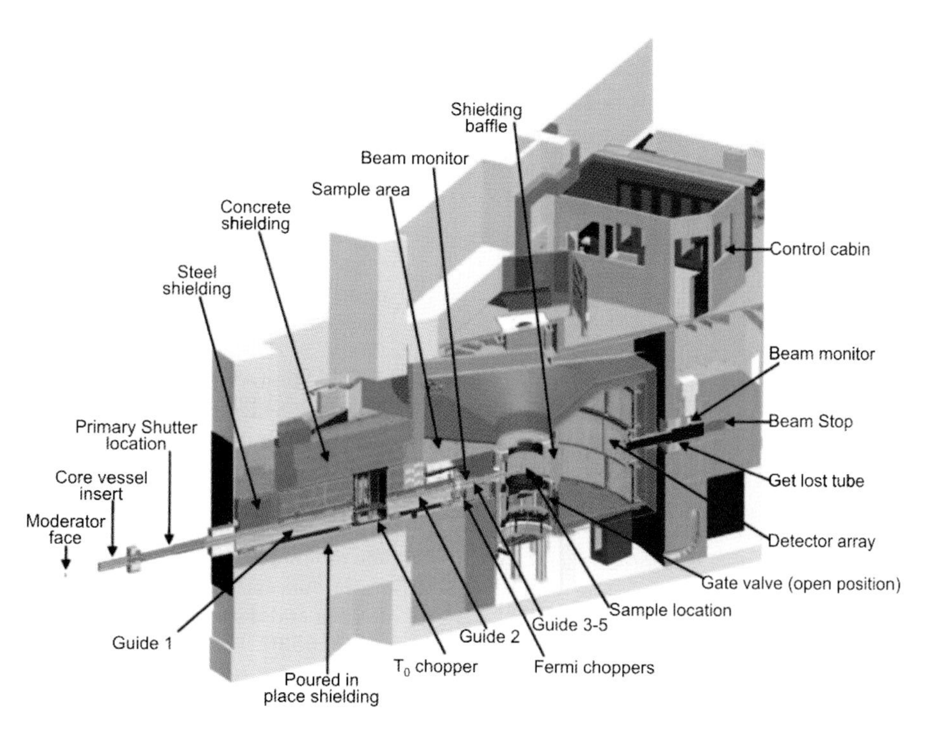

Fig. 14. An example of a neutron spectrometer, ARCS at the SNS, after Abernathy *et al.* [54].

6.4. *Neutron instruments for fundamental and particle physics*

Precision tests of quantum mechanics or the Standard Model of particle physics are performed using beams of cold or ultracold neutrons. The particle physics experiments designed for these purposes are typically set up to study the neutron itself: its electric dipole moment or its β decay.

7. Conclusion

We have presented a summary of select methods in the handling of neutron beams that are of particular relevance to modern neutron experimental facilities, along with a short description of production and experimental uses of the neutrons at the end stations of the beams. This is an exciting time for neutron optics, as we are witnessing the creation of neutron optical systems on a par with the very best of photonic systems and concepts, at the same time that generational leaps in both neutron production and computational methods have taken place.

Acknowledgments

The authors are extremely grateful for useful discussions and support from S. Peggs. C. P. Cooper-Jensen is supported by The Cluster of Research Infrastructures for Synergies in Physics (CRISP), cofunded by the partners and the European Commission under the 7th Framework Programme Grant Agreement 283745.

References

[1] J. Chadwick, Possible existence of a neutron, *Nature*, **129**, 312 (1932).
[2] H. Maier-Leibnitz and T. Springer, The use of neutron optical devices on beam–hole experiments, *Reactor Sci. Technol.* **17**, 217 (1963).
[3] L. Meitner and O. R. Frisch, Disintegration of uranium by neutrons: A new type of nuclear reaction, *Nature* **143**, 239 (1939).
[4] G. T. Seaborg, The interaction of fast neutrons with lead. PhD thesis, University of California, Berkeley (1937).
[5] A Second Target Station at ISIS (2000).
[6] R. L. Kustom, An overview of the spallation neutron source, arXiv, p. 0008212 (2000).
[7] Technical Design Report of Spallation Neutron Source Facility in J-PARC. Technical Report, JAEA, JAEA-Technology 2011-035 (2011).
[8] S. Peggs, ESS Technical Design Report. Technical Report, European Spallation Source ESS AB, ESS-doc-274-v15 (2013).
[9] K. N. Klausen, Fission, spallation or fusion-based neutron sources, *Pramana–J. Phys.* **71**, 623 (2008).
[10] Institut Laue Langevin Annual Report (2012).
[11] http://www.isis.stfc.ac.uk
[12] J. Penfold and R. K. Thomas, The application of the specular reflection of neutrons to the study of surfaces and interfaces, *J. Phys. Condens. Matter* **2**, 1369 (1990).
[13] E. Fermi and W. H. Zinn, *Phys. Rev.* **70**, 103 (1946).
[14] E. Fermi and L. Marshall, Interference phenomena of slow neutrons, *Phys. Rev.* **71**, 666 (1947).
[15] F. Mezei, Novel polarized neutron devices: Supermirror and spin component amplifier, *Commun. Phys.* **1**, 81 (1976).
[16] http://www.swissneutronics.ch
[17] P. Böni, Supermirror-based beam devices, *Physica B* **234–236**, 1038 (1997).
[18] J. B. Hayter and H. A. Mook, Discrete thin-film multilayer design for X-ray and neutron supermirrors, *J. Appl. Cryst.* **22**, 35 (1989).
[19] I. Carron and V. Ignatovich, Algorithm for preparation of multilayer systems with high critical angle of total reflection, *Phys. Rev. A* **67**, 043610 (2003).
[20] L. G. Parratt, Surface studies of solids by total reflection of X-rays, *Phys. Rev.* **95**(2) 359 (1954).
[21] R. Maruama, D. Yamazaki, T. Ebisawa and K. Soyama, Effect of interfacial roughness correlation on diffuse scattering intensity in a neutron supermirror, *J. Appl. Phys.* **105**, 083527 (2009).
[22] O. Elsenhans, P. Böni, H. P. Friedli, H. Grimmer, P. Buffat, K. Leifer and I. Anderson, Thin films for neutron optics, *SPIE* **1738**, 130 (1992).
[23] M. Ay, C. Schanzer, M. Wolff and J. Stahn, New interface solution for NI/TI multilayers, *Nucl. Instrum. Methods Phys. Res. A* **562**, 389 (2006).
[24] H. Takenaka, H. Ito, K. Nagai, Y. Muramatsu, E. Gullikson and R. C. C. Perera, Soft X-ray reflectivity and structure evaluation of Ni/C/Ti/C multilayer X-ray mirrors for water-window region, *Nucl. Instrum. Methods Phys. Res. A* **467–468**, 341 (2001).
[25] A. E. Munter, B. J. Heuser and K. M. Skulina, Neutron reflectivity measurements of titanium–beryllium multilayers, *Physica B* **221**, 500 (1996).
[26] T. Ebisawa, N. Achiwa, S. Yamada, T. Akiyoshi and S. Okamoto, Neutron reflectivities of Ni–Mn and Ni–Ti multilayers for monochromators and supermirrors, *J. Nucl. Sci. Technol.* **16**, 647 (1979).
[27] J. Saroun and J. Kulda, Neutron ray-tracing simulations and data analysis with RESTRAX, *Proc. of SPIE* **5536**, 124 (2004).
[28] D. Wechsler, G. Zsigmond, F. Streffer and F. Mezei, VITESS: virtual instrumentation tool for pulsed and continuous sources, *Neutron News* **11**(4), 25 (2000).

[29] P. Willendrup, E. Farhi and K. Lefmann, MCSTAS 1.7, a new version of the flexible Monte Carlo neutron scattering package, *Physica B* **350**, 735 (2004).

[30] P. A. Seeger and L. L. Daemen, The Neutron Instrument Simulation Package, NISP, in *Proc. SPIE* **5536**, 109 (2004).

[31] W.-T. Lee, X.-L. Wang, J. L. Robertson, F. Klose and Ch. Rehm, Ideas — a Monte Carlo simulation package for neutron-scattering instrumentation, *Appl. Phys. A* **74**, s1502 (2002).

[32] V. Hugouvieux, E. Farhi, M. R. Johnson, F. Juranyi, P. Bourges and W. Kob, Structure and dynamics of L-GE: Neutron scattering experiments and *ab initio* molecular dynamics simulations, *Phys. Rev. B* **75**, 104208 (2007).

[33] E. Klinkby, B. Lauritzen, E. Nonbol, P. K. Willendrup, U. Filges, M. Wohlmuther and F. X. Gallmeier, Interfacing MCNPX and MCSTAS for simulation of neutron transport, *Nucl. Instrum. Methods Phys. Res. A* **700**, 106 (2013).

[34] J. M. Carpenter and D. F. R. Mildner, Neutron guide tube gain for a remote finite source, *Nucl. Instrum. Methods* **196**, 341 (1982).

[35] D. F. R. Mildner, Neutron intensity gains for converging guide systems, *Nucl. Instrum. Methods Phys. Res. A* **301**, 395 (1991).

[36] J. R. D. Copley, Transmission properties of short curved neutron guides. Part 1: Acceptance diagram analysis and calculations, *Nucl. Instrum. Methods Phys. Res. A* **355**, 469 (1995).

[37] L. D. Cussen, Resolution calculations for novel neutron beam elements, *J. Appl. Cryst.* **33**, 1393 (2000).

[38] P. M. Bentley and K. H. Andersen, Accurate simulation of neutrons in less than one minute. Part 1: Acceptance diagram shading, *Nucl. Instrum. Methods Phys. Res. A* **602**, 564 (2008).

[39] F. Mezei, The *raison d'etre* of long pulse spallation sources, *J. Neutron Res.* **6**, 3 (1997).

[40] C. Schanzer, P. Böni, U. Filges and T. Hils, Advanced geometries for ballistic neutron guides, *Nucl. Instrum. Methods Phys. Res. A* **529**, 63 (2004).

[41] P. Böni, New concepts for neutron instrumentation, *Nucl. Instrum. Methods Phys. Res. A* **586**, 1 (2008).

[42] K. H. Klenø, K. Lieutenant, K. H. Andersen and K. Lefmann, Systematic performance study of common neutron guide geometries, *Nucl. Instrum. Methods Phys. Res. A* **696**, 75 (2012).

[43] J. Stahn, U. Filges and T. Panzner, Focusing specular neutron reflectometry for small samples, *Eur. Phys. J. Appl. Phys.* **58**, 11001 (2012).

[44] H. Wolter, Spiegelsysteme streifenden einfalls als abbildende optiken für röntgenstrahlen, *Annalen der Physik* **445**, 94 (1952).

[45] M. C. Weisskopf, B. Brinkman, C. Canizares, G. Garmire, S. Murray and L. P. van Speybroeck, An overview of the performance and scientific results from the Chandra X-ray observatory, *Publ. Astron. Soc. Pac.* **114**, 1 (2002).

[46] D. F. R. Mildner and M. V. Gubarev, Wolter optics for neutron focusing, *Nucl. Instrum. Methods Phys. Res. A* **634**, S7 (2011).

[47] Y. S. Bagdasarova, Wolter mirror microscope: Novel neutron focussing and imaging optic. PhD thesis, MIT (2010).

[48] P. M. Bentley and K. H. Andersen, Optimisation of focusing neutronic devices using artificial intelligence techniques, *J. Appl. Cryst.* **42**, 217 (2009).

[49] P. M. Bentley, M. Boehm, I. Sutton, C. D. Dewhurst and K. H. Andersen, Global optimisation of an entire neutron guide hall. Submitted to *J. Appl. Cryst.* (2010).

[50] E. Farhi, Y. Debab and P. Willendrup, IFIT: A new data analysis framework. Applications for data reduction and optimization of neutron scattering instrument simulations with MCSTAS, *J. Neutron Res.* **17**, in press.

[51] M. Bertelsen and K. Lefmann, Guide_bot: Automated optimization of neutron guides, *J. Appl. Cryst.*, in preparation.

[52] A. Hilger, N. Kardjilov, M. Strobl, W. Treimer and J. Banhart, The new cold neutron radiography and tomography instrument Conrad at HMI Berlin, *Physica B* **385–386**, 1213 (2006).

[53] A. C. Hannon, Results on disordered materials from the general materials diffractometer, GEM, at ISIS, *Nucl. Instrum. Methods Phys. Res. A* **551**, 88 (2005).

[54] D. L. Abernathy, M. B. Stone, M. J. Loguillo, M. S. Lucas, O. Delaire, X. Tang, J. Y. Y. Lin and B. Fultz, Design and operation of the wide angular-range chopper spectrometer arcs at the Spallation Neutron source, *Rev. Sci. Instrum.* **83**, 015114 (2012).

Phillip M. Bentley is the group leader for Neutron Optics and Shielding at the European Spallation Source in Lund, Sweden, and is a co-opted researcher at the University of Uppsala, Sweden. Prior to this, he was responsible for designing neutron optical systems at the Australian Nuclear Science and Technology Organisation (ANSTO) in Australia, and at the Institut Laue-Langevin, Grenoble, France; following on from work at HZB (formerly the Hahn-Meitner Institut) in Berlin on neutron instrumentation, and a PhD in metallic magnetism at the University of Leeds. His research interests include the design and optimization of complex optical systems and neutron beam delivery geometries, and novel neutron beam components.

Carsten P. Cooper-Jensen works in the Neutron Optics and Shielding Group at the European Spallation Source in Lund, Sweden. Previously he worked for 12 years at DTU-Space, Denmark, where he developed multilayer coatings for hard X-ray astronomy and coated optics for the NASA satellite Nuclear Spectroscopic Telescope Array (NuSTAR). His research interests lie in multilayer coatings for neutrons and hard X-rays. He is a co-opted researcher at the University of Uppsala, Sweden.

Ken H. Andersen is Head of the Neutron Instruments Division at the European Spallation Source in Lund, Sweden. He was previously in charge of the Neutron Optics lab at the Institut Laue-Langevin (ILL) in Grenoble, France and has worked as a neutron instrument scientist at both ILL and ISIS in the UK. His research interests center around the design and optimization of neutron instruments for both steady-state and pulsed neutron sources. He is an adjunct professor at the Niels Bohr Institute at Copenhagen University, Denmark.

Reviews of Accelerator Science and Technology
Vol. 6 (2013) 275–290

DOI: 10.1142/S1793626813300132

Beam–Materials Interactions

Nikolai V. Mokhov
Accelerator Physics Center,
Fermilab, MS 220, P.O. Box 500,
Batavia, IL 60510, USA
mokhov@fnal.gov

Challenging applications related to high-intensity beam–materials interactions are described along with consequences of controlled and uncontrolled impacts of intense beams on components of medium- and high-energy accelerators, beamlines, target stations, beam collimators and absorbers, detectors, shielding, and the environment. Requirements on simulation code capabilities for such applications are derived. The principal classes of deleterious effects in materials under irradiation are described with real-life examples, modeling techniques, and analyses of uncertainties in simulations.

Keywords: Intensity frontier; radiation effect classes; Monte-Carlo simulations and uncertainties.

1. Introduction

The next generation of medium- and high-energy accelerators for megawatt proton, electron, and heavy-ion beams moves us into a completely new domain of extreme energy deposition density up to 0.1 MJ/g and power density up to 1 TW/g in beam interactions with matter [1]. The consequences of controlled and uncontrolled impacts of such high-intensity beams on components of accelerators, beamlines, target stations, beam collimators and absorbers, detectors, shielding, and the environment can range from minor to catastrophic. Challenges also arise from the increasing complexity of accelerators and experimental setups, as well as from design, engineering, and performance constraints. All these put unprecedented requirements on the accuracy of particle production predictions, the capability and reliability of the codes used in planning new accelerator facilities and experiments, the design of machine, target and collimation systems, detectors, and radiation shielding and minimization of their impact on the environment. This leads to research activities involving new materials and technologies, as well as code developments whose predictive power and reliability are absolutely crucial.

2. Challenging Applications and Demands

Particle transport simulation tools and the physics models and calculations required in developing relevant codes are all driven by application. The most demanding applications are the high-power accelerators (e.g. spallation neutron sources, heavy-ion machines, and neutrino factories), accelerator-driven systems (ADSs), high-energy colliders, and medical facilities. Here are a few examples of demanding applications and corresponding issues addressed in the beam–materials interaction simulations [2]:

- *Beam collimation.* Operationally, high-power accelerators are limited by beam losses, not current limitations. Conventional radiation shielding can be bulky, costly, and not easily implemented. Only with a very efficient beam collimation system can one reduce uncontrolled beam losses in the machine to an allowable level, protecting personnel and components against excessive irradiation, maintain operational reliability over the life of the machine, provide acceptable hands-on maintenance conditions, and reduce the impact of radiation on the environment, under both normal operation and accident conditions [3–5].

- *High-power targetry.* The principal issues include: production and collection of maximum numbers of particles of interest; suppression of background particles in the beamline; target and beam window operational survival and lifetime (compatibility, fatigue, stress limits, erosion, remote handling, and radiation damage); protection of focusing systems, including provision for superconducting coil quench stability; heat loads, radiation damage, and activation of components; thick shielding and spent beam handling with respect to prompt radiation and groundwater activation. For further details, see Ref. 6.
- *Absorbers* for misbehaved beams along the beamlines, abort beam dumps and those downstream of the production targets and interaction regions at colliders are other challenging systems in the megawatt accelerators. These should be able to withstand an impact of beams of up to full power — say, 0.2–20 MW — without destruction over a designed lifetime (at least a few years), fully contain the beam energy, and execute the initial shielding functions. The major absorber technology for high-intensity beams is a core built up of many thin graphite slabs encapsulated in an aluminum shell with cooling water channels. It has been proven in more than 20 years of operational experience at the Tevatron, with the peak instantaneous temperature rise of ~1000°C per pulse. The core is contained in steel shielding surrounded by concrete. A similar design is used at the Large Hadron Collider (LHC), with the beam swept in a spiral during the abort. Other technologies for high-power beam absorbers include a stationary beryllium, aluminum, or nickel wall liquid-cooled dump, a water-cooled aluminum shell rotating drum, and a water vortex beam absorber considered for an 18 MW electron beam at the International Linear Collider (ILC). In the ILC case, the beam is rastered with dipole coils to avoid water boiling. The entrance beam window and catalytic recombination are of serious concern in such a design.
- *High-energy colliders.* These include proton (LHC), heavy-ion (LHC and the Relativistic Heavy-Ion Collider — RHIC), e^+e^- (ILC and the Compact Linear Collider — CLIC), and muon colliders. The principal issues address overall machine and interaction region design; accelerator and detector component protection against beam-induced radiation load (superconducting magnets) and damage (heating, material integrity, and component lifetime); electronics soft errors; and detector backgrounds. All particle interactions and transport need to be accurately treated to predict machine and detector performance, radiation damage, residual radiation (hands-on maintenance), air, soil, and groundwater activation, and prompt radiation on surface and in underground experimental halls.

3. Interactions of Fast Particles with Matter

Electromagnetic interactions, decays of unstable particles, and strong inelastic and elastic nuclear interactions all affect the passage of high-energy particles through matter. The physics of these processes is described in detail in numerous books, handbooks, and reviews (see, for example, Refs. 7–9). At high energies the characteristic feature of the phenomenon is creation of hadronic cascades and electromagnetic showers (EMSs) in matter due to multiparticle production in electromagnetic and strong nuclear interactions. Because of consecutive multiplication, the interaction avalanche rapidly accrues, passes the maximum, and then dies as a result of energy dissipation between the cascade particles and due to ionization energy loss. Energetic particles are concentrated around the projectile axis forming the shower core. Neutral particles (mainly neutrons) and photons dominate with a cascade development when energy drops below a few hundred MeV.

The length scale in hadronic cascades is a nuclear interaction length λ_I (16.8 cm in iron), while in EMS it is a radiation length X_0 (1.76 cm in iron); see Refs. 7–9 for definitions and values of these quantities in other materials. The hadronic cascade longitudinal dimension is 5–10 λ_I, while in EMS it is 10–30 X_0. It grows logarithmically with primary energy in both cases. Transversely, the effective radius of the hadronic cascade is about λ_I, while for EMS it is about $2r_M$, where r_M is a Moliere radius (see Refs. 7–9). At the same time, low-energy neutrons coupled to photons propagate a much larger distance in matter around the cascade core, both longitudinally and transversely, until they dissipate their energy in a region of a fraction of an electron volt.

Muons — created predominantly in pion and kaon decays during the cascade development — can travel hundreds of thousands of meters in matter along the cascade axis. Neutrinos — the usual muon partners in such decays — propagate even farther, hundreds of thousands of kilometers, until they exit the Earth's surface. A rather unusual problem arising from neutrino–materials interactions is a potential radiation hazard to a general public at very large distances from a high-luminosity multi-TeV muon collider [10]. This will require placing such a machine at a depth of a few hundred meters along with a well-thought-out collider layout design and special muon beam manipulation techniques.

4. Simulations

As stated in Sec. 2, the demanding applications at high-intensity accelerators put unprecedented requirements on the accuracy of particle production predictions, and the capability and reliability of the simulation codes used. The challenge is detailed and accurate (to a % level) modeling of all particle interactions with 3D system components (up to tens of kilometers of the accelerator lattice in some cases) in the energy region spanning up to 15 decades as a basis of accelerator, detector, and shielding designs and their performance evaluation, for both short-term and long-term effects.

The current versions of five general-purpose, all-particle codes are capable of this: FLUKA [11], GEANT4 [12], MARS15 [13], MCNP6 [14], and PHITS [15]. These are used extensively worldwide for accelerator applications. A substantial amount of effort (up to several hundreds of man-years) has been put into development of these codes over the last few decades. The user communities for the codes reach several thousands of people worldwide. All five codes can handle a very complex geometry, have powerful user-friendly built-in graphical user interfaces (GUIs) with magnetic field and tally viewers, and variance reduction capabilities. Tallies include volume and surface distributions (1D to 3D) of particle flux, energy, reaction rate, energy deposition, residual nuclide inventory, prompt and residual dose equivalent, displacement per atom (DPA)

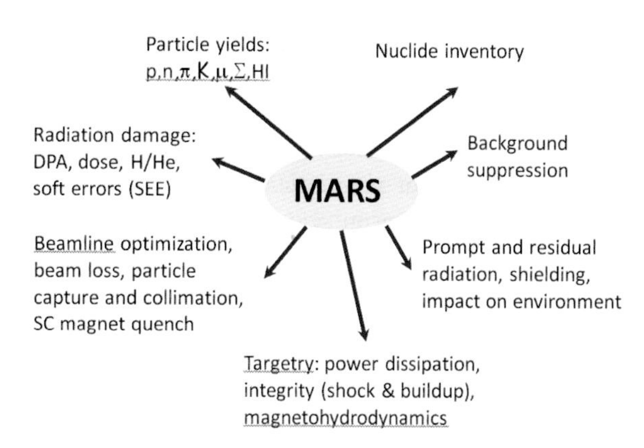

Fig. 1. Major applications at the intensity frontier of particle–matter interaction MARS15 code.

for radiation damage, event logs, intermediate source terms, etc. All the aspects of beam interactions with accelerator system components are addressed in sophisticated Monte Carlo simulations benchmarked — wherever possible — with dedicated beam tests. Figure 1 shows the energy and intensity frontier applications where the codes listed above are used.

As an example, the advanced features in the current MARS15 code [13, 16] — created by accelerator developmental needs — include:

- Reliable description of cross-sections and particle yields from a fraction of eV to many TeV for hadron, photon, and heavy-ion projectiles on nuclei.
- Precise modeling of leading particle production and low-momentum transfer processes (elastic, diffractive, and inelastic), crucial for beam-loss and collimation studies.
- Reliable modeling of π^0 production (electromagnetic showers), $K^{\pm}$ and K^0 production (neutrino and kaon rare decay experiments), proton–antiproton annihilation, and stopped hadrons and muons.
- Nuclide inventory, residual dose, displacement per atom (DPA), and hydrogen and helium production.
- Precise modeling of multiple Coulomb scattering with projectile and target form factors included.
- Reliable and CPU-efficient modeling of low-energy electromagnetic showers and electromagnetic interactions of particles and heavy ions down to 1 keV/A in compounds (energy deposition, radiation damage, and backgrounds) with

bremsstrahlung and direct pair production by heavy particles at high energies.
- Hadron/muon photo- and electroproduction.
- Accurate particle transport in arbitrary geometry in the presence of magnetic and electrical fields with objects ranging in size from microns to kilometers.
- Variance reduction techniques, crucial for modeling rare processes and thick shielding.
- Enhanced tagging of the origin of a given signal/tally — geometry, process, and phase space — invaluable for source term and sensitivity analyses.
- User-friendly geometry description and visual editing.
- Interfaces to MAD, ANSYS, and hydrodynamics codes.

Most of the processes in MARS15, such as electromagnetic showers, hadron–nucleus interactions, decays of unstable particles, emission of synchrotron photons, photohadron production and muon pair production, can be treated exclusively (analogously), inclusively (with corresponding statistical weights), or in a hybrid mode. The choice of method is left for the user to decide — via the input settings — what is the most appropriate and computationally efficient for the considered physics case.

Inclusive particle production is based on a comprehensive set of phenomenological formulas tuned to data in all the important phase space regions. An example of modeling of nontrivial behavior of the invariant proton production cross-section is shown in Fig. 2 for the $p + Be \to p + X$ inclusive reaction in comparison with data [17–19]. The color-coded lines represent three distinct physics mechanisms simulated: resonance and diffractive dissociation (blue), fragmentation (green), and central region (black). Quasielastic scattering and Fermi motion are modeled in addition, supplied with a phenomenological model for cascade and evaporation nucleon production.

The LAQGSM module is based on the quark–gluon string model above 10 GeV and intranuclear cascade, pre-equilibrium, and evaporation models at lower energies [20]. It is used in MARS15 for photon, hadron, and heavy-ion projectiles at projectile energies from a few MeV/A to 1 TeV/A. This provides a power of full, theoretically consistent modeling of exclusive and inclusive distributions of secondary particles, spallation, fission, and fragmentation products. It was recently modified to improve its performance in the crucial-for-the-intensity-frontier energy region of 0.7–12 GeV. Figure 3 shows results on neutron inclusive production calculated with this model in comparison with data [21] in interactions of a

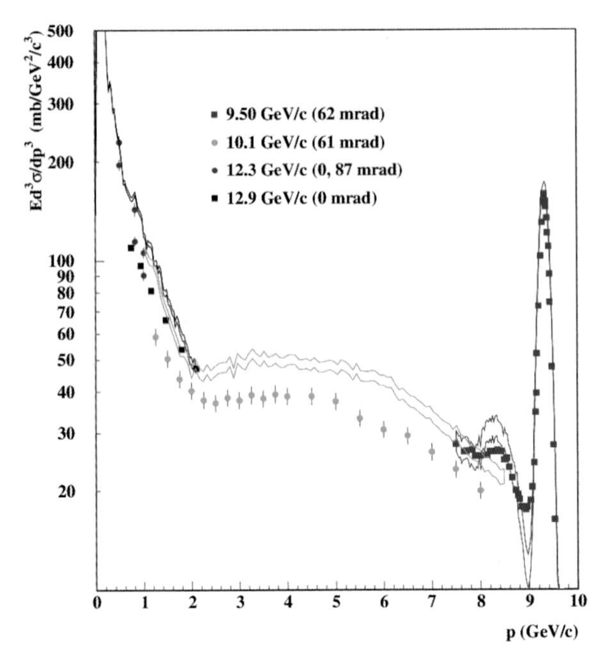

Fig. 2. Proton production cross-section at indicated angles for protons of labeled momenta on a beryllium nucleus calculated with MARS15 (lines) and compared to data [17–19].

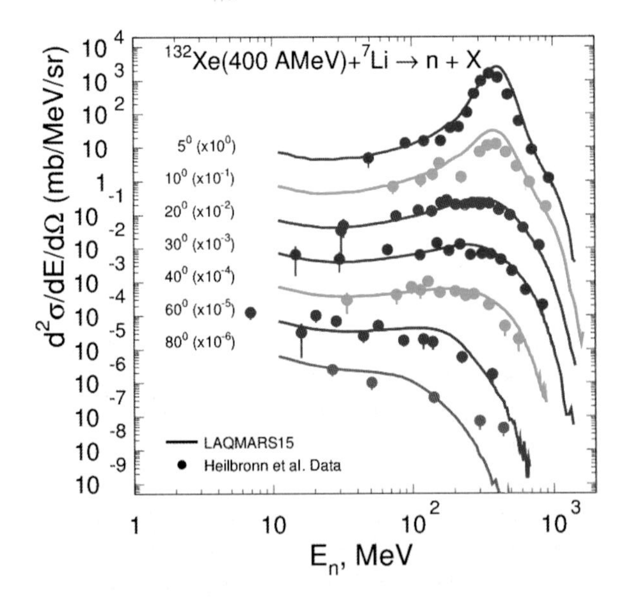

Fig. 3. Double-differential neutron production cross-section for 400 MeV/A xenon on a lithium nucleus calculated with MARS15 (LAQGSM [20] mode) and compared to data [21].

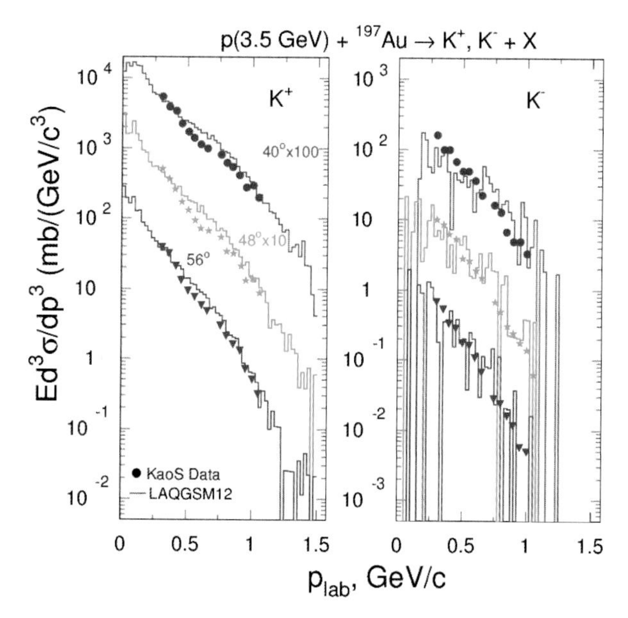

Fig. 4. Kaon production cross-section for 3.5 GeV protons on a gold nucleus calculated with MARS15 (LAQGSM) and compared to data [22].

400 MeV/A xenon projectile with a lithium nucleus. Comparison of the model with data [22] in Fig. 4 reveals good agreement for $K^{\pm}$ large-angle production for the 3.5 GeV proton interactions with the gold nucleus.

5. Nuclide Production and Residual Activation

As mentioned in Sec. 2, one of the fundamental operational limitations at high-power accelerators is the beam loss rate, which corresponds to the tolerable residual dose levels on the machine components and corresponding nuclide production rates. A worldwide-spread "1 W/m rule" was developed at the brainstorming workshop [3] used nowadays as one of the primary guides in high-intensity accelerator design considerations. Based on a thorough analysis of a world experience with high-power machines and related calculation results, the 1 W/m beam loss rate was derived as a universal design goal applicable to any proton accelerator or beamline with proton energy above about 200 MeV. In common conditions, this continuous beam loss rate results in a contact residual dose rate of 50–100 mrem/h on an outer surface of a typical massive accelerator magnet after 30 days of irradiation and 1 day of cooling. Observations and numerous calculations over 15 years have confirmed the applicability and usefulness of this simple rule.

Three approaches are used with the codes described in the previous section to calculate 3D distributions of the residual dose rate in an arbitrary configuration of accelerator setups:

(1) Calculate the production rates of all nuclides generated in the object/region of interest; solve the Bateman equations governing the decay and transmutation of nuclides using transmutation trajectory analysis for predefined irradiation and cooling times; convert the calculated activities to individual doses at a distance using specific gamma ray constants or run corresponding Monte Carlo for the emitted photons.
(2) Calculate the spatial distribution of the residual dose rate using built-in ω factors which relate the density of inelastic nuclear interactions to a contact dose rate; correct for a small object size; apply Monte-Carlo-based distance correction.
(3) Start as in item 2, then scale from surface disk sources to a remote point.

Although quite complex and time-consuming, the first method is fully consistent. It also provides a nuclide inventory — a detailed distribution of nuclides and activities produced in the accelerator and its system components. The second method is easier and faster, while the third is good for engineering estimates. Figure 5 (from Ref. 16) shows comparisons between measured and predicted residual activity in a copper target irradiated with a 500 MeV/A uranium beam. The MARS15 performance is quite impressive.

6. Materials Under Irradiation

Depending on the material, the level of energy deposition density and its time structure, one can face a variety of effects in materials under irradiation. Figure 6 shows two classes of effects — related to beam-induced heat dissipation and changes in material properties — that can be observed in superconducting magnets under irradiation.

The most damaging in a typical high-intensity accelerator environment are the following:

- Thermal shocks and quasi-instantaneous damage;

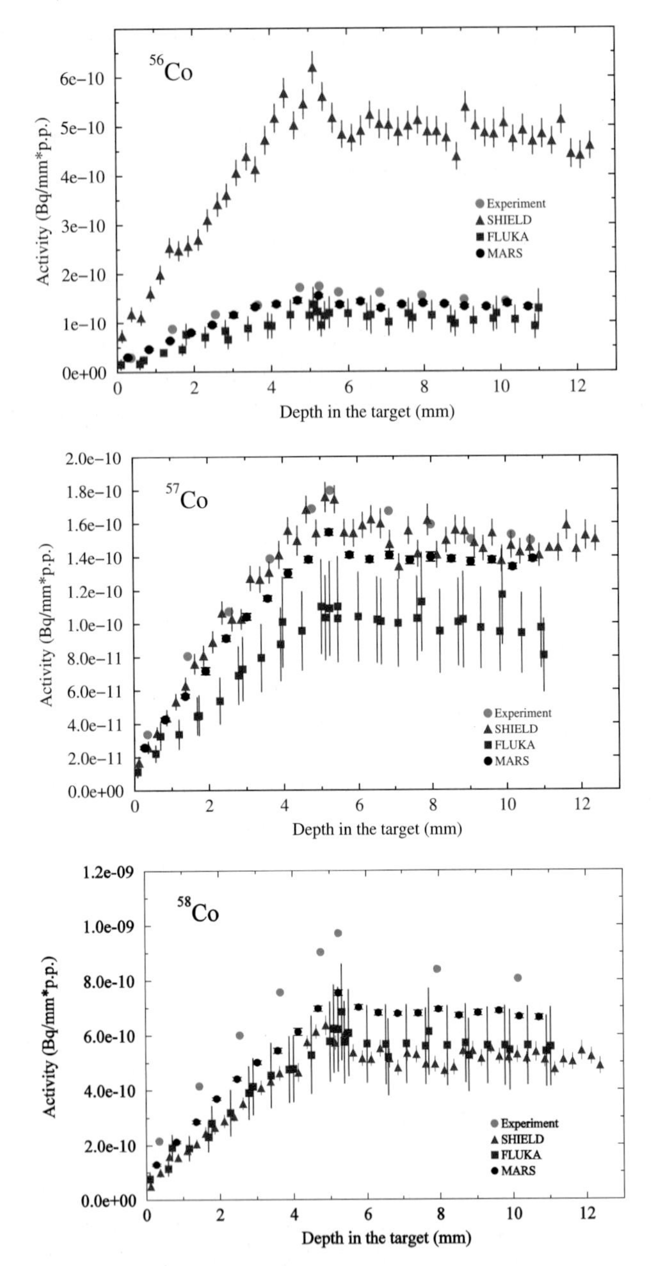

Fig. 5. Measured [23] and calculated distributions of specific residual activity of cobalt isotopes generated and stopped in a copper target irradiated with a 500 MeV/u uranium beam, 11 mm in diameter. The target length was twice the range of the uranium ions and the transverse target size was 50 mm. It was assembled using copper disks, and the activation foils were inserted between the disks. Statistical errors for MARS15 correspond to 2.5 million histories; statistical errors for SHIELD and FLUKA are taken from Ref. 23.

- Organic insulation property deterioration due to dose buildup;
- Radiation damage to metals, ceramic and other inorganic materials due to atomic displacements (DPA) as well as helium and hydrogen production.

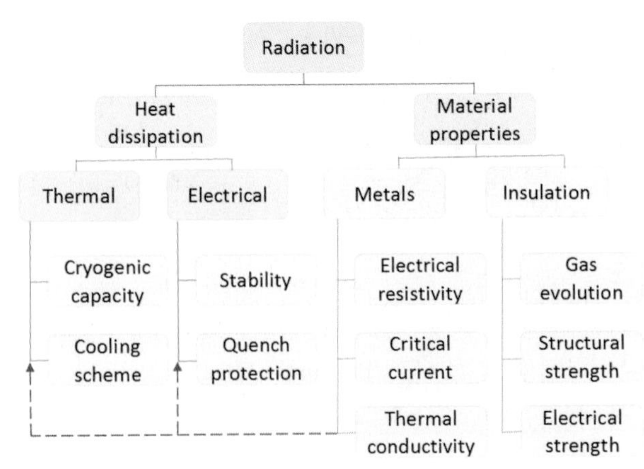

Fig. 6. Types of beam-induced deleterious effects in superconducting magnets. (Courtesy of V. V. Kashikhin.)

6.1. *Short pulses*

Short pulses with energy deposition density in the range from 0.2 kJ/g (W), 0.6 J/g (Cu), and 1 kJ/g (Ni and Inconel) to about 15 kJ/g cause thermal shocks resulting in fast material ablation and slower structural changes. The latter are shown in Fig. 7 for the Fermilab antiproton production target irradiated by a 120 GeV proton beam (rms beam spot size of 0.2 mm) with 3×10^{12} protons per pulse.

Fig. 7. Beam-induced damage to the Fermilab antiproton production target — a stack of 10-cm-diameter nickel and Inconel disks.

Fig. 8. A hole indicated as created in the Tevatron's 5-mm-thick primary tungsten collimator.

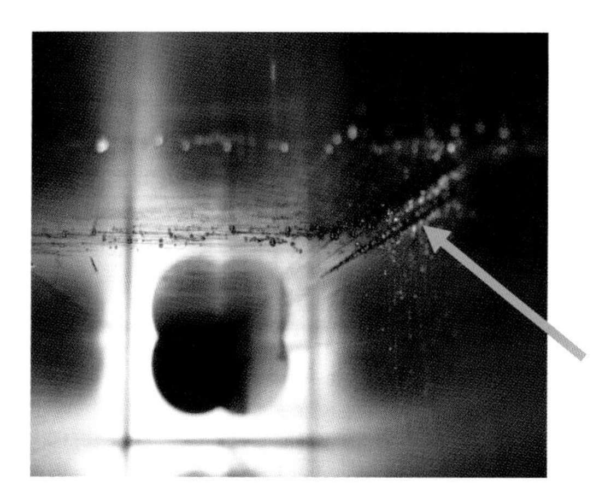

Fig. 9. A 25-cm-long groove indicated as created in the Tevatron's secondary stainless steel collimator.

MARS simulation explained target damage, and reduction in antiproton yield, and justified better target materials.

An outstanding example of the fast material ablation at accelerators is destruction of the Tevatron primary (Fig. 8) and secondary (Fig. 9) collimators caused by an accidental loss of the 980 GeV beam in 2003 [24]. The damage was induced by a failure in the CDF Roman Pot detector positioning at the end of a 980 × 980 GeV proton–antiproton colliding beam store. The dynamics of this failure over the first 1.6 ms, including excessive halo generation and superconducting magnet quenching, were studied via realistic simulations using the STRUCT [25] and MARS codes. It was shown that the interaction of a misbehaved proton beam with the superconducting magnets and collimators resulted in rapid local heating. Detailed consideration was given to the ablation process for the collimator material taking place in high vacuum. It was shown that ablation of the tungsten primary collimator resulted in the creation of the hole in it, while a groove was created in the stainless steel secondary collimator jaw surface with parameters fully agreeing with the postmortem observations.

Beam pulses with energy deposition density in excess of 15 kJ/g bring materials to the hydrodynamic regime. It was first shown in studies for the SSC's 20 TeV proton beam (400 MJ, 300 μs spill) on a graphite beam dump [26] and later for the collider superconducting magnets, steel collimators, and tunnel-surrounding Austin chalk. Since the beam duration was comparable to the characteristic time of expected hydrodynamic motions, the static energy deposition capability of the MARS code has been combined with the two- and three-dimensional hydrodynamics of the LANL's MESA and SPHINX codes. It was found in simulations that a hole was drilled by the beam in the graphite dump at the rate of 7 cm/μs, with pressures of a few kbar generated. Later these effects were studied in detail for the SPS and LHC targets and beam dumps using coupling of the FLUKA (energy deposition) and BIG2 (hydrodynamics) codes [27, 28]. Figure 10 shows the calculated physical state of the solid tungsten target at the end of the SPS proton pulse (rms beam spot size of 0.088 mm) at 7.2 μs. It is seen that within the inner 2 mm radius, a strongly coupled (SC) plasma state exists, which is followed by an expanded hot (EH) liquid. The melting front is seen propagating outward.

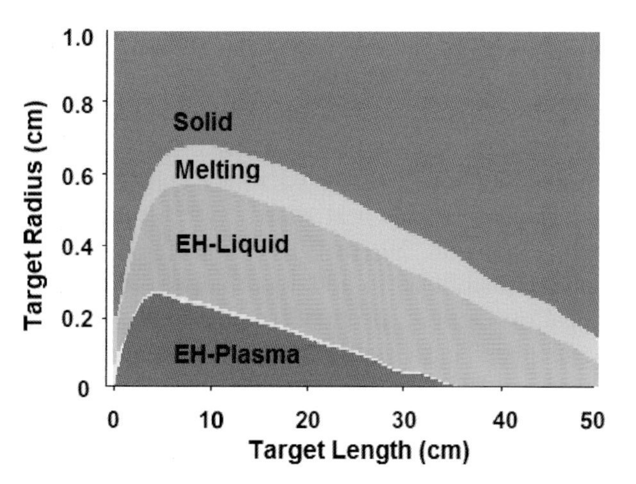

Fig. 10. The tungsten target's physical state after the SPS beam pulse [27].

6.2. *Organic materials*

Contrary to the MeV-type accelerators with their insulators made mostly from ceramics or glass, the majority of insulators in high-energy accelerator equipment are made from organic materials: epoxy, G11, polymers, etc. Apart from electronics and optical devices, the organic materials are the ones most sensitive to radiation. A large number of radiation tests have been made on these materials, and the results are extensively documented (see Ref. 29 for references). The impact of radiation on organic materials is a three-step process [30]:

(1) Production of free radicals by radiation.
(2) Reaction of free radicals: crosslinking, chain scission, formation of unsaturated bonds (C=C, etc.), oxidation, and gas evolution.
(3) Change of molecular structure: modification and degradation affected by the irradiation temperature and atmosphere, as well as by the presence of additives.

The findings for organic materials under irradiation are [30]:

- Degradation is enhanced at high temperatures.
- Radiation oxidation in the presence of oxygen accelerates degradation.
- Radiation oxidation is promoted in the case of a low dose rate.
- Additives can improve radiation resistance. For example, 1% by the weight of the antioxidant in polyethylene can prolong its lifetime by 5–10 times.

Dose limits on insulators are usually defined for a certain level of changes in the material properties critical to the application. For example, 10% degradation of ultimate tensile strength is a typical criterion for epoxy, CE/epoxy resins, and G11. Similar changes in electrical resistivity are often used as the criterion. For the given insulator and irradiation conditions, its radiation damage is proportional to the peak energy deposition density or dose accumulated in the hottest region. Radiation damage thresholds based on the results of dedicated radiation tests [29], experience, or indirect evidences are used worldwide as a basis for design and an estimate of the component lifetime or the operation time prior to replacement. For example, the dose limit used for the LHC superconducting magnet insulators is 25–40 MGy (2.5–4 Grad). Other projects utilizing superconducting magnet technologies assume a lower limit of 7–10 MGy.

It is worth noting here that energy deposition — responsible for damage in insulators and, for example, for cable quench stability in superconducting magnets — is modeled in accelerator applications quite accurately. In the majority of cases, FLUKA and MARS15 results on energy deposition coincide within 10% and agree with data.

7. DPA and Gas Production

The dominant mechanism of structural damage to inorganic materials is displacement of atoms from their equilibrium position in a crystalline lattice due to irradiation with formation of interstitial atoms and vacancies in the lattice. The resulting deterioration of material critical properties is characterized — in the most universal way — as a function of displacements per target atom (DPA). DPA is a strong function of the projectile type, energy, and charge, as well as material properties including the temperature.

7.1. *DPA model*

Three major codes — FLUKA, MARS15, and PHITS — use very similar implementation of the NRT model [31, 32] to calculate DPA. A primary knock-on atom (PKA) created in nuclear collisions can generate a cascade of atomic displacements. This is taken into account via damage function $\nu(T)$. DPA is expressed in terms of damage cross-section σ_d:

$$\sigma_d(E) = \int_{T_d}^{T_{\max}} \frac{d\sigma(E,T)}{dT} \nu(T)\, dT,$$

where E is the kinetic energy of the projectile, T the kinetic energy transferred to the recoil atom, T_d the displacement energy, and $T_{\max}$ the highest recoil energy according to kinematics. In a modified Kinchin–Pease model [31], $\nu(T)$ is zero at $T < T_d$, unity at $T_d < T < 2.5T_d$, and $k(T)E_d/2T_d$ at $2.5T_d < T$, where E_d is "damage" energy available for generating atomic displacements by elastic collisions. T_d is an irregular function of atomic number (∼40 eV). The displacement efficiency, $k(T)$, introduced as a result of simulation studies on evolution of atomic displacement cascades [33], drops from 1.4 to 0.3 once the PKA energy is increased from 0.1 to 100 keV, and exhibits a weak dependence on target material and temperature.

The implementation of this model in MARS15 [34] includes electromagnetic elastic (Coulomb) scattering, the Rutherford cross-section with Mott corrections, and nuclear form factors (a factor-of-2 effect). Resulting displacement cross-sections due to Coulomb scattering are shown in Fig. 11 for various projectiles on a carbon target. For elementary particles, energy dependence of σ_d disappears above 2–3 GeV, while it continues to higher energies for heavy ions. For projectiles heavier than a proton, σ_d grows with a projectile charge z as z^2/β^2 at $\gamma\beta > 0.01$, where β is a projectile velocity. All products of elastic and inelastic nuclear interactions, as well as Coulomb elastic scattering of transported charged particles (hadrons, electrons, muons, and heavy ions) from 1 keV to 10 TeV, contribute to DPA in the model.

DPA for neutrons from 10^{-5} eV to 20–150 MeV is described in MARS15 using the NJOY99+ENDF-VII [35, 36] database for 393 nuclides [16]. A corresponding output is shown in Fig. 12. Such results

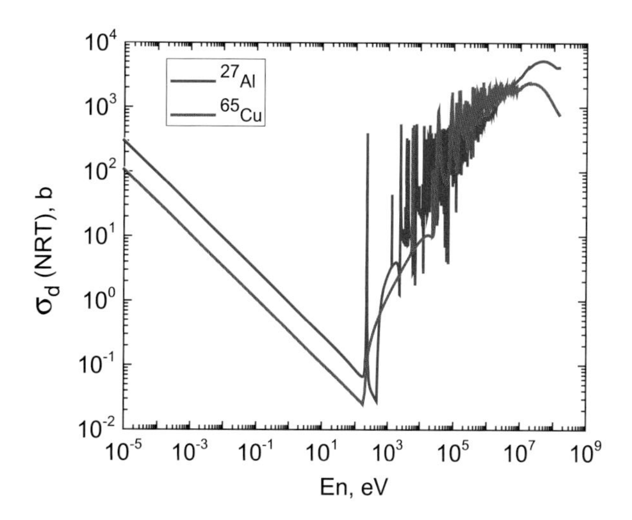

Fig. 12. NRT neutron defect production cross-sections.

are then corrected using experimental defect production efficiency η, a ratio of the number of single interstitial atom vacancy pairs (Frenkel pairs) produced in a material to the number of defects calculated using the NRT model. The values of η have been measured [37] for many important materials in the reactor energy range and are presented in Fig. 13.

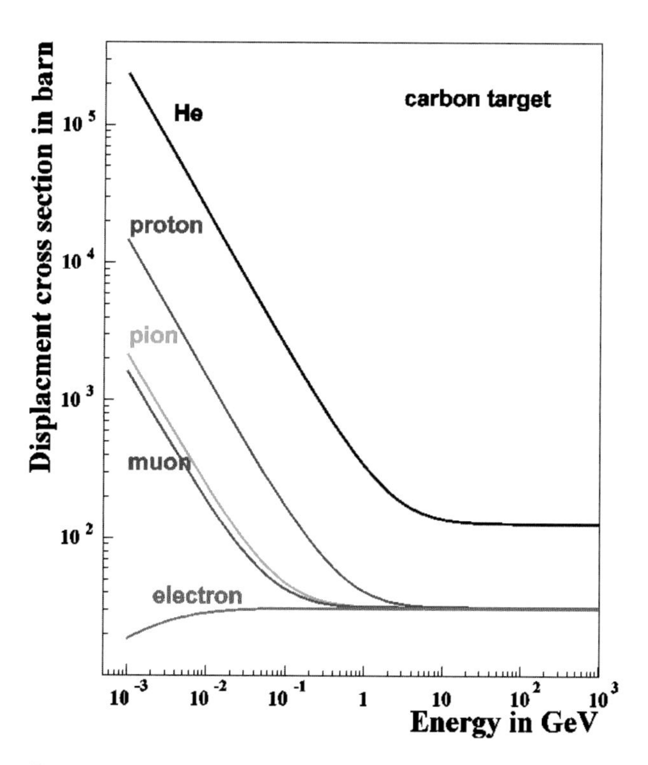

Fig. 11. Displacement cross-section in carbon for various charged projectiles.

7.2. *DPA modeling verification*

There is no direct way to measure DPA. Therefore, DPA model realizations have recently been tested via thorough code intercomparisons. MARS15

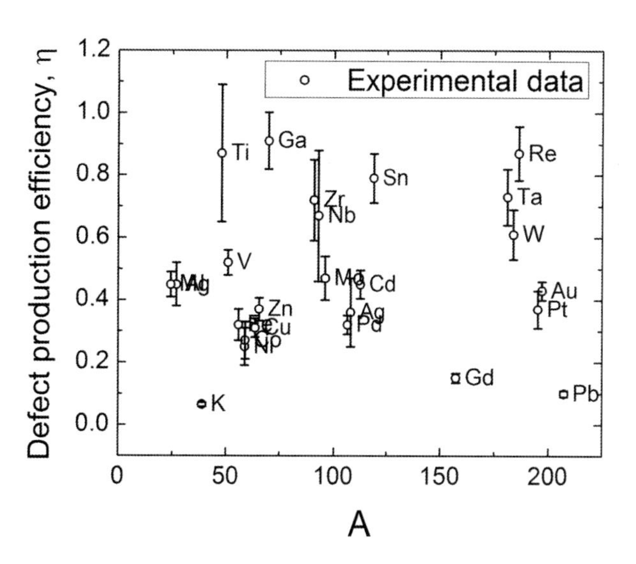

Fig. 13. Measured [37] defect production efficiency.

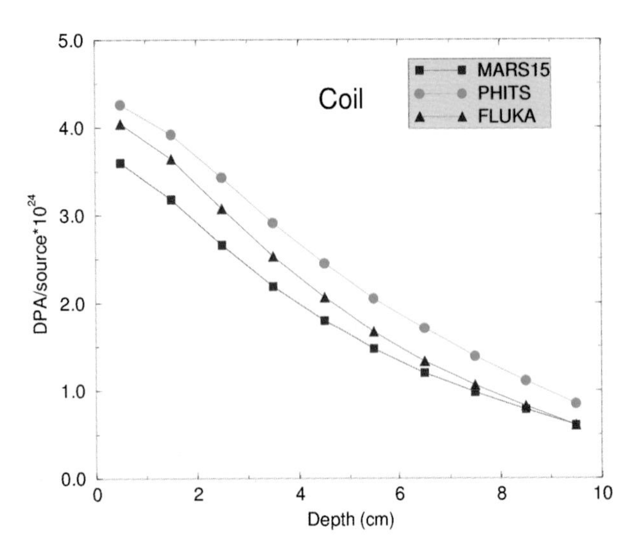

Fig. 14. The DPA rate as a function of the thickness of the Mu2e production solenoid superconducting coil, as calculated with the three codes.

calculations are in perfect agreement with results of detailed studies [38, 39]. Three major codes, FLUKA, MARS15, and PHITS, agree in their DPA calculations within 15% for proton and heavy-ion beams for a variety of irradiation conditions. As an example, Fig. 14 shows such a comparison for a superconducting coil of the Mu2e pion production solenoid for an 8 GeV proton beam on a tungsten target.

Despite substantial progress in the field of radiation damage over the last several years, there are still several misses and open questions:

- A desperate need for radiation damage measurements at cryogenic temperatures;
- Measurements with charged particle beams and their relation to neutron data;
- Annealed versus nonannealed defects;
- Low-energy neutron DPA in compounds;
- A consistent link of calculated DPA to observed changes in material properties.

7.3. *Hydrogen and helium production*

At accelerators, radiation damage to structural materials is amplified by increased hydrogen and helium gas production for high-energy beams. In the Spallation Neutron Source (SNS)–type beam windows, the ratio of He/atom to DPA is about 500 of that in fission reactors. These gases can lead to grain boundary embrittlement and accelerated swelling. In the simulation codes analyzed here, uncertainties on production of hydrogen are about 20%, while for helium these could be as high as 50%.

8. Beam Loss and Shielding

At high-intensity accelerators, the deleterious effects of controlled and uncontrolled beam loss on components of beamlines, target stations, beam absorbers, shielding, and the environment can be so severe that the cost of the systems to prevent or mitigate such an impact can be a substantial fraction of the entire facility cost. Two examples of dealing with these issues [40] from the beam–materials point of view are given in this section.

8.1. *LBNE experiment*

The Long-Baseline Neutrino Experiment (LBNE) will explore the interactions of the world's highest-intensity neutrino beam by sending it from Fermilab more than 1000 km through the Earth's mantle to a large liquid argon detector. Consequences of accidental and operational losses of a 120 GeV, 2.3 MW proton beam in the LBNE beamline (Fig. 15) are simulated with the MARS15 and STRUCT codes. The tolerable beam loss limits are derived with respect to the beamline component integrity and the impact on the environment.

The main criteria which have guided the design of the primary beamline are transmission of high-intensity beam with minimum losses and precision of targeting, keeping activation of components and groundwater below the regulatory limits in normal and accidental conditions.

STRUCT and MARS simulations have evaluated the impact of a localized full beam loss at any location along the beamline and a sustained small fractional loss. The proton beam energy considered in these studies was 120 GeV. In the first case, a peak beam pipe temperature twice the melting point for stainless steel is reached with a single lost full beam of 1.6×10^{14} protons per pulse (ppp). At an initial intensity of 4.9×10^{13} ppp, beam pipe failure is probable after 4–5 lost full beam pulses. Therefore, large beam loss for even a single pulse needs to be robustly prevented via an Integrated Beam Permit System (IBPS). This is a common practice nowadays when the IBPS is programmed to the maximum number of pulses that might be lost after the commissioning is

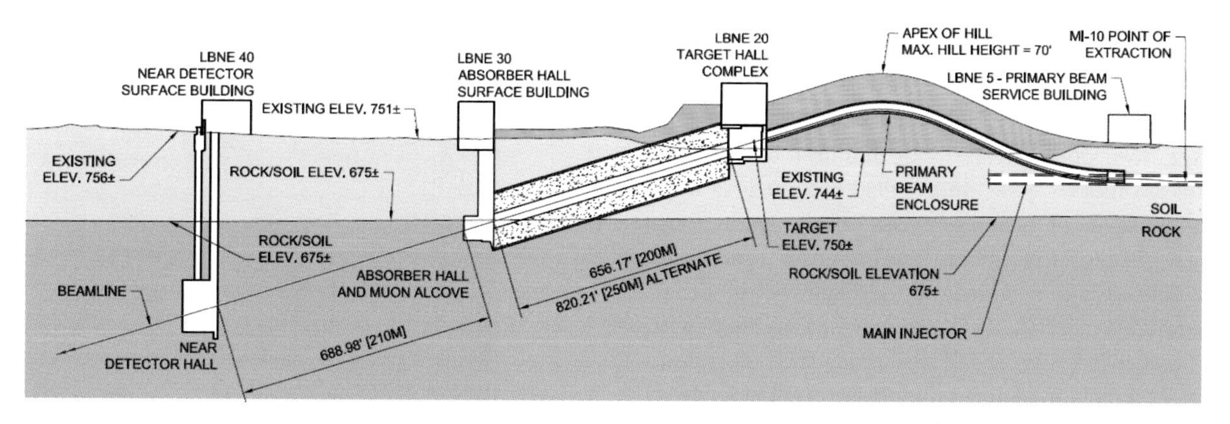

Fig. 15. The LBNE target hall and beamline shielding. Elevations and height are shown in feet. Lengths of the decay channel and dirt shielding between muon alcove and near detector hall are shown in feet and meters. At beam transport, beam losses are highest at the apex. The radiation levels are highest in the target hall near the target, with about 85% of the protons interacting with it.

done with a very-low-intensity beam and everything is tuned.

In the second case, the magnitude of the beam loss is chosen as a value which is within accuracy limitations for intensity monitors at the beginning and end of the beamline, and which might not produce a vacuum failure. The STRUCT simulations have shown that the highest loss rate takes place in the quadrupole magnet and two adjacent dipole magnets located right at the apex of the primary beamline (Fig. 15). Under this condition, for a scenario where there is accidental beam loss at 0.3% of the beam for 30 continuous days, calculated with MARS15 the peak contact dose after 24 h of cooldown is: (1) for tunnel walls, 5 mSv/h over a 20 ft ($\sim$6 m) region of the tunnel, and >1 mSv/h over a 50 ft ($\sim$15 m) tunnel region; (2) for the hottest magnet, 500 mSv/h over $\sim$1 m of magnet steel, and >100 mSv/h over most of a 3 m magnet.

Even after waiting six months with no beam, a magnet would still be at >30 mSv/h in the hottest region. The Fermilab limit of "0.5 mSv/h on contact to safely permit all necessary maintenance" dictates sustained localized beam loss to be a factor of 1000 less than considered above, in good agreement with the requirements of the current NuMI experiment at Fermilab. IBPS will again take care of this.

Radiological requirements for the design of the beamlines and experimental facilities are described in the *Fermilab Radiological Control Manual* [41]. The 2.3 MW LBNE proton beam is produced in the form of 1.6×10^{14} protons every 1.333 s. Based on the experience with operation of the NuMI beamline, it was concluded that for the radiological calculation a continuous beam loss rate of 10^{-5} and an accidental loss of 2 pulses per hour will be used.

The beamline passes through the aquifer regions, and therefore radiation requirements are quite stringent and vary from region to region. The calculated soil shielding required for 2.3 MW beam, for unlimited occupancy classification (1 mSv/year), is 6.4 m for continuous fractional beam loss of the 10^{-5} level and 7 m for two localized full beam pulses lost per hour. To reduce the accidental muon dose at the site boundary to <10 μSv/year, 122 m of the soil path of the muons is required.

Prompt radiation is one of the main issues in the above-grade target option. There are two contributions to the prompt dose rate at both onsite and offsite locations: direct radiation outside the shielding and sky-shine, which is primarily neutron backscattered radiation from the air. The primary beam transport line, the target hall, and the decay pipe (Fig. 15), as sources, contribute to the dose in areas accessible to members of the public. The LBNE beamline is pointed toward the nearest site boundary. Therefore, the direct muon dose adds up to the prompt doses from other sources, at the nearest site boundary. Based on the MARS calculations, both the annual direct and sky-shine doses are calculated for both offsite and onsite locations. A direct accidental muon dose at the apex of the transport line is also included in the offsite dose. Besides that, detailed consideration is given to surface and groundwater contamination and air activation.

8.2. *Mu2e experiment*

The Mu2e experiment is devoted to studies of the charged lepton flavor violation, which up to now has never been observed and can manifest itself as the conversion of μ^- to e^- in the field of a nucleus without emission of neutrinos. One of the main parts of the Mu2e experimental setup is its production solenoid (PS), in which negative pions are generated in interactions of the primary proton beam with the target (see Fig. 16). These pions then decay into muons, which are delivered by the transport solenoid to the detectors. The off-axis 8 GeV proton beam will deliver 6×10^{12} protons per second to the heavy metal target, placed at the center of the PS bore. The constraints in the PS shielding insert (absorber) design are quench stability of the superconducting coils, low dynamic heat loads to the cryogenic system, a reasonable lifetime of the coil components, acceptable hands-on maintenance conditions, compactness of the absorber that should fit into the PS bore, and an aperture large enough to not compromise pion collection efficiency, cost, weight, and engineering requirements.

Thorough optimization of the absorber design was performed with MARS15. The following quantities were focused on: dynamic heat load, peak power density, number of displacements per atom (DPA) in the helium-cooled solenoid coils, peak absorbed dose, and peak neutron flux in the coils. As an example, neutron flux isocontours are shown in Fig. 17.

Limits on the radiation quantities were set [42] based on: quench protection requiring that the peak coil temperature should not violate the allowable value of 5 K with a 1.5 K thermal margin for the peak power density, 10% degradation of the ultimate tensile strength for the absorbed dose, RRR (residual resistivity ratio) degradation from ~1000 to ~100 in an Al stabilizer for DPA, and requirements from the particular cooling system designed for the dynamic heat load. In the current design all the quantities analyzed, peak power density 18 (30) μW/g, peak DPA/year 4 $(6)10^{-5}$, peak absorbed dose/lifetime 1.7 (7) MGy, and dynamic heat load 20 (100) W satisfy the limits shown in parentheses. The optimized design meets these limits.

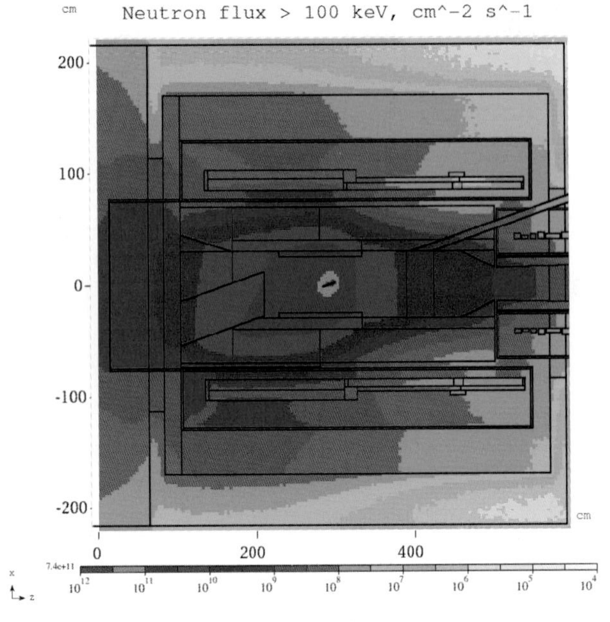

Fig. 17. Neutron flux in the Mu2e pion production solenoid.

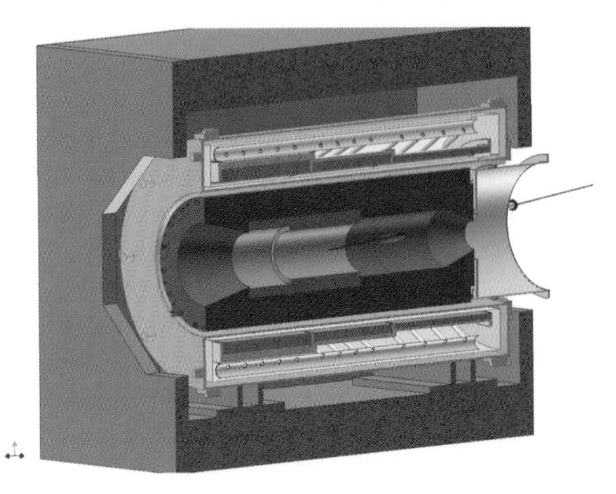

Fig. 16. The Mu2e pion production solenoid with a bronze absorber.

Another feature of the current approach in beam–materials studies is coupling of the energy deposition codes described above with the finite element system, such as ANSYS [43] and COMSOL [44] for detailed thermal and stress analyses. In the Mu2e case, the 3D thermal analysis was performed for the radiation heat load in the case of the optimized absorber design. The dynamic heat load map in the coil and the support structure generated by the MARS15 code were applied to all parts of the cold mass. The finite element model created by COMSOL Multiphysics was discretized to the level of individual layers and the interlayer insulation/conducting sheets. The maximum temperature of 4.8 K is found in the middle of the inner surface of the thickest coil section (Fig. 18). As in the LBNE

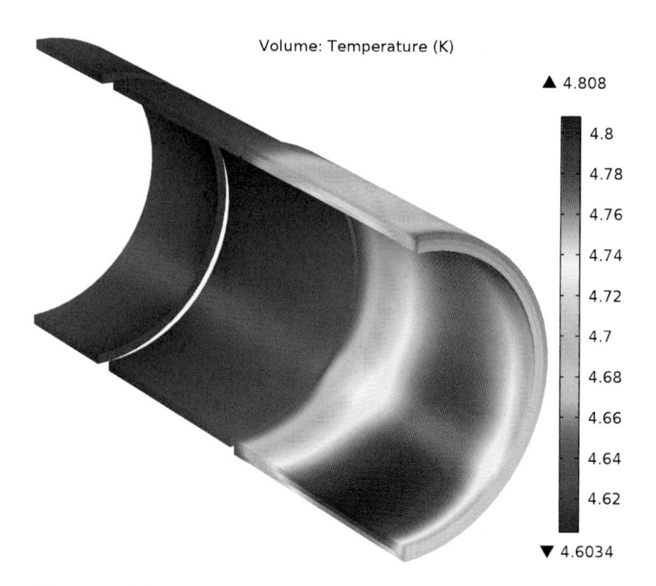

Fig. 18. Temperature distribution in Mu2e solenoid inner coil.

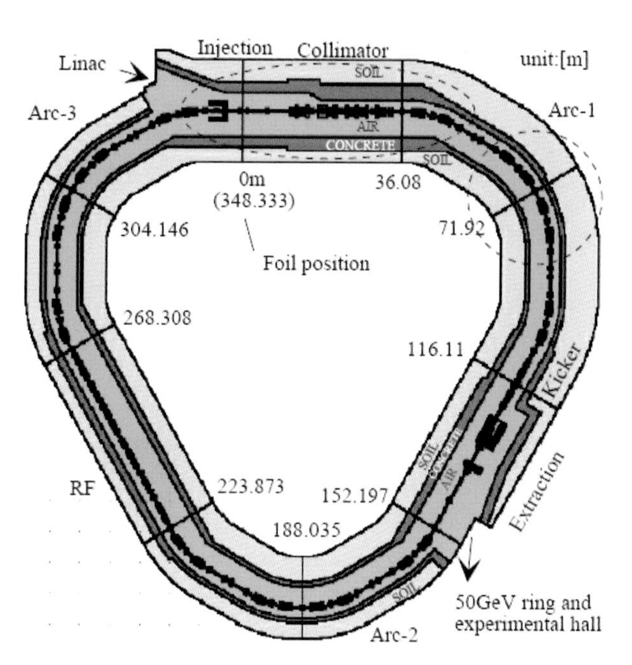

Fig. 19. The MARS computer model of the entire J-PARC 3 GeV accelerator.

case, thorough consideration is given to the radiation environment in the Mu2e experiment.

9. Geometry and Visualization

In a modern approach to accelerator complex upgrade and design, a realistic model of the whole machine for multiturn beam loss, energy deposition, activation, and radiation shielding studies is built by reading in MAD lattices directly and creating a complete geometry and magnetic field model in the framework of such codes as FLUKA, MARS, and GEANT. In MARS it is done via the MAD-MARS Beamline Builder (MMBLB) [45]. Such realistic modeling takes time and substantial effort, but experience confirms enormous benefits and insight. An example is shown in Fig. 19. The entire 3 GeV ring of the Japan Proton Accelerator Research Complex (J-PARC) — with its injection, extraction, and collimation systems — was built into the model with all the magnetic elements, materials, magnetic fields, and shielding, thus allowing thorough optimization of the design and performance.

A geometry module capable of highly accurate particle tracking in arbitrary 3D accelerator complex systems, comprising many thousands and millions of elements in the presence of arbitrary magnetic and electrical fields providing a user-friendly way for description, debugging and visualization and being computing-efficient, is the central element of the simulation codes used for the current energy and intensity frontier applications. The codes considered in this article basically provide these capabilities. A representative example is the powerful ROOT geometry and visualization module [46] implemented in MARS15. Geometry models created for MARS15 can be used with other Monte Carlo codes (e.g. Geant4), and one can use the ROOT models created for Geant4 with MARS15. ROOT provides a large set of geometrical elements along with a possibility of producing composite shapes and assemblies as well as 3D visualization. Examples of ROOT models built from scratch and imported to MARS15 are shown in Figs. 20 and 21, respectively.

MMBLB [45] was recently redesigned for use with the ROOT geometry packages. Usually, a beamline consists of elements of several predefined types: magnets (dipoles, quadrupoles, etc.), correctors, and collimators. A user has to describe how to build 3D geometry models for elements of each type. Using the information on the elements including a magnetic field presented in a selected optics file, the beamline builder will generate a corresponding 3D geometry model. An example of such a model of the Fermilab Booster is shown in Fig. 22. For scoring information on radiation fields inside the beamline elements, one can define histograms in a local coordinate system of each beamline element.

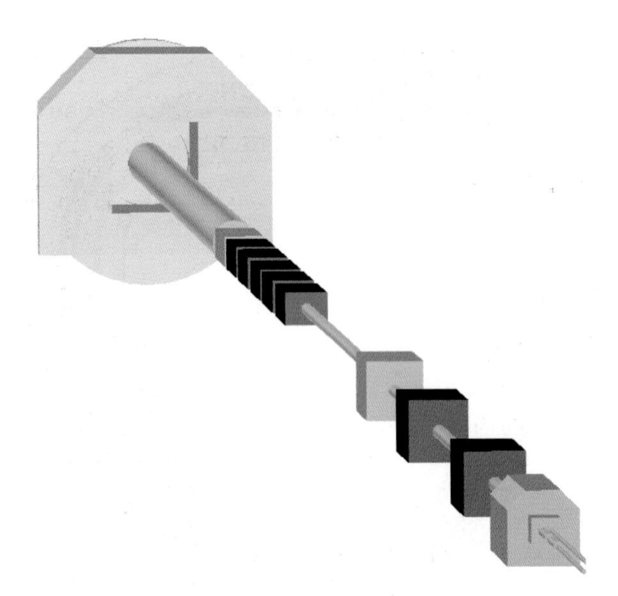

Fig. 20. The ROOT model of the LHC IR5 used in MARS15 for optimization of the inner triplet and simulation of machine-related backgrounds in the CMS detector.

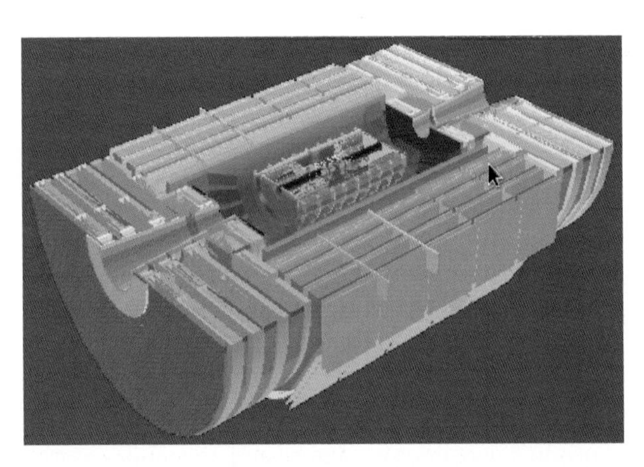

Fig. 21. A 3D cutaway of the CMS detector imported to MARS15 from a ROOT source tree.

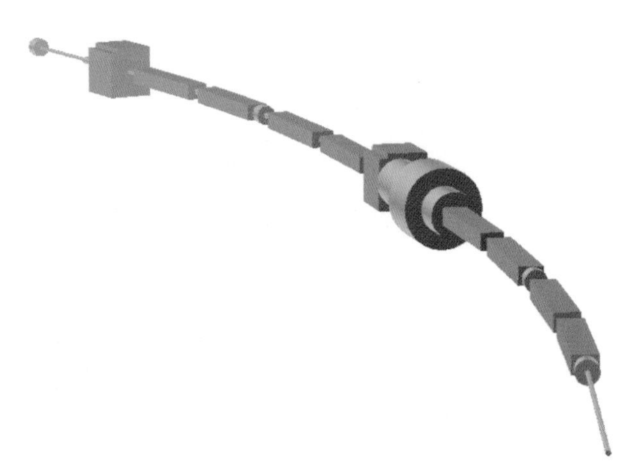

Fig. 22. 3D view of a fragment of the Fermilab Booster MARS15 model.

10. Uncertainties in Simulations

The predictive power, capabilities, and reliability of the major particle–matter interaction codes used in accelerator applications are quite high. At the same time, analysis of the status and uncertainties in modeling of radiation effects of high-power beams has revealed some issues. The most fundamental one is particle production in nuclear interactions, which is the heart of all such simulations and the key to collimator, target, and other machine component design as well as fixed target and collider experiment planning. Overall, the situation is quite good for beam energies below 1 GeV and above 10 GeV, with the accuracy of predictions being at the 20% level in most cases. At intermediate energies — most interesting for the intensity frontier — there are substantial theoretical difficulties. Moreover, the experimental data contradict each other at these energies. The main problem is with the low-energy pion production, which is crucial, for example, to all the Project X experiments. The accuracy of beam-induced macroscopic effect predictions in high-energy accelerator applications today is:

- Energy deposition effects (instantaneous and accumulated) $<15\%$.
- DPA calculations by the latest versions of the FLUKA, MARS15, and PHITS codes coincide within 20%. We still need a better link of the calculated DPA to the observed changes in material properties.
- Hydrogen gas production $<20\%$ and helium gas production $<50\%$.
- Beam loss generation and collimation: quite good in FLUKA and MARS15 (Tevatron, J-PARC, LHC).
- Radiological issues (prompt and residual): a factor of 2 for most radiation values if all details of geometry, material composition, and source term are taken into account.

11. Future Developments and Data Needs

The areas which require further development in Monte Carlo codes and the dedicated experiments to thoroughly benchmark include:

- Further advances in physics models for heavy ions, including their physics and CPU performance; more data to benchmark the models.

- Particle production event generators: pion, kaon, and antiproton yields; consistent data to benchmark these for 2–8 GeV proton beams; pre-equilibrium emission of heavy fragments; multifragmentation, photonuclear, muon, and neutrino-induced reactions, and delayed particles.
- Reliable experimental data on longitudinal and lateral energy deposition profiles in fine-segmented setups with combination of low-Z and high-Z composite materials for primary beam (heavy ions) and hadron, electron, and low-energy-dominated cases.
- Further refinements and justification of DPA models.
- Materials–beam tests: cryogenic temperatures, high-energy protons, annealing and oxidation effects.
- Moving from the calculated dose and DPA to changes in material properties: ready for coupling shower simulation codes and "materials" modeling codes.
- Further refinements of models predicting single-event upsets in front-end electronics verified against experimental data.
- Refined and user-friendly integrated systems for coupled energy deposition and hydrodynamics calculations.
- Practical direct linking of CAD models and particle shower simulation codes.
- Beam tests at the LHC to reveal the high potential of the hollow electron beam collimation.

Acknowledgments

The author is grateful to his numerous collaborators and coauthors, partially listed in the references below. This work was supported by the Fermi Research Alliance, LLC, under contract DE-AC02-07CH11359 with the US Department of Energy.

References

[1] N. V. Mokhov, Dealing with megawatt beams, in *Proc. Workshop on Shielding Aspects of Accelerators, Targets and Irradiation Facilities — SATIF-10* (Geneva, Switzerland, 2010), pp. 105–111.

[2] N. V. Mokhov, Advances and future needs in particle production and transport code developments, in *Symp. Accelerators for America's Future* (Washington, DC, 2009), Fermilab-Conf-09-638-APC.

[3] Beam halo and scraping, in *Proc. 7th ICFA Workshop on High-Intensity High-Brightness Hadron Beams*, eds. N. V. Mokhov and W. Chou (1999).

[4] N. V. Mokhov, Beam collimation at hadron colliders, *AIP Conf. Proc.* **693**, 14 (2003).

[5] N. V. Mokhov *et al.*, *JINST* **6**, T08005 (2011).

[6] E. Noah, Targets and secondary beam extraction. This volume.

[7] A. N. Kalinovskii, N. V. Mokhov and Yu. P. Nikitin, *Passage of High-Energy Particles Through Matter* (American Institute of Physics, New York, 1989).

[8] *Handbook of Accelerator Physics and Engineering*, 2nd edn., eds. A. W. Chao, K. H. Mess, M. Tigner and F. Zimmermann (World Scientific, 2013).

[9] J. Beringer *et al.* (Particle Data Group), *Phys. Rev. D* **86**, 010001 (2012).

[10] N. V. Mokhov and A. Van Ginneken, Neutrino radiation at muon colliders and storage rings, in *Proc. 9th Int. Conf. Rad. Shielding*; *J. Nucl. Sci. Technol.* **S1**, 172 (2000).

[11] A. Ferrari, P. Sala, A. Fasso and J. Ranft, FLUKA: a multi-particle transport code. CERN-2005-010 (2005); http://www.fluka.org

[12] S. Agostinelli *et al.*, *Nucl. Instrum. Methods A* **506**, 250 (2003); J. Allison *et al.*, *IEEE Trans. Nucl. Sci.* **53**, 270 (2006); http://geant4.cern.ch

[13] N. V. Mokhov, *The MARS Code System User's Guide*, Fermilab-FN-628 (1995); N. V. Mokhov and S. I. Striganov, MARS15 overview, *AIP Conf. Proc.* **896**, 50 (2007); http://www-ap.fnal.gov/MARS

[14] T. Goorley *et al.*, *Nucl. Technol.* **180**, 298 (2012); http://mcnp.lanl.gov

[15] K. Niita *et al.*, *Radiat. Meas.* **41**, 1080 (2006); http://phits.jaea.go.jp

[16] N. Mokhov, P. Aarnio, Yu. Eidelman, K. Gudima, A. Konobeev, V. Pronskikh, I. Rakhno, S. Striganov and I. Tropin, MARS15 code developments driven by the intensity frontier needs, Fermilab-Conf-12-635-APC (2012), to be published in *Prog. Nucl. Sci. Technol.* (2013).

[17] I. A. Vorontsov, V. A. Ergakov, G. A. Safronov, *et al.*, Measurement of inclusive cross-sections pi−, pi+, p, H-2, H-3, He-3 at angle of 3.5 degrees in the interaction of 1.1-GeV/c protons with Be, Al, Cu, Ta and comparison with the fusion models. Report ITEP-83-085 (1983).

[18] L. Z. Barabash, A. E. Buklei, V. B. Gavrilov, *et al.*, *Sov. J. Nucl. Phys.* **26**, 90 (1982).

[19] G. J. Marmer, K. Reibel, D. M. Schwartz, *et al.*, *Phys. Rev.* **179**, 1294 (1969).

[20] S. G. Mashnik, K. K. Gudima, R. E. Prael, A. J. Sierk, M. I. Baznat and N. V. Mokhov, CEM03.03 and LAQGSM03.03 event generators for the MCNP6, MCNPX and MARS15 transport codes. LANL Report LA-UR-08-2931 (2008); arXiv:0805.0751v1 [nucl-th], 6 May 2008.

[21] L. Heilbronn *et al.*, *Nucl. Sci. Eng.* **157**, 142 (2007).

[22] W. Scheinast *et al.*, *Phys. Rev. Lett.* **96**, 072301 (2006).

[23] E. Mustafin, H. Iwase, E. Kozlova, D. Schardt, A. Fertman, A. Golubev, R. Hinca, M. Pavlovic, I. Strasik and N. Sobolevsky, Measured residual activity induced by U ions with energy of 500 MeV/u in Cu target, in *Proc. EPAC 2006* (Edinburgh, Scotland), TUPLS141.

[24] A. I. Drozhdin, N. V. Mokhov, D. A. Still and R. V. Samulyak, Beam-induced damage to the Tevatron collimators: analysis and dynamic modeling of beam loss, energy deposition and ablation. Report Fermilab-FN-751 (2004).

[25] A. I. Drozhdin and N. V. Mokhov, *STRUCT User's Reference Manual*, SSCL-MAN-0034 (1994); http://www.ap.fnal.gov/~drozhdin/STRUCT

[26] D. C. Wilson, C. A. Wingate, J. C. Goldstein, R. P. Godwin and N. V. Mokhov, in *IEEE Proc. Part. Accel. Conf.* (1993), pp. 3090–3092.

[27] N. A. Tahir, R. Schmidt, M. Bruger, *et al.*, *New J. Phys.* **10**, 073028 (2008).

[28] N. A. Tahir, J. B. Sancho, A. Shutov, R. Schmidt and A. R. Piriz, *Phys. Rev. STAB* **15**, 051003 (2012).

[29] H. Schonbacher, Ref. 8, pp. 793–797.

[30] A. Idesaki, Irradiation effects of gamma-rays on cyanate ester/epoxy resins, in *Second Workshop on Radiation Effects in Superconducting Magnet Materials* (KEK, Tsukuba, Japan, 2013).

[31] G. H. Kinchin and R. S. Pease, *Rep. Prog. Phys.* **18**, 1 (1955).

[32] M. J. Norgett, M. T. Robinson and I. M. Torrens, *Nucl. Eng. Design* **33**, 50 (1975).

[33] R. E. Stoller, *J. Nucl. Mat.* **276**, 22 (2000).

[34] N. V. Mokhov, I. L. Rakhno and S. I. Striganov, Simulation and verification of DPA in materials, in *Proc. Workshop on Appl. High Intensity Proton Accel.* (World Scientific, 2010), pp. 128–131.

[35] M. B. Chadwick *et al.*, *Nucl. Data Sheets* **107**, 2931 (2006).

[36] R. E. MacFarlane *et al.*, LANL Report LA-12740-M (1994).

[37] C. H. M. Broeders and A. Yu. Konobeyev, *J. Nucl. Mat.* **328**, 197 (2004).

[38] I. Jun *et al.*, *IEEE Trans. Nucl. Sci.* **56**(6), 3229 (2009).

[39] M. J. Boschini *et al.*, arXiv:1011.4822v6 [physics. space-ph] (2011).

[40] N. Mokhov, S. Childress, A. Drozhdin, V. Pronskikh, D. Reitzner, I. Tropin and K. Vaziri, Beam-induced effects and radiological issues in high-intensity high-energy fixed target experiments. Fermilab-Conf-12-634-APC (2012), to be published in *Prog. Nucl. Sci. Technol.* (2013).

[41] *Fermilab Radiological Control Manual (FRCM)*, http://esh.fnal.gov/xms/FRCM

[42] V. S. Pronskikh, R. Coleman, V. V. Kashikhin and N. V. Mokhov, Radiation damage to Mu2e apparatus, in *Proc. 14th Int. Workshop on Neutrino Factories, Super Beams and Beta Beams (NuFact2012)* (Williamsburg, USA, 2012).

[43] ANSYS, http://www.ansys.com

[44] COMSOL Multiphysics user interface, http://www.comsol.com/products/multiphysics

[45] M. A. Kostin, O. E. Krivosheev, N. V. Mokhov and I. S. Tropin, An improved MAD-MARS beam line builder: user's guide. Fermilab-FN-0738-rev (2004).

[46] http://root.cern.ch/drupal

Nikolai V. Mokhov received his Ph.D. in Theoretical and Mathematical Physics from the Moscow Physical Engineering Institute, Russia, in 1976. He has been Senior Scientist and Department Head at Fermilab since 1992. He has 44 years of experience in Monte Carlo physics simulation code developments, and is the author of the MARS code system. He has been involved in various fields of radiation physics; design of numerous accelerator systems in Russia, the USA, Europe, and Japan; the pioneer scheme for protection of superconducting magnets against irradiation; and design of various detector and experimental setup components. He codiscovered the top quark at Fermilab in 1995. Dr. Mokhov is the spokesman of the Crystal Collimation experiment at Tevatron and of the JASMIN US–Japan Collaboration. He has been an APS Fellow since 2002. He is the author of 642 scientific papers and the book *Passage of High-Energy Particles Through Matter* (AIP, 1989).

Reviews of Accelerator Science and Technology
Vol. 6 (2013) 291–310

DOI: 10.1142/S1793626813300144

John Adams and CERN: Personal Recollections

G. Brianti
Formerly CERN, 5 Chemin des Tulipiers,
1208 Geneva, Switzerland
g_brianti@bluewin.ch

D. E. Plane
CERN, 1211 Geneva 23, Switzerland
david.plane@cern.ch

By any standards, John Adams had a most remarkable career. He was involved in three important, emerging technologies, radar, particle accelerators and controlled fusion, and had an outstanding impact on the last two. Without a university education, he attained hierarchical positions of the highest level in prestigious national and international organizations. This article covers the CERN part of his career, by offering some personal insights into the different facets of his contributions to major accelerator projects, from the first strong-focusing synchrotron, the PS, to the SPS and its conversion to a proton–antiproton collider. In particular, it outlines his abilities as a leader of an international collaboration, which has served as an example for international initiatives in other disciplines.

Keywords: Accelerators; CERN; PS; SPS; John Adams; international organizations.

1. Introduction

The eminent role played by John Adams in the field of particle accelerators on the international scene, in particular at CERN, has been well documented in a number of excellent publications. In his honor, CERN introduced a series of annual lectures, the John Adams Memorial Lectures, which have continued to this very day. The inaugural lecture in 1985, covering the whole of his career, was delivered by the visionary Prof. Edoardo Amaldi [1], who appointed Adams in 1953. It was followed 24 years later (2009) by E. J. N. Wilson [2], who centered his talk on his personal experience of working with Adams. Michael Crowley-Milling published in 1993 [3] a complete biography of this exceptional personality and, in 1977, M. Goldsmith and E. Shaw wrote *Europe's Giant Accelerator*, describing the design and construction of the SPS complex [4].

While referring to them, we intend to complement these publications by giving an account of our own personal experience with him, starting from the working ambiance in the early CERN during the design and construction of the PS and, later on, of the SPS. We then extend the review of his achievements to developments, such as the proton–antiproton project [5], which led to the discovery of the W and Z bosons, and later LEP and the LHC [6]. In addition, a section is devoted to the SPS experimental areas, which have served a large community of particle physicists for more than 30 years.

Giorgio Brianti, the primary author, wrote the sections dealing with the early days of CERN and the construction of the PS (Secs. 2–4), the ones dealing with the SPS story and construction (Secs. 5 and 6), and the one on further developments (Sec. 8). David Plane, the second author, wrote Sec. 7, on the SPS experimental areas.

John Adams also made major contributions to thermonuclear and plasma research — these are well described in Refs. 1–3.

2. CERN...upon Arve[a]

On a sunny, late September afternoon in 1954, a few hours after my arrival in Switzerland, I reached

[a]Arve, the second Geneva river, is near the Physics Institute, where the early PS Division was housed.

the terrace of the Physics Institute in downtown Geneva for the classical five o'clock tea with some new colleagues. CERN, I learned, had just become an official organization after the ratification of the founding treaty by a sufficient number of European parliaments. Someone guided me toward a tall, handsome and distinguished-looking man with a pipe in his hand, whispering in my ear, "Follow me, I'll introduce you to the boss." John at that time was only 34, but had a very natural authority. To say that we had a conversation would be an exaggeration because of my still uncertain English, but I understood that I was assigned to the Magnet Group. After this first encounter, I moved down to the barracks, where the entire PS Division was housed, waiting for the Meyrin site to became available. The job was to build the 25 GeV Proton Synchrotron (PS), the initial *raison-d'être* of CERN.

In the preparatory phase (1952–1953), Odd Dahl, a Norwegian professor from Bergen, had been the PS project leader, while Frank Goward, who had come from Harwell with John Adams and Mervyn Hine, was his deputy. Finally, Dahl renounced going to Geneva and Goward took on the job. In the spring of 1954, Goward died suddenly and John became the PS leader. John and Hildred Blewett from BNL also joined the PS team to help with the design [3]. Most of the other members were very young applied physicists or engineers coming straight from university, who in general had heard the word "synchrotron" for the first time in connection with CERN.

I soon found out that the Magnet Group had the task of designing and producing the magnetic system, which would guide and focus the particles around the 628 m ring.

The Group Leader was Colin Ramm, an able Australian, who worked closely with Cornelius (Kees) Zilverschoon, a clever and extroverted Dutchman who was Head of the Engineering Group. The second-hand barracks, acquired from the Swiss army as war remnants, included offices and a small electronics laboratory, in addition to which a small assembly hall was available, where magnet and other sizeable models were installed for measurements and development work. At that time, Lorenzo Resegotti, Claude Germain, Bastian de Raad, Dirk Neet and Jean Pierre Stroot were also members of the Group.

The initial plan was to build a weak-focusing synchrotron of 10 GeV, a scaled-up version of the Cosmotron, but the recent invention of the strong-focusing principle by Courant, Livingston and Snyder (1952) [7] prompted CERN to take the bold decision to construct a 25 GeV (later 28 GeV) machine based on the new principle, but at the same cost. One of the great novelties of strong-focusing synchrotrons was of course the magnet system, so it was not surprising that it received so much attention. The other novelty was the appearance of the transition energy, requiring a sudden jump in the radio frequency phase during acceleration, when beam stability is lost.

There were also major doubts about the influence of non-linear resonances caused by magnet imperfections. It seemed that this would lead to almost-impossible construction tolerances. E. J. N. Wilson [2] quotes an account of a meeting in Harwell at the end of 1952, prior to the arrival of John Adams in Geneva:

"It was John's job to help resolve the many doubts there still were about this decision to change to alternating-gradient focusing. John Lawson had warned of the dangers of non-linear resonances and Kjell Johnsen had to be persuaded that transition would not be a problem. John and Mervyn Hine studied the non-linear resonances driven by magnet imperfections using ACE, one of the first computers available in the UK. It seemed that because of the high field gradient (n-value) of the first design, magnet construction tolerances would need to be unrealistically tight to avoid these resonances. Hine writes: 'I remember at the end of the Harwell meeting John summarized and took over. He stepped into the authority position and wrote a summary on the blackboard in his wonderfully clear left-hand writing.' "

At the time of my arrival, there were still traces of this debate in the form of magnet models with very narrow and sharp pole pieces (looking like two opposing knives) corresponding to $n = (R/B) * (dB/dR) = 4000$, with gradients of about 50 T/m, as in strong quadrupole magnets of a modern separate function FODO lattice. However, in a FODO lattice the quadrupole magnets cover no more than 10–15% of the total magnetic length, while with combined function magnets the gradient is present over the entire length! The magnet weight was estimated to be only 800 tons for a 30 GeV machine, which, in all probability, would not have worked. Finally, more realistically, the n value adopted was 288 for a total magnet iron weight of 3300 tons.

The above reference gives me the opportunity to introduce two other very important persons of the early CERN, Mervyn Hine and Kjell Johnsen. Mervyn, also a British physicist, forming with John the so-called "Harwell twins," was John's "alter ego," especially on scientific and technical grounds, while the Norwegian Kjell was originally an assistant of Prof. Dahl. Both were very competent in many fields. John ran the project by leading two types of regular meetings: the staff meeting attended by all academic colleagues, where progress reports and general information were on the agenda, including of course extensive discussion; and the more restricted Parameter Committee, in which only Group Leaders participated. Being too junior, I participated in only the first type of meeting, where I was very impressed by the intellectual power and the competence of John, Mervyn and Kjell. However, John always had the final word at the end of the meeting by summing up the discussion in a concise and clear way, and with a very natural authority. He gave the impression of being able to promptly appreciate fully the pros and cons of any technical or managerial issue and reach the most appropriate decision.

From the very beginning of CERN, John and his leading team wanted to have complete in-house technical expertise for all machine systems, not only at the academic level but also at all other levels. This coincided with the view of Prof. Edoardo Amaldi [1], who was the Secretary-General (Director-General) of the provisional organization from 1953 to 1954 and who held the view that the development of CERN should be based on very solid technical expertise, by recruiting not only physicists but also engineers. Indeed, he did just that for a number of Italian engineers, including Franco Bonaudi (the first CERN employee), Mario Morpurgo, Guido Petrucci, Lorenzo Resegotti and myself. Other engineers and technicians were recruited from all over Europe. Locally, almost the entire staff of the mechanical workshop (a dozen or so people) of the neighboring Physics Institute of the University of Geneva were also recruited by CERN. Although this was not popular at the Institute, it provided us with a quite exceptional team of mechanical workers.

Returning to the early work in a common hall, I would like to mention a particular magnet model

Fig. 1. The adjustable pole model for PS magnet development.

(Fig. 1) used to find the most appropriate profile for the magnet laminations. In a time when computer programs were not yet available, it consisted of a stack of about 1 cm C-shaped plates perpendicular to the actual final block plane, which could be moved independently of each other by appropriate long screws. Varying the profile of the pole pieces in this way allowed us to determine the optimal profile of the pole pieces. The trouble was that the screws were very hard to move — providing an excellent way to develop our young muscles!

Among the exotic objects in the hall was a mechanical machine with a rotating shaft for cycling current through the magnet model. The rotation of the shaft produced, at appropriate times, the opening and closing of a set of switches. One day Albert Picot, a minister of the local government, who was a strong supporter of CERN, was visiting our provisional quarters, and he walked around the hall accompanied by John. Just when they arrived in front of this machine, which happened to be in operation, something went wrong and a switch suddenly opened under full current, producing a terrifying discharge, which sounded like an explosion. Picot disappeared very quickly, but continued to support CERN... from a distance.

All the machine components were developed in this single assembly hall and our daily work there helped us grow in mutual understanding and promoted friendships. The differences in our national cultures were not obstacles to our working together, but rather a source of mutual enrichment. Most of us were still single, so often we continued to stay together outside the normal working hours for outings and weekend trips around Geneva.

The Magnet Group rapidly became stronger and stronger, with the arrival of Simon van der Meer [8], Gunther Plass, Helmut Reich and others. The first job of Simon, who arrived with a high reputation, was to calculate the pole-face windings to be installed on the pole surface for correcting the magnetic field imperfections. It was considered to be a difficult problem, which I and other earlier arrivals were reluctant to tackle. Of course, he was able to solve it in a rather short time using only a mechanically hand-driven calculator.

The well-trained professional mechanical workers were not only skilled according to the best Swiss watch-making tradition, but also inventive and very enthusiastic in their work. They were able to produce masterpieces, particularly for magnetic measurements. I would like to cite two examples: a small plastic turbine driven by compressed air, reaching almost 1000 turns/min for rotating coils used to measure magnetic fields, and a permeameter (Fig. 2) for measuring magnetic steel properties. The latter consisted of three toroidal concentric coils split in the median plane embedded in a plastic structure, which could open and close for the insertion of the steel rings to be measured. It has since become a standard instrument for all subsequent accelerators.

My initial job was the choice of the steel laminations. The choice fell on classical car body steel, which was cheap and, at that time, was produced by many companies in Europe. However, a problem had to be solved: the usual car body steel had too high a coercive force, which would cause too high a magnetic remnant field at injection; moreover, this property varied widely from batch to batch. The high coercive force was due to the very small metallurgical grains, which were required in order to avoid the so-called "orange skin" when laminations were deep-drawn to produce car bodies. I found out in the literature that a rather light final cold work (skin pass) on the laminations, followed by appropriate heat treatment, would increase enormously the grain size and hence reduce considerably the coercive force and the remnant field. An additional important precaution was to stock all the steel laminations in batches, and then select laminations from each batch to produce a block.

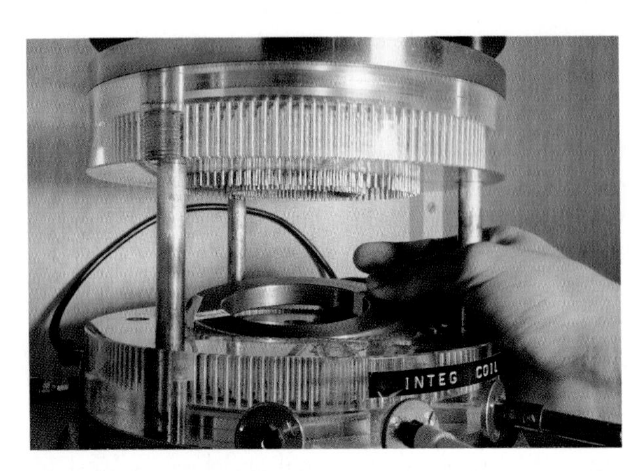

Fig. 2. The original permeameter, developed to measure magnetic steel properties.

In 1956 we moved to the Meyrin site, which at that time was entirely on Swiss territory but adjacent to the French–Swiss border. By choosing this location, the local authorities thought that any possible extension would have to be in France, which indeed happened ten years later for the construction of the Intersecting Storage Rings (ISR) and then, more extensively, for the SPS, LEP and LHC.

3. PS Construction

The intense prototype and model work progressed rather rapidly toward a final design, consisting in a magnet system composed of 100 units made up of 10 blocks each (see Fig. 3), mounted on a sturdy common girder and with common coils.

The 1 mm laminations, punched by very precise dies, were electrically insulated by a sheet of paper glued by a new thermal setting epoxy resin, araldite. After assembly, the blocks had to be heated and then cleaned of the excess polymerized glue by metallic brushes: a terrible job!

Figures 4–6 show the first complete PS magnet unit, a section of the PS ring and a view inside the completed ring.

I mention these technical details to illustrate the care that John wanted to be applied to all aspects of the technical solutions before adopting them. He was aware that building the first European research accelerator based on the brand-new and untested strong-focusing principle was already a considerable risk, which had to be attenuated as much as possible by a very careful and prudent design. Among many, one example of this prudent approach is the circular concrete beam, supported by pillars going down to the

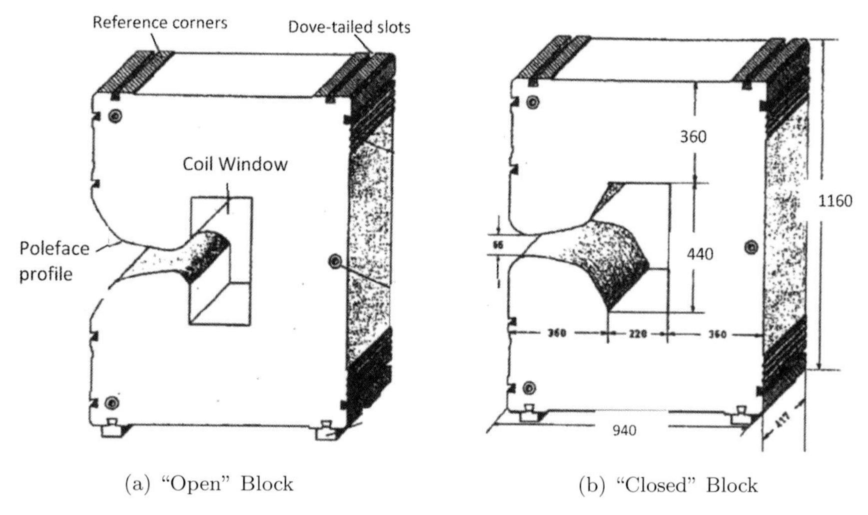

(a) "Open" Block (b) "Closed" Block

Fig. 3. PS "open" and "closed" magnet blocks.

Fig. 4. The first PS magnet unit and its proud builders. The author (G. B.) is in the back row, second from the right.

rock, on which the machine is installed. The temperature of the tunnel is controlled in order to insure a very accurate alignment [3].

This was considered by some people to be "overdesigned" and this opinion tainted the reputation of the CERN accelerator community for some time, especially after the construction of the second large machine, the ISR. Later on John himself showed a much less conservative technical approach when constructing his new machine, the SPS, and, in particular, when he approved and encouraged the adventure of the proton–antiproton project.

Returning to my personal story, I would like to underline another characteristic of John's managerial skills. I was asked to supervise both the production of the steel laminations and of the thousand magnet blocks on the premises of the two chosen firms in Genoa. After an intense preparation, the production of the magnet blocks started, but the first blocks were out of tolerance due to hasty and careless procedures. I warned my bosses in Geneva and I was expecting that they would come to discuss the situation with the firm. Instead I was told, "Write an order on CERN's behalf to stop the production and

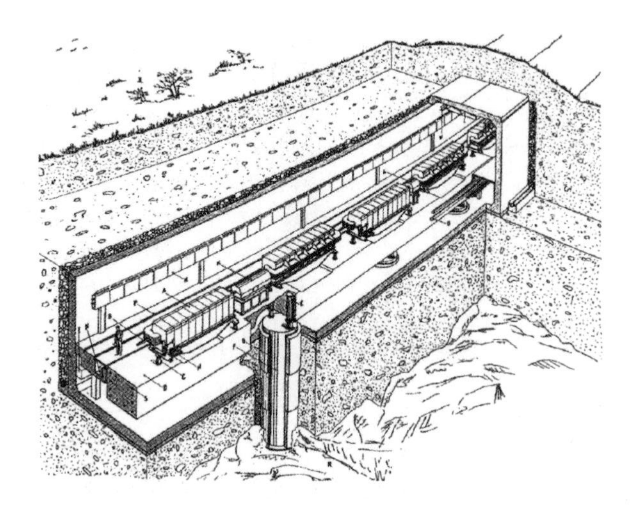

Fig. 5. A section of the PS ring.

go away. Leave a telephone number to call you back when they will intend to follow your instructions." After a few days they called me back and the production restarted following the correct procedure. This confidence in a young person at the beginning of his career marked my ensuing professional life.

The production of the blocks continued according to schedule, so that, at the beginning of 1958, I could return to Geneva. John gave me a new job, this time in the Controls Group, with the responsibility of completing the main control room, which was well behind schedule. In the end we installed it more or less on time, but it certainly did not look like the modern computer-based control centers. All important elements were installed in racks along the walls, with forests of cables connecting them on the front faces. On the central desk there were only telephones, a microphone, and two meters — a voltmeter giving the energy and an ampere meter giving the current pulse by pulse.

The commissioning of the machine and its final success are well documented in the references. The evening of 24 November 1959 will remain forever in our memories as a milestone in our professional lives. We were particularly proud of the fact that the PS started operation six months ahead of the sister machine, the AGS in Brookhaven. Figures 7 and 8 are photographs taken at that time.

Under the guidance of John, we all learned that adopting very sound engineering of any single component and paying extreme attention to tolerances was of paramount importance for the success of the PS. We were taught to tackle technical design and construction on the basis of an attitude which was one of the facets of John's personality, namely "constructive pessimism," just the opposite of "blind optimism." Indeed, John was a pessimist not in a negative way, but in the sense that he believed that nature had no reason to make gifts to accelerator designers. Therefore the correct attitude consisted in understanding the finest details of each problem in order to make a design which left nothing to chance

Fig. 6. The PS ring.

Fig. 7. PS startup, 24 November 1959. From left: Adams, Geibel, Hildred Blewett, Laslett, Schmelzer, Schnell, Pierre Germain.

Fig. 8. Mervyn Hine and Kjell Johnsen in the PS control room.

on the way to success. Some people confused this with conservatism and overcautiousness. Later on, based on his experience with the PS, he became more confident about using smaller safety factors and was able to take significant risks in transforming the SPS into a proton–antiproton collider.

It is also interesting to note that John did not publish many articles in well-known scientific or technical magazines. His decisive influence on projects was essentially through decisions after thorough discussion in staff meetings or personal contacts with people responsible for given machine systems. These discussions and decisions are recorded in innumerable minutes of meetings and in a collection of his handwritten personal notebooks. Of course, he also made many reports to official CERN bodies, like the Council and the Scientific Policy Committee.

4. CERN at a Turning Point

Already, before the completion of the PS, John decided to go back to England, and in August 1959 he was appointed Director Designate of the newly constituted Culham Laboratory for fusion research. The appointment was to become effective during 1960, but the tragic accidental death of Director-General Bakker in an air crash in April 1960 changed the plan. John was named Acting Director-General and soon after Director-General until 1 August 1961, when Viktor Weisskopf could take over as Director-General.

After a slow start the experimental program around the PS moved ahead, and reflection on future projects got underway. In 1963 the European Committee for Future Accelerators (ECFA) was formed, on which all European physicists were represented, with the scope of deciding which new projects should be adopted. Prof. Edoardo Amaldi was its first Chairman. Two main lines were considered: a proton collider fed by the PS, and a synchrotron of 300 GeV [2, 3, 9]. The construction of the ISR (Intersecting Storage Rings, with 30 GeV proton beams) was approved in December 1965 for completion in 1970, together with the intermediate injector for a tenfold increase in the PS intensity, which later became the Booster. The decision to build the 300 GeV accelerator was postponed, for at least two reasons. The first was financial, in order not to cumulate its cost with that of the ISR, but the second was more fundamental. In fact, it was judged impossible to build it in, or adjacent to, the Geneva site with the classical cut-and-fill method for the construction of the 7-km-long tunnel. It was deemed necessary to establish a new laboratory somewhere in Europe. The Council launched a call to the Member States for a new site, which had considerable quantitative success: in a short time, sites were proposed in almost each country, with even more than one in some countries. Of course, this raised several difficulties, such as the cost of the new laboratory, the role of the existing one in Geneva and the actual choice of the new site [2, 3, 9].

Meanwhile, in the US at the end of 1966, it was decided to establish the new National Accelerator Laboratory (NAL) in Batavia near Chicago and build the so-called 200 BeV machine. Soon after, in 1967, R. R. Wilson was appointed as Director and transformed the project into a 200–400 BeV machine based on a separated function focusing scheme [3].

Two years later, in 1969, the CERN Council took the bold decision to call back John Adams and appointed him as Director-General Designate of the new laboratory, yet to be approved [2–4, 9]. There are no better words than those of John himself to summarize the difficulties:

"Looking back, I think one can discern a number of reasons why our Member States hesitated to reach a decision on the 300 GeV Programme in the form it was presented at that time. In the first place the economic situation in 1969 for science in general and nuclear physics in particular was very different from the ebullient years around 1964 and 1965 when the 300 GeV Programme was first put forward. It was evident that several Member States of CERN and possibly all of them found the cost of the Programme too high compared with their other investments in science and with the growth rates in their total science investments, which had dropped from figures around 15% per annum in 1965 to a few per cent per annum in 1969. In the second place, the idea of constructing a second European laboratory for nuclear physics remote from the existing one, which had seemed attractive in 1965, looked inappropriate in 1969, particularly since it implied running down the existing CERN laboratory when the new one got under way. In the third place, so many delays had occurred in the 300 GeV Programme and the American machine was coming along so fast that an eight-year Programme to reach experimental exploitation seemed too long.

Fourthly, it turned out that choosing one site amongst five technically possible sites presented non-trivial political problems for the Member States of CERN." [10].

5. Preparation of the 300 GeV Project

Initially with a very small staff (only Ted Wilson was full-time), John started immediately to work on two lines: how to reduce the cost of the project and how to solve the site problem. The existing project was essentially a scaled-up version of the PS with combined function magnets. This design implied that the field on the central orbit could only be about 1.3 T because the field in the narrower part of the gap cannot exceed 1.8–2.0 T. In a separated function machine, the much higher central field leads to a smaller circumference and hence to a lower cost, or to a higher energy for the same cost. E. J. N. Wilson has given an excellent account of this phase [2]. John adopted the new separated function design, but the problem of the new site remained, including not only the choice among the five technically suitable sites of the final shortlist but also the fate of the existing laboratory in Geneva. At this point, the idea put forward earlier by Colin Ramm of a synchrotron installed in an underground tunnel, albeit of lower energy, was seriously considered. It turned out that the land between the Meyrin site and the Jura mountains at a sufficient depth was made of molasse, a sedimentary rock composed of sandstone and marl, which was relatively easy to excavate by means of a full-size boring machine [3]. The extension of this molasse bed was easily sufficient for a machine diameter of 2200 m, adequate for the wanted energy, at the reasonable depth of about 40 m.

These changes led to a cheaper solution making full use of the existing installations (injectors, a large experimental area) and assuring the future development of the Geneva Laboratory. They also led to the approval of the 300 GeV project, which turned out to be crucial for the subsequent, but unforeseen at that stage, larger projects, LEP and the LHC. Indeed, the question of another laboratory away from Geneva never came up again! Figure 9 shows the President of the CERN Council on the day the 300 GeV project was approved.

I believe that the installation of the laboratory on this unique site over the last 60 years was, and continues to be, a great asset to CERN and has been crucial for its success, not only because, from the

Fig. 9. Professor Edoardo Amaldi, President of the CERN Council, 19 February 1971. The 300 GeV program was approved.

material point of view, it allowed the use and reuse of existing installations, but also because, from the staff point of view, it ensured the continuity and the harmonious evolution of technical expertise.

6. 300 GeV Program

The 300 GeV program had a duration of eight years, but with the proviso of starting physics experiments at least in the existing West Area in 1976. In addition, there was the idea of the so-called "missing magnet" scheme, consisting in installing all the quadrupole magnets but only half of the dipole magnets in order to start the physics earlier at 200 GeV. Later on, one would install the rest of the dipole magnets, realign the machine and reach full energy.

Immediately after the approval, John (appointed Director-General of CERN Laboratory II) started to set up the team for the detailed design and construction of the machine and its experimental areas [1–4]. Of course, he considered accelerator experts not only from CERN but also from other European national laboratories.

Among the CERN staff, the first to be appointed was Bastian de Raad for beam transfer (injection and extraction), followed by Simon van der Meer [8] for the power supplies and Hans Horisberger for mechanical engineering. John also called me to discuss my possible task. At that time I was building the PS Booster, the new 800 MeV injector for the PS, consisting of four superimposed synchrotrons, with the purpose of a tenfold increase in the PS intensity, due for completion in 1972. John asked me to take up the responsibility of the magnet system, which, being one of the most urgent tasks, would have required abandoning immediately the Booster. I felt that this was not appropriate. He understood and proposed to me the less urgent job of the experimental areas, provided that I would participate immediately in the management meetings. Finally, the magnet system was entrusted to Roy Billinge, returning from FNAL.

Other important persons came from outside, namely Hans-Otto Wüster from DESY, who became John's Deputy, Michael Crowley-Milling from Daresbury for Controls, and Robert Lévy-Mandel from Saclay for supervising civil engineering and general installation work. Figures 10 and 11 are photographs taken during the SPS construction phase, and Figs. 12–14 show John in the company of close colleagues.

John adopted the classical tree structure for the management, consisting of groups for the various systems, for the machine layout and for some general tasks. Together John, the Group Leaders, Hans-Otto Wüster, E. J. N. Wilson (for parameters), B. Milman (for planning and budgets) and R. Florent (for

Fig. 10. John Adams in front of an early 300 GeV layout.

Fig. 11. John Adams at work in 1975.

Fig. 13. John Adams with Giorgio Brianti.

Fig. 12. John Adams with his deputy, Hans-Otto Wüster, on their way to the SPS inauguration ceremony.

Fig. 14. J. Adams, R. Billinge and M. Crowley-Milling in 1976.

purchasing) formed the so-called Parameter Committee, which was in fact the management structure for the project [4].

John, who was also an excellent architect, designed himself the office and laboratory buildings at the new French Prévessin site, about 3 km away from the Meyrin site. It consisted of four three-floor cross-shaped buildings and a large central assembly hall. However, not all people taking part in the program moved to Prévessin; John had the wisdom not to duplicate all functions by making use of some services of Laboratory I. I had mixed feelings concerning this new site. If, on one hand, it strengthened the team spirit for the realization of the project, on the other hand it provoked a certain separation among the staff of the two CERN laboratories.

In addition to the weekly Parameter Committee meeting, John ran the project with another weekly meeting on civil engineering, with variable participation according to the subjects. Of course, the civil engineering was very important not only for the novel aspect of the deep underground work, but also because John was convinced that strict control of it, with the aim of minimizing surfaces and volume, would be one of the best ways to limit expenditures. Facing John during this type of meeting was quite an experience. The person presenting the request of a given civil engineering work had to defend its need down to almost the last cubic centimeter! John challenged how solid and justified the request was by looking the defender firmly in the eyes, putting a lot of questions and sometime proposing alternatives. At the end, if the defender came out with a success, even if partial, he could be sure to have worked for the benefit of the project.

Another facet of the effectiveness of John's project direction was to follow very closely with his own eyes not only the paperwork, but also the realization in terms of models, prototypes and, above all, civil engineering progress. He was aware of every detail. When Robert Lévy-Mandel took him around the site, usually on Saturday, he often realized that John already knew everything. The tunnel boring was completed in 1974 (Fig. 15), to the delight of all.

Coming back to technicalities, the project departed from previous CERN machines in a number of components. One was the very compact magnets with coils insulated by glass fiber impregnated with epoxy resin, but without the customary extra layers of mica used for all other previous CERN magnets.

Fig. 15. Completion of the SPS tunnel excavation in 1974.

Roy Billinge's idea was that this simplified insulation was not only sufficient but enabled the uncovering of fabrication faults. The aperture of the vacuum chambers was also considerably reduced with respect to the previous project by reducing the space allowed for orbit distortion, alignment and magnet errors and counting on orbit correction after the first turn. Hence the magnet cross-sections were notably smaller compared to previous projects.

Models and prototypes confirmed the soundness of the solution but, when the fabrication was already well advanced, a number of magnets already installed in the tunnel showed defective coil insulation. This was a big blow, with the fear of repeating at CERN the FNAL problems [3]. The novel type of insulation was thought to be responsible and all kinds of theories explaining the accident were formulated, until Roy himself found out that, in the factory, the coil ends were cleaned with phosphoric acid, which damaged the insulation. This concerned in total 280 magnets. An emergency program was established to restore the situation and, in the end, this incident had little or no effect on the date of completion of the machine. During the entire crisis, called the "black January," John remained extraordinarily calm, with no reproaches or accusations toward anybody — a distinctive mark of his superior quality as a leader and a manager. Another very important area where the project was very innovative was the control system, the task of Michael Crowley-Milling, based for the first time on a distributed network of computers and on innovative equipment for the control desk.

A view inside the newly completed tunnel is shown in Fig. 16.

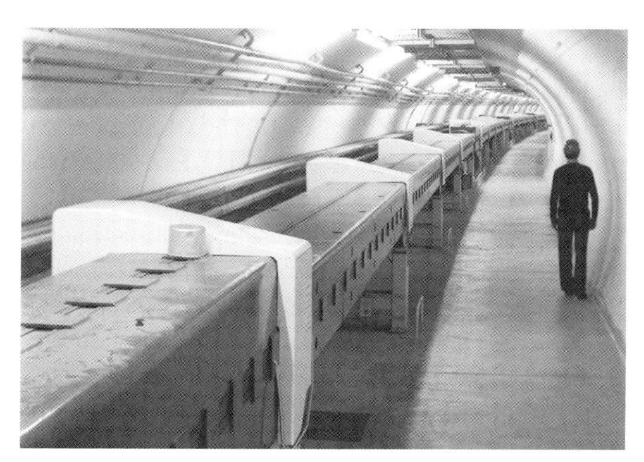

Fig. 16. The SPS ring in 1976.

It is also worth mentioning that during the construction some pressure was felt from external laboratories to adopt the "missing magnet" scheme, namely to install initially only one half of the classical room temperature magnets and eventually use superconducting magnets to complete the machine. John did not like the idea at all, considering that the risk was too high. Finally, he decided to complete the machine in one go, but asked me to collaborate with Saclay on the design of a superconducting dipole prototype for future use, if not in the machine, for the secondary beams. This successful collaboration was very profitable later for the LHC.

The construction continued with no other major problems and on the morning of 17 June 1976 the beam reached the energy of 300 GeV as defined in the program, enabling John to announce it to the Committee of Council. In the afternoon, during the Council session, John requested permission from the Delegates to go to a higher energy and, having obtained it, finally announced that the energy of 400 GeV had been reached. At that time I thought that this two-step procedure was *coquetterie* on John's behalf, but indeed some Delegates had wanted to limit the energy, fearing a higher cost [3].

It was during this period that my contact with John became very frequent. What I can say is that, if John's performance was quite remarkable during group meetings, the experience of personal contact with him was also amazing and instructive. When you presented a proposal he looked at you in a very firm way, over his half-spectacles, to assess how convinced you were of the proposal. Then he started to use one of his preferred managerial tools — the pipe. He would take out various small tools from a little bag and begin to carefully clean the pipe. Meanwhile, I am sure, he was weighing carefully in his mind all aspects and consequences of what you said, together with possible alternatives, and when the pipe was properly cleaned, he arrived at the decision or the answer, often absolutely final.

While the SPS was approaching completion, an important managerial decision had to be taken by the Council. The term of office of Willy Jentschke as Director-General of Laboratory I was ending in December 1975, while that of John for Laboratory II extended for another three years. It was generally admitted that the two laboratories had eventually to be reunited either in 1976 or three years later. The first alternative prevailed, but who should be appointed as Director-General of the reunified laboratory? The obvious choice was John, because of his abilities and international reputation, but there was some reluctance from the community of physicists to accept an "engineer" as Director-General. After many discussions, the final decision was to appoint John as Executive Director-General and Leon Van Hove as Research Director-General for the classical duration of five years (1976–1980). The two newly named Director-Generals, plus the outgoing D-G and the Council President, are shown in Fig. 17. To ease his continuing task for the completion of the SPS, John changed the internal SPS organization by appointing Bas de Raad as Head of the accelerator, Robert Lévy-Mandel for site and general services, and myself for the experimental areas. In 1979, I succeeded M. Crowley-Milling as SPS Division Leader.

My impression is that, for John, this last period of leadership was not the happiest of his life. He was used to being the only captain on the ship and therefore had some difficulties in having to always reach an agreement with Leon on practically all matters of importance, including managerial and administrative issues. However, goodwill prevailed on both sides and, as we shall see, important decisions were taken in those years for the future of CERN.

The SPS era was particularly propitious for advancing international collaboration, a constant theme during John's career at CERN, and indeed an aim pursued by all CERN managements. Already back in 1952, Amaldi spent several hours with John and was impressed by his interest in creating a new

Fig. 17. Van Hove, Jentschke, Levaux (President of the Council) and John Adams in 1975.

European laboratory [1]. This is all the more remarkable since large nations alone could still undertake ambitious accelerator projects, and not all research physicists supported the idea of a multinational organization.

During the PS construction period, in addition to the close collaboration with the Brookhaven Laboratory mentioned above, first contacts were established with Russian physicists at the International Conference on High Energy Physics in Geneva in 1956. Later on, the PS Division received the first team of Russian physicists at the beginning of the sixties and constructed in 1969 the extraction system for the Serpukhov accelerator, which was the highest energy accelerator at that time [3].

It was during the SPS construction that John developed contacts with China, resulting in an exchange of visits. He was invited to visit China in 1977, and there he was received by the Chinese Premier Teng Hsiao-ping, who at that time made the historical announcement that China would open up to the world and to the use of the best of Western technology [3]. In parallel, a collaboration was established with KEK in Japan in view of the construction of LEP and for the development of superconducting magnets.

Last but not least, the CERN model was at the origin of other European scientific organizations, like ESRO and ELDO, unified later on in ESA (European Space Agency) and ESO (European Southern Organization, for astronomical research in the southern hemisphere). ESO was actually founded at CERN, and had its head-quarters at CERN for a long time.

Concerning space research, the idea of European collaboration dates back to a letter written by Prof. Edoardo Amaldi to Prof. Luigi Crocco in Princeton, dated 16 December 1958. This event was celebrated recently by giving the name *Prof. Edoardo Amaldi* to the third space Automated Transfer Vehicle (ATV-3) of ESA, which was launched on 23 March 2012 in a mission to the *International Space Station* carrying copies of this famous letter.

7. SPS Experimental Areas

John Adams' first major CERN project, the PS, suffered a very late start of the physics research program, through no responsibility of his own. He clearly understood the complexities of performing experiments at high energy machines and, during his first term as Director-General, set up the NPRC (Nuclear Physics Research Committee) to oversee the experimental program. Learning from this regrettable experience, he encouraged every effort to facilitate the rapid start of the SPS physics research program.

To this end, the SPS Experimental Areas Group under the leadership of one of the authors (Giorgio Brianti) provided, from the very outset, a wide range of secondary and attenuated primary beams, which were fully instrumented with detectors for beam tuning, particle identification and beam spectrometry. John particularly appreciated the provision — quite exceptional at the time — of standardized but state-of-the-art instrumentation which avoided much duplication of effort and which would have been a headache to maintain and to accommodate in the beam lines had there been many different types of devices.

A West Area Preparation Committee was set up well before the first beam, in order to inform users about the facilities and infrastructures and how to use them. Liaison physicists helped to install the experiments, and the group provided many other services (e.g. gas supplies, drawing office, power supplies). In order to ensure that the facilities were in line with the needs of the users, John appointed James Allaby, a member of the Experimental Physics Division, as SPS Physics Coordinator.

This part of the report will describe very briefly the facilities as initially installed for the first round of SPS experiments; more details can be found in App. A.

7.1. *Beam instrumentation*

The beamlines were well equipped with the basic instrumentation required for beam tuning and diagnostics, including scintillator counters, analog wire chambers for measurement of beam position and size, and scintillator filament scanners to optimize the beam divergence. All were remotely removable from the beam to limit multiple Coulomb scattering.

Momentum measurements of individual beam particles were provided by four digital wire chambers, arranged in pairs surrounding beam magnets [11].

Particle identification was provided by a combination of threshold Cerenkov and CEDAR

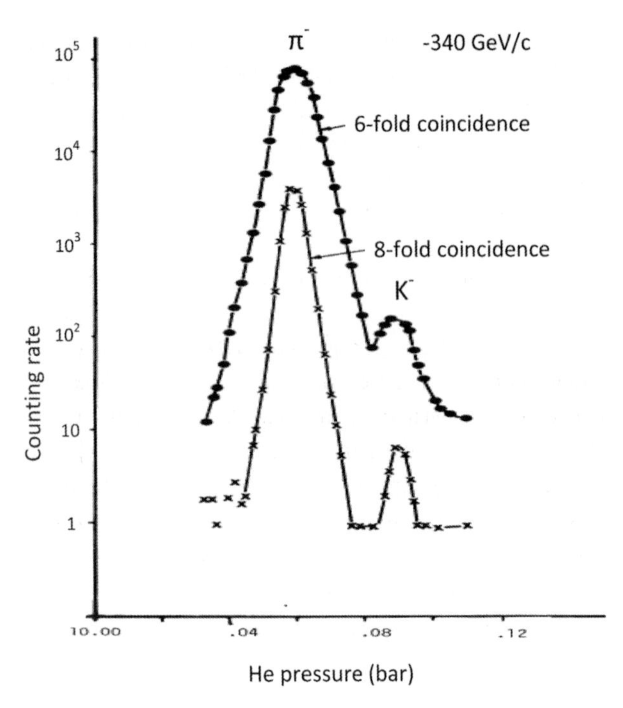

Fig. 18. Pressure curve for a CEDAR-N, showing good separation between π^- and K^- at 340 GeV/c.

(CErenkov Differential with Achromatic Ring focus) counters [12]. Figure 18 presents a pressure curve for a North Area CEDAR, showing good separation between π^- and K^- at 340 GeV/c.

7.2. *West experimental area*

This existing area was the first to come into operation. The primary proton beam was split into three parts at the level of the West Area, but was limited in energy to 250 GeV/c due to the background muon dose rate at the CERN fence downstream of the area.

A large number of experiments were already installed before the first extracted SPS beams became available in late 1976.

7.3. *North experimental areas*

Physicists eagerly awaited the higher energy beams planned for the North Areas. Unlike the West Hall, the North Experimental Area buildings and the transfer tunnels leading from the SPS to the target areas had to be conceived, designed, constructed and equipped with all the necessary infrastructure. Beamlines were commissioned in accordance with the priorities of the physics program and the readiness of the experiments to set up and take data.

The first area to receive beam was EHN2 (Experimental Area North 2) in early 1978, housing two muon experiments. Shortly after, also in 1978, the beams to the large experimental hall EHN1 (Experimental Hall North 1) came into operation, serving nine experiments.

Two years later, there followed two beams in the shielded, underground high intensity facility ECN3 (Experimental Cave North 3) [13], where two large experiments were installed.

8. Further Developments

8.1. *SPS as proton–antiproton collider*

During the period 1976–1980, John set up the "Accelerator Club," where the foremost accelerator experts of the Laboratory met regularly to discuss improvements to the existing machines and ideas for new developments. The most interesting one was to use the SPS as a proton–antiproton collider, an idea put forward very strongly in 1976 by Carlo Rubbia [5]. This required having proton and antiproton beams of comparable intensity, constituting the major problem to be overcome. The only way forward was to produce antiprotons by the 26 GeV protons of the PS (a production rate of one antiproton for one million protons) and then store them in an accumulator ring prior to their injection into the SPS. Like all storage rings, the accumulator ring had limited acceptance in the three dimensions, whereas the antiprotons emerging from the target had a large spread in production angles and momentum. The only solution was to concentrate the beam by either electron or stochastic cooling. Since stochastic cooling had been invented by Simon van der Meer [8] in 1972 and tested successfully in the ISR in 1974, the proposal was based on this method.

Van der Meer's original report on stochastic cooling was published in 1972 and the first successful tests were conducted in the ISR in 1974 and in 1976 by him and various other people, among them W. Schnell and L. Thorndahl. In the same period, ideas were put forward for the accumulation of antiprotons in storage rings by D. Möhl, P. Strolin and L. Thorndahl, and independently by P. MacIntyre. A similar proposal was made at Fermilab again by C. Rubbia, D. Cline, P. MacIntyre and F. Mills [5].

Modifications to the SPS were also needed, including the insertion of low beta sections around the collision points, a significant decrease in the vacuum pressure and, of course, the construction of huge (for the time) underground experimental areas for mobile experiments (UA1 on a platform, UA2 on air cushions). The mobility of these experiments in and out of the SPS ring was necessary in order to allow periods of fixed target operation at least once a year.

Leon Van Hove, the Research Director-General, supported the project from the beginning, while the accelerator community was initially skeptical, but was eventually won over after further feasibility studies were made, and by the prospects of undertaking a very challenging enterprise. Prior to the final design of the Antiproton Source, a test synchrotron called ICE (Initial Cooling Experiment) was quickly assembled by G. Petrucci using the refurbished magnets of the g-2 experiment in order to test both electron and stochastic cooling. The stochastic cooling method obtained a brilliant confirmation. A committee chaired by F. Bonaudi finalized the accelerator project.

The scheme consisted in using the PS at the maximum beam intensity concentrated over one quarter of the circumference, in order to match the circumference of the Antiproton Accumulator (AA). This was obtained by extracting the beam from the Booster in 10 bunches, instead of the usual 20, by recombining vertically the bunches of pairs of Booster rings, and by further reducing the 10 bunches to 5 in the PS by RF gymnastics. The beam was then extracted from the PS at 26 GeV and directed to the target at the entrance of the AA. The antiprotons were collected at 3.5 GeV by a magnetic horn (another important invention of S. van der Meer [8]).

R. Billinge and S. van der Meer [8] directed the design and construction of the AA (Fig. 19). Despite the great sophistication and the number of elements, the ring was constructed and tested successfully in less than three years. The formation of a full antiproton stack took two to three days, or 100,000 PS pulses. A question much debated at the time was what to do with the antiproton stack: direct injection into the SPS at 3.5 GeV or postacceleration in the PS to 26 GeV in order to inject into the SPS above the transition energy. Since there was no agreement among Bonaudi's Committee on this point, John, a convinced supporter of the project after his initial hesitation, took it upon himself to study thoroughly the question and decided in favor of postacceleration of the antiproton beam in the PS [3]. It was a wise decision, which undoubtedly facilitated the reliable operation of the collider. In Fig. 20, John Adams,

Fig. 19. The Antiproton Accumulator.

Fig. 20. John Adams and Carlo Rubbia.

equipped with a pipe and associated equipment, and Carlo Rubbia are having an animated discussion.

In 1978 the project was approved, and the first proton–antiproton collisions occurred on 10 July 1981. The first real period of physics exploitation was in 1982, with initial luminosities in the low 10^{29} cm^{-2} s^{-1} and an integrated luminosity of 28 nb^{-1} (sufficient for the discovery of W's). The year 1983 saw the collected integrated luminosity increased to 153 nb^{-1} and the discovery of the Z boson.

A few years later, a substantial improvement of the Antiproton Source was obtained by separating the function of collection and accumulation/cooling of antiprotons. This implied the addition of a second ring (Antiproton Collector, AC) around the original AA. Consequently, the luminosity went well above 10^{30} cm^{-2} s^{-1}, the record being $6 \cdot 10^{30}$ cm^{-2} s^{-1}.

Looking back to the early 1980s, one nontechnical but very important fallout of the proton–antiproton undertaking was the daily working together of the experimental teams and of the accelerator people. We all remember with nostalgia the animated discussions of the five o'clock meetings in the SPS Control Room to decide the course of action for the following day on the basis of the status of the antiproton stack in the AA. But it worked well in the end!

8.2. *LEP*

In 1977 a meeting of ECFA reached the conclusion that the next big project should be an electron–positron collider of at least 100 GeV beam energy. Studies by accelerator experts showed that, in order to reach such a high energy, a ring of about 30 km would be necessary. Moreover, only the use of superconducting radio frequency cavities would deliver this energy.

John, who was more inclined to a proton machine using superconducting magnets, became convinced that the physicists wanted the electron–positron collider and participated actively in its study with the help of Michael Crowley-Milling, who was then the Directorate Member for the accelerator program. However, John remained of the opinion that a proton machine would be required after LEP and so, in a page of his handwritten notebook of 1977, he advocated a LEP tunnel of a sufficient width (3.5 m) to accommodate also a proton ring using superconducting magnets of 4.5 T and a beam energy of 3 TeV. He called this SPEC (Super Proton Electron Complex) [14].

A tunnel of 30 km could be built at CERN, but about one third of it would not be in the good molasse rock but would penetrate the problematic limestone of the Jura mountains, which certainly contained cavities with water at high pressure.

LEP was finally approved in 1981, with Herwig Schopper as Director-General, who called in a well-known physicist, Emilio Picasso, as Project Leader. Emilio formed a team of experienced accelerator people and was able to mobilize the large required effort throughout the laboratory. In order to minimize the problems of excavation in the limestone, the tunnel circumference was reduced to about 27 km and the machine plane tilted in order to reduce the maximum depth of access shafts to about 180 m.

The execution of the project does not belong here, because John had no part in it, but I would like to stress his vision of a proton machine in the same tunnel mentioned above. In fact, I started the work on the future Large Hadron Collider already in 1982 with a group of the best CERN experts on a highly part-time basis, since the main effort was concentrated on LEP. Thinking back to that period, I remember with emotion John in the Accelerator Club taking part in the discussion of the initial project, of the next proton machine [6], which was to

Fig. 21. The LEP collider.

Fig. 22. The LHC collider.

materialize much later as the Large Hadron Collider (LHC)! Installations in the LEP and LHC tunnel are show in Figs. 21 and 22.

9. Conclusion

John Adams was an exceptional personality. Without a formal academic education but provided with a wide range of natural abilities, he became an extraordinary engineer and a charismatic leader. In remembering him with emotion, a number of distinctive qualities come to mind, like wide technical knowledge, rationality, organization, vision, authority, capacity of analysis and political abilities. But he was also rather reserved about his deep thoughts, feelings and emotions, so that it is not easy to rank these exceptional natural and developed gifts in a given order. It was a mix of all these qualities which made him an *extraordinary engineer*, capable of designing and constructing very complex machines involving a wide range of technologies in time and within budget; a *charismatic leader*, able to assemble, direct and orient international teams toward well-defined goals; and the *right man at the right time* for launching CERN on the track of a very successful international scientific organization. Later on, he also directed and oriented all other further developments of what has become the foremost laboratory for particle physics. Once, he said that he was very fortunate to have started his professional life at the emergence of three new technologies: radar during the war, accelerators at Harwell and CERN, and controlled fusion at Culham.

Clearly, a combination of fortunate circumstances helped to determine his destiny, but an inkling of his approach to life can be obtained from his statement that "if there are not enough events or they are too infrequent, then they can often be stirred up by various means, but there is no substitute for a clear sense of direction. An aimless person in a sea of random events takes a very long time indeed to reach any goal." [15]

Certainly, John Adams was a person with a mission in life and a "clear sense of direction"; he achieved much more than he, or anybody else, could have expected at the beginning of his professional life. CERN and the entire community of particle physics owe him the long-lasting gratitude due to a founding father.

Acknowledgments

The authors acknowledge the help of Niels Doble and Lau Gatignon, who provided information about the SPS' early particle beams, and thank Alfredo Placci for his knowledge of the CEDAR devices.

Appendix A. SPS Experimental Areas

Responding to the demands of the experimental community, intense beams of hadrons, electrons, tagged photons, and muons were provided, including a superconducting RF-separated and a polarized muon beam. A group of liaison physicists, including one of the authors (D. E. P.), designed the beamlines and the layout of the experiments, coordinated their installation, and helped the users profit from the facilities — they also modified the beamlines and layouts as required.

Appendix A.1. *Beam instrumentation*

Both analog and digital wire chambers were provided, the former for beam profiles and the latter for beam spectrometry. Almost all had 1 mm wire spacing and could give profiles for beam intensities ranging from 10^4 to 10^{11} particles/s [16].

Most secondary, and later tertiary, beams were equipped with four digital wire chambers, arranged in pairs surrounding bending magnets, in order to provide momentum measurements for individual particles. A resolution on one pion mass resulted in most cases [11].

Particle identification was provided by threshold Cerenkov and CEDAR counters. Two types of differential Cerenkov counters were provided: CEDAR-W for the West Area, providing π–K separation up to 150 GeV/c, and flagging protons above 12 GeV/c; CEDAR-N for the higher energy North Area, providing π–K separation up to 340 GeV/c (see Fig. 18) and capable of tagging protons above 60 GeV/c [12].

The sophisticated optical system of the CEDAR counters consisted of a quartz Mangin mirror (back surface reflection) and a chromatic corrector. Procurement of the Mangin mirror was no easy matter, and Claude Bovet and the team which designed and built the CEDARs profited from the presence on the CERN site of the optical experts from the European Southern Observatory project.

Command of all instrumentation and electronics was under user-friendly computer control, and all the electronics was provided and serviced by the SPS Experimental Areas Group.

Appendix A.2. *West Area*

This existing area was the first to come into operation.

Figure 23 is a schematic drawing of the West Area as initially installed. The labels have the following meanings: T for "target," H for "hadron," E for "electron," S for "separated," P for "attenuated proton," Y for "hyperon," N for "neutrino beams" and WAn for "West Area experiment number 'n'."

A particularly novel feature was the provision of a superconducting-RF (S band)–separated beam to the general purpose Ω spectrometer.

Note the large number of experiments installed for the start of SPS operation.

Appendix A.3. *North Areas*

The first area to be served was EHN2 (Experimental Area North 2) in early 1978, housing two muon experiments. In designing the M2 muon beam [17], special care was taken to avoid the muon halo problem which has troubled the Fermilab beam. A sophisticated muon tracking program, HALO [18], helped to identify technical solutions, and special 5-m-long magnetic collimators and magnetized shielding blocks downstream of the hadron absorber proved to be efficient in limiting the muon halo at the experiment to an acceptable level.

Shortly after, also in 1978, the beams to the large experimental hall EHN1 (Experimental Hall North 1) came into operation. An innovative feature of beams to EHN1 was the introduction of "wobbling stations" [19] in the production target zone, which allowed, for each target, two beams to be produced

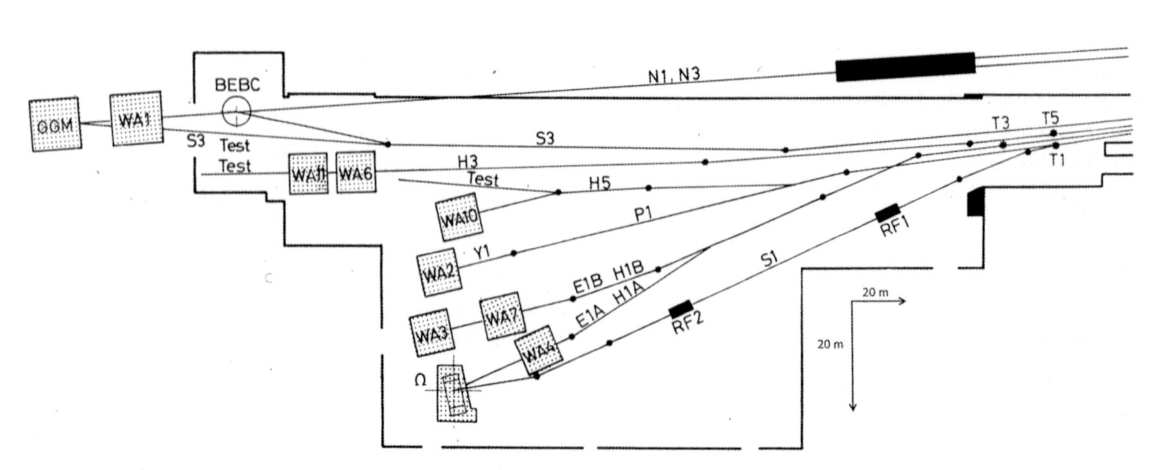

Fig. 23. Layout of the West Area beams and experiments in 1976. Note the different scales in the horizontal and vertical directions.

Table 1. Characteristics of the North Area beams as initially installed.

Production target	Experimental hall	Beam	Beam properties
T2	EHN1	H2	High energy, high resolution secondary beam, for charged hadrons, electrons, polarized protons from Λ decay, and attenuated primary protons. Maximum momentum 400 GeV/c, momentum acceptance $\pm 2\%$.
T2	EHN1	H4	High energy, high resolution secondary beam, for charged hadrons, electrons, polarized protons from Λ decay, and attenuated primary protons. Maximum momentum 400 GeV/c, momentum acceptance $\pm 1.4\%$.
T4	EHN1	H6	High energy, secondary beam, for charged hadrons. Maximum momentum 200 GeV/c, momentum acceptance $\pm 1.5\%$.
T4	EHN1	H8	High energy, high resolution secondary beam, for charged hadrons, electrons and attenuated primary protons. Maximum momentum 400 GeV/c, momentum acceptance $\pm 3\%$.
T6	EHN2	M2	Muon beam–high backward/forward polarization for $P_{\mu^+} \approx 0.9/0.6 P_{\pi^+}$. Maximum momentum 300 GeV/c; momentum acceptance $\pm 10\%$ for parent particles, $\pm 6\%$ for muons.
T8	ECN3	H10	Intense π beam, up to 10^{10} pions/pulse. Maximum momentum 400 GeV/c, momentum acceptance $\pm 12\%$.
T10	ECN3	E12	Broadband (momentum acceptance $\pm 28\%$) e–γ beam. Maximum momentum of e 300 GeV/c.

at zero or very small production angles, with only loose coupling in their momenta.

Fierce radiation levels prevailed in the West and North Area target regions. To limit the effects of radiation damage the bending magnets immediately downstream of the targets were equipped with mineral insulated (high alumina concrete), radiation-resistant coils [20], and the upstream quadrupoles were installed on "plug-in" bases which assured automatic alignment, powering and cooling of replacement elements.

Characteristics of the North Area beams, as initially installed, are summarized in Table 1.

References

[1] E. Amaldi, John Adams and his times. First John Adams Memorial Lecture, CERN 86-04.

[2] E. J. N. Wilson, Sir John Adams: His legacy to the world of particle accelerators. J. Adams Memorial Lecture 2009, CERN-2011-001.

[3] M. Crowley-Milling, *John Bertram Adams, Engineer Extraordinary: A Tribute*, Gordon and Breach Science (1993).

[4] M. Goldsmith and Edwin Shaw, *Europe's Giant Accelerator*, Taylor and Francis (London, 1977).

[5] R. Cashmore, L. Maiani and J.-P. Revol, *Prestigious Discoveries at CERN, 1973 Neutral Currents, 1983 W & Z Bosons* (Springer) — G. Brianti, CERN's contribution to accelerators and beams (pp. 25–40).

[6] *Large Hadron Collider in the LEP tunnel: Proc. ECFA-CERN Workshop CERN 84-10* (Mar. 1984).

[7] E. D. Courant, M. S. Livingston and H. S. Snyder, *Phys. Rev.* **88**, 1190 (1952).

[8] V. C. Chohan, Simon van der Meer (1925–2011): A modest genius of accelerator science, *Rev. Accel. Sci. Technol.* **4**, 279 (2011).

[9] *CERN Courier*, Vol. 53, archive page 14 — The 300 GeV project.

[10] J. B. Adams, The European 300 GeV accelerator, in *Proc. 8th Int. Conf. on High-Energy Accelerators* (CERN, Geneva, Switzerland, 1971), pp. 25–31.

[11] G. Dubois, A. Placci, L. Pregernig, M. Rabany, B. Skalli and G. Vismara, *IEEE Trans. Nucl. Sci.* **25**(1), 808 (1978).

[12] C. Bovet, R. Maleyran, A. Placci and M. Placidi, *IEEE Trans. Nucl. Sci.* **NS25**, 572 (1978).

[13] G. Brianti and N. Doble, The SPS North Area High Intensity Facility: NAHIF. CERN/SPS/EA 77-2, CERN/SPSC/77-72, SPSC/T-18 (1977).

[14] J. Lawson and G. Brianti, 50 years of synchrotrons. 12th J. Adams Memorial Lecture, CERN 97-04.

[15] J. B. Adams, "Take a risk" in "Letters to a Young Scientist," *New Scientist* **45**, 2 (1970).

[16] P. Dreesen and G. Vismara, *Nucl. Instum. Methods* **156**, 325 (1978).

[17] R. Clifft and N. Doble, CERN/SPSC/74-12 (CERN/Lab. II/EA/74-2).

[18] C. Iselin, HALO, a computer program to calculate muon halo. CERN 74-17.

[19] H. Atherton, G. Brianti and N. Doble, Towards a layout of the 300 GeV North Experimental Area. CERN/Lab. II/EA/72-1 (July 1972).

[20] R. L. Keizer and M. Mottier, Mineral insulated magnets. CERN/SPS/EMA/77-3, presented at 6th Int. Conf. Magnet Technology (Bratislava, 1977).

Giorgio Brianti was appointed directly by Prof. E. Amaldi and spent his whole professional career at CERN from 1954 to 1995. He took part in the construction of PS from 1954 to 1959 and was Head of the PS Operation Group from 1960 to 1964. He was also leader of the Synchro-cyclotron Division from 1964 to 1967 of the Synchrotron Injector (Booster) Division from 1968 to 1972; Head of the SPS Experimental Areas Group from 1972 to 1978; Leader of the SPS Division from 1979 to 1980; Director from 1981 to 1993 in charge of accelerators, proton-antiproton collider and preparation of the LHC project.

David E. Plane earned his PhD on the 10-inch hydrogen bubble chamber at the Liverpool synchro-cyclotron, following which he moved to CERN. After working on hyperon resonances and fits to baryon decay rates within SU(3) and SU(6), he joined the SPS project which was under the leadership of John Adams. In the mid-1980s he returned to CERN's Research Sector where he joined OPAL, a LEP experiment. In 1998 he became spokesman of the experiment. He was awarded the Rutherford Medal of the UK Institute of Physics, and elected Fellow of the Institute. He is also a member of the Bologna Academy of Science. Retired, he is currently an Honorary CERN Staff Member.

Reviews of Accelerator Science and Technology
Vol. 6 (2013) 311

DOI: 10.1142/S179362681392001X

Erratum

Industrialization of Superconducting RF Accelerator Technology

[*Reviews of Accelerator Science and Technology*, Vol. 5 (2012) pp. 265–283]

Michael Peiniger,* Michael Pekeler† and Hanspeter Vogel‡
RI Research Instruments GmbH, Friedrich-Ebert-Straße 1,
51429 Bergisch Gladbach, Germany
*michael.peiniger@research-instruments.de
†michael.pekeler@research-instruments.de
‡hanspeter.vogel@research-instuments.de

The following errors were discovered after publication:

1. On page 271, Section 4.3, the third sentence currently reads:

   "The cavities for CESR were produced by ACCEL and those for KEK-B by Mitsubishi Heavy Industries."

   It should read:

   "The cavities for CESR were produced by ACCEL and those for KEK-B by Mitsubishi Electric Company."

2. On page 277, Table 2, the 8th column of the table for manufacturer of the KEK-B cavities currently reads "MHI", it should read "MELCO".

3. On page 287, Table 3, the 4th column of the table for Industry of the BEP-II SRF modules currently reads "MHI", it should read "MELCO".

4. On page 278, Figure 16 caption currently reads:

   "Two 500 MHz SRF accelerator modules produced by Mitsubishi Heavy industries and KEK for the BEB-II collider at IHEP, China (courtesy of KEK)".

   It should read:

   "Two 500 MHz SRF accelerator modules produced by Mitsubishi Electric Company and KEK for the BEB-II collider at IHEP, China (courtesy of KEK)".